Alternaria Diseases of Crucifers: Biology, Ecology and Disease Management

Govind Singh Saharan • Naresh Mehta
Prabhu Dayal Meena

Alternaria Diseases of Crucifers: Biology, Ecology and Disease Management

Springer

Govind Singh Saharan
Plant Pathology
CCS Haryana Agricultural University
Hisar, Haryana, India

Naresh Mehta
Plant Pathology
CCS Haryana Agricultural University
Hisar, Haryana, India

Prabhu Dayal Meena
Crop Protection Unit
ICAR
Bharatpur, Rajasthan, India

ISBN 978-981-10-9064-6 ISBN 978-981-10-0021-8 (eBook)
DOI 10.1007/978-981-10-0021-8

Springer Singapore Heidelberg New York Dordrecht London

Softcover re-print of the Hardcover 1st edition 2016

Printed on acid-free paper

Springer Science+Business Media Singapore Pte Ltd. is part of Springer Science+Business Media (www.springer.com)

Dedicated to Beloved Dr Prithwi Raj Verma

(10 January 1940–07 Feb 2015)

Foreword

डा. एस. अय्यप्पन
सचिव एवं महानिदेशक
Dr. S. AYYAPPAN
SECRETARY & DIRECTOR GENERAL

भारत सरकार
कृषि अनुसंधान और शिक्षा विभाग एवं
भारतीय कृषि अनुसंधान परिषद
कृषि मंत्रालय, कृषि भवन, नई दिल्ली 110 001

GOVERNMENT OF INDIA
DEPARTMENT OF AGRICULTURAL RESEARCH & EDUCATION
AND
INDIAN COUNCIL OF AGRICULTURAL RESEARCH
MINISTRY OF AGRICULTURE, KRISHI BHAVAN, NEW DELHI 110 001
Tel.: 23382629; 23386711 Fax: 91-11-23384773
E-mail: dg.icar@nic.in

Brassica species including *B. campestris*, *B. juncea*, *B. napus* and *B. carinata* are an important group of oilseed crops, constituting almost 13.2 % of the world edible oil requirement. Together these occupy about 29.39 million hectares of area with an annual production of 53.01 million tons in the world. These have wide adaptability and are often grown under varied agroclimatic conditions throughout the world. During the last two decades, the area and production of these crops have increased substantially and the total production has almost trebled.

Considerable potential exists for improving production and productivity of oilseed brassicas, as well as cruciferous vegetables through breeding improved varieties that are resistant to both biotic and abiotic stresses. Among biotic factors limiting the productivity, the diseases like *Alternaria* blight, white rust, downy mildew, *Sclerotinia* stem rot and powdery mildew are the serious one. Presently, *Alternaria* diseases are due to four species, viz. *A. brassicae* (Berk.) Sacc., *A. brassicicola* (Schwein.) Wiltsh., *A. raphani* Groves and Skolko and *A. alternata* (Fr.) Kiessl, causing heavy yield losses annually in cruciferous crops.

To have an in-depth knowledge of *Alternaria* diseases of crucifers and to evolve effective management strategies, it is necessary that a comprehensive review of literature relating to pathogen and its taxonomy, process of infection, pathogenesis, fine structure, biochemistry of host–pathogen interaction, phytotoxins, pathogenic variability, distribution, yield losses, disease cycle, epidemiology and forecasting, resistance, and sources; various techniques for germ-plasm screening; in vitro studies; biotechnological approaches; and disease management practices is available for the workers. Presently such a comprehensive document is lacking. This book "*Alternaria Diseases of Crucifers: Biology, Ecology and Disease Management*" is an outcome of sincere efforts of the authors focusing on oilseed crops' diseases. The suggestions by the authors on priority areas of research will benefit the *Brassica* researchers to plan their research in a better way.

I congratulate the authors for bringing out this long-awaited publication and believe that this book will be of immense use to the scientists, teachers, students, extension specialists and all those interested in the production of crucifer crops.

New Delhi, India
22 April 2015

S. Ayyappan

Preface

The aim of this book *Alternaria Diseases of Crucifers: Biology, Ecology and Disease Management* is to present a comprehensive information available in literature on fundamental and applied knowledge of *Alternaria* species infecting Brassicaceae crops and weeds. Since the first publication of *Monograph on Alternaria Diseases of Crucifers* (Verma and Saharan, 1994), voluminous valuable research data have been generated and published that encouraged the authors to update the information in the form of a book. The four species of *Alternaria*, viz. *A. brassicae* (Berk.) Sacc., *A. brassicicola* (Schwein.) Wiltsh., *A. raphani* Groves and Skolko. and *A. alternata* (Fr.) Kiessl., are most widely distributed and cause severe quantitative and qualitative losses in crucifers where these crops are grown in the world. *Brassica* crops are grown for high-quality edible (rapeseed–mustard, canola and other rapes) and industrial (*Crambe*) oil, common vegetables (cabbage, cauliflower, radish, kohlrabi, broccoli, Brussels sprouts, kales and other *Brassica* vegetables) and a few weeds. For convenience of the readers and coherence of the text, the information has been arranged in 13 chapters with several subsections. The arrangement includes the disease, geographical distribution, symptoms on different hosts, host range, yield losses and disease assessment method, while detailed description on pathogen includes taxonomic position, phylogeny, variability, sporulation, perpetuation and spore germination, host–parasite interactions in the form of seed infection, disease cycle, process of infection and pathogenesis, epidemiology, forecasting, fine structures, biochemical changes and phytotoxins, host defence mechanism, techniques to study host–parasite relationships, management practices (including cultural, chemical, biological control practices) and deployment of host resistance. The last section deals with gaps in our understanding, and knowledge about management of these diseases, and offers suggestions for future research priorities. The subject matter has been vividly illustrated with photographs, graphs, figures, histogram, tables and coloured plates, which makes it stimulating, effective and easy to comprehend by the readers. The headings and subheadings of each chapter have been arranged in numbered series to make the subject matter contiguous. All important and relevant references have been included for further consultations by the researchers, teachers and students.

We believe that this book will be immensely useful to researchers, teachers, extension specialists, students and all others who are interested in the diagnosis and management of *Alternaria* diseases of crucifers. Suggestions for the improvement are always welcomed.

Hisar, India Govind Singh Saharan
Naresh Mehta
Bharatpur, India Prabhu Dayal Meena

Acknowledgements

Authors are highly grateful to the following persons/scientists/publishers/societies/journals/institutes/websites and all others whose valuable materials such as photographs (macroscopic, microscopic, electron micrographs, scanning electron micrographs), drawings, figures, histograms, graphs, tables, flow charts, etc., have been used through reproduction in the present document. The address of the author(s)/source(s) from where the material is adapted can be obtained from the reference which has been cited in the reference section of the book.

A. Persons/Scientists
A. Tsuneda
B. M. Pryor
B. P. H. J. Thomma
Claudia A. Jasalavich
D. P. Lawrence
E. G. Simmons
F. M. Humpherson-Jones
J. H. C. Woudenberg
J. P. Tewari
M. Soledade C. Pedras
P. R. Verma
P. Parvatha Reddy
P. S. Bains
Philippe Simonea
R. P. Awasthi
R. Y. Parada
S. J. Kolte
S. K. Shrestha
Yangrae Cho

B. Journal
African Journal of Biotechnology
Australian Journal of Agricultural Research
Botany
Brazilian Phytopathological Society
Canadian Journal of Botany

Canadian Journal of Plant Pathology
Crop Protection
European Foundation for Plant Pathology
Fungal Genet & Biology
Indian Phytopathology
International Journal of Agriculture, Environment & Biotechnology
Journal of Mycology and Plant Pathology
Journal of Oilseed Brassica
Journal of Phytopathology
Molecular Microbiology
Molecular Plant Microbe Interaction
Molecular Plant Pathology
Molecules
Mycologia
Mycological Progress
Mycological Research
Phytochemistry
Plant Molecular Biology
Plant Pathology
Plant Pathology
PLoS Pathogens
Proceedings of the National Academy of Sciences USA
Review of Plant Pathology
Seed Science & Technology Journal
The Canadian Journal of Plant Pathology
The Plant Pathology Journal
Transaction of British Mycological Society
Tropical Plant Pathology

Website
http://cals.arizona.edu/PLP/pryorlab/alternaria.html
http://nt.ars-grin.gov/fungaldatabases/
http://www.mycobank.org
http://www.marinespecies.org
www.elsevier.com/locate/yfgbi
www.sciencedirect.com

Publishers
CABI, UK
CSIRO Publishing
Elsevier
John Wiley & Sons, Inc.
Springer, Netherlands
Studium Press LLC, USA
Taylor & Francis Group

Institutions

Canadian Phytopathological Society
Indian Phytopathological Society
Indian Society of Mycology and Plant Pathology
Korean Society of Plant Pathology
Mycological Society of America
Pryor Laboratory, University of Arizona School of Plant Sciences, USA
Society for Rapeseed-Mustard Research
The American Phytopathological Society
The Australasian Plant Pathology Society
The British Society for Plant Pathology
The Korean Society of Plant Pathology

Databases

MycoBank, International mycological Association
Systematic Mycology and Microbiology Laboratory Fungal Database, US Department of Agriculture

Contents

About the Authors

Govind Singh Saharan Renowned Educationist and Plant Pathologist has contributed in the diverse fields of Plant Pathology including standardization of artificial inoculation techniques, identification of resistance sources, pathogenic variability, genetics of host–parasite interaction, epidemiology and management of several diseases. He has about 250 publications in peer-reviewed journals and 10 books and supervised over 11 MSc and PhD students. Professor Saharan is internationally recognized a visiting Professor at the University of Alberta, Edmonton, Canada (1991 and 1994); Saskatoon Research Station, Canada (1991, 1994, 1997); and Rothamsted Research, UK (1994, 1997). He has been on the panel of experts of SAUs, ICAR, CSIR, UGC and DBT in India.

Naresh Mehta Associate Dean, Professor (Plant Pathology), Gold medalist, Educationist and Plant Pathologist, has contributed in diverse field of plant pathology covering pathogenic variability, genetics of host–pathogen interaction, epidemiological studies, identification of resistant sources, biochemical/genetical basis for resistance, residual analysis of fungicides and disease management. He has published more than 96 research papers in peer-reviewed journals, 2 books, 10 manuals, 8 review articles, 18 book chapters, 18 lead lectures in the international and national conferences and 79 research papers presentations in the international/national conferences. He has supervised seven MSc and PhD students. Professor Mehta has been admitted as fellow of prominent Plant Pathological Society of India; councillor (North Zone) and editor-in-chief, ISMPP; and member of editorial board of INSOOP and Indian Phytopathological Soc. (IPS), New Delhi, India. Professor Mehta has been a visiting scientist at the University of Alberta, Edmonton, Canada, and invited to deliver lecture in the 9th International Congress of Plant Pathology at Torino, Italy. Professor Mehta is on the panel of experts of SAUs.

Prabhu Dayal Meena is presently working as Principal Scientist (Plant Pathology) at ICAR-Directorate of Rapeseed-Mustard Research, Bharatpur (Rajasthan), India. He has made important contributions to the various aspects including resistance, epidemiology, forecasting and biocontrol for the management of rapeseed–mustard diseases. He has published about 60 research papers, two books and four bulletins and has supervised nine MSc and co-supervised a PhD student. He has been honoured with fellow of ISMPP, PPAI, ISOR and Dr PR Kumar Outstanding Brassica Scientist Award (2011) of SRMR and Best Scientist Award of DRMR for 2012–2013. Meena is internationally recognized and has been working at Rothamsted Research, UK (2007).

List of Figures

List of Tables

List of Plates

1 Introduction

1.1 Status of Genus *Alternaria*

Alternaria is an ubiquitous fungal genus that includes saprobic, endophytic and pathogenic species. It is associated with a wide variety of substrates including seeds, plants, agricultural products, animals, soil and atmosphere. Species of *Alternaria* are known as serious plant pathogens, causing major losses on a wide range of crops. There are over 4000 *Alternaria*/host associations recorded in the USDA Fungal Host Index ranking the genus 10th among nearly 2000 fungal genera based on the total number of host records. While very few *Alternaria* species appear to have a sexual stage (*Pleospora*) to their life cycle, the most lack sexuality altogether. Several species of *Alternaria* with their taxonomic and morphological characteristics, pathogenic nature, and synonymous along with distribution have been described extensively (Elliott 1917; Neergaard 1945; Ellis 1971, 1976; Simmons 1967, 2002, 2007).

Several taxa are also important postharvest pathogens and causative agents of phaeohyphomycosis in immunocompromised patients or airborne allergens. Because of the significant negative health effects of *Alternaria* on plants and their surroundings, a correct and rapid identification of *Alternaria* species would be of great value to agriculturists, researchers, medical mycologists and the public alike (Woudenberg et al. 2013).

Alternaria was originally described by Nees (1816) based on *A. tenuis* as the only species. Characteristics of the genus included the production of dark-coloured phaeodictyospores in chains and a beak of tapering apical cells. Von Keissler (1912) synonymized both *A. tenuis* and *Torula alternata* (Fries 1832) with *Alternaria alternata*, due to ambiguities in Nees' description of *A. tenuis*. Description of two additional genera, *Stemphylium* (Wallroth 1833) and *Ulocladium* (Preuss 1851) in phaeodictyosporic hyphomycetes, further complicated the taxonomic resolution in this group of fungi. Several redescriptions and revised criteria of these genera (Saccardo 1886; Elliot 1917; Wiltshire 1933, 1938; Joly 1964) resulted in a growing number of new species. Results of lifetime study on *Alternaria* taxonomy based upon morphological characteristics were summarized in a classical identification manual by Simmons (2007), in which 275 *Alternaria* species were recognized. One species was transferred to the genus *Prathoda* and three new genera, *Alternariaster*, *Chalastospora* and *Teretispora*, were segregated from *Alternaria*. Molecular studies revealed multiple non-monophyletic genera within the *Alternaria* complex, and *Alternaria* species clades, which do not always correlate to species group upon morphological characteristics described (Pryor and Gilbertson 2000; Chou and Wu 2002; de Hoog and Horre 2002; Pryor and Bigelow 2003; Hong et al. 2005; Inderbitzin

G.S. Saharan et al., *Alternaria Diseases of Crucifers: Biology, Ecology and Disease Management*,
DOI 10.1007/978-981-10-0021-8_1

et al. 2006; Pryor et al. 2009; Runa et al. 2009; Wang et al. 2011; Lawrence et al. 2012). The *A. alternata*, *A. brassicicola*, *A. infectoria*, *A. porri* and *A. radicina* species group were strongly supported by these studies; two new species groups, *A. sonchi* (Hong et al. 2005) and *A. alternantherae* (Lawrence et al. 2012), and three new genera, *Crivellia* (Inderbitzin et al. 2006), *Undifilum* (Pryor et al. 2009) and *Sinomyces* (Wang et al. 2011), were described. The latest molecular revision of *Alternaria* (Lawrence et al. 2013) introduced two new species groups, *A. panax* and *A. gypsophilae*, and elevated eight species groups to sections within *Alternaria*. The sexual phylogenetic *Alternaria* lineage, the *A. infectoria* species group, did not get the status of section, in contrast to the eight asexual phylogenetic lineages. The *Alternaria* complex currently comprises the genera *Alternaria*, *Chalastospora* (Simmons 2007), *Crivellia*, *Embellisia*, *Nimbya*, *Stemphylium*, *Ulocladium*, *Undifilium* and the recently described *Sinomyces* together with eight sections of *Alternaria* and the *A. infectoria* species group.

Woudenberg et al. (2013) conducted a study to delineate the phylogenetic lineages within *Alternaria* and allied genera to create a robust taxonomy. Phylogenetic inferences were drawn on sequence data of parts of the 18S nrDNA (SSU), 28S nrDNA (LSU), the internal transcribed spacer regions 1 and 2 and intervening 5.8S nrDNA (ITS), glyceraldehydes-3-phosphate dehydrogenase (GAPDH), RNA polymerase second largest subunit (*RPB2*) and translation elongation factor 1-alpha (*TEF1*) gene regions of ex-type and reference strains of *Alternaria* species and all available allied genera.

The most common and destructive diseases of Brassicaceae crops worldwide are those caused by four species of *Alternaria*, viz. *A. brassicae* (Berk.) Sacc., *A. brassicicola* (Schwein.) Wiltsh., *A. raphani* Groves and Skolko and *A. alternata* (Fr.) Keissl. Although attack by *Alternaria* at the seedling stage can lead to death of young plants, infection on leaves, stems and siliquae generally results in heavy losses in seed yield and quality (Table 1.1).

1.2 *Brassica* Crops

Brassica oilseed crops occupy over 29.39-million hectares of the world's agricultural lands with 53.01 mt total production yielding on an average of 1700 kg/ha. Their ability to survive and grow at low temperatures enables the oilseed *Brassica* to be cultivated successfully in cool agricultural regions, at high elevations and as winter crops in the subtropics. The small round seeds of *Brassica* oilseed crops contain over 40 % oil on a dry weight basis and, after the extraction of oil, provide a meal containing over 40 % high-quality protein. In the Western countries, the meal is utilized exclusively as a feed for livestock and poultry, but in many Asian countries, it is used as an organic fertilizer for field crops.

The Brassicaceae family, to which the genus *Brassica* belongs, contains approximately 3500 species in 350 distinct genera of many important crop plants yielding high-quality, edible and industrial oils, common vegetables and weeds. Based on the evidence that some vegetable types were in common use in the Neolithic age (Chang 1968; Hyams 1971), and direct reference to oilseed rape and mustard in the ancient Indian Sanskrit writings of 2000–1500 BC (Singh 1958), the *Brassica* vegetables and oilseeds may well have been among the earliest plants domesticated by man. Greek, Roman and Chinese writings of 500–200 BC also mention these crops and their medicinal value (Prakash and Hinata 1980). Oilseed rape was introduced in China and Japan around the time of Christ (Hougen and Stefansson 1983). Although its cultivation began in the thirteenth century in Europe, its industrial use was not widespread until its superior qualities as lubricant oil were recognized (Shahidi 1990). Its use as an edible vegetable oil in Western countries is very recent. Unlike most other oilseeds, rapeseed comes from several species of genus *Brassica* (Shahidi 1990) including *B. napus* L., *B. rapa* L. (*B. campestris* L.) and *B. juncea* (L.) Czern. & Coss., which are known as rape, turnip rape and leaf mustard, respectively. Common names for *B. napus* are rape, rapeseed, oil rape, colza, oilseed rape, swede rape and Argentine rape; for *B. rapa*

Table 1.1 Yield losses in crucifer crops due to *Alternaria* diseases in different countries of the world

Crop/cultivar	Country (location)	Yield losses (%)	References
Rapeseed	Germany	75	Klemm (1938), Raabe (1939)
Rapeseed	Germany	20–50	Daebeler et al. (1986)
Rapeseed	England	60	Smith et al. (1988)
Rapeseed	Lithuanian	37–100	Brazauskiene and Petraitiene (2006)
Brassica napus	Canada	42	Degenhardt et al. (1974)
Brassica napus	India (Kangra)	17	Kumar (1997)
Brassica rapa	Canada	70	Degenhardt et al. (1974)
Brassica carinata	India (Kangra)	11	Kumar (1997)
Brassica juncea cv. Prakash	India (Hisar)	18	Saharan (1984)
Brassica juncea cv. Varuna	India (Hisar)	17	Saharan (1984)
Brassica juncea cv. Varuna	India (Pantnagar)	34	Saharan (1984)
Brassica juncea cv. RLM-514	India (Ludhiana)	49	Saharan (1984)
Brassica juncea cv. Varuna	India (Kanpur)	21–22	Prasad et al. (2003)
Brassica juncea cv. Varuna	Nepal	32–57	Shrestha et al. (2005)
Brassica campestris var. Brown Sarson cv. BSH-1	India (Hisar)	26	Saharan (1984)
Brassica campestris var. Yellow Sarson cv. YSPb-24	India (Hisar)	35	Saharan (1984)
Brassica campestris var. Yellow Sarson cv. YS-151	India (Pantnagar)	45	Saharan (1984)
Brassica campestris var. Yellow Sarson	India (Kangra)	28	Kumar (1997)
Eruca sativa	India	56	Jain (1992)
Cabbage, cauliflower and broccoli	USA	50	Ramsey and Smith (1961)
Cauliflower	Germany	50	Stoll (1948)
Cabbage	Germany	70–90	Domsch (1957)
Cabbage	Germany	80–100	Gorshkov (1976)
Brassica oleracea	England	80	Smith et al. (1988)
Radish	India	18	Suhag et al. (1983)

are rapeseed, oil turnip and Polish rape; and for *B. juncea* are brown mustard, oriental mustard, Indian mustard and rapeseed. In China, all three species are grown, but winter-grown rape constitutes the major source of rapeseed. In India, turnip rape and mustard may be considered rapeseed, and in North America and Europe, rape and turnip rape are regarded as rapeseed. The other oilseed crops of the Brassicaceae include *B. rapa* L. var. *toria* (turnip rape, toria), *B. rapa* L. var. Brown Sarson (turnip rape, Brown Sarson), *B. rapa* L. var. Yellow Sarson (turnip rape, Yellow Sarson), *B. nigra* (L.) Koch (black mustard), *B. hirta* Moench (*Sinapis alba* L.) (white mustard), *B. carinata* A. Braun (Abyssinian mustard, Ethiopian mustard), *B. tournefortii* Gouan (wild turnip), *Eruca sativa* Mill. (*E. vasicaria* spp. *sativa* (Mill.) Thell.) (garden rocket, taramira), *Camelina sativa* Crantz. (false flax, Dutch flax, gold-of-pleasure), *Crambe abyssinica* Hochst. ex. O.E. Schulz and *C. hispanica* L. The genome designations of varietal or subspecific taxa of agriculturally important brassicas and radish are given in Table 1.2 (Anonymous 1985).

The traditional varieties of rapeseed that are being grown in the Asian countries contain 22–60 % erucic acid in their oil and high percentage of glucosinolate in their defatted meal. The presence of erucic acid compromises the nutritional value of oil,

Table 1.2 Genomic designations of varietal or subspecific taxa of agriculturally important Brassicas and radish (Anonymous 1985)

Brassica sp. (n)	spp. or var.	2n genome descriptor	Common name
nigra (8)	–	bb	Black mustard
oleracea (9)	–	cc	Cole crops
	acephala	cc. a	Kales
	alboglabra	cc.al	Chinese kale, kai-lan
	botrytis	cc.b	Cauliflower, heading broccoli
	capitata	cc.c	Cabbage
	costata	cc.co	Portuguese cabbage
	gemmifera	cc.g	Brussels sprouts
	gongylodes	cc.go	Kohlrabi
	italic	cc.i	Broccoli, calabrese
	medullosa	cc.m	Marrow stem kale
	palmifolia	cc.p	Kale (Jersey kale)
	ramosa	cc.ra	Thousand-head kale
	sabauda	cc.s	Savoy cabbage
	sabellica	cc.sa	Collards
	selensia	cc.se	Borecole
campestris (10) syn. *rapa*	–	aa	–
	chinensis	aa.c	Pak choi
	narinosa	aa.na	–
	nipposinica	aa.n	–
	oleifera	aa.o	Turnip rape, toria
	parachinensis	aa.pa	Choy sum
	pekinensis	aa.p	Chinese cabbage, pechay
	perviridis	aa.pe	Tender green, komatsuna, mustard spinach
	rapifera	aa.r	Turnip
	trilocularis	aa.t	Sarson
	utilis	aa.u	–
carinata (17)	–	bbcc	Ethiopian mustard
juncea (18)	–	aa.bb	–
	capitata	aabb.c	Head mustard
	crispifolia	aabb.cr	Cut leaf mustard
	faciliflora	aabb.f	Broccoli mustard
	lapitata	aabb.l	Large petiole mustard
	multiceps	aabb.m	Multishoot mustard
	oleifera	aabb.o	Oilseed mustard, raya
	rapifera	aabb.r	Root mustard
	rugosa	aabb.ru	Leaf mustard
	spicea	aabb.sp	Mustard
	tsa-tsai	aabb.t	Big stem mustard
napus (19)	–	aacc	Fodder rape
	oleifera	aacc.o	Oil rape
	rapifera	aacc.r	Swede, rutabaga

(continued)

Table 1.2 (continued)

Brassica sp. (n)	spp. or var.	2n genome descriptor	Common name
Raphanus sativus (9)	–	rr	Radish
	radicola	rr.r	Radish, daikon
	oleifera	rr.o	Oil radish
	caudatus	rr.c	Rat tail radish

and glucosinolate reduces the feeding value of the meal. Since the late 1970s, both *B. napus* and *B. rapa* Canadian varieties have been genetically modified to contain low erucic acid and glucosinolates and named these 'double-low' cultivars as 'canola' in 1979. The term canola thus refers to a rapeseed cultivar that contains less than 30 μmol/g of one or any combination of the four known aliphatic glucosinolates (gluconapin, progoitrin, glucobrassicanapin and napoleiferin) in its defatted meal, and less than 2 % of the fatty acyl content of the oil is erucic acid. Canada recently has also developed canola-type *B. juncea* cultivars.

The main groups of cultivated *Brassica* vegetables are kales (*B. oleracea* L. var. *acephala*) including kitchen kale, green kale, dwarf Siberian kale, narrow stem kale, collards and trochunda; cabbages (*B. oleracea* L. var. *captitata*, var. *sabauda*, var. *bullata*) including headed cabbages, Brussels sprouts and savoy cabbage; kohlrabi (*B. oleracea* L. var. *gongylodes*); inflorescence kales (*B. oleracea* L. var. *botrytis*, var. *italica*) including cauliflower, broccoli and sprouting broccoli; branching bush kales (*B. oleracea* L. var. *fruticosa*) including cow kale, borecole, thousand-headed kale); Chinese kale (*B. alloglabra* L.) (Snogerup 1980); and radishes (*Raphanus sativus* L.).

Crambe is a newly emerging oilseed crop with increasing commercial acreage in the USA, Poland and some other countries of the world. Crambe oil is a potential raw material for rubber and plastic industries. The prospective species for crambe cultivation are *Crambe abyssinica*, *C. hispanica* and some other annuals.

1.3 The Disease

The disease is known by more than 40 names in the world on the basis of symptoms produced on different parts of cruciferous crops. All the four *Alternaria* species cause symptoms in the seedling stage on cotyledons and in the adult stage on leaves, leaf petiole, stem, inflorescence, siliquae and seeds. In general, symptoms are similar on all infected host species in the form of lesions. There may be variations in shape, size, colour, formation of concentric rings, yellow halo around the lesions under different agro-ecological zones, host genotypes, nutritional status of soil and pathotypes involved. In rapeseed, effects on cell membrane, chloroplast and mitochondria have been recorded (Verma and Saharan 1994; Tewari 1991b). The disease is known to be distributed all over the world wherever host pathogens interact under suitable environmental conditions. Host range of all the four *Alternaria* species is very wide infecting oilseed *Brassica*, cruciferous vegetables, wild cruciferous hosts and weeds. Quantitative and qualitative losses in yield of oilseeds and vegetables range from 11 to 100 % (Table 1.1) depending upon time of infection, prevailing environmental conditions after infection and strategies used for its control (Verma and Saharan 1994; Czyzewska 1969; Kadian and Saharan 1983, 1984; Kolte 1985; Tewari and Conn 1993; Seidle et al. 1995; Kumar 1997; Gupta et al. 1998; Meah et al. 2002; Prasad et al. 2003; Mondal et al. 2007; Brazauskiene and Petraitiene 2006; Smith et al. 1988; Gorshkov 1976; Stoll 1948; Domsch 1957). The disease has been assessed using several methods including descriptive keys (Mayee and Datar 1986; Saharan 1991; Horsfall and Barratt 1945; Fontem et al. 1991; Redman et al. 1967), standard area diagrams (Conn et al. 1990; James 1974), incidence–severity relationships (Seem 1984), inoculum disease intensity relationships (Saharan and Kadian 1983b; Dueck and Degenhardt 1975), infection type and host reaction as resistant and susceptible (Krishnia et al. 2000a) and disease stress tolerance attributes (Gupta et al. 2002).

1.4 The Pathogen

Alternaria is an ubiquitous fungal genus that includes saprobic, endophytic and pathogenic species. Four species of *Alternaria*, viz. *A. alternata*, *A. brassicae*, *A. brassicicola* and *A. raphani* are pathogenic on cruciferous crops. *Alternaria brassicae* is a major pathogen of oil-yielding *Brassica*, while other three are more common on vegetable crops. As early as 1836, Berkeley identified the causal fungus on Brassicaceae as *Macrosporium brassicae*, which was later renamed as *A. brassicae*. Phylogeny, taxonomy, morphology, classification, infection process, identification characteristics and synonyms have been well documented (Verma and Saharan 1994; Woudenberg et al. 2013; Simmons 2007; Thomma 2003; Ellis 1968a, b, 1971; Neergaard 1945). Identification, cloning and sequencing of virulence genes of *Alternaria* species infecting crucifers will resolve many doubts about their relationship with crucifer's hosts (Cramer et al. 2006; Jasalavich et al. 1995; Kim et al. 2007). Pathogenicity factors and transcription factor *Amr1* have been identified in *A. brassicicola* (Mamgain et al. 2013; Cho et al. 2012). A non-ribosomal peptide synthase gene (*AbNPS2*) is important for cell wall integrity, conidial viability and virulence of aged spores of *A. brassicicola* (Kim et al. 2007). More than 100 genes have been functionally analysed through various techniques like gene knockout and overexpression making *A. brassicicola* the species of choice for functional genomic research (Oide et al. 2006; Cho et al. 2006, 2007, 2009; Kim et al. 2007; Mamgain et al. 2013).

Twenty-four liquid and solid media for growth and sporulation of *Alternaria* species pathogenic on crucifers have been reported (Verma and Saharan 1994); *Alternaria brassicae*, *A. brassicicola* and *A. raphani* grow well on most carbon sources (Taber et al. 1968). The optimum temperatures for growth in culture of *A. brassicae*, *A. brassicicola* and *A. raphani* are between 20 and 25 °C along with 95–100 % relative humidity for good sporulation (Ansari et al. 1989; Taber et al. 1968; Changsri and Weber 1960, 1963). The optimum pH requirement for growth and sporulation of all the three species is 6–8. Maximum growth and sporulation of *A. brassicae* occurs with alternating light and darkness (Ansari et al. 1989; Verma and Saharan 1994; Taber 1964). The pathogen survives and perpetuates through infected seeds, diseased plant debris and pathogen propagules in the soil and other crucifers/weed hosts in a particular agroecosystem (Chupp and Sherf 1960; Dixon 1981; Ellis 1968a, b; Verma and Saharan 1994). Optimum temperature for spore germination is 20–25 °C at 90 % or more relative humidity. Continuous light completely inhibits sporulation (Verma and Saharan 1994; Singh and Suhag 1983). All four species of *Alternaria* are seed borne in crucifers. The pathogen has been detected from all the parts of infected seeds. Its transmission through seed is very high under congenial temperature conditions (Atkinson 1950; Vannacci and Pecchia 1988; Sivapalan and Browning 1992; Kubota et al. 2006; Shrestha et al. 2000). The disease cycle starts from the primary infections initiated directly from infected seed, spores on crops residues and on cruciferous hosts, weeds or possibly from microsclerotia and chlamydospores produced on infected debris. Spores are produced abundantly in wet weather and are dispersed locally by rain splash and wind. Under congenial weather conditions, lesions develop and produce windborne spores, which cause secondary infections during crop season. The cycle continues throughout the season when conditions are favourable to infect seed and other parts of plants, which become the source of survival of the pathogen (Kolte 1985; Saharan 1992; Verma and Saharan 1994; Mehta et al. 2005).

The process of infection and pathogenesis of four *Alternaria* species on cruciferous hosts has been very well understood. The role of external and internal factors during host–pathogen interactions, enzymes, toxins and genes governing pathogenesis has been determined. Identification of *A. brassicicola* genes, *AbVF 19* and *Amr1*, makes this pathogen as an efficient and successful facultative parasite of crucifers (Verma and Saharan 1994; Cho et al. 2007, 2009, 2012; Giri et al. 2013; Mamgain et al. 2013).

1.5 Epidemiology and Forecasting

Alternaria blight of crucifers develops in epidemic form when temperature ranges from 15 to 25 °C, relative humidity is >90 % and wind velocity is 2–5 km/h and during the presence of intermittent rains (Ansari et al. 1988; Saharan 1991; Awasthi and Kolte 1989b; Verma and Saharan 1994; Yadav and Brar 2003). Closer spacing (30×15 cm), high doses of nitrogen (80 Kg N/ha) and frequent irrigation rapidly increase severity of disease in rapeseed–mustard (Saharan 1991; Stankova 1972; Verma and Saharan 1994). Various models have been developed to measure the temporal progression of disease under field conditions when crops are sown on different dates under varied environmental conditions. Disease forecasting models have also been developed (Verma and Saharan 1994; Awasthi and Kolte 1994; Dang et al. 2006; Magarey et al. 2005; Mehta et al. 2002, 2008; Kumar et al. 2013; Mahapatra and Das 2014; Mehta 2014) after taking into account the leaf wetness period, minimum and maximum temperature, relative humidity, date of sowing, crop age and variety and species of *Brassica* crops under different agro-ecological conditions. Models based on weekly weather data beginning from week of sowing till up to 6 weeks of crop growth can be used for reliable forewarning of *Alternaria* blight; reliable forewarning regarding the crop age at first appearance of disease, peak severity of disease and maximum severity of disease in different crop varieties is possible (Kumar et al. 2013).

1.6 Pathogenic Variability

Pathogenic variability in the form of pathotypes/races/strains/variants has been identified in four species of *Alternaria* causing blight and black spot disease in crucifers. The pathogenic variability in *Alternaria* species is governed by determinant attributes, viz. pathological, symptomatological, morphological, cultural, nutritional, biochemical, genetical, molecular and proteome level and both thermo- and fungicidal sensitivity. Initial observations on variations in cultural characteristics and pathogenesis of different isolates of *Alternaria* were made by Stoll (1952), in *A. brassicicola*–vegetables, by Van Schreven (1953) in *A. brassicae–Brassica* and by Atkinson (1953) in *A. raphani–radish*–host pathosystem. *Alternaria alternata* strains showed differences in their physiological and pathological characteristics isolated from crambe. Strain B is most virulent, strain A is moderately virulent, and strain C is least pathogenic on crambe (Czyzewska 1969, 1971). On rapeseed–mustard group of crops, three races of *A. brassicae*, viz. RM-1, RM-2 and V-3, are virulent. Race RM-1 isolated from rapeseed–mustard is avirulent on *B. oleracea* var. *capitata*, Race RM-2 isolated from *B. rapa* is avirulent on both *B. oleracea* var. *capitata* and *B. oleracea* var. *botrytis*, and Race V-3 is most virulent infecting all the test host differentials. This race was isolated from vegetable host species including radish, cabbage and cauliflower (Saharan and Kadian 1983a). Thirteen isolates of *A. brassicae* tested on selected cultivars of winter rape differed in their virulence (Mridha 1983). Stoll (1952) characterized three pathotypes of *A. brassicae* infecting siliquae of cauliflower as highly aggressive, less aggressive and non-pathogenic. Atkinson (1953) classified *A. raphani* isolates in to two races as 'wild type' and 'variant type'. Three isolates of *A. brassicae* designated as A, C and D differ in their morphology, growth, sporulation and cultural characteristics along with virulence on *B. carinata* (Kolte et al. 1989, 1991; Awasthi and Kolte 1989a; Vishwanath and Kolte 1997). Four pathotypes of *A. brassicae* from *B. juncea* were identified and designated as Bj-4, Bj-5, Bj-6 and Bj-7 by Gupta et al. (2004), on the basis of host differentials and symptomatological variations. Pathotypes DLK, RSR-1 and GDP of *A. brassicae* were identified by Mehta et al. (2003). Some researchers have mentioned only the number of isolates exhibiting differential reactions without designating pathotypes (Kumar et al. 2003; Mehta et al. 2003; Sangwan and Mehta 2007). In the absence of host differentials, other parameters have been used to identify pathogenic variability in *Alternaria* species infecting crucifers.

Out of these criteria, symptomatology (Gupta et al. 2004; Kolte et al. 1991; Goyal et al. 2013), morphology (Kolte et al. 1989, 1991), genetics (Sharma et al. 2013; Priyanka et al. 2014) and proteome level have been correlated with virulence of pathotypes.

1.7 Fine Structures

Fine structures of *A. brassicicola* have been studied through electron microscopy. The sequence of events in the production and maturation of spores has been described. The changes in the internal organelles of hyphae, conidia and conidiophore have been observed. Conidiophores have a similar structure to mature hyphae, except that after spore production, they have a pore in the tip and an annule. There are variations in the number of nuclei in the cells of vegetative hyphae, conidiophores and conidia with six chromosomes in dividing nuclei of vegetative hyphae (Campbell 1970a, b, 1972; Knox-Davis 1979).

1.8 Biochemistry

During *Alternaria*–crucifers host–pathogen interactions, a number of biochemical changes take place in the host as well as in the pathogen. These biochemical changes produce various kinds of primary and secondary metabolites, which influence the host defence system and pathogen virulence. *Alternaria brassicicola* produces compounds like anti-tumouric depudecin, antibiotic complex brassicicolin and phytotoxic brassicicenes. The production of glucosinolates and phytoalexins has been correlated with host resistance (Verma and Saharan 1994; Atwal et al. 2003; Mathpal et al. 2011; Sharma et al. 2010; Doughty et al. 1996; Jung et al. 2002).

1.9 Resistance

Host resistance in crucifers against *Alternaria* species has various components and it is multilayered. Inheritance of resistance in inter- and intraspecific crosses of *B. juncea* and *B. carinata* to *A. brassicae* is governed by additive genes, dominant genes, additive x additive-type epistatic genes, additive x dominance and dominance x dominance type of non-allelic interaction genes (Singh and Singh 1989; Krishnia et al. 2000b). Inter-mating between resistance plants helps in increasing the level of resistance against *A. brassicae* through pyramiding of resistant genes. High level of horizontal resistance in genotypes of oilseed *Brassica* has been recorded (Saharan and Kadian 1983b; Saharan and Krishnia 2001). *Brassica* genotypes, Rajat, Kranti, RH-781 and RL-1359, have been identified with disease tolerance attributes (Gupta et al. 2002). Genotypes PR-8988 and PR-9024 show higher degree of partial resistance or slow blighting (Kumar and Kolte 2001). Epicuticular wax (Candle, Tobin, Altex, Midas, Tower) and low number and narrow stomatal aperture (Tower, RC-781) provide resistance to *Alternaria* infection in *Brassica* species (Saharan and Kadian 1983b; Conn et al. 1984; Tewari 1991a, b). The concentration of phenolic compounds, activation of polyphenol oxidase and catalase are higher in tolerant genotype of mustard (Gupta et al. 1990; Kiran et al. 2002). Chitinase-modifying proteins, *Cmps*, are secreted by fungal pathogens of crucifers, which interfere with fungalysin *Cmp* activity to improve plant resistance to multiple fungal diseases (Naumann and Wicklow 2013). *GLIP1* in association with ethylene signalling may be a critical component in plant resistance (*A. thaliana*) to *A. brassicicola* (Oh et al. 2005). *Brassica juncea* plants transformed with chitinase gene tagged with an overexpressing promoters, 35S *CaMV*, give defence response by degrading the cell walls of invading fungi (Mondal et al. 2003). Increased level of PAL, PPO and peroxidase plays an important role in the defence mechanisms of *B. juncea* genotypes against *Alternaria* pathogenesis (Parihar 2012). Treatment by β-aminobutyric acid leads to proper balance of oxidant and antioxidants suitable for expression of resistance in *B. carinata* against *A. brassicae* by curtailing pathogens ingress during early stages of colonization (Chavan et al. 2013). Zeatin, a cytokinin, up-regulates plant immunity via an elevation of

MAPK-4 and antagonizes the effects of *A. brassicae* (Marmath et al. 2013a, b). Transgenic expression of hevein, the rubber tree lectin in *B. juncea* cv. RLM-198, confers protection against *A. brassicae* (Kanrar et al. 2002); β-aminobutyric acid pretreatment of *B. juncea* plants induces *A. brassicae* resistance mediated through an enhanced expression of pathogenesis-related protein genes, independent of SA and JA accumulation (Kamble and Bhargava 2007). Hypersensitive response gene (*hsr 203J*) like homologues of *Brassica* plays important role in differential defence response against *A. brassicae* (Mishra et al. 2010). The cDNA encoding *Pm-AMP1* has been successfully incorporated into the genome of *B. napus*, and its in planta expression confers greater protection against *A. brassicae.* A cysteine-rich antimicrobial peptide, *Pm AMP1*, was isolated from *Pinus monticola* (Verma et al. 2012). Combined expression of a barley class II chitinase and type I ribosome-inactivating protein in transgenic *B. juncea* provides resistance against *A. brassicae* (Chhikara et al. 2012). Transcriptional responses to exposure to the brassicaceous defence metabolites camalexin and allyl isothiocyanate in *A. brassicicola* have been recorded (Sellam et al. 2007).

Elicitation and accumulation of phytoalexins in crucifers after exposure to *Alternaria* and their role in disease resistance have been demonstrated (Verma and Saharan 1994). Calcium sequestration property of *A. brassicae* can be used in enhancing resistance to this pathogen in rapeseed by soil or foliar application of calcium compounds (Tewari 1991a, b). Numerous sources of resistance to *Alternaria* species have been identified from different *Brassica* species and their near and distantly related coenospecies, but very few have been utilized to develop resistant cultivars (Verma and Saharan 1994; Saharan 1992, 1997; Saharan et al. 2003; Sharma et al. 2002). Sources of multiple disease resistance in *B. juncea*, *B. rapa*, *B. carinata* and *B. napus* have also been identified. Reports of strong and positive correlation in the increased level of resistance against different *Brassica* pathogens will facilitate improvement in accumulation of resistance to multiple disease resistance (Kumar and Saharan 2002; Mitchell-olds et al. 1995). In spite of several bottlenecks in the development of resistance cultivars, various methods/techniques including conventional as well as biotechnological approaches are being utilized to incorporate desired traits in cruciferous crops against *Alternaria* disease (Saharan et al. 2003; Nowicki et al. 2012; Aneja and Agnihotri 2013).

1.10 Phytotoxins

Alternaria species pathogenic to crucifers produce host-specific and non-host-specific toxins, which facilitate their pathogenic process to become successful pathogen. Prior to colonization, necrotrophs must kill their host cells at a distance by producing both toxins and lytic enzymes often by triggering genetically programmed apoplastic pathways or by directly causing cell damage resulting in necrosis. *Alternaria brassicae* and *A. brassicicola* pathogenic to crucifers produce a number of toxins and metabolites belonging to chemical groups containing terpenoids, pyranones, steroids and nitrogen. Effects of toxins on plants at physiological, biochemical and molecular level have been investigated. The role of toxins in the process of infection, their biosynthesis, mode of action, chemical structure, role in host defence and transformation into phytoalexins has been understood (Verma and Saharan 1994; Bains and Tewari 1989; Buchwaldt and Green 1992; Lou et al. 2013; Chen et al. 2005; Li et al. 2008; Rotem 1994; He et al. 1998; Pedras et al. 1998, 2001, 2002, 2009; Thomma 2003; Marmath et al. 2013a, b).

1.11 Disease Management

Common cultural practices like the use of clean, bold, healthy and treated seed of recommended cultivars, long crop rotation (3–4 years), sanitation, weed control, shallow (2-cm depth) planting at recommended time, use of balanced nutrients, proper plant density (45 × 30 cm), proper drainage in the field, plant debris management, use of tolerant/resistant cultivars, application of chemicals/bioagents at a proper time with adequate foliage coverage and education of farmers about the importance of appropriate practices have been advocated to manage *Alternaria* diseases of *Brassica* crops (Verma and Saharan 1994; Kolte 1985; Saharan and Chand 1988; Saharan 1992, 1997; Peruch and Michereff 2007). Seed treatment at 50 °C for 20 min is highly effective in controlling the seed-borne inoculum. Numerous chemicals and bioagents have been recommended to control seed-borne infection of *Alternaria* (Verma and Saharan 1994; Vannacci and Harman 1987; Latif et al. 2006). In vitro and in vivo testing of a large number of chemicals against *Alternaria* species infecting crucifers has been found very effective in controlling the disease and increasing the yield under field conditions. The number of sprays, optimum doses, optimum crop growth stage, spraying time, compatibility of different fungicides with insecticides, residual toxicity, nature of persistence, interval of sprays and cost–benefit ratio of the most effective fungicides have been determined (Verma and Saharan 1994; Singh and Singh 2005; Meena et al. 2004; Khan et al. 2007; Mondal et al. 2007; Sultana et al. 2009; Saharan 1991, 1992; Marshall and Haris 1984; Bonin and Fratczak 1987; Brazauskiene and Peteraitiene 2004; Davies et al. 1986). Indiscriminate use of high doses of fungicides may affect pollen biology of the crop resulting into adverse effect on seed yield (Williams and Pink 1987; Jain et al. 2000). Fungicides like iprodione, procymidone and fludioxonil have shown resistance against isolates of *A. brassicicola*, which may affect their efficacy under field conditions to control the disease (Huang and Levy 1995; Iacomi-Vasilescu et al. 2004). Some of the plant extracts and biocontrol agents have efficacy as high as fungicides in the control of *Alternaria* diseases of crucifers along with higher yield. Antagonistic mechanisms of biocontrol have been studied in a few selected host–parasite systems. Application of biocontrol agents like *Trichoderma harzianum*, *Pseudomonas fluorescence* and *Bacillus subtilis* initiates a number of biochemical changes in *B. juncea*, which trigger plant defence response (Verma and Saharan 1994; White et al. 1990; Tsuneda and Skoropad 1980; Danielsson et al. 2006; Sharma et al. 2010). The use of resistant cultivars is the most easy, economical, environmental friendly and safest way of plant disease control. However, worldwide efforts are being made to develop and release resistant cultivars of crucifers to mange *Alternaria* disease (Verma and Saharan 1994; Saharan et al. 2003).

Integration of all plant disease control strategies, viz. cultural, chemical, biological, nutritional, host resistance, biotechnological and genetic engineering including pest management, is the best way to manage *Alternaria* disease of crucifers (Verma and Saharan 1994; Saharan and Mehta 2002; Kolte 2005; Mehta 2014).

1.12 Techniques

Twenty-four standardized, reproducible methods, procedures and tools have been described, which will help the researchers to refine and reconfirm their findings in the future (Verma and Saharan 1994).

The last chapter of the book highlights some priority areas of research to be conducted by future researchers to manage and understand *Alternaria*–crucifer pathosystem in a better way.

References

Aneja JK, Agnihotri A (2013) Alternaria blight of oilseed brassicas: epidemiology and disease control strategies with special reference to use of biotechnological approaches for attaining host resistance. J Oilseed Bras 4:1–10

Anonymous (1985) Crucifer genetics cooperative (GrGc). Department of Plant Pathology, University of Wisconsin, Madison, USA, Section G, pp 1–2

Ansari NA, Khan MW, Muheet A (1988) Effect of temperature and relative humidity on spore germination of *Alternaria brassicae* and development of Alternaria blight on oilseed crucifers. Rev Trop Plant Pathol 5:79–84

Ansari NA, Khan MW, Muheet A (1989) Effect of some factors on growth and sporulation of *Alternaria brassicae* causing Alternaria blight of rapeseed and mustard. Acta Bot Ind 17:49–53

Atkinson RG (1950) Studies on the parasitism and variation of *Alternaria raphani*. Can J Res Sect C 28:288–317

Atkinson RG (1953) Survival and pathogenicity of *Alternaria raphani* after five years in dried soil cultures. Can J Bot 31:542–547

Atwal AK, Ramandeep, Munshi SK, Mann APS (2003) Biochemical changes in relation to Alternaria leaf blight in Indian mustard. Plant Dis Res 19:57–59

Awasthi RP, Kolte SJ (1989a) Variability in *Alternaria brassicae* affecting rapeseed and mustard. Indian Phytopathol 42:275. (Abstr.)

Awasthi RP, Kolte SJ (1989b) Effect of some epidemiological factors on occurrence and severity of Alternaria blight of rapeseed and mustard. In: Proceedings of IDRC (Canada) oil crops: meetings held at Pantnagar and Hyderabad, India, 4–17 January, pp 49–55

Awasthi RP, Kolte SJ (1994) Epidemiological factors in relation to development and prediction of Alternaria blight of rapeseed and mustard. Indian Phytopathol 47:395–399

Bains PS, Tewari JP (1989) Bioassay and mode of action of an *Alternaria brassicae* toxin. Can J Plant Pathol 11:184–189

Berkeley MJ (1836) Fungi. In: Smith JE (ed) The English flora. vol 5, pt. II. Longman, London, p 339

Bonin K, Fratczak E (1987) Control of fungal diseases in winter oil seed rape. In: Proceedings of the 7th international rapeseed congress held during May 11–14, 1987, Poznan, Poland, pp 1281–1287

Brazauskiene I, Petraitiene E (2004) Effects of fungicide application timing on the incidence and severity of Alternaria blight (*Alternaria brassicae*) and on the productivity of spring oilseed rape (*Brassica napus* L. ssp. *oleifera annua* Metzg.). Agron Res 2:121–133

Brazauskiene I, Petraitiene E (2006) The occurrence of *Alternaria* blight (*Alternaria* spp.) and Phoma stem canker (*Phoma lingam*) on oilseed rape in central Lithuania and pathogenic fungi on harvested seed. J Plant Prot Res 46:295–311

Buchwaldt L, Green H (1992) Phytotoxicity of destruxin B and its possible role in the pathogenesis of *Alternaria brassicae*. Plant Pathol 41:55–63

Campbell R (1970a) An electron microscope study of exogenously dormant spores, spore germination, hyphae and conidiophores of *Alternaria brassicicola*. New Phytol 69:287–293

Campbell R (1970b) Ultra-structure of an albino strain of *Alternaria brassicicola*. Trans Br Mycol Soc 54:309–313

Campbell R (1972) Changes in volume of reproductive structures during spore production by *Alternaria brassicicola*. Trans Br Mycol Soc 59:153–156

Chang K (1968) Archeology of ancient China. Science 162:519–526

Changsri W, Weber GF (1960) Studies of *Alternaria* spp. pathogenic on Cruciferae. Phytopathology 50:631. (Abstr.)

Changsri W, Weber GF (1963) Three *Alternaria* species pathogenic on certain cultivated crucifers. Phytopathology 53:643–648

Chavan V, Bhargava S, Kamble A (2013) Temporal modulation of oxidant and anti-oxidative responses in *Brassica carinata* during β-aminobutyric acid-induced resistance against *Alternaria brassicae*. Physiol Mol Plant Pathol 83:35–39

Chen LY, Price TV, Silvapulle MJ (2005) Dark leaf spot (*Alternaria brassicicola*) on Chinese cabbage: spatial patterns. Aust J Agr Res 56:699–714

Chhikara S, Chaudhury D, Dhankher OP, Jaiswal PK (2012) Combined expression of a barley class II chitinase and type I ribosome inactivating protein in transgenic *Brassica juncea* provides protection against *Alternaria brassicae*. Plant Cell Tissue Organ Cult 108:83–89

Cho Y, Davis JW, Kim KH, Wang J, Sun QH, Cramer RA Jr, Lawrence CB (2006) A high throughput targeted gene disruption method for *Alternaria brassicicola* functional genomics using linear minimal element (LME) constructs. Mol Plant Microbe Interact 19:7–15

Cho Y, Cramer R, Kwang-Hyung K, Davis J, Mitchell T, Figuli P, Pryor B, Lemasters E, Lawrence CB (2007) The *Fus3/Kss1* MAP kinase homolog *Amk1* regulates the expression of hydrolytic enzyme genes in the fungus *Alternaria brassicicola*. Fungal Genet Biol 44:543–553

Cho Y, Kim KH, La Rota M, Scott D, Santopietro G, Callihan M, Mitchell TK (2009) Identification of virulence factors by high throughput targeted gene deletion of regulatory genes in *Alternaria brassicicola*. Mol Microbiol 72:1316–1333

Cho Y, Srivastava A, Ohm RA, Lawrence CB, Wang KH, Grigoriev IV, Marahatta SP (2012) Transcription factor *Amr1* induces melanin biosynthesis and suppresses virulence in *Alternaria brassicicola*. PLoS Pathog 8:e1002974. doi:10.1371/journal.ppat.1002974

Chou HH, Wu WS (2002) Phylogenetic analysis of internal transcribed spacer regions of the genus *Alternaria* and the significance of filament-beaked conidia. Mycol Res 106:164–169

Chupp C, Sherf AF (1960) Chapter 8: Crucifer diseases. In: Vegetable diseases and their control. Ronald Press Company, New York, pp 237–288

Conn KL, Tewari JP, Hadziyev D (1984) The role of epicuticular wax in canola in resistance to *Alternaria brassicae*. Phytopathology 74:851. (Abstr.)

Conn KL, Tewari JP, Awasthi RP (1990) A disease assessment key for Alternaria black spot in rapeseed and mustard. Can Plant Dis Surv 70:19–22

Cramer RA, Rota CM, La Cho Y, Thon M, Craven KD, Knudson DL, Mitchell TK, Lawrence CB (2006) Bioinformatic analysis of expressed sequence tags derived from a compatible *Alternaria brassicicola*–*Brassica oleracea* interaction. Mol Plant Pathol 7:113–124

Czyzewska S (1969) Alternaria blight of *Crambe abyssinica*. Acta Mycol 5:175–211

Czyzewska S (1971) The pathogenicity of *Alternaria* spp. isolated from *Crambe abyssinica*. Acta Mycol 7:171–240

Daebeler F, Amelung D, Riedel V (1986) Leaf spot damage caused by *Alternaria* sp. in winter rape. Wissenschaftliche Zeitschrift der Wilhelm-Pieck-

Universitat Rostock, Naturwissens chaftliche Reihe 35:52–54

Dang JK, Sangwan MS, Mehta N, Sharma OP, Dhandapani A (2006) Development of prediction model for Alternaria blight caused by *Alternaria brassicae* of rapeseed and mustard. Plant Dis Res 21:199–201

Danielsson J, Reva O, Meijer J (2006) Protection of oilseed rape (*Brassica napus*) toward fungal pathogens by strains of plant-associated *Bacillus amyloliquefaciens*. Microb Ecol 54:134–140

Davies JML, Ann DM, Dobson SC, Jones OW, Gladders P, Melville SC, Mc Pherson GM, Popham MD, Price RJ, Wafford JD (1986) Fungicide spray timing on Brussels sprouts. Asp Appl Biol 12:21–28

de Hoog GS, Horré R (2002) Molecular taxonomy of the *Alternaria* and *Ulocladium* species from humans and their identification in the routine laboratory. Mycoses 45:259–276

Degenhardt KJ, Skoropad WP, Kondra ZP (1974) Effects of Alternaria blackspot on yield, oil content and protein content of rapeseed. Can J Plant Sci 54:795–799

Dixon GR (1981) Pathogens of crucifer crops. In: Vegetable crop diseases. AVI Publ. Company Inc, Westport, pp 119–123

Domsch KH (1957) The blackening of rape and cabbage siliquae. Z Pflanzenkr Pflanzen schutz 64:65–79

Doughty KJ, Blight MM, Bock CH, Fieldsend JK, Pickett JA (1996) Release of alkenyl isothiocyanates and other volatiles from *Brassica rapa* seedlings during infection by *Alternaria brassicae*. Phytochemistry 43:371–374

Dueck J, Degenhardt K (1975) Effect of leaf age and inoculum concentration on reaction of oilseed *Brassica* spp. to *Alternaria brassicae*. Proc Am Phytopathol Soc 2:59. (Abstr.)

Elliott JA (1917) Taxonomic characters of the genera *Alternaria* and *Macrosporium*. Am J Bot 4:439–476

Ellis MB (1968a) *Alternaria brassicae*. Commonwealth Mycological Institute. Descriptions of Pathogenic Fungi and Bacteria No. 162. CMI, Kew

Ellis MB (1968b) *Alternaria brassicicola*. Commonwealth Mycological Institute. Descriptions of Pathogenic Fungi and Bacteria. No. 163. CMI, Kew

Ellis MB (1971) Dematiaceous hyphomycetes. Commonwealth Mycological Institute, Kew, No. 608, pp 464–497

Ellis MB (1976) More dematiaceous hyphomycetes. Commonwealth Mycological Institute, Kew, pp 1–507

Fontem DA, Berger RD, Weingartner DP, Bartz JA (1991) Progress and spread of dark leaf spot in cabbage. Plant Dis 75:269–274

Fries E (1832) Systema mycologicum, vol 3. E. Moritz, Greifswald, 500 p

Giri P, Taj G, Kumar A (2013) Comparison of artificial inoculation methods for studying pathogenesis of *Alternaria brassicae* (Berk.) Sacc. on *Brassica juncea* (L.) Czern. (Indian mustard). Afr J Biotechnol 12:2422–2426

Gorshkov AK (1976) Infectious damping-off of cabbage seedlings under glass. Trudy Vses s-kh Inst Zaoch Obrazovaniya 117:25–29

Goyal P, Chattopadhyay C, Mathur AP, Kumar A, Meena PD, Datta S, Iquebal MA (2013) Pathogenic and molecular variability among *Brassica* isolates of *Alternaria brassicae* from India. Ann Plant Prot Sci 21:349–359

Gupta SK, Gupta PP, Yadav TP, Kaushik CD (1990) Metabolic changes in mustard due to Alternaria leaf blight. Indian Phytopathol 43:64–69

Gupta V, Chawla HKL, Dhawan K, Mehta N (1998) Alterations in total phenols and glucosinolates in *Brassicae juncea* leaves during interaction with *Alternaria brassicae*. Cruciferae News Lett 20:83–85

Gupta K, Saharan GS, Singh D (2002) Effective selection criteria to assess disease tolerance in Indian Mustard (*Brassica juncea* L. Czern and Coss). J Mycol Plant Pathol 32:72–80

Gupta K, Saharan GS, Mehta N, Sangwan MS (2004) Identification of pathotypes of *Alternaria brassicae* from Indian mustard [*Brassica juncea* (L.) Czern. and Coss]. J Mycol Plant Pathol 34:15–19

He C, Rusu AG, Poplawski AM, Irwin JA, Manners JM (1998) Transfer of a supernumerary chromosome between vegetatively incompatible biotypes of the fungus *Colletotrichum gloeosporioides*. Genetics 150:1459–1466

Hong SG, Cramer RA, Lawrence CB, Pryor BM (2005) Alt a1 allergen homologs from *Alternaria* and related taxa: analysis of phylogenetic content and secondary structure. Fungal Genet Biol 42:119–129

Horsfall JG, Barratt RW (1945) An improved grading system for measuring plant disease. Phytopathology 35:655 (Abstr.)

Hougen FW, Stefansson BR (1983) Rapeseed. In: Advances in cereal science and technology, vol 5. American Association of Cereal Chemists, St. Paul, pp 261–289

Huang R, Levy Y (1995) Characterization of iprodione-resistant isolates of *Alternaria brassicicola*. Plant Dis 79:828–833

Hyams E (1971) Plants in the service of man. J. M. Dent and Sons, London, pp 33–61

Iacomi-Vasilescu B, Avenot H, Bataille-Simoneau N, Laurent E, Guenard M, Simoneau P (2004) *In vitro* fungicide sensitivity of *Alternaria* species pathogenic to crucifers and identification of *Alternaria brassicicola* field isolates highly resistant to both dicarboximides and phenylpyrroles. Crop Prot 23:481–488

Inderbitzin P, Shoemaker RA, O'Neill NR, Turgeon BG, Berbee ML (2006) Systematics and mating systems of two fungal pathogens of opium poppy: the heterothallic *Crivellia papaveracea* with a *Brachycladium penicillatum* asexual state and a homothallic species with a *Brachycladium papaveris* asexual state. Can J Bot 84:1304–1326

Jain SC (1992) Assessment of losses in taramira (*Eruca sativa* Lam.) due to Alternaria blight disease. Madras Agric J 79:57–58

Jain AK, Briegleb BP, Minschwaner K, Wuebbles DJ (2000) Radiative forcings and global warming potentials of thirty-nine greenhouse gases. J Geophys Res 105(20):773–790

James WC (1974) Assessment of plant diseases. Annu Rev Phytopathol 12:27–48

Jasalavich CA, Morales VM, Pelcher LE, Seguin-Swartz G (1995) Comparison of nuclear ribosomal DNA sequences from *Alternaria* species pathogenic to crucifers. Mycol Res 99:604–614

Joly P (1964) Le genre Alternaria. Encyclopédie Mycologique Paris France 35:174–179

Jung DS, Na YJ, Ryu KH (2002) Phylogenic analysis of *Alternaria brassicicola* producing bioactive metabolites. J Microbiol 40:289–294

Kadian AK, Saharan GS (1983) Symptomatology, host range and assessment of yield losses due to *Alternaria brassicae* infection in rapeseed and mustard. Indian J Mycol Plant Pathol 13:319–323

Kadian AK, Saharan GS (1984) Studies on spore germination and infection of *Alternaria brassicae* of rapeseed and mustard. J Oilseed Res 1:183–188

Kamble A, Bhargava S (2007) β–aminobutyric acid-induced resistance in *Brassica juncea* against the necrotrophic pathogen *Alternaria brassicae*. J Phytopathol 155:152–158

Kanrar S, Venkateswari JC, Kirti PB, Chopra VL (2002) Transgenic expression of hevein, the rubber tree lectin, in Indian mustard confers protection against *Alternaria brassicae*. Plant Sci 162:441–448

Khan MM, Khan RU, Mohiddin FA (2007) *In vitro* studies on the variation of different temperature and culture media on the growth of *Alternaria brassicae* (Berk.) Sacc. infecting rapeseed-mustard. Ann Plant Prot Sci 15:526–527

Kim KH, Cho Y, La Rota M, Cramer R, Lawrence C (2007) Functional analysis of the *Alternaria brassicicola* non-ribosomal peptide synthetase gene *AbNPS2* reveals a role in conidial cell wall construction. Mol Plant Pathol 8:23–39

Kiran DHR, Mehta N, Sangwan MS (2002) Evaluating relative resistance of Brassicas to Alternaria blight through tissue culture technique. J Mycol Plant Pathol 32:231–233

Klemm M (1938) The most important diseases and pests of colza and rape. Dtsch Landw Pr, lxv 19:239; 20:251–252

Knox-Davies PS (1979) The nuclei of *Alternaria brassicicola*. Trans Br Mycol Soc 72:81–90

Kolte SJ (1985) Diseases of annual edible oilseed crops, vol 2. CRC Press, Boca Raton, p 135

Kolte SJ (2005) Tackling fungal diseases of oilseed Brassicas in India. Brassica 7:7–13

Kolte SJ, Awasthi RP, Vishwanath (1989) Disease problems in Brassicas and research activities at Pantnagar. In: IDRC (Canada) oil crops: proceedings of the three meetings held at Pantnagar and Hyderabad, India, 4–17 January 1989, pp 43–48

Kolte SJ, Bardoloi DK, Awasthi RP (1991) The search for resistance to major diseases of rapeseed mustard in India. In: Proceedings of the 8th International Rapeseed Congress, 9–11 July 1991, Saskatoon, Canada 60 p. (Abstr.), vol 1, pp 219–225

Krishnia SK, Saharan GS, Singh D (2000a) Genetics of Alternaria blight resistance in inter and intra-specific crosses of *Brassica juncea* and *Brassica carinata*. Ann Biol 16:212–216

Krishnia SK, Saharan GS, Singh D (2000b) Genetic variation for multiple disease resistance in the families of inter-specific cross of *Brassica juncea x Brassica carinata*. Cruciferae News Lett 22:51–53

Kubota M, Abiko K, Yanagisawa Y, Nishi K (2006) Frequency of *Alternaria brassicicola* in commercial cabbage seeds in Japan. J Gen Plant Pathol 72:197–204

Kumar A (1997) Assessment and economics of avoidable yield losses due to Alternaria blight in oilseed Brassicas. Plant Dis Res 12:152–156

Kumar B, Kolte SJ (2001) Progression of Alternaria blight of mustard in relation to components of resistance. Indian Phytopathol 54:329–331

Kumar S, Saharan GS (2002) Sources of multiple disease resistance in *Brassica* spp. J Mycol Plant Pathol 32:184–188

Kumar S, Sangwan MS, Mehta N, Kumar R (2003) Pathogenic diversity in isolates of Alternaria brassicae infecting rapeseed-mustard. J Mycol Plant Pathol 33:59–64

Kumar A, Agrawal R, Chattopadhyay C (2013) Weather based forecast models for diseases in mustard crop. Mausam 64:663–670

Latif MA, Abu Kaoser MS, Khan MAI, Rahman H, Hossain MA (2006) Efficacy of some plant extracts in controlling seed-borne fungal infections of mustard. Bangladesh J Microbiol 23:168–170

Lawrence DP, Park MS, Pryor BM (2012) *Nimbya* and *Embellisia* revisited, with nov. comb for *Alternaria celosiae* and *A. perpunctulata*. Mycol Prog 11:799–815

Lawrence DP, Gannibal PB, Peever TL, Pryor BM (2013) The sections of *Alternaria*: formalizing species-groups concepts. Mycologia 105:530–546

Li HY, Xiao S, Chye ML (2008) Ethylene and pathogen inducible *Arabidopsis* acyl-COA binding protein 4 interacts with an ethylene-responsive element binding protein. J Exp Bot 59:3997–4006

Lou J, Fu L, Peng Y, Zhou L (2013) Metabolites from *Alternaria* fungi and their bioactivities. Molecules 18:5891–5935

Magarey RD, Sutton TB, Thayer CL (2005) A simple generic infection model for foliar fungal plant pathogens. Phytopathology 95:92–100

Mahapatra S, Das S (2014) Effect of meteorological factors on progression of Alternaria leaf blight of mustard and comparison of Logistic and Gompertz growth models in predicting disease severity. Indian Phytopathol 67:155–158

Mamgain A, Roychowdhury R, Jagatpati T (2013) *Alternaria* pathogenicity and its strategic controls. Res J Biol 1:1–9

Marmath KK, Giri P, Sharma S, Taj G, Kumar A (2013a) *In-silico* interaction studies of *Alternaria brassicae* toxin destruxin B and potential partners of *MAPK4* cascade. Int J Agric Environ Biotechnol 6:203–210

Marmath KK, Giri P, Taj G, Pandey D, Kumar A (2013b) Effect of zeatin on the infection process and expression of *MAPK-4* during pathogenesis of *Alternaria brassicae* in non-host and host *Brassica* plants. Afr J Biotechnol 12:2164–2174

Marshall J, Harris RI (1984) Broad spectrum disease control in oilseed rape with prochloraz. In: Proceedings of the British crop protection conference, pest and diseases during 19–22 November 1984, Brighton Metropole, England, Vol 2, pp 729–734

Mathpal P, Punetha H, Tewari AK, Agrawal S (2011) Biochemical defense mechanism in rapeseed-mustard genotypes against Alternaria blight disease. J Oilseed Brass 2:87–94

Mayee CD, Datar VV (1986) Host range and disease assessment scales. In: Phytopathometry. Marathwada Agricultural University, Prabhani, pp 110–111

Meah MB, Hau B, Siddiqua MK (2002) Relationships between disease parameters of Alternaria-blight (*Alternaria brassicae*) and yield of mustard. Zeitschrift für Pflanzenkrankheiten und Pflanzenschutz – J Plant Dis Protec 109:243–251

Meena PD, Meena RL, Chattopadhyay C, Kumar A (2004) Identification of critical stage for disease development and biocontrol of Alternaria blight of Indian mustard (*Brassica juncea*). J Phytopathol 152:204–209

Mehta N (2014) Epidemiology and forecasting for the management of rapeseed-mustard diseases. J Mycol Plant Pathol 44:131–147

Mehta N, Sangwan MS, Srivastava MP, Kumar R (2002) Survival of *Alternaria brassicae* causing Alternaria blight in rapeseed-mustard. J Mycol Plant Pathol 32:64–67

Mehta N, Sangwan MS, Srivastava MP (2003) Morphological and pathological variations in rapeseed-mustard isolates of *Alternaria brassicae*. Indian Phytopathol 56:188–190

Mehta N, Sangwan MS, Saharan GS (2005) Fungal diseases of rapeseed mustard. In: Saharan GS, Mehta N, Sangwan MS (eds) Diseases of oilseed crops. Indus Publishing Company, New Delhi, pp 15–86

Mehta N, Sangwan MS, Kumar R, Niwas R (2008) Progression of Alternaria blight on different varieties of rapeseed- mustard in relation to weather parameters. Plant Dis Res 23:28–33

Mishra A, Pandey D, Goel A, Kumar A (2010) Molecular cloning and *in-silico* analysis of functional homologues of hypersensitive response gene(s) induced during pathogenesis of Alternaria blight in two genotypes of Brassica. J Proteomic Bioinform 3:244–248

Mitchell-Olds TR, James V, Palmer MJ, Williams PH (1995) Genetics of *Brassica rapa* (syn. *Campestris*) L. multiple disease resistance to three fungal pathogens; *Peronospora parasitica*, *Albugo candida* and *Leptosphaeria maculans*. Heredity 75:362–369

Mondal KK, Chatterjee SC, Viswakarma N, Bhattacharya RC, Grover A (2003) Chitinase-mediated inhibitory activity of *Brassica* transgenic on growth of *Alternaria brassicae*. Curr Microbiol 47:171–173

Mondal KK, Bhattacharya RC, Koundal KR, Chatterjee SC (2007) Transgenic Indian mustard (*Brassica juncea*) expressing tomato glucanase leads to arrested growth of *Alternaria brassicae*. Plant Cell Rep 26:247–252

Mridha MAU (1983) Virulence of different isolates of *Alternaria brassicae* on winter oilseed rape cultivars. In: Proceedings of the 6th international rapeseed conference, Paris, France, pp 1025–1029

Naumann TA, Wicklow DT (2013) Chitinase modifying proteins from phylogenetically distinct lineages of *Brassica* pathogens. Physiol Mol Plant Pathol 82:1–9

Neergaard P (1945) Danish species of *Alternaria* and *Stemphylium* taxonomy, parasitism, economical significance. Oxford University Press, Oxford, 559 p

Nees von Esenbeck CGD (1816) Das System der Pilze und Schwamme. Stahelschen Buchhandlung, Wurzburg, 329 p

Nowicki M, Nowakowska M, Niezgoda A, Kozik EU (2012) Alternaria black spot of crucifers: symptoms, importance of disease, and perspectives of resistance breeding. Veg Crop Res Bull 76:5–19

Oh IS, Park AR, Bae MS, Kwon SJ, Kim YS, Lee JE, Kang NY, Lee S, Cheong H, Ohkmae KP (2005) Secretome analysis reveals an *Arabidopsis* lipase involved in defense against *Alternaria brassicicola*. Plant Cell 17:2832–2847

Oide S, Moeder W, Krasnoff S, Gibson D, Haas H, Yoshioka K, Gillian Turgeon B (2006) NPS6, encoding a non-ribosomal peptide synthetase involved in siderophore-mediated iron metabolism, is a conserved virulence determinant of plant pathogenic ascomycetes. Plant Cell 18:2836–2853

Parihar PS (2012) Changes in metabolites of *Brassica juncea* (Indian mustard) during progressive infection of *Alternaria brassicae*. Nat Sci 10:39–42

Pedras MSC, Kuhn AQ, Taylor JL (1998) The phytoalexin camalexin is not metabolized by *Phoma lingam*, *Alternaria brassicae*, or phytopathogenic bacteria. Plant Sci 139:1–8

Pedras MSC, Zaharia IL, Gai Y, Zhou Y, Ward DE (2001) In planta sequential hydroxylation and glycosylation of a fungal phytotoxin: avoiding cell death and overcoming the fungal invader. Proc Natl Acad Sci U S A 98:747–752

Pedras MSC, Zaharia LI, Ward DE (2002) The destruxins: synthesis, biosynthesis, biotransformation, and biological activity. Phytochemistry 59:579–596

Pedras MS, Chumala PB, Jin W, Islam MS, Hauck DW (2009) The phytopathogenic fungus *Alternaria brassicicola*: phytotoxin production and phytoalexin elicitation. Phytochemistry 70:394–402

Peruch LAM, Michereff SJ (2007) Saprophytic survival of *Alternaria brassicicola* and management of broccoli leaf debris. Ciência Rural 37:13–18

Prakash S, Hinata K (1980) Taxonomy, cytogenetics and origin of crop Brassicas- a review. Opera Bot 55:1–57

Prasad R, Saxena D, Chandra S (2003) Yield losses by Alternaria blight in promising genotypes of Indian mustard. Indian Phytopathol 56:205–206

Preuss CGT (1851) Übersicht untersuchter pilze, besonders aus der Umgegend von Hoyerswerda. Linnaea 24:99–153

Priyanka PG, Tasleem M, Taj G, Mal R, Kumar A (2014) Morphological, cultural, pathogenic and molecular variability amongst Indian mustard isolates of *Alternaria brassicae* in Uttarakhand. Afr J Biotechnol 13:441–448

Pryor BM, Bigelow DM (2003) Molecular characterization of *Embellisia* and *Nimbya* species and their relationship to *Alternaria*, *Ulocladium* and *Stemphylium*. Mycologia 95:1141–1154

Pryor BM, Gilbertson RL (2000) Molecular phylogenetic relationships amongst *Alternaria* species and related fungi based upon analysis of nuclear ITS and *mt SSU* rDNA sequences. Mycol Res 104:1312–1321

Pryor BM, Creamer R, Shoemaker RA, McLain-Romero J, Hambleton S (2009) *Undifilum*, a new genus for endophytic *Embellisia oxytropis* and parasitic *Helminthosporium bornmuelleri* on legumes. Botany 87:178–194

Raabe A (1939) Investigations on parasitic fungal diseases of colza and rape. Zbl Bakt Abt 2:35–52

Ramsey GB, Smith MA (1961) Market diseases of cabbage, cauliflower, turnips, cucumbers, melons and related crops. US Dept Agric, Agric Handbook 184:49

Redman CE, King EP, Brown JF (1967) Tables for converting Barratt-Horsfall rating scores to estimate mean percentages. Eli Lilly Co., Indianpolis, 100 p

Rotem J (1994) The genus *Alternaria*: biology, epidemiology and pathogenicity. APS Press, St. Paul, 326 p

Runa F, Park M, Pryor B (2009) *Ulocladium* systematics revisited: phylogeny and taxonomic status. Mycol Prog 8:35–47

Saccardo PA (1886) Hyphomycetese. In: Sylloge fungorum omnium hucusque cognitorum, vol 4. Patavii, Sumptibus Auctoris, Typis Seminarii, New York, p 807

Saharan GS (1984) A review of research on rapeseed-mustard pathology in India. Paper presented in the annual rabi oilseed workshop held at Jaipur, 6–10 August, 1984

Saharan GS (1991) Assessment of losses, epidemiology and management of black spot disease of rapeseed-mustard. In: Proceedings of GCIRC 8th International Rapeseed Congress, 9–11 July, Saskatoon, Canada, p 84 (Abstr.), vol 2, pp 465–470

Saharan GS (1992) Disease resistance. In: Labana KS, Banga SS, Banga SK (eds) Breeding oilseed brassicas. Narosa Publishing House, New Delhi, pp 181–200

Saharan GS (1997) Disease resistance. In: Kalia HR, Gupta SK (eds) Recent advances in oilseed brassicas. Kalyani Publishers, Ludhiana, pp 233–259

Saharan GS, Chand JN (1988) Chapter 3: Diseases of rapeseed and mustard. In: Diseases of oilseed crops (in Hindi). Directorate of Publication, Haryana Agricultural University Press, Hisar, pp 84–91

Saharan GS, Kadian AK (1983a) Physiologic specialization in *Alternaria brassicae*. Cruciferae News Lett 8:32–33

Saharan GS, Kadian AK (1983b) Analysis of components of horizontal resistance in rapeseed and mustard cultivars against *Alternaria brassicae*. Indian Phytopathol 36:503–507

Saharan GS, Krishnia SK (2001) Multiple disease resistance in rapeseed and mustard. In: Nagarajan S, Singh DP (eds) Role of resistance in intensive agriculture. Kalyani Publications, New Delhi, pp 98–108

Saharan GS, Mehta N (2002) Fungal diseases of rapeseed-mustard. In: Gupta VK, Paul YS (eds) Diseases of field crops. Indus Publishing Company, New Delhi, pp 193–201

Saharan GS, Mehta N, Sangwan MS (2003) Nature and mechanism of disease resistance to Alternaria blight in rapeseed-mustard system. Ann Rev Plant Pathol 2:85–128

Sangwan MS, Mehta N (2007) Pathogenic variability in isolates of *Alternaria brassicae* (Berk.) Sacc. from different agro-climatic zones of India. Plant Dis Res 22:101–107

Seem RC (1984) Disease incidence and severity relationships. Annu Rev Phytopathol 22:133–150

Seidle E, Rude S, Petrie A (1995) The effect of Alternaria black spot of canola on seed quality and seed yield and studies on disease control. Agriculture and Agri-Food Canada, Saskatoon, 41 p

Sellam A, Dongo A, Guillemette T, Hudhomme P, Simoneau P (2007) Transcriptional responses to exposure to the brassicaceous defence metabolites camalexin and allyl-isothiocyanate in the necrotrophic fungus *Alternaria brassicicola*. Mol Plant Pathol 8:195–208

Shahidi F (1990) Rapeseed and Canola: global production and distribution. In: Shahidi F (ed) Canola and rapeseed: production, chemistry, nutrition and processing technology. Van Nostrand Reinhold (Pub.), New York, 1, pp 3–13

Sharma G, Kumar VD, Haque A, Bhat SR, Prakash S, Chopra VL (2002) *Brassica coenospecies*: a rich reservoir for genetic resistance to leaf spot caused by *Alternaria brassicae*. Euphytica 125:411–417

Sharma N, Rahman MH, Liang Y, Kav NNV (2010) Cytokinin inhibits the growth of *Leptosphaeria maculans* and *Alternaria brassicae*. Can J Plant Pathol 32:306–314

Sharma M, Deep S, Bhati DS, Chowdappa P, Selvamani R, Sharma P (2013) Morphological, cultural, pathogenic and molecular studies of *Alternaria brassicae* infecting cauliflower and mustard in India. Afr J Microbiol Res 7:3351–3363

Shrestha SK, Mathur SB, Munk L (2000) Transmission of *Alternaria brassicae* in seeds of rapeseed and mustard, its location in seeds and control. Seed Sci Technol J 28:75–84

Shrestha SK, Munk L, Mathur SB (2005) Role of weather on Alternaria leaf blight disease and its effect on yield and yield component of mustard. Nepal Agric Res J 6:62–72

Simmons EG (1967) Typification of *Alternaria, Stemphylium* and *Ulocladium*. Mycologia 59:67–92
Simmons EG (2002) *Alternaria* themes and variations (305–309). Mycotaxon 83:127–145
Simmons EG (2007) Alternaria: an identification manual, CBS Biodiversity series no. 6. Utrecht, The Netherlands, p 775
Singh DP (1958) Rape and mustard. Indian Central Oilseed Committee, Hyderabad, 105 p
Singh H, Singh D (1989) Studies on genetic control of resistance to *Alternaria brassicae* in Indian mustard. In: Proceedings of XII Eucarpia Congress, 11–13, Abstr
Singh RB, Singh RN (2005) Fungicide management of foliar disease of mustard in mid-eastern India. Indian Phytopathol 58:51–56
Singh R, Suhag LS (1983) Some pathological studies on *Alternaria alternata* causing leaf and pod blight of radish in Haryana. Indian Phytopathol 36:174–176
Sivapalan A, Browning JW (1992) Incidence of *Alternaria brassicicola* (Schw.) Wiltsh. on *Brassica oleracea* seeds. Aust J Exp Agric 32:535–537
Smith IM, Dunez J, Phillips DH, Lelliott RA, Archer SA (1988) European handbook of plant diseases. Blackwell Scientific Publications, Oxford
Snogerup S (1980) The wild forms of *Brassica oleracea* group (2n = 18) and their possible relations to the cultivated ones. In: Tsunoda S, Hinata K, Gomez-campa C (eds) Brassica crops and wild allies, vol 7. Japan Scientific Society, Press, Tokyo, pp 121–132
Stankova J (1972) Varietal variability of winter rape in its susceptibility to dark leaf spot and the factors influencing the development of the disease. Rostlinna Vyroba 18:625–630
Stoll K (1948) On the Alternaria blackening of Brassicas. NachrBl Dtsch PflSch Dienst Berl N F 2:174–178
Stoll K (1952) The origin, injurious effect and control of Brassica blackening. NachrBl Dtsch Pflsch Dienst Berl N F 6:81–85
Suhag LS, Singh R, Malik YS (1983) Assessment of losses caused by *Alternaria alternata* on radish seed crop and its control by chemicals. Indian Phytopathol 36:758–760
Sultana NA, Khan MAH, Islam MN, Kamran N (2009) Evaluation of appropriate time for the application of rovral against alternaria blight incidence and yield of mustard. Int J Sustain Agric 1:20–23
Taber RA (1964) A study of pathogenicity and *in vitro* growth of *Alternaria* spp. isolated from *Brassica* spp. with special reference to *A. raphani* from rape. M.Sc. thesis, University of Saskatchewan, Saskatoon, Canada
Taber RA, Vanterpool TC, Williard AT (1968) A comparative nutritional study of *Alternaria raphani, A. brassicae* and *A. brassicicola* with special reference to *A. raphani*. Phytopathology 58:609–616
Tewari JP (1991a) Structural and biochemical bases of the black spot disease of crucifers. Adv Struct Biol 1:325–349
Tewari JP (1991b) Current understanding of resistance to *Alternaria brassicae* in crucifers. In: Proceedings of GCIRC 8th International Rapeseed Congress, 9–11 July, Saskatoon, Canada, p 84 (Abstr.), vol 2, pp 471–476
Tewari JP, Conn KL (1993) Reaction of some wild crucifers to *Alternaria brassicae*. Bull OILB SROP 16:53–58
Thomma BPHJ (2003) Pathogen profile-*Alternaria* spp. from general saprophyte to specific parasite. Mol Plant Pathol 4:225–236
Tsuneda A, Skoropad WP (1980) Interactions between *Nectria inventa*, a destructive mycoparasite, and fourteen fungi associated with rapeseed. Trans Br Mycol Soc 74:501–507
Van Schreven DA (1953) *Alternaria, Stemphylium* en *Botrytis* aantasting bij Koolzaad (*Brassica napus*). Tijdschr Plantenziekt 59:105–136
Vannacci G, Harman GE (1987) Biocontrol of seed-borne *Alternaria raphani* and *A. brassicicola*. Can J Microbiol 33:850–856
Vannacci G, Pecchia S (1988) Location and transmission of seed-borne *Alternaria raphani* Groves and Skolko in *Raphanus sativus* L.: a case study. Archivfur Phytopathologie und Pflanzenschutz 24:305–315
Verma PR, Saharan GS (1994) Monograph on *Alternaria* diseases of crucifers. Saskatoon Research Centre, Technical Bulletin 1994–6E, Agriculture and Agri-Food Canada, Saskatoon, SK, Canada, 162 p
Verma SS, Yajima WR, Rahman MH, Shah S, Liu JJ, Ekramoddoullah AK, Kav NN (2012) A cysteine-rich antimicrobial peptide from *Pinus monticola* (*PmAMP1*) confers resistance to multiple fungal pathogens in canola (*Brassica napus*). Plant Mol Biol 79:61–74
Vishwanath K, Kolte SJ (1997) Variability in *Alternaria brassicae*: response to host genotypes, toxin production and fungicides. Indian Phytopathol 50:373–381
Von Keissler K (1912) Zur Kenntnis Pilzflora Krains. Beihefte zum Botanischen Zentralblatt 29:434
Wallroth CFW (1833) Flora Cryptogamica Germaniae Sectio 2. J.L. Schrag, Nürnberg
Wang Y, Geng Y, Ma J, Wang Q, Zhang XG (2011) *Sinomyces*: a new genus of anamorphic *Pleosporaceae*. Fungal Biol 115:188–195
White JG, Linfield CA, Lahdenpera ML, Uoti J (1990) Mycostop- a novel bio-fungicide based on *Streptomyces griseoviridis*. In: Proceedings of British crop protection conference, pests and diseases, Brighton, 1, pp 221–226
Williams PH (1985) Crucifer genetics cooperative (GrGc). Cruciferae Newsl 10:1–2
Williams J, Pink DAC (1987) Development of an *in vitro* screen in Brussels sprout for resistance to *Alternaria brassicae*. Cruciferae News Lett 12:91
Wiltshire SP (1933) The foundation species of *Alternaria* and *Macrosporium*. Trans Br Mycol Soc 18: 135–160
Wiltshire SP (1938) The original and modern conceptions of *Stemphylium*. Trans Br Mycol Soc 21:211–239
Woudenberg JHC, Groenewald JZ, Binder M, Crous PW (2013) *Alternaria* redefined. Stud Mycol 75: 171–212
Yadav MS, Brar KS (2003) Relationship between meteorological factors and incidence of Alternaria blight and white rust of Indian mustard in South–western Punjab. Plant Dis Res 18:80–82

The Disease

2

2.1 Introduction

Alternaria blight caused by four species of *Alternaria* is known by several names (synonyms) on the basis of colour, shape, size and location of lesions on different parts of cruciferous host species. General symptoms are similar on all plant hosts with minor variations influenced by environmental conditions, host species, genotypes, nutritional conditions and pathotypes. Pathotype-specific symptoms can be observed on *Brassica* genotypes. The disease is widely distributed all over the world wherever the congenial environmental conditions and host–pathogen interaction occur. Its host range is very wide ranging from oil-yielding *Brassica* to vegetables and a large number of cruciferous weeds. The disease causes heavy yield losses in the form of quality and quantity of cruciferous crops (11–100 %). The disease incidence and intensity can be accessed through various keys, scales and diagrams and using modern techniques including remote sensing, video image and PCR. Host resistance and categorization of *Brassica* genotypes for breeding resistance varieties can be done through internationally acceptable procedures and methods suggested in this chapter. Sometimes, under field conditions, symptoms of mixed infections of *Alternaria* blight, powdery mildew and white rust can be observed on *B. juncea* leaves.

2.2 Synonyms

The diseases of Brassicaceae crops caused by four species of *Alternaria* are known by many names: *Alternaria* leaf spot; *Alternaria* pod spot; *Alternaria* leaf and pod spot; *Alternaria* blight; *Alternaria* leaf blight; *Alternaria* pod blight; *Alternaria* leaf and pod blight; *Alternaria* dark leaf spot; black spot; brown spot; black ring spot; blight pod spot; stem streak; seed rot; siliquae mould; black mould; grey leaf spot; grey leaf mould of rapeseed, mustard and crucifers; rape black; brassicaceous black; black spot of cabbage; brown rot of cauliflower; *Alternaria* storage rot of vegetables; *Alternaria* blight of vegetables; *Alternaria* sooty spot; black pod blotch; brown rot; curd blight; curd drying; root and foot rot; and damping-off (Anonymous 1985a, b, c, d; Czyzewska 1969, 1970, 1971; Eddins 1952; Ellis 1968a, b; Harter and Jones 1923; Kolte 1985a, b; Leppik 1973; Neergaard 1945; Petrie 1975; Pound 1946; Pound *et al.* 1951; Sherf and Macnab 1986; Subramanian 1971; Tewari 1985; Van Schreven 1953; Walker

G.S. Saharan et al., *Alternaria Diseases of Crucifers: Biology, Ecology and Disease Management*,
DOI 10.1007/978-981-10-0021-8_2

1952; Weber 1973; Weimer 1924; Wiltshire 1947; Yoshii 1933).

2.3 Symptomatology

2.3.1 Rapeseed–Mustard

On oil-yielding *Brassica* crops, all three species of *Alternaria* (*A. brassicae*, *A. brassicicola* and *A. raphani*) are reported to cause symptoms at the seedling stage on cotyledons (Plate 2.1a, b) and at the adult stage on leaves, leaf petiole, stem, inflorescence, siliquae and seeds. There may be variations in shape, size, colour and intensity of lesions on different host tissues and species under different agro-ecological zones. In general, lesions produced by *A. brassicae* are grey compared to black sooty velvety by *A. brassicicola.* All three species produce distinct lesions with yellow halos around them. The disease first appears on the cotyledons and hypocotyls in the form of small light brown lesions which soon turn black due to the appearance of spore masses (under humid conditions) which act as a source of infection for the other parts of the plant. Damping-off of rapeseed seedlings due to *A. brassicicola* is common in Finland (Tahvonen 1979).

The initial infection by *A. brassicae* on the lower leaves starts as minute brown to blackish lesions, which multiply rapidly and later spread to the upper leaves, stem and siliquae (Plate 2.1c–d). On the leaves, lesions may vary in size from 1 to 20 mm increases and cover more areas (Plate 2.2a–b). In some *Brassica* species, formation of concentric rings in the lesions, and a zone of yellow halo around the lesions, is very prominent (Kadian and Saharan 1983). Several lesions on the leaves coalesce to cause blighting and defoliation under humid weather (Plate 2.2a–b). The lesions on the stem are at first linear and then expand but remain usually elongated with pointed ends (Plate 2.2c–d). In severe attacks, the upper part of the stem and siliquae withers. Siliquae may show sunken, dark-brown to black circular lesions. Deep lesions on the siliquae cause infection in the seed (Plate 2.3a–b). On siliquae of *B. rapa* var. *toria* and Yellow and Brown Sarson, lesions are more prominent, enlarged and dark brown to black in colour than on *B. juncea* where they are light coloured with distinct grey centres. The diseased seeds just beneath the black spot are small, shrivelled and grey to brown in colour. In years of severe outbreaks, infection of the stem may be sufficiently intense to cause the whole plant to wither before many of the pods have even been matured or formed (Kolte 1985a, b; Saharan and Chand 1988; Vasudeva 1958). Detailed symptomatology of *Brassica* species infected with *A. brassicae* has been described by Conn et al. (1990) and Kadian and Saharan (1983).

In *A. brassicae* infections in rapeseed, the chloroplasts are more severely affected than the mitochondria. Changes in chloroplasts include swelling and eventual disintegration. The mitochondria, although, show swelling of intra-cristal spaces, but without disintegration. In severely affected cells, vesiculated plasma membranes with electron-dense deposits, intact grana from disintegrated chloroplasts and mitochondria with swollen intra-cristal spaces are recognizable (Tewari 1991a, b).

On Brassicaceae, the lesions on leaves due to *A. brassicicola* are sooty black, velvety and copiously covered with black conidiophores and conidia, whereas those caused by *A. brassicae* are grey, dense and sparsely covered with brown conidiophores and conidia (Changsri 1961).

The lesions caused by *A. raphani* are small with raised margins and are surrounded by a yellow translucent halo. Isolate specific lesions have been recorded on *B. juncea* genotypes which vary in size, shape, colour, number of concentric rings in each spot and presence and absence of yellow halo around the spots (Gupta et al. 2004). Under field conditions on *B. juncea* leaves, symptoms of mixed infections of *Alternaria* blight, powdery mildew and white rust can also be observed.

Alternaria alternata causes root rot of rape (*R. rapa*) in Canada (Berkenkamp and Vaartonou 1972) and root and foot rot of *Eruca sativa* in India (Bhargava et al. 1980). The first evidence of the disease in the field is poor germination of seeds. The infected cotyledons first become chlorotic and then necrotic. The necrotic lesions extend on the stem up to 6–7 cm above the soil level. The root system of severely infected plants is completely destroyed (Bhargava et al. 1980).

Plate 2.1 *Alternaria* blight symptoms on various rapeseed–mustard plant parts; (**a**) *Alternaria* infection on cotyledons leaves; (**b**) *Alternaria* brown lesion on cotyledons leaves; (**c**) *Alternaria* infection on first leaves; (**d**) *Alternaria* spots on leaf

2.3.2 Taramira (*Eruca sativa*)

The disease appears on pods as small lesions, circular and blackish in colour. As the disease progresses, several lesions coalesce and cover large areas, resulting in the death of the entire pod. Similar lesions are also noticed on stems where they occur as circular yellow areas enlarging into concentric circles of black sooty colour.

2.3.3 Crambe (*Crambe abyssinica*)

Alternaria brassicae infects mainly leaves and stems, *A. brassicicola* is a major pathogen of stem and siliquae, *A. alternata* infects all above-ground plant parts, and *A. raphani* has not been recorded on this crop (Czyzewska 1971). Generally, symptoms caused by all three species are similar. *Alternaria*-infected seeds when sown cause

Plate 2.2 *Alternaria* blight symptoms on various rapeseed–mustard plant parts. (**a**) Lesions increase in size and cover larger area. (**b**) Concentric rings in the spot. (**c**) *Alternaria* lesions on stem. (**d**) Severe *Alternaria* infection on the lower portion of the plant

pre- and post-emergence damping-off of seedlings. In dry seasons, diseased cotyledons turn yellow, shrivel and dehisce prematurely. During humid weather, initial spots on the infected cotyledons are small, dark brown, which quickly become large and cover all the tissues. Such cotyledons generally become slimy with mycelium, conidiophores and conidia leading to quick seedling death.

Plate 2.3 *Alternaria* blight symptoms on various rapeseed–mustard plant parts. (**a**) *Alternaria* infection on siliquae. (**b**) Deep *Alternaria* infection on siliquae. (**c**) Severely infected siliquae. (**d**) *Alternaria*-infected seeds (*right*)

Leaf lesions are round (0.5–2 mm dia.), slightly depressed with a smooth surface and brown to dark brown in colour. The tissues around the spots turn lighter in colour and then become yellow. Sometimes, the brown dead tissues in the central part of the spot crumble and fray at the leaf edge in the form of 0.5-mm-width brown tissue border. The spots caused by *A. brassicae* are larger (1–10 mm) and appear isolated with visible infection centre. The concentric rings in the lesion are more sparsely arranged usually at a distance of 1–2 mm; two to three rings appear on one spot. Larger spots are lighter brown in colour and have a slight violet tint.

Spots on the infected stems are dark brown to black, mostly elongated in the form of streaks and somewhat recessed. In damp weather, infected stems become constricted causing the seedlings to topple. The infected collar region appears recessed with elongated black to olive lesion of up to 2 cm encircling the stem partially or completely. On siliquae, initial lesions are small (0.3–3 mm), round and olive brown in colour which later turn blackish brown after sporulation.

Infected siliquae are small and deformed and produce small, discoloured, shrivelled seeds, which are non-viable (Czyzewska 1969, 1971; Holcomb and Newman 1970; Leppik 1973).

2.3.4 Garden Stock (*Matthiola incana*)

The initial symptoms appear on the lower leaves first, which gradually spread upward onto the leaves, stem and flowers. The lesions first appear round and then elongate to 3–15 mm in size. Older lesions are pale or greyish green with concentric zones with brown centres. Such lesions become dark brown to black during sporulation. On stem, and flowers, lesions with water-soaked margins are also common (Davis et al. 1949).

2.3.5 Vegetable Crops (Cruciferous Vegetables)

The symptoms produced by *A. brassicae* and *A. brassicicola* are similar and often indistinguishable. Seed infection by either fungus may cause severe damping-off or a stunting of young plants. On infected seedlings, the pathogen produces dark-brown necrotic areas on the cotyledons and similar coloured streaks on the hypocotyl. On older plants, all above-ground parts are attacked including leaves, stems, Brussels sprouts buttons, cauliflower curds, inflorescence and siliquae including seeds in seed crop.

In cruciferous vegetables, on leaves, symptoms first appear as minute dark-brown to black spots each surrounded by a halo of chlorotic tissues. Leaf spots that vary in size from pinpoints to 2–3 inches in diameter are common on old lower leaves. The enlargement of the spots may be in concentric circles. Older lesions are circular, often zonate with a papery, thin centre, and may be covered with a mat of spores, which are yellow in the case of *A. brassicae* and dark olive brown in the case of *A. brassicicola.* The centres of the lesions may fall out to give a shot-hole effect. Severe defoliation by *A. brassicae* has been reported in some cultivars of stubble turnips. The lesions are linear on stem, petiole and siliquae (Chupp 1925; Chupp and Sherf 1960).

Cauliflower and broccoli heads show a browning, beginning at the margin of the individual flower and flower clusters. On plants grown for seed, dark necrotic lesions occur on the main axis, the inflorescence branches and on the siliquae. These lesions coalesce rapidly and cause premature ripening and splitting of the siliquae resulting in high levels of seed infection. Infected seeds are small, shrunken, discoloured or covered with fungal growth and have low viability. Cankers may form just below the nearly mature cabbage heads resulting in stump rot and death of plants.

On turnips and rutabaga, when foliage is infected, the roots may also become infected and develop symptoms, especially during storage. The leaf spots are nearly circular, often zonate, and are of various shades of brown to black. Dark spores may cover the spot if temperature is high during storage (Sherf and Macnab 1986). *Alternaria raphani* is common on radish plants kept for seed purpose. Leaf lesions are raised, spherical to elliptical and up to 1 cm in diameter; black sporulation may be seen on the lesions. The centre soon dries and may drop off (Singh 1987). The natural infection of radish seeds with *A. raphani* may result in a poor germination, pre- or post-emergence blight, appearance of distinctive lesions on cotyledons and hypocotyls, presence of scab-like lesions on table radish and spotting and blighting of leaves, stalks and siliquae (Atkinson 1950).

On kohlrabi (*Brassica oleracea* var. *gongylodes* L.) plants, disease symptoms first appear as small dark brown, which turn black on leaves; spots vary in size from pinpoints to 1 cm in diameter. As the spots enlarge, a definite zonation, or concentric rings, becomes evident and slightly zonate giving a target spot appearance to the lesion. As spores are produced, spots in the centre become darker, and the dead centre might later tear off, or partially drop away, giving the leaf spots a shot-hole appearance. As the disease progresses, numerous spots and spotted leaves turn yellow and die prematurely. The pathogen isolated and identified is *A. brassicicola* (Schwein) Wiltshire (El-Mohamedy 2007).

2.3.6 Weeds

A brassicaceous perennial, exotic, rangeland weed (*Lepidium draba* L.) shows tiny black spots on leaves enlarging over time to become circular to irregular and cream coloured around the initial black spots and sometimes with dark-brown borders or chlorotic holes (Caesar and Lartey 2009).

2.4 Geographical Distribution

Among the different *Alternaria* species reported to infect brassicaceous hosts, *A. alternata* is an extremely common saprophyte and a weak pathogen found on many plants and other substrata. It is cosmopolitan and reported to be widespread on all kinds of brassicaceous plants including rapeseed–mustard, *Crambe* and *Brassica* vegetables (Ellis 1971; Verma and Saharan 1993, 1994).

Alternaria brassicae (CMI map 353) and *A. brassicicola* (CMI map 457) have been reported from almost every continent on Brassicaceae hosts (Anonymous 1969). *Alternaria brassicae* is considered most destructive on oil-yielding brassicas, and both are common on vegetable crucifers. They are known to occur in Africa, Argentina, Australia, Austria, Bangladesh, Brazil, Britain, Bhutan, Bulgaria, Burma, Canada, Chile, China, Cyprus, Czechoslovakia, Denmark, Egypt, England, Ethiopia, Finland, France, Germany, Ghana, Guinea, Holland, Hong Kong, Hungary, India, Iran, Ireland, Italy, Jamaica, Japan, Kenya, Libya, Malawi, Malaya, Mauritius, Morocco, Mozambique, Nepal, Netherland, New Guinea, New Zealand, Nicaragua, Nigeria, Norway, Pakistan, the Philippines, Poland, Rhodesia, Romania, Russia, Sabah, Saudi Arabia, Scotland, Sierra Leone, Singapore, Spain, South Africa, Sri Lanka, Sudan, Sweden, Switzerland, Taiwan, Tanganyika, Tanzania, Trinidad, Turkey, Uganda, the USA, USSR and Zambia (Anonymous 1980, 1981, 1984, 1985a, b, c, d; Ellis 1971; Kolte 1985a; Saharan 1992b; Verma and Saharan 1993, 1994; Nourani et al. 2008).

Alternaria cheiranthi (Lib.) Bolle. is common on wallflowers and occasionally recorded on Brassicaceae. It is recorded from Belgium, Denmark, France, Germany, Great Britain, Holland, Ireland and Italy (Ellis 1971).

Alternaria raphani, which occurs on various brassicaceous hosts, has been recorded from Canada, Denmark, Egypt, Greece, India, Iran, Japan, the Netherlands and the USA (Ellis 1971). It is most common on radish, but also occurs on other *Brassica* vegetables and oil-yielding crops (Verma and Saharan 1993, 1994; Nourani et al. 2008).

Alternaria alternata has been reported from Canada, Denmark, France, Poland, Latvian SSR, Leningrad USSR and Maryland, USA (Czyzewska 1969, 1970; Leppik 1973; Neergaard 1945). The chronological records of *Alternaria* species causing diseases of brassicaceous hosts reported from various countries of the world are given in Table 2.1. The names of various hosts in this table are as reported in the original papers.

2.5 Host Range

Alternaria species occur worldwide and attack a large number of brassicaceous crops. Among the common hosts of economic importance are rapeseed–mustard, cabbage, Chinese cabbage, cauliflower, broccoli, horse radish, turnip, radish, Brussels sprouts, collards, kohlrabi, rutabaga and swedes. Apart from these, the inventory of hosts infected by *A. brassicae*, *A. brassicicola*, *A. raphani* and *A. alternata* is given in Table 2.2 (Ansari et al. 1990; Atkinson 1950; Chupp 1925; Czyzewska 1969; Ellis 1968a, b, 1971; Kadian and Saharan 1983; Kolte 1985a, b; McDonald 1959; Putnam et al. 1972; Rai and Sinha 1963; Saharan et al. 1982; Tewari and Conn 1993; Verma and Saharan 1993, 1994; Walker 1927; Weber 1932).

2.6 Yield Losses

2.6.1 Rapeseed–Mustard

Heavy infections on leaves, stems and siliquae adversely influence both quantity and quality of yield of *Brassica* crops (Butler 1918). In India,

Table 2.1 Records of *Alternaria* species on Brassicaceae (Verma and Saharan 1993, 1994, updated 2014)

Alternaria species	Host	Disease	Location	Recording year	Reference
A. alternata (*A. tenius*)	Cauliflower	Pox	Italy	1932	Verona, (1932)
	Cabbage	Seed crop, black spot	USSR	1959	Nelen, (1959)
	B. campestris and *B. napus*	Black spot	Canada	1963	Taber and Vanterpool (1963)
	Brassica spp.	Black spot	Canada	1967	Conners (1967)
	C. abyssinica	Black spot	Poland	1969	Czyzewska (1969)
	B. campestris	Root rot	Canada	1971	Berkenkamp and Vaartonou (1972)
	Garden rocket	Black spot	Saudi Arabia	1978	Sheir et al. (1981)
	Brassica spp.	Black spot	USA	1978	Babadoost and Gabrielson (1979)
	Eruca sativa	Root and foot rot	India	1979	Bhargava et al. (1980)
	Radish	Leaf and pod blight	India	1980	Suhag et al. (1983)
A. brassicae (*Macrosporium brassicae; A. herculea*)	Cabbage	Leaf spot	USA	1909	Fawcett (1909)
	Collards	Black mould	USA	1913	Higgins (1917)
	Cauliflower	Leaf spot and brown rot	USA	1918	Weimer (1924)
	Cabbage	Black leaf spot	USA	1918	Harter and Jones (1923)
	Cabbage	Leaf spot	USA	1922	Chupp (1923)
	Cabbage	Leaf spot	USA	1922	Milbraith (1922)
	Cabbage and radish	Leaf spot	Russia	1922	Estifeyeff (1925)
	Cabbage	Black leaf spot	Trinidad	1922	Stell (1922)
	Brassica spp.	Leaf spot	Holland	1924	Bolle (1924)
	Brassica spp.	Black leaf speck	USA	1926	Nelson (1926)
	Cabbage	Leaf spot	China	1926	Porter (1926)
	Brassicaceous vegetables	Leaf spot	USA	1926	Weimer (1926)
	B. pekinensis	Leaf spot	USA	1927	Gardner (1929)
	Brassica spp.	Leaf spot	Dominican Republic	1927	Ciferri and Gonzalez (1927)
	Cauliflower	Brown rot	USA	1927	Walker (1927)
	Cauliflower	Black leaf spot	Italy	1928	Gardner (1929)
	Cauliflower	Black spot	Colombia	1929	Toro (1929)
	Cabbage	Leaf spot and storage rot	North Caucasus	1930	Kikoina (1930)
	Cabbage	Seed infection	Holland	1930	Voisenat (1930)
	Cabbage	Leaf spot	Italy	1931	Pollacci (1932)
	Cabbage and radish	Leaf spot	Japan	1933	Yoshii (1933)
	Brassicaceous vegetables	Leaf spot	UK	1933	Moore (1944)

(continued)

Table 2.1 (continued)

Alternaria species	Host	Disease	Location	Recording year	Reference
	Cauliflower and broccoli	Black leaf spot and grey leaf spot	Canada	1934	Conners (1935)
	Chinese cabbage	Leaf spot	USA	1934	Davis (1934)
	Brassica spp.	Seed infection	Germany	1934	Juhans (1934)
	Brassicaceous vegetables	Leaf spot	Trinidad, Philippines	1934	Fajardo and Palo (1934)
	Matthiola incana	Leaf spot	Great Britain	1935	Ware (1936)
	Turnip	Root rot	USA	1935	Chupp (1935)
	Cabbage	Leaf spot	Morocco	1937	Malencon and Delecluse (1937)
	B. oleracea	Leaf spot	Africa	1937	Roger and Mallamaire (1937)
	Colza and rape	Leaf spot	Germany	1938	Klemm (1938)
	Cauliflower head	Grey rot	Brazil	1938	Arruda (1938)
	Horse radish	Leaf spot	Germany	1938	Boning (1938)
	Iberis umbellata	Leaf spot	Denmark	1938	Neergaard (1939)
	Horse radish	Leaf spot	USA	1940	Kadow and Anderson (1940)
	Cabbage	Leaf spot	China	1940	Teng (1940)
	Turnip	Leaf spot	Australia	1941	Anonymous (1941)
	Cabbage	Black spot	USA	1941	Godfrey (1941)
	Cabbage	Storage rot and ball head	USA	1942	Myers (1942)
	Cabbage and cauliflower	Leaf spot	USA	1942	Snyder and Baker (1943, 1945)
	Radish and turnip	Leaf spot	Argentina	1943	Marchionatto (1947)
	Turnip	Leaf spot	UK	1943	Moore (1948)
	Colza and white mustard	Black spot	Sweden	1944	Bjorling (1944)
	Cabbage	Seed infection	USA	1944	Anonymous (1944)
	Cauliflower and broccoli	Head browning	USA	1944	Rangel (1945)
	Cabbage and turnip	Brown rot	Tanganyika	1945	Wallace and Wallace (1945)
	Turnip, kohlrabi, cabbage and radish	Leaf spot	Norway	1945	Jorstad (1945)
	Brassicaceous vegetable, *M. incana* and *I. umbellata*	Leaf spot	Denmark	1945	Neergaard (1948)
	Broccoli	Dark leaf spot	UK	1946	Moore (1948)
	Radish	Leaf spot	UK	1946	Moore (1948)
	Cabbage and mustard	Leaf spot	Ceylon	1947	Bond (1947)

(continued)

Table 2.1 (continued)

Alternaria species	Host	Disease	Location	Recording year	Reference
	Rape	Seedling blight	Canada	1948	Vanterpool (1950)
	C. maritima	Grey leaf spot	France	1949	Darpoux and Faivre-Amiot (1949)
	Cabbage, mustard and turnip	Leaf spot	Holland	1949	Flik and Saaltink (1950)
	Turnip	Leaf spot	New Zealand	1949	Brien and Dingley (1953)
	Brassica spp.	Leaf spot	Holland	1950	Anonymous (1951b)
	Lunaria annua	Leaf spot	USA	1950	Baker and Davis (1950)
	Cabbage	Leaf spot	Nyasaland	1950	Gillman (1952)
	Sugar beet	Leaf spot	USA	1950	McFarlane et al. (1954)
	Brassica spp.	Leaf spot	India	1950	Mehta et al. (1950)
	B. napus	Dark leaf	Holland	1950	Van Schreven (1953)
	Turnip	Leaf spot	Kenya	1951	Anonymous (1951a)
	Cabbage	Leaf spot	New Caledonia	1951	Bugnicourt et al. (1951)
	Turnip	Leaf spot	Sudan	1951	Tarr (1954)
	Chinese cabbage	Leaf spot	USSR	1952	Nelen (1966)
	Candytuft	Stem spot	Tanganyika	1952	Wallace (1954)
	C. *abyssinica*	Grey leaf	Latvian SSR	1954	Leppik (1973)
	Brassica spp.	Dark leaf spot	Ireland	1956	McKay (1956)
	Rape	Grey leaf spot	Canada	1956	McDonald (1959)
	Cabbage	Seedling blight	USSR	1956	Tupenevich and Shirko (1956)
	Cabbage and turnip	Leaf spot	Nicaragua	1957	Litzenberger and Stevenson (1957)
	Colza	Black spot	France	1958	Louvet (1958)
	Rape	Siliquae and seed infection	Poland	1958	Czyzewska (1958)
	Cabbage	Leaf spot	Greece	1959	Koleva-Sekutkovska (1959)
	Turnip and *Sisymbrium orientale*	Leaf spot	Australia	1959	Anonymous (1959)
	Cabbage	Leaf spot	Primorskii Krai, USSR	1959	Nelen (1959)
	L. annua and *L. rediviva*	Leaf spot	Romania	1959	Negru (1959)
	Rape	Grey leaf spot	Canada	1961	Downey and Bolton (1961)

(continued)

Table 2.1 (continued)

Alternaria species	Host	Disease	Location	Recording year	Reference
	Turnip	Root rot and leaf spot	USA	1961	Ramsey and Smith (1961)
	Radish	Leaf spot	Canada	1963	Taber and Vanterpool (1963)
	Cruciferous	Leaf spot	New Zealand	1964	Morton (1964)
	Brassica	Leaf spot	New Zealand	1964	Morton (1964)
	Thlaspi arvense and *Lepidium* spp.	Leaf spot	Canada	1966	Petrie and Vanterpool (1966)
	Sinapis alba	Seed infection	Denmark	1967	Jorgensen (1967)
	Radish	Leaf spot	Romania	1967	Barbu and Dinescu (1969)
	Taramira (*Eruca sativa*)	Black spot	India	1969	Prasada et al. (1970)
	B. alboglabra	Leaf spot	Singapore	1969	Seow and Lim (1969)
	C. abyssinica	Grey leaf spot	Poland	1970	Czyzewska (1970)
	C. abyssinica	Grey leaf spot	USSR	1970	Leppik (1973)
	Rape	Black spot	Western Australia	1971	Bokor (1972)
	Radish	Leaf spot	India	1977	Rao (1977)
	Turnip rape	Black spot	Spain	1979	Romero and Jimenez (1979)
	Turnip	Leaf spot	Bulgaria	1979	Khristov (1979)
	B. napus	Leaf spot	Mexico	1980	Ponce and Mendoza (1983)
	Rape	Leaf spot	Sweden	1981	Doughty et al. (1991)
	Anagallis arvensis, *Convolvulus aruensis*	Leaf spot	India	1982	Saharan et al. (1982)
	Rape	Leaf spot	UK	1982	Anonymous (1983a, b)
	Brassica spp.	Dark leaf spot	Ireland	1982	Anonymous (1983a, b)
	Charlock weed	Leaf spot	Ireland	1982	Anonymous (1983a, b)
	Cabbage, Brussels sprouts and cauliflower curds	Curds spot	UK	1982	Anonymous (1983a, b)
	Brassica spp.	Leaf spot	Ireland	1983	Ryan et al. (1984)
	Swede rape	Leaf spot	Russia	1983	Vakhurusheva (1983)
	B. napus and *B. campestris*	Leaf spot	Italy	1985	Tosi and Zazzerini (1985)

(continued)

Table 2.1 (continued)

Alternaria species	Host	Disease	Location	Recording year	Reference
	Oilseed rape	Dark leaf spot	UK	1985	Anonymous (1985d)
	Candytuft (*Iberis* sp.)	Leaf spot	India	1987	Gurha and Dhar (1987)
	B. napus and *B. carinata*	Leaf spot	Ethiopia	1987	Kidane and Bekele (1987)
	Rape	Dark leaf spot	Germany	1988	Daebeler and Amelung (1988)
	Chinese cabbage		Brazil	2003	Reis and Boiteux (2010)
	Pak choi		Brazil	2003	Reis and Boiteux (2010)
	Turnip		Brazil	2003	Reis and Boiteux (2010)
	Chinese cabbage		Brazil	2003	Reis and Boiteux (2010)
	Leaf mustard		Brazil	2003	Reis and Boiteux (2010)
	Salad rocket		Brazil	2004	Reis and Boiteux (2010)
	Oilseed rape		Brazil	2004	Reis and Boiteux (2010)
	Wild mustard		Brazil	2004	Reis and Boiteux (2010)
	Wild radish		Brazil	2004	Reis and Boiteux (2010)
	Radish		Brazil	2004	Reis and Boiteux (2010)
	Crambe abyssinica	Leaf spot	Australia	2004	You *et al.* (2005)
	Stem radish		Brazil	2005	Reis and Boiteux (2010)
	Leaf mustard		Brazil	2005	Reis and Boiteux (2010)
	Chinese cabbage	Leaf spot	Brazil	2005	Michereff *et al.* (2012)
	Lepidium draba	Leaf spot	America	2007	Caesar and Lartey (2009)
	Brassica napus	Leaf spot	Iran	2007	Nourani *et al.* (2008)
A. brassicicola	Cabbage	Leaf spot	Holland	1924	Bolle (1924)
	Cabbage and cauliflower	Damping-off and dark spot	Burma	1934	Su (1934)
	Cauliflower	Inflorescence blight	Mauritius	1943	Coombes and Julien (1949)
	Chinese cabbage	Leaf spot	Jamaica	1944	Anonymous (1945a, b)

(continued)

Table 2.1 (continued)

Alternaria species	Host	Disease	Location	Recording year	Reference
	Godetia hybrida	Seed infection	Denmark	1945	Anonymous (1945a, b)
	C. abyssinica	Black spot	Denmark	1945	Neergaard (1945)
	Vegetable *Brassica* and weeds	Leaf spot	Denmark	1945	Neergaard (1945)
	Cabbage, cauliflower and kale	Black spot	UK	1947	Green (1947)
	Kohlrabi	Leaf spot	Ceylon	1947	Bond (1947)
	Cabbage	Lcaf spot	Nyasaland	1950	Gillman (1952)
	Turnip, cauliflower and radish	Leaf spot	Sudan	1951	Tarr (1951, 1954)
	Cabbage	Leaf spot	New Zealand	1952	Brien and Dingley (1955)
	Cabbage, Brussels sprouts and radish	Leaf spot and blackening	Germany	1952	Stoll (1952)
	Cabbage and cauliflower	Damping-off	Finland	1952	Linnasalmi (1952)
	Cabbage	Seed infection	Turkey	1954	Gobelez (1956)
	Brassica spp.	Dark leaf spot	Ireland	1956	McKay (1956)
	Cabbage	Dark spot	USSR	1959	Nelen (1959)
	B. rapa	Leaf spot	Malaya	1963	Jamil (1966)
	Brassica spp.	Leaf spot	Brunei	1963	Johnston (1964)
	Rape	Leaf spot	Chile	1963	Bertossi (1963)
	Crucifers *Brassica*	Leaf spot	New Zealand	1964	Morton (1964)
	Cabbage	Leaf spot	Saudi Arabia	1967	Anonymous (1967)
	Brassica spp.	Leaf spot	Canada	1967	Conners (1967)
	C. abyssinica	Black spot	Poland	1969	Czyzewska (1969)
	C. abyssinica	Black spot	USA	1969	Holcomb and Newman (1970)
	Brassica spp.	Leaf spot	Sarawak	1972	Kueh (1972)
	White cabbage	Storage rot	UK	1973	Geeson (1979)
	Crucifers	Leaf spot	India	1977	Rao (1977)
	Cabbage	Leaf spot	Saudi Arabia	1978	Sheir *et al.* (1981)
	Brassica spp. seed crop	Leaf spot	USA	1978	Babadoost and Gabrielson (1979)
	Cabbage	Storage rot	Finland	1981	Tahvonen (1981)
	Cabbage	Leaf spot	Brazil	1982	Bolkan et al. (1983)
	Brassica spp.	Leaf spot	UK	1982	Anonymous (1983a, b)
	Brassica spp.	Leaf spot	Ireland Republic	1983	Ryan et al. (1984)

(continued)

Table 2.1 (continued)

Alternaria species	Host	Disease	Location	Recording year	Reference
	E. sativa	Leaf blight	India	1984	Sharma (1985)
	Brassica spp. oilseeds	Black spot	Spain	1984	Galvez and Romero (1988)
	Rape	Dark leaf spot	Germany	1988	Daebeler and Amelung (1988)
	Cabbage	Sooty spot	Japan	1998	Kubota and Abiko (1998)
	Chinese cabbage		Brazil	2003	Reis and Boiteux (2010)
	Chinese cabbage		Brazil	2003	Reis and Boiteux (2010)
	Kohlrabi	Leaf spot	Egypt	2004	El-Mohamedy (2007)
	Oilseed rape		Brazil	2004	Reis and Boiteux (2010)
	Wild radish		Brazil	2005	Reis and Boiteux (2010)
	Stem turnip		Brazil	2005	Reis and Boiteux (2010)
	Leaf mustard		Brazil	2005	Reis and Boiteux (2010)
	Cabbage	Leaf spot	Brazil	2005	Michereff et al. (2012)
	Cauliflower	Leaf spot	Brazil	2005	Michereff et al. (2012)
	Broccoli	Leaf spot	Brazil	2005	Michereff *et al.* (2012)
	Salad rocket		Brazil	2006	Reis and Boiteux (2010)
	Turnip		Brazil	2006	Reis and Boiteux (2010)
	Wild mustard		Brazil	2006	Reis and Boiteux (2010)
	B. napus	Leaf spot	Iran	2007	Nourani et al. (2008)
	Radish		Brazil	2008	Reis and Boiteux (2010)
A. raphani	*Brassica* spp.	Leaf spot	Denmark	1945	Neergaard (1948)
	Garden stock	Leaf spot	California	1946	Davis et al. (1949)
	Radish	Seed, leaf, stem, pods and root infection	USA	1947	Mclean (1947)
	Matthiola spp.	Leaf spot	Canada	1953	Conners (1954)
	B. oleracea var. *capitata*	Leaf spot	Greece	1953	Critopoulos (1953)
	Cabbage		USSR	1959	Nelen (1959)

(continued)

Table 2.1 (continued)

Alternaria species	Host	Disease	Location	Recording year	Reference
	Brassica spp.	Leaf spot	Canada	1963	Taber and Vanterpool (1963)
	Radish and *T. arvense*	Leaf spot	Canada	1966	Petrie and Vanterpool (1966)
	Radish	Leaf spot	Saudi Arabia	1978	Sheir et al. (1981)
	I. amara	Leaf spot	India	1981	Narian et al. (1982)
	Turnip	Leaf spot	USA	1982	Cotty and Alcorn (1984)
	B. campestris var. *rapa*	Leaf spot	India	2002	Sangwan et al. (2002)
	B. campestris var. *toria*	Leaf spot	India	2002	Sangwan et al. (2002)
	B. carinata	Leaf spot	India	2002	Sangwan et al. (2002)
	B. napus	Leaf spot	India	2002	Sangwan et al. (2002)
	B. campestris var. Yellow Sarson	Leaf spot	India	2002	Sangwan et al. (2002)
	B. campestris var. Brown Sarson	Leaf spot	India	2002	Sangwan et al. (2002)
	B. chinensis	Leaf spot	India	2002	Sangwan et al. (2002)
	B. juncea	Leaf spot	India	2002	Sangwan et al. (2002)
	B. tournefortii	Leaf spot	India	2002	Sangwan et al. (2002)
	B. pekinensis	Leaf spot	India	2002	Sangwan et al. (2002)
	B. alba	Leaf spot	India	2002	Sangwan et al. (2002)
	Eruca sativa	Leaf spot	India	2002	Sangwan et al. (2002)
	B. napus	Leaf spot	Iran	2007	Nourani et al. (2008)

yield losses in *Alternaria*-infected plants increase considerably after winter rains (Dey 1948). Shrivelling of seeds and reduction in quantity of oil content is the major effect in severe infections (Chahal and Kang 1979; Chohan 1978; Kaushik et al. 1984; Milbraith 1922; Nijhawan and Hussain 1964; Vasudeva 1958). The seed production of brassicas is significantly reduced by *Alternaria* infections, which invade siliquae and penetrate the seeds besides damaging the assimilatory tissues of the leaves and stems (Bandyopadhya et al. 1974; Nielsen 1933). *Alternaria brassicae* infection is also known to affect chemical composition of seed including protein, total carbohydrates and ash (Degenhardt et al. 1974; Nijhawan and Hussain 1964).

In Canada, Degenhardt et al. (1974) reported that the combined effect of *A. brassicae* and *A. raphani* infection under artificially inoculated field conditions resulted in 70 and 42 % losses in yield of *B. campestris* (*B. rapa*) and *B. napus*, respectively. According to their estimates, *A. brassicae* alone can cause 63 % loss in *B. campestris* and 42 % in *B. napus*; yield reductions due

Table 2.2 Host species susceptible to *Alternaria brassicae*, *Alternaria brassicicola*, *Alternaria raphani* and *Alternaria alternata* (Verma and Saharan 1994, updated 2014)

Alternaria species	Host	Common name	Reference
A. brassicae	*Armoracia rusticana*	Horse radish	Tewari and Conn (1993)
	A. lapathifolia		Bolle (1924)
	Anagallis arvensis		Saharan et al. (1982)
	Brassica oleracea var. *gongylodes*	Kohlrabi	Weiss (1960)
	B. oleracea var. *viridis*	Kale	Weiss (1960)
	B. napobrassica	Rutabaga	Weiss (1960)
	B. oleracea var. *gemmifera*	Brussels sprouts	Weiss (1960)
	B. oleracea var. *botrytis*	Cauliflower	Weiss (1960)
	B. oleracea var. *capitata*	Cabbage	McDonald (1959)
	B. napus ssp. *oleifera*	Rapeseed	McDonald (1959)
	B. tournefortii	Wild mustard	Kadian and Saharan (1983)
	B. nigra	Black mustard	Kadian and Saharan (1983)
	B. chinensis	Chinese cabbage	Kadian and Saharan (1983)
	B. pekinensis	Chinese cabbage	Kadian and Saharan (1983)
	B. rugosa		Kadian and Saharan (1983)
	B. rapa var. Brown Sarson		Kadian and Saharan (1983)
	B. rapa var. Yellow Sarson		Kadian and Saharan (1983)
	B. rapa var. *toria*		Kadian and Saharan (1983)
	B. hirta	White mustard	Kadian and Saharan (1983)
	B. kabar	White mustard	Weiss (1960)
	B. rapa ssp. *oleifera*	Canola	Tewari and Conn (1993)
	B. rapa ssp. *rapifera*	Turnip	Tewari and Conn (1993)
	B. carinata	Abyssinian mustard	Tewari and Conn (1993)
	B. orientalis		Bolle (1924)
	B. japonica		Banga et al. (1984)
	B. rapa		Rai and Sinha (1963)
	B. juncea	Mustard	Mehta et al. (1950)
	B. alboglabra		Seow and Lim (1969)
	Chenopodium album		Tripathi and Kaushik (1984)
	Cleome ciliata		Lapis and Ricaforte (1974)
	Convolvulus arvensis		Saharan et al. (1982)
	Citrus aurantium		Rao (1977)
	Cyamopsis psoraloides	Guar	Siddiqui (1963)
	Crambe abyssinica	Crambe	Czyzewska (1970)
	C. maritima	Crambe	Leppik (1973)

(continued)

Table 2.2 (continued)

Alternaria species	Host	Common name	Reference
	Cheiranthus cheiri	Wall flower	McDonald (1959)
	Eruca sativa	Taramira	Kadian and Saharan (1983)
	Erysimum cheiranthoides	Wormseed mustard	Tewari and Conn (1993)
	Iberis amara	Candytuft	McDonald (1959)
	Iberis umbellata		Neergaard (1939)
	Isatis tinctoria		Bolle (1924)
	Lallemantia iberica		Darpoux (1945)
	Lepidium sativum	Garden cress	McDonald (1959)
	L. latifolium		McDonald (1959)
	Lactuca sativa	Lettuce	Lapis and Ricaforte (1974)
	Lunaria annua	Honesty	Baker and Davis (1950)
	Matthiola incana	Stock	Ware (1936)
	Portulaca oleracea		Lapis and Ricaforte (1974)
	Raphanus sativus	Radish	McDonald (1959)
	Sinapis alba	White mustard	Tewari and Conn (1993)
	S. arvensis	Wild mustard	Tewari and Conn (1993)
	Sisymbrium officinale	Hedge mustard	Putnam et al. (1972)
	S. altissimum	Tumbling weed	Putnam et al. (1972)
	S. *orientale*		Anonymous (1959)
	Thlaspi arvense	Stink weed	Petrie and Vanterpool (1966)
A. brassicicola	*B. oleracea* var. *capitata*	Cabbage	Rai and Sinha (1963)
	B. oleracea var. *botrytis*	Cauliflower	Rai and Sinha (1963)
	B. oleracea var. *caulorapa*		Rai and Sinha (1963)
	B. juncea	Mustard	Bolkan et al. (1983)
	B. pekinensis	Chinese cabbage	Bolkan et al. (1983)
	B. rapa	Turnip	Jamil (1966)
	B. carinata	Abyssinian mustard	Galvez and Romero (1988)
	C. abyssinica	Crambe	Czyzewska (1969)
	E. sativa	Taramira	Sharma (1985)
	Godetia hybrida		Anonymous (1946)
	I. umbellata		Neergaard (1948)
	Lunaria rediviva		Negru (1959)
	R. sativus	Radish	Rao (1977)
A. raphani	*B. rapa* var. *toria*		Rai and Sinha (1963)
	B. juncea	Indian mustard	Rai and Sinha (1963)
	B. oleracea var. *botrytis*	Cauliflower	Atkinson (1950)
	B. oleracea var. *capitata*	Cabbage	Atkinson (1950)
	B. napabrassica	Swedes	Atkinson (1950)
	Cheiranthus cheiri	Wall flower	Atkinson (1950)
	E. sativa	Taramira	Rai and Sinha (1963)
	I. amara	Candytuft	Atkinson (1950)
	I. armena		Narain *et al.* (1982)

(continued)

Table 2.2 (continued)

Alternaria species	Host	Common name	Reference
	Lactuca sativa	Lettuce	Atkinson (1950)
	M. incana	Stock	Baker and Davis (1950)
	T. arvense	Stink weed	Petrie and Vanterpool (1966)
	B. campestris var. *rapa*		Sangwan *et al.* (2002)
	B. campestris var. *toria*		Sangwan *et al.* (2002)
	B. carinata		Sangwan *et al.* (2002)
	B. napus		Sangwan *et al.* (2002)
	B. campestris var. Yellow Sarson		Sangwan *et al.* (2002)
	B. campestris var. Brown Sarson		Sangwan *et al.* (2002)
	B. chinensis		Sangwan *et al.* (2002)
	B. juncea		Sangwan *et al.* (2002)
	B. tournefortii		Sangwan *et al.* (2002)
	B. pekinensis		Sangwan *et al.* (2002)
	B. alba		Sangwan *et al.* (2002)
	Eruca sativa		Sangwan *et al.* (2002)
A. alternata	*B. rapa* var. Brown Sarson		Singh and Suhag (1983)
	B. rapa var. *toria*		Singh and Suhag (1983)
	B. juncea	Indian Mustard	Singh and Suhag (1983)
	B. oleracea var. *capitata*	Cabbage	Singh and Suhag (1983)
	B. oleracea var. *botrytis*	Cauliflower	Rai and Sinha (1963)
	B. oleracea var. *caulorapa*		Rai and Sinha (1963)
	Beta vulgaris	Spinach	Singh and Suhag (1983)
	Convolvulus arvensis	Hiran Khuri	Singh and Suhag (1983)
	C. abyssinica	Crambe	Czyzewaska (1969)
	E. sativa	Taramira	Rai and Sinha (1963)
	1. amara		Rai and Sinha (1963)
	Lepidium sativum		Rai and Sinha (1963)
	Lycopersicon esculentum	Tomato	Singh and Suhag (1983)
	R. sativus	Radish	Rai and Sinha (1963)
	Trianthema monogyna	Santhi	Singh and Suhag (1983)

to *A. raphani* alone were 42 % and 34 %, respectively. In 1955 and 1956 crop seasons in Canada, about 20 % yield losses in rapeseed were attributed to *A. brassicae* infection (McDonald 1959). However, Tewari and Conn (1988) estimated an average 30 % yield loss from central region of Alberta. According to Daebeler et al. (1986), *Alternaria* leaf spot damage in winter rapeseed ranged from 20 to more than 50 % in the German Democratic Republic.

Kanwar and Khanna (1979) reported considerable deterioration in seed quality and quantity due to *Alternaria* infection. The losses in yield (Tables 2.3, 2.4, 2.5, 2.6 and 2.7) have been reported from 35 to 45 % in the case of Yellow Sarson (Table 2.3; Saharan 1984), 25–45 % in Brown Sarson cv. BSH-1 (Table 2.3; Chahal and Kang 1979) and 17–48 % in raya (mustard) (Table 2.3; Saharan 1984). Kolte (1982) reported 17–60 % loss in yield of Rai and Sarson. Yield losses ranging from 10 to 75 % have been reported in different oil-yielding crops from India (Saharan 1992a). Yield losses are heavier in Yellow Sarson (38–45 %) followed by Brown

Table 2.3 Assessment of yield losses in rapeseed–mustard due to *Alternaria* (Saharan 1984)

Cultivar	Location	Disease intensity (%)		Reduction in disease intensity (%)	Reduction in yield (%)	Reduction in oil content (%)
		Sprayed*	Unsprayed			
Prakash (raya)	Hisar	18.0	71.1	74.6	17.9	3.4
Varuna (raya)	Hisar	20.6	67.4	67.9	16.8	2.5
BSH-1 (Brown Sarson)	Hisar	19.7	62.7	68.5	25.6	–
YSPb-24 (Yellow Sarson)	Hisar	19.7	69.9	71.5	35.4	3.5
RLM-514 (raya)	Ludhiana	41.5	62.5	33.8	48.5	–
YS-151 (Yellow Sarson)	Pantnagar	11.0	14.9	26.6	45.0	–
Varuna (raya)	Pantnagar	9.2	15.2	40.0	34.1	–

Difolatan @ 2-g product/L water, Blitox @ 2.5-g product/L water and Dithane M-45 @ 2-g product/L water were sprayed at Hisar, Ludhiana and Pantnagar, respectively.
*Significant @5%

Table 2.4 Influence of *Alternaria* pod infection on yield components of raya cultivar Parkash (Kadian and Saharan 1983)

Yield components						
Category of infection	Pod length[a] (cm)	No. of seeds/pod	No. of infected seeds/pod	1000-seed weight (g)	Seed germination (%)	Oil content (%)
0. Healthy pods	7.2	19.4	0.0	2.3	93	40.5
1. Superficial lesions on pod	7.1	19.4	0.0	2.3	93	40.2
2. One to two deep lesions/pod	7.1	19.4	0.9	2.3	91	37.9
3. Three to five deep lesions/pod	6.4	17.4	2.4	2.2	82	37.7
4. More than five deep lesions/pod	6.4	18.1	3.6	2.1	72	36.6
Correlation (r)				0.92	0.97	0.89

[a]Average of 1000 pods

Sarson (26 %) and mustard (17–18 %) (Saharan 1984, 1992a). There is a reduction in oil content from 1 to 10 % in the infected seeds (Tables 2.4, 2.5, 2.6 and 2.7). Deep lesions on the *Brassica* siliquae increased the percentage of seed infection and decreased pod length, seeds per pod, 1000-seed weight, percent seed germination and percent oil content (Tables 2.3, 2.4, 2.5 and 2.6; Bandyopadhya *et al.* 1974; Chahal and Kang, 1979; Kadian and Saharan 1983; Nijhawan and Hussain 1964; Kolte *et al.* 1987; Saharan 1984; Singh and Bhowmik 1981; Tripathi *et al.* 1987). Daebeler and Amelung (1988) and Saharan (1991) correlated disease intensity on foliage and/or siliquae with the components of yield losses of rapeseed–mustard. The increase in

Table 2.5 Influence of *Alternaria* pod infection on yield components of Brown Sarson (Kadian and Saharan 1983)

Yield components						
Category of infection	Pod length[a] (cm)	No. of seeds/pod	No. of infected seeds/pod	1000-seed weight (g)	Seed germination (%)	Oil content (%)
0. Healthy pods	5.9	18.4	0.0	2.3	97.5	43.4
1. Superficial lesions on pod	4.8	18.2	0.5	2.2	96.5	43.2
2. One to two deep lesions/pod	4.8	18.1	2.8	2.1	85.5	42.5
3. Three to five deep lesions/pod	5.0	17.9	3.9	2.0	81.0	41.3
4. More than five deep lesions/pod	4.9	17.9	4.8	2.0	70.0	38.9
Correlation (r)				0.98	0.92	0.87

[a]Average of 1000 pods

Table 2.6 Influence of *Alternaria* pod infection on yield components of Yellow Sarson (Kadian and Saharan 1983)

	Yield components						
	Category of infection	Pod length[a] (cm)	No. of seeds/pod	No. infected seeds/pod	1000-seed weight (g)	Seed germination (%)	Oil content (%)
0.	Healthy pods	7.3	19.3	0.0	3.0	95.0	45.7
1.	Superficial lesions on pod	7.3	19.3	0.0	3.0	95.0	45.6
2.	One to two deep lesions/pod	7.2	17.6	1.5	2.9	90.0	44.3
3.	Three to five deep lesions/pod	7.0	17.8	1.8	2.6	84.0	43.6
4.	More than five deep lesions/pod	7.6	19.6	3.5	2.2	69.5	40.3
	Correlation (r)				0.91	0.86	0.90

[a]Average of 1000 pods

number of deep lesions on pods causes a steep rise in the number of infected seeds per pod, and sharp decline in percent seed germination, and percent oil content of rapeseed–mustard crops (Saharan 1991; Table 2.7). According to Ansari et al. (1988), the loss in oil content of the seed from rapeseed-diseased plants, over the seeds from healthy plants, ranged between 14.6 and 36.0 %. Under Nepal conditions, *Alternaria* blight of mustard causes 32–57 % loss in yield along with 4.2 to 4.5 % losses in oil content (Shrestha et al. 2005).

Jain (1992) reported yield loss up to 56 % in certain *Eruca sativa* cultivars, and in Yellow Sarson, the loss in yield went up to the extent of 70 %. Verma and Saharan (1994) reported yield losses ranging from 10 to 75 % in oilseed *Brassica* in India. In a field experiment, Barman and Bhagwati (1995) found that severely infected plants yield fewer seeds as compared to the healthy plants.

Alternaria infection generally increases green seed counts in *B. rapa* and causes seed shrivelling and substantial yield reductions in *B. rapa* and *B. napus* (Seidle et al. 1995). Pod infection results in increased losses compared to the infection on leaves. Deep lesions on the pods and increased seed infection reduce pod length, seed/pod, 1000-seed weight, seed germination and oil content. Losses due to pod infection have been found to be heavier in *B. campestris* than other

Table 2.7 Effect of black spot on number of infected seeds per pod (a) per cent seed germination (b) and per cent oil content (s) of rapeseed–mustard (Saharan 1991)

Infection category	*B. juncea*			*B. campestris* var. Brown Sarson			*B. campestris* var. Yellow Sarson		
	a	b	c	a	b	c	a	b	c
0	0	95.5	40.5	0	96.5	43.5	0	95.0	45.7
1	0	95.0	40.0	0.54	96.0	43.2	0.85	94.0	45.0
2	0.85	91.5	37.8	2.76	82.5	42.5	2.56	90.5	44.0
3	2.40	83.0	37.0	3.90	80.0	40.3	2.96	79.5	43.2
4	3.62	71.5	36.5	4.95	69.5	36.9	3.69	65.0	39.0
5	6.56	50.5	32.0	5.50	42.0	34.9	4.50	41.5	34.5
	–	−0.97	−0.89	–	−0.92	−0.87	–	−0.86	−0.90

0 = Healthy pods, 1 = minute superficial lesions on pods, 2 = one to two deep lesions on pods, 3 = three to five deep lesions on pods, 4 = five to eight deep lesions on pods, 5 = more than nine deep lesions on pods

Brassica spp. *Alternaria brassicae* causes maximum average yield loss of 27.53 % in Yellow Sarson, 17.16 % in *B. napus* and 10.72 % in *B. carinata* in Himachal Pradesh (Kumar 1997). Gupta et al. (1998) in their detailed studies on *Alternaria* leaf blight-induced changes in fatty acid composition of siliquae wall and seeds of mustard (*Brassica juncea* L.) observed that erucic acid decreases, while linoleic acid increases in the siliquae walls and seeds of *A. brassicae*- and *A. brassicicola*-infected seeds while linoleic acid remained unchanged in the seeds. Disease increased palmitic acid and oleic acid in the siliquae. However, it does not affect palmitic acid, but oleic acid is lower in diseased than in healthy seeds. Ram and Chauhan (1998) estimated the loss in seed yield from 28.6 to 71.4 %. Seed germination is adversely affected in Indian mustard, and *B. napus* at infection levels 3 to 5, while seed germination is adversely affected in *Brassica campestris* at levels 2 to 5 (Kumar 2001).

Meah et al. (2002) observed quantitative relationships between yield of mustard (*Brassica campestris* cv. BINA 2) and disease parameters of *Alternaria* blight in five field trials conducted under natural conditions in Bangladesh. Yield (Y) measured as seed weight in g/m² is linearly related to disease severity (DS), assessed at 65-day-age growth stage (1997–1998):

$$\text{Y} = 203.78 - 3.38\,\text{DS}\left(r^2 = 0.47\right).$$

DS significantly reduced Y and this relation is influenced by seed quality (1998–1999).

$$\text{Y healthy seeds} = 217.13 - 15.53\,\text{DS}\left(\text{r}^2 = 0.875\right) \text{ and}$$

$$\text{Y diseased seeds} = 136.57 - 7.10\,\text{DS}\left(\text{r}^2 = 0.844\right).$$

However, the percent yield loss caused by 1 % DS is 1.66 % in 1997–1998, but 7.2 and 5.2 %, respectively, for healthy and diseased seeds in 1998–1999.

Prasad et al. (2003) assessed yield losses due to *Alternaria* blight in (1999–2001) Indian mustard genotypes PAB 9534, PAB 9511, JMM 915, RN 490 and Varuna under protected and unprotected conditions using Varuna and PAB 9511 as the susceptible and resistant controls, respectively. The protected plots were sprayed with 0.25 % mancozeb starting from 40 days after sowing and 3 subsequent sprays at 15-day intervals. The disease appeared 45 days after sowing. The highest disease intensity was recorded at flowering and pod formation stage. Treatment with mancozeb reduced disease incidence in all the genotypes. The highest disease intensity (46.1 and 41.95 %) was recorded in Varuna for both years. In both years, the highest reduction (72.6 and 59.0 %) in disease severity was recorded in RN 490 and lowest (17.8 and 16.1 %) in the protected plots compared with the unprotected plots (39.6 and 32.5 %). The highest seed yield loss (20.8 and 21.9 %) was observed in Varuna under unprotected conditions; however, it also gives the highest seed yield (20.3 and 19.5 q/ha) followed by RN 490 (18.5 and 18.3 q/ha) in the protected plots. Pooled analysis of data revealed that

Varuna produced the highest disease intensity (22.0 and 44.0 %) and yield performance (19.9 and 15.7 q/ha) in protected and unprotected plots, respectively. The 1000-g seed weight of RN 490 in protected (5.2 g) and unprotected (4.8 g) plots was similar to Varuna. In West Bengal, *Alternaria* blight of rapeseed–mustard causes loss in yield of about 47 % (Mondal et al. 2007).

In Europe, epidemics occur on rape about two years in every five, and in these years losses may be as high as 60 % in individual crops. In rape, and other brassicaceous seed crops, *A. brassicae* causes yield loss due to premature ripening and shedding of seed before harvest and by reducing the 1000 grain weight. In England, the disease first caused serious losses in rape in 1980–81, and since then an estimated £3–4 million have been spent annually on fungicidal control (Smith et al. 1988). In Germany, *A. brassicae* caused 75 % losses in rape (Klemm 1938; Raabe 1939). In Lithuanian, *Alternaria* blight damaged 37.2 to 100 % rape siliquae with disease severity of 6.66 to 7.24 % (Brazauskiene and Petraitiene 2006).

Table 2.8 Influence of *Alternaria* infection on siliquae of *Crambe abyssinica* (Czyzewska 1971)

Infection categories	Seed germination (%)	Healthy seedlings (%)	1000-seed weight (g)
1. None	96.0	67.0	7.9
2. Slight	95.0	62.5	7.7
3. Moderate	94.5	45.5	6.9
4. Strong	82.0	45.0	3.2
5. Very Strong	66.2	39.5	1.7

1. No infection
2. Slight infection: siliquae normal in size, several small spots on the fruit pod surface, normally developed seeds
3. Moderate infection: siliquae somewhat smaller or normal in size, numerous spots on fruit hull, seeds smaller
4. Strong infection: siliquae smaller in size, almost the whole surface of the fruit pod covered with black spots, seeds small
5. Very strong infection: siliquae deformed, the whole surface of the fruit pod covered with black spots, seeds shrunken, small, sometimes dry and black, in which case only the seed pod is left

2.6.2 Crambe

In Poland during 1953 and 1954, *Alternaria* blight of *C. abyssinica* caused very poor germination, and 80 % of the planting had to be ploughed in or sown again (Czyzewska 1969). Severely diseased plants produced small, wrinkled, deformed siliquae with small shrivelled and discoloured seeds having poor viability. Diseased seeds cause damping-off of seedlings leading to lower plant population in the field. The weight of 1000 siliquae can drop three times depending on the degree of infection, and the weight of 1000 seeds can drop more than four times. The decrease in germination and the number of healthy seedlings can be more than 30 % (Table 2.8) (Czyzewska 1969, 1971; Holcomb and Newman 1970; Leppik 1973).

2.6.3 Vegetable Crops (Cruciferous Vegetables)

In vegetable crops, losses occur from damping-off of seedlings, and spotting of lower leaves, and heads of cabbage, turnip and rutabaga. The disease can be destructive in seedbeds, especially in cabbage, cauliflower and Brussels sprouts. Spotting and browning of cauliflower, broccoli and cabbage heads reduce quality and market value of these crops (Sherf and Macnab 1986). *Alternaria brassicicola* is the most important pathogen of *B. oleracea* seed crops. Reduction in seed yield may be as high as 80 %, and the pathogen may severely depress germination to the extent that infected seeds may be unsalable (Smith et al. 1988). *Alternaria brassicicola* has been recognized as an important cause of deterioration of white cabbage in cold storage (Kear et al. 1977). According to Gorshkov (1976), damping-off of cabbage by *A. brassicae* may result into 80–100 % losses. In wet seasons in the USA, *A. brassicae* leaf spot on cabbage, cauliflower, broccoli and other brassicas has been known to reduce the yield by more than 50 %. The market losses are due to decay, which develops in transit and storage (Ramsey and Smith 1961). In Magdeburg district of Germany, *Alternaria* caused seed losses of up to 50 % in cauliflower seed plants (Stoll 1948) and 70 to 90 % in rape and seed cabbage plantings (Domsch 1957).

Alternaria alternata infection in radish reduces pod length, number of seeds per pod and thousand-seed weight leading to reduction in seed yield as much as 18 % (Suhag et al. 1983).

2.7 Disease Assessment Keys/ Severity Charts

Alternaria diseases can be assessed by various methods described below:

2.7.1 Visual Assessment Methods

There are two well-documented visual assessment methods.

2.7.1.1 Descriptive Keys

In diagrams, the plants with varying amounts and kinds of disease symptoms are categorized with an accompanying description. The descriptive keys for assessment of disease severity used by different workers are given in Table 2.9. These keys measure the disease in a scoring scale of 0–5 or 0–9. The zero (0) score indicates no symptoms on any part of the host, and 1–5 or 1–9 scores indicate the presence of different degrees of disease symptoms on various host parts. A score of 1 is indicative of the least amount of disease, and scores 5 or 9 indicate the maximum disease (75 % or more) of the infected host. These descriptive keys are being successfully used for estimating disease severity of plants with differentiated disease resistance or of plants subjected to different environmental conditions or cultural procedures. Host resistance is measured by grouping into resistant (0–1), moderately resistant (1.1–3), moderately susceptible (3.1–5), susceptible (5.1–7) and highly susceptible (7.1–9). Quantitative estimates of the disease indices are calculated using numerical rating based on the severity of the disease (McKinney 1923):

$$\text{Disease Index}(\%) = \frac{\sum \text{sample frequency} \times \text{numerical rating 1}) + \ldots \text{sample frequency} \times \text{numerical rating 5 or 9})}{\text{Total no. samples} \times \text{maximum numerical rating 5 or 9}} \times 100$$

Disease assessments recorded at different growth stages of the crop will help produce temporal disease progress curves. Growth stage key of rapeseed is given in Table 2.10.

An improved grading system for assessing plant diseases proposed by Horsfall and Barratt (1945) has been successfully used by Fontem et al. (1991) to measure the temporal progress and spread of dark leaf spot in cabbage. This grading is based on the principle that according to the Weber–Fechner law, the human eye distinguishes according to the logarithm of the light intensity. Hence, the grades should be based on the ability to distinguish, rather than the extent of the disease. Below 50 %, the eye sees the amount of diseased tissue. Above 50 %, it sees the amount of diseased free tissue. This scoring system (1–12) is based on 50 % as a midpoint. The grades differ by a factor of two in either direction (Table 2.9). Several plants (20 or more) at random are graded.

$$\text{Mean grade} = \frac{\text{grade reading}}{\text{number of reading}}$$

A calibration curve is set up with grade numbers on the x-axis and percentage disease on a special semilog. Y-axis with one and one-half phases from either end up to 50 %. This scheme has been very useful in testing the efficacy of fungicides and varietal resistance and in surveys of plant disease. Fontem et al. (1991) used the conversion table of Redman et al. (1967) to convert the Horsfall–Barratt rating score to disease proportions in black spot of cabbage.

2.7.1.2 Standard Area Diagrams

Pictorial representation of the host plant with known and graded amounts of disease is compared with

Table 2.9 Keys for the assessment of *Alternaria* disease severity on crucifers (Mayee and Datar 1986; Verma and Saharan 1994)

Scoring scale	Description
0.	No symptoms on leaf
1.	Small, round, black dots covering 1 % or less of the leaf area
3.	Grey, circular spots containing concentric rings to cover 1–10 % area
5.	Lesions enlarge, grey circular spots with concentric rings and a black border to cover 11–25 % of the leaf area
7.	Lesions enlarge and coalesce with each other to cover 26–50 % of the leaf area
9.	Lesions coalesced with concentric rings and black border to cover 50 % or more of leaf area; defoliation occurs
Saharan, 1991	
0.	Healthy pods
1.	Minute superficial lesions per pod
2.	One to two deep lesions per pod
3.	Three to five deep lesions per pod
4.	Five to eight deep lesions per pod
5.	More than nine deep lesions per pod

Horsfall and Barratt (1945), Fontem *et al.* (1991)			
Values to convert back to 0%			
Midpoint Elanco Formula %[a]			
1.	No symptoms	0	0
2.	0–3 % area diseased	1.50	2.34
3.	3–6 % area diseased	4.50	4.68
4.	6–12 % area diseased	9.00	9.37
5.	12–25 % area diseased	18.50	18.75
6.	25–50 % area diseased	37.50	37.50
7.	50–75 % area diseased	62.50	62.50
8.	75–87 % area diseased	81.50	81.25
9.	87–94 % area diseased	91.00	90.63
10.	94–97 % area diseased	96.50	96.31
11.	97–100 % area diseased	98.50	97.66
12.	100 % area diseased	100	100

[a]Redman et al. (1967)

Braverman (1971)	
0.	Zero percent leaf area affected
0.5	Traces to 10 % leaf area affected
1.0	11–20 % leaf area affected
1.5	21–30 % leaf area affected
2.0	31–40 % leaf area affected
2.5	41–50 % leaf area affected
3.0	51–60 % leaf area affected
3.5	61–70 % leaf area affected
4.0	71–80 % leaf area affected
4.5	81–90 % leaf area affected
5.0	91–100 % leaf area affected

Alternaria blight leaf phase, scoring using 0–5 modified scale (Saharan 1997; Krishnia et al. 2000)

Scoring scale: *Alternaria* blight leaf phase

Grade	**Description**		**Disease reaction**
0	No symptoms on leaf		Resistant
1	Small, round, black dots covering up to 1 % of the leaf area		
2	Grey circular spots containing concentric rings to cover 1.1–10 % of leaf area		

(continued)

Table 2.9 (continued)

Grade	Description	Disease reaction
3	Lesions enlarge, grey circular spots with concentric rings and black boarder to cover 10.1–25 % of the leaf area	Susceptible
4	Lesions enlarge and coalesce with each other to cover 25.1–50 % leaf area	
5	Lesions coalesced, with concentric rings and black boarder to cover 50 % of the leaf area	

The *Alternaria* blight infection scoring on siliquae using 0–5 modified disease rating scale

Scoring scale: *Alternaria* blight siliquae phase

Grade	Description	Disease reaction
0	Healthy siliquae	Resistant
1	Minute superficial lesions per siliquae	
2	One to two deep lesions per siliquae	
3	Three to five deep lesions per siliquae	Susceptible
4	Six to nine deep lesions per siliquae	
5	More than nine deep lesions per siliquae	

Table 2.10 Growth stage key for oilseed rape (NIAB 1985)

Growth stage	Decimal code	Characteristics
Germination	0.0	Dry seed
Emergence	0.8	Cotyledons emerged
Leaf production	1.0	Cotyledons unfolded
	1.01	First true leaf emerged
	1.02 to 1.09	Second true leaf emerged
Stem extension	2.00	No internodes (rosette)
	2.01	One internode
	2.02 to 2.09	Two internodes
Flower and bud development	3.0	Leaf buds only
	3.3	Green flower buds visible
	3.7	Yellow flower buds visible
Flowering	4.1	First flower opened
	4.2	20 % buds on terminal racemes flowered
	4.3 to 4.9	30 % buds flowered
Pod development	5.1	Lowest pods 2 cm long
	5.2	20 % pods 2-cm long
	5.3 to 5.9	30 % pods 2-cm long
Seed development	6.1	Seeds present
	6.2	Seeds green
	6.5	Seeds brown
	6.7	Seeds black but soft
	6.9	All seeds black and hard

diseased leaves and/or siliquae to allow estimation of disease severity. It consists of a set of pictures giving a schematized illustration of the grades distinguished as in Fig. 2.1 (Conn et al. 1990).

In contrast to descriptive keys, standard area diagrams allow estimation of intermediate levels of disease severity by comparing a diseased plant with diagrams showing both more and less disease. To calculate the disease severity, the leaf and/or siliquae to be assessed are matched to one of the diagrams of the black areas, or area damaged, shown (representing 1, 5, 10, 20, 30 and 50 %) for each leaf and/or siliquae covered by the actual lesions. Disease severity is calculated using the following formula:

$$\text{Disease severity (Area)}\,\% = \frac{\text{Area of the plant tissue affected by disease}}{\text{Total area}} \times 100$$

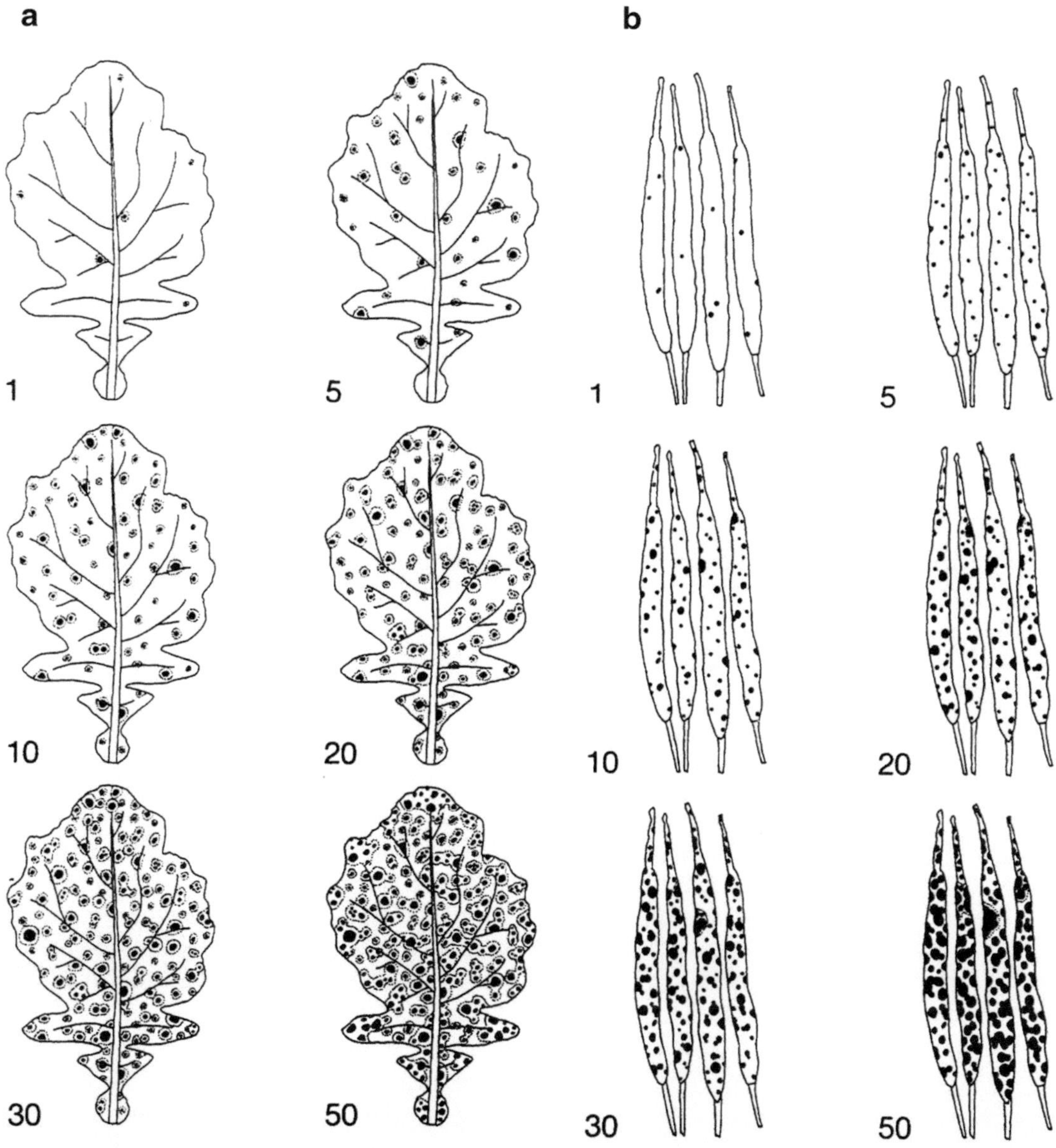

Fig. 2.1 A schematized drawings of (**a**) leaves and (**b**) siliquae of crucifers showing *Alternaria* infection grades (Conn et al. 1990)

Although only a few representative percentage infections (1, 5, 10, 20, 30, 50) are given in the diagrams for assessment of severity, interpolations for arriving at the middle-level percentage such as 2, 3, 15, 30, 40, 60 etc. can be easily practised and recorded. The extent of interpolation will be dictated by the ability of the observer to detect particular differences. Once the severity percentages are decided upon, the visual scales/keys can be used for categorization. However, once the observer becomes familiar with the diagrams, the ranking numbers can be used as such for comparison between the genotypes on 0–5 or 0–9 scale. It is expected that 10 % of the representative population be considered for assessment. The percentage severity data as such can be transformed for any subsequent epidemiological analysis by using the disease index formula given earlier. The standard area diagrams constructed by James (1974) account for the logarithmic decrease in the activity of the eye in estimating disease severities approaching 50 % in their selection of representative keys. Estimations of disease severity intermediate between two keys are often made by careful interpolation. As logistic transformation fits many disease progress curves, equal interval grades on the logit scale are being often used.

2.7.2 Incidence–Severity Relationships

The relationship between disease incidence and disease intensity (severity) (I–S relationship) is an epidemiologically significant concept. Since incidence is easier to measure than severity, any quantifiable relationship between the two measures permits estimation of severity based upon incidence data, which are more precise and easily acquired. Where resource limitations prevent the collection of accurate severity measurements, estimation of severity based upon incidence data will be highly beneficial in disease and yield loss assessments. The I–S relationship can be used through the analysis of correlation, and regression, multiple infection methods, and measurement of aggregation (Seem 1984). There is enormous scope in exploring the I–S relationship especially in understanding the type of disease spread in a sampling unit in *Alternaria* diseases.

2.7.3 Inoculum–Disease Intensity Relationships

An alternate method of disease assessment, especially in studies of host resistance, is based on the number of spores produced per lesion. While analysing components of horizontal resistance in rapeseed–mustard cultivars against *A. brassicae*, Saharan and Kadian (1983) indicated spore count per lesion as one of the most important parameters. An optimum inoculum concentration must be determined to obtain differences in susceptibility among cultivars/genotypes (Dueck and Degenhardt 1975).

2.7.4 Remote Sensing Method

Aerial infrared photography using remote sensing procedure is commonly utilized to detect plant diseases. The detection of diseased plant tissue on false-colour infrared film is due to its greater reflection of near infrared light (700–950 nm) compared to healthy tissue. Colour infrared photographs have been analysed with micro-densitometers or other types of electronic scanning devices to quantify the disease severity. This technique is useful to employ in areas where the same crop is contiguous over large areas. The use of this technique in assessing incidence of *Alternaria* diseases has not yet been reported.

2.7.5 Video Image Analysis

This method can be easily employed for assessing *Alternaria* diseases of brassicaceous plants since its lesions and healthy tissue have different colours. Recent advances in electronic and computer technology allow video cameras to interface directly with a microcomputer. Rapid, automated, nonsubjective estimates of disease severity are made possible by computer-controlled analysis of video images. The accuracy of estimates of disease severity, obtained using video analysis, is high and appears to be independent of both the complexity of the host–pathogen system and disease severity.

2.7.6 Stress Tolerance Attributes

Gupta et al. (2002) assessed stress tolerance attributes for the performance of cultivars. Genotypes can be categorized in the following four groups based on their performance in disease and no disease stress environment: group A (genotypes expressing uniform superiority in both disease and no disease stress environment), group B (genotypes performing favourably only in no disease stress environment), group C (genotypes yielding relatively higher only in disease stress environment) and group D (genotypes performing poorly in both disease and no disease stress environment).

The optimal selection criteria should distinguish group A from the other three groups.

Let

Y_P = the potential yield of a given genotype in no disease stress environment

Y_S = the yield of a given genotype in disease stress environment

Y_{P^-} = mean yield in no disease stress environment

Y_{S^-} = mean yield in disease stress environment

The following stress tolerance attributes were defined from these four yield measurements:

$$\text{Disease Stress Intensity}(\text{DSI}) = \frac{\mathbf{Y_{S^-}}}{\mathbf{1 - Y_{P^-}}}$$

It ranges between 0 and 1, and the larger the value of DSI, the more severe is the stress intensity.

$$\text{Mean productivity}(\text{MP}) = \frac{\mathbf{Y_S + Y_P}}{\mathbf{2}}$$

This index favours higher yield potential and lower disease stress tolerance. Rosielle and Hamblin (1981) showed that under most yield trials, the correlations between MP and Y_P and MP and Y_S would be positive. Thus, selection based on MP generally increases the average performance in both disease and no disease environments. However, MP fails to distinguish the group A and group B genotypes.

$$\text{Disease Tolerance}(\text{TOL}) = \mathbf{Y_p - Y_s}$$

A larger value of TOL represents relatively more sensitivity to disease stress; thus, a smaller value of TOL is favoured. Selection based on TOL favours genotypes with low yield potential under no disease stress conditions and high yield under disease stress conditions. Under most yield trails, the correlations between TOL and Yp would be positive. Thus, TOL fails to distinguish between group C and group A.

$$\text{Disease stress susceptibility index}(\text{DSSI}) = \frac{1 - \mathbf{Y_s / Y_p}}{\mathbf{DSI}}$$

The smaller the value of DSSI, the greater is the disease stress tolerance. Under most yield trials, TOL and DSSI are positively correlated. Selection based on DSSI favours genotypes with low yield potential and high yield stress conditions. Thus, DSSI also fails to distinguish group A from group C.

$$\text{Geometric mean productivity}(\text{GMP}) = \sqrt{\mathbf{Y_s \times Y_p}}$$

GMP is based on the arithmetic means, and therefore, it has an upward biasness due to a relatively larger difference between Yp and Ys, whereas the geometric mean is less sensitive to large extreme values. Thus, GMP is a better indicator than MP in separating group A from other groups.

$$\text{Disease stress tolerance index}(\text{DSTI}) = \frac{(\mathbf{Y_p})(\mathbf{Y_s})}{(\mathbf{Y_{p^-}})^2}$$

DSTI is estimated based on GMP and thus the rank correlation between DSTI and GMP is equal to 1. The higher the value of DSTI for a genotype, the higher are its disease tolerance and yield potential. The disease stress intensity value is also incorporated in the estimation of DSTI. Thus, DSTI is expected to distinguish group A from group B and group C (Fernandez 1992).

References

Anonymous (1941) Plant diseases. Notes contributed by the biological branch. Agric Gazette NSW Hi 7:369–371; 384:435–438

Anonymous (1944) Divisions of plant pathology and seed investigations. Report of New York State Agriculture Experimental Station 1942–1943, 34–43, 1944, 53–58

Anonymous (1945a) Annual report of the Department of Agriculture, Jamaica, for the year ending 31st March 1944, 16 p

Anonymous (1945b) Plant pathology section. Report of the Department of Agriculture Canada, 1944–45, 29–34.

Anonymous (1946) Plant diseases in Denmark in 1945. Annual survey of data collected by the state Phytopathological Experiment Station. Tidsskr Planteavl, Ii (3):373–437

Anonymous (1951a) Some diseases and pests of cultivated crops in 1950 and their control. Versl. PlZiekt. Dienst Wageningen 118, 120

Anonymous (1951b) Department of Agriculture, Kenya, Annual Report 1949. Vol II- Record of Investigations, 89 p

Anonymous (1959) Plant disease survey for the twelve months ending 30th June 1959. In: Twenty-ninth annual report N.S.W. Department of Agriculture. Biological Branch-Division of Science Services, 54 p

Anonymous (1967) Report of the second session of the Near East Plant Protection Commission, F.A.O. held at Tripoli, Libya, 6–13 May 1967. F.A.O. Meeting Report PL: 1967/M/3, 36 p

Anonymous (1969) CMI distribution map of *Alternaria brassicicola*. Commonwealth Mycological Institute, Kew

Anonymous (1980) Diseases of rape. Commonwealth Agricultural Bureau-Annotated Bibliography No. M3, 38 p

Anonymous (1981) *Brassica* diseases. Annual report Edinburgh School Agriculture, 1981, Edinburgh, UK, pp 131–132, 137–138

Anonymous (1983a) Diseases and pests of horticulture crops. Annual report Edinburgh School Agriculture, 1983, Edinburgh, UK, pp 41–42, 95–96

Anonymous (1983b) Diseases of vegetables. Research report 1982, Horticulture. Dublin Irish Republic; An Foras Taluntais, pp 44–45

Anonymous (1984) Oilseed rape-diseases and pests. Annual report Edinburgh School Agriculture, 1983. Edinburgh, UK, pp 39–40

Anonymous (1985a) Diseases of vegetables. Annual report Edinburgh School Agriculture, 1984, Edinburgh, UK, pp 54, 57–58

Anonymous (1985b) Foliar diseases of Brassicas. Research report 1984, Horticulture. Dublin, Ireland; An Foras Taluntais, pp 45–46

Anonymous (1985c) Oilseed rape and other Brassicas. Annual report Edinburgh School Agriculture, 1985, Edinbourgh, UK, pp 57–59, 102–105

Anonymous (1985d) Oilseed rape and other Brassicas. Annual report Edinburgh School Agriculture, 1985, Edinbourgh, UK, pp 57–59, 102–105

Ansari NA, Khan MW, Muheet A (1988) Effect of Alternaria blight on oil content of rapeseed and mustard. Curr Sci 57:1023–1024

Ansari NA, Khan MW, Muheet A (1990) Host range of *Alternaria brassicae*. Acta Bot Ind 18:104–105

Arruda SC (1938) Grey rot of cauliflower. Biologico 4:343–344

Atkinson RG (1950) Studies on the parasitism and variation of *Alternaria raphani*. Can J Res Sect C 28:288–317

Babadoost M, Gabrielson RL (1979) Pathogen causing *Alternaria* diseases of *Brassica* seed crops in Western Washington. Plant Dis Rep 63:815–820

Baker KF, Davis LH (1950) Some diseases of ornamental plants in California caused by species of *Alternaria* and *Stremphylium*. Plant Dis Rep 34:403–413

Bandyopadhya DC, Saba GN, Mukherjee D (1974) Note on variations in quantitative composition of seeds of "B-9" variety of Yellow Sarson caused by Alternaria blight. Indian J Agric Sci 44:406–407

Banga SS, Labana KS, Medhi BN (1984) *Alternaria* incidence in some alloplasmic lines of Indian mustard [*Brassica juncea* (L.) Czern & Coss.]. Theor Appl Genet 67:195–196

Barbu V, Dinescu I (1969) Studies in some Alternarioses of cultivated plants. Phytopathol Z 64:344–354

Barman B, Bhagwati R (1995) Assessment of yield loss due to *Alternaria brassicae* in mustard. Plant Health 1:80–82

Berkenkamp B, Vaartonou H (1972) Fungi associated with rape root rot in Alberta. Can J Plant Sci 52:973–976

Bertossi EO (1963) Fungus phytopathology of Valdivia. 4th contribution. Riv Univ Santiago 48:41–56

Bhargava SN, Shukla DN, Singh N (1980) Root and foot-rot of Eruca sativa caused by Alternaria alternata (Fr) Keissler. Curr Sci 49:452

Bjorling K (1944) Some fungus diseases of Colza and white mustard. Vaxtskyddsnotiser Vaxtskyddsanstalt Stockh 8:73–77

Bokor A (1972) Diseases of rape. J Agric West Aust 13:45–48

Bolkan HA, Ribeiro WRC, Dianese JC (1983) Occurrence and pathogenicity of *Alternaria brassicicola* in Brazil. Plant Dis 67:825–827

Bolle PC (1924) Plant diseases caused by the blackening fungi (Phaeodictyae). Meded Phytopath. Lab. "Willie Commelin Scholten", Baam [Holland] 7:77

Bond TET (1947) Notes on Ceylon fungi and plant diseases. Part 1. (1–15). Ceylon J Sci Sect 12:171–193

Boning K (1938) The most important diseases and pests of horseradish. Nachr Schadl-Bekampf Leverkusen 13:62–87

Brazauskiene I, Petraitiene E (2006) The occurrence of Alternaria blight (*Alternaria* spp.) and Phoma stem canker (*Phoma lingam*) on oilseed rape in central Lithuania and pathogenic fungi on harvested seed. J Plant Prot Res 46:295–311

Brien RM, Dingley JM (1953) First supplement to "A revised list of plant diseases recorded in New Zealand", 1949–1951. N Z J Sci Technol Sect. A 34:556–562

Brien RM, Dingley JM (1955) Second supplement to "A revised list of plant diseases recorded in New Zealand", 1952–1955. N Z J Sci Technol Sect A 37:27–35

Bugnicourt F, Cohic F, Dadant R (1951) A catalogue of the animal and plant parasites of the cultivated plants of New Caledonia. Noumea, Institute Francais d'oceania, 120 p

Butler EJ (1918) Fungi and diseases in plants. Thacker Spink & Co., Calcutta, 298 p

Caesar AJ, Lartey RT (2009) First report of a leaf spot caused by *Alternaria brassicae* on the invasive weed *Lepidium draba* in North America. Plant Dis 93:846

Chahal AS, Kang MS (1979) Different levels of Alternaria blight in relation to grain yield of brown sarson. Indian J Mycol Plant Pathol 9:260–261

Changsri W (1961) Studies of *Alternaria* spp. pathogenic on crucifereae. Disease Abstracts 21:1698

Chohan IS (1978) Diseases of oilseed crops, future plans and strategy for control under small holdings. Indian Phytopathol 31:1–15

Chupp C (1923) Black mold or leaf spot caused by *Alternaria brassicae* (Berk.) Sacc. In: Diseases of field and vegetable crops in the United States in 1922, U.S. Dept. Agric. Bur. Plant. Indus. Plant Dis Surv Bull 26(Suppl):107

Chupp C (1925) Alternaria leaf spot and blight of crucifers. In: Manual of vegetable garden diseases. The Mac Millan Co., New York, pp 146–150

Chupp C (1935) *Macrosporium* and *Colletotrichum* rots of turnip roots. Phytopathology 25:269–274

Chupp C, Sherf AF (1960) Crucifer diseases. In: Vegetable diseases and their control. Ronald Press Company, New York, pp 237–288

Ciferri R, Gonzalez RF (1927) Parasitic and saprophytic fungi of the Dominican Republic. (11th. series). Bol R Soc Espanola Hist Nat 27:267–280

Conn KL, Tewari JP, Awasthi RP (1990) A disease assessment key for Alternaria blackspot in rapeseed and mustard. Can Plant Dis Surv 70:19–22

Conners IL (1935) Fourteenth annual report of the Canadian Plant Disease Survey, 8 + 116 p

Conners IL (1954) Thirty third annual report of the Canadian Plant Disease Survey 1953. 15 + 124 p

Conners IL (1967) An annotated index of plant diseases in Canada and fungi recorded on plants in Alaska, Canada and Greenland. Canadian Department of Agriculture, Publication, 1251, Queen's Printer, Ottawa, pp 61–64

Coombes GAN, Julien JH (1949) The production of vegetable seeds in Mauritius (1943–1946). Bull Depart Agric Mauritius 50, 38 p

Cotty PI, Alcorn SM (1984) *Alternaria raphani* on turnip in Arizona. USA Plant Dis 68:732

Critopoulos PD (1953) A contribution to the fungus flora of Greece. Bull Torrey Bot Club 80:325–341

Czyzewska S (1958) Phytopathological and mycological studies on seeds of rape (*B. napus* var. oleifera). Roczn Naukrol 78 (Ser. A) 2:283–307

Czyzewska S (1969) Alternaria blight of *Crambe abyssinica*. Acta Mycol 5:175–211

Czyzewska S (1970) Effect of temperature on the growth and sporulation of *Alternaria* species isolated from *Crambe abyssinica*. Acta Mycol 6:261–276

Czyzewska S (1971) The pathogenicity of *Alternaria* spp. isolated from *Crambe abyssinica*. Acta Mycol 7:171–240

Daebeler F, Amelung D (1988) Occurrence and importance of Alternaria dark leaf spot in winter rape. Nachrichtenblatt fur den Pflanzen schutz in der DDR 42:196–199

Daebeler F, Amelung D, Riedel V (1986) Leaf spot damage caused by *Alternaria* sp. in winter rape. Wissenschaftliche Zeitschrift der Wilhelm-Pieck-Universitat Rostock, Naturwissens chaftliche Reihe 35:52–54

Darpoux H, Faivre-Amiot A (1949) The antagonistic action of Actinomycete 105 towards some microorganisms, mostly phytopathogenic. Experiments on their application in the control of plant diseases. Ann Inst Pasteur 76:72–74

Darpoux H (1945) A contribution to the study of the disease of oleaginous plants in France. Ann Epiphyt 11:71–103

Davis LH, Sciaroni RH, Pritchard F (1949) Alternaria leaf spot of garden stock in California. Plant Dis Rep 33:432–433

Davis WH (1934) *Alternaria brassicae* as a parasite of Chinese cabbage. Phytopathology 24:1379–1380

Degenhardt KJ, Skoropad WP, Kondra ZP (1974) Effects of Alternaria blackspot on yield, oil content and protein content of rapeseed. Can J Plant Sci 54:795–799

Dey PK (1948) Plant pathology. Administrative report, Agriculture Department of UP, 1946–1947, pp 39–42

Domsch KH (1957) The blackening of rape and cabbage siliquae. Z Pflanzenkr Pflanzen schutz 64:65–79

Doughty KJ, Porter AJR, Morton AM, Kiddie G, Bock CH, Wallsgrove R (1991) Variation in the glucosinolate content of oilseed rape (*Brassica napus* L.) leaves. II. Response to infection by *Alternaria brassicae* (Berk.) Sacc. Ann Appl Biol 118:469–477

Downey RK, Bolton JL (1961) Production of rape in Western Canada. Publ Depart Agric Can 1021, 19 p

Dueck J, Degenhardt K (1975) Effect of leaf age and inoculum concentration on reaction of oilseed *Brassica* spp. to *Alternaria brassicae*. Proc Am Phytopathol Soc 2, 59 (Abstr.)

Eddins AH (1952) Diseases, deficiencies and injuries of cabbage and other crucifers in Florida, Bulletin 492. Florida Agriculture Experiment Station, Gainesville

Ellis MB (1968a) *Alternaria brassicae*. Commonwealth Mycological Institute. Descriptions of Pathogenic Fungi and Bacteria, No. 162. CMI, Kew

Ellis MB (1968b) *Alternaria brassicicola*. Commonwealth Mycological Institute. Descriptions of Pathogenic Fungi and Bacteria, No. 163. CMI, Kew

Ellis MB (1971) Dematiaceous hyphomycetes. Commonwealth Mycological Institute, Kew, No. 608, pp 464–497

El-Mohamedy RSR (2007) First record of Alternaria leaf spot on Kohlrabi in Egypt. Egypt J Phytopathol 35:87–88

Estifeyeff PG (1925) Diseases of cultivated and wild plants in the Djetyssouy region in the period 1922–1924. Pamphlet of the Djetyssouy Plant Protection Station, Alma-Ata [verni], 14 p

Fajardo TG, Palo MA (1934) A serious leaf spot of Chinese celery cabbage, wongbok and other cruciferous plants in Trinidad Valley, Mountain Province Luzon. Philip J Agric 5:143–153

Fawcett HS (1909) Cabbage diseases. Florida Agricultural Experimental Station Annual Reporter, 59–60

Fernandez GCJ (1992) Effective selection criteria for assessing plant stress tolerance. In: Proceedings of the international symposium on adaptation of vegetables and other food crops in temperature and water stress, Shanhua, Taiwan, 13–16 August 1992, pp 257–270

Flik HM, Saaltink GJ (1950) The most important diseases, pests and injuries of agricultural crops in 1949. Maandb LandbVoor 7:241–250

Fontem DA, Berger RD, Weingartner DP, Bartz JA (1991) Progress and spread of dark leaf spot in cabbage. Plant Dis 75:269–274

Galvez RC, Romero MF (1988) *Alternaria brassicicola*, a new causal agent of black spot of crucifers in Spain. In: Communicaciones dell 3 Congreso Nacional de Fitopatologia. Puerto de la cruz, 29 October–2 November, pp 82–86

Gardner MW (1929) Indiana plant diseases, 1927. Proc Indiana Acad Sci 38:143–157

Geeson JD (1979) The fungal and bacterial flora of stored white cabbage. J Appl Bacteriol 46:189–193

Gillman H (1952) Report of the acting director of agriculture. Report Department of Agriculture, Nyasald, 36

Gobelez M (1956) Research work on the varieties and areas of spread of bacterial and parasitic diseases affecting and contaminating the seeds of cultivated plants grown in certain provinces of Central Anatolia as well as the approximate degree of damage caused by such diseases. Zir Fak Yayinl 107:131

Godfrey GH (1941) An outbreak of cabbage blackspot disease in the lower Rio Grande Valley of Texas. Plant Dis Rep 25:119–120

Gorshkov AK (1976) Infectious damping-off of cabbage seedlings under glass. Trudy Vses s-kh Inst Zaoch obrazovaniya No 117:25–29

Green DE (1947) Black spot disease of Brassica seed crops. Gardners' Chron Ser 3 122(3149):178–180

Gupta V, Chawla HKL, Dhawan K and Mehta N (1998) Alterations in total phenols and glucosinolates in *Brassicae juncea* leaves during interaction with *Alternaria brassicae*. Cruciferae Newsl, pp 83–85

Gupta A, Meyer JM, Goel R (2002) Development of heavy metal-resistant mutants of phosphate solubilizing *Pseudomonas* sp. NBRI 4014 and their characterization. Curr Microbiol 45:323–327

Gupta K, Saharan GS, Mehta N, Sangwan MS (2004) Identification of pathotypes of *Alternaria brassicae* from Indian mustard [*Brassica juncea* (L.) Czern & Coss]. J Mycol Plant Pathol 34:15–19

Gurha SN, Dhar V (1987) A new leaf spot disease of candy tuft. Indian J Mycol Plant Pathol 17:238–239

Harter LL, Jones LR (1923) Cabbage disease. U.S. Department of Agriculture Farmers' Bulletin, 1351, 28

Higgins BB (1917) Notes on some diseases of collards. Georgia Agric Exp Stat Annu Rep 29:21–27

Holcomb GE, Newman BE (1970) *Alternaria circinans* and other fungal pathogens on *Crambe abyssinica* in Louisiana. Plant Dis Rep 54:28

Horsfall JG, Barratt RW (1945) An improved grading system for measuring plant disease. Phytopathology 35:655 (Abstr.)

Jain SC (1992) Assessment of losses in Taramira (*Eruca sativa* L.) due to Alternaria blight disease. Madras Agric J 79:57–58

James WC (1974) Assessment of plant diseases. Annu Rev Phytopathol 12:27–48

Jamil M (1966) Annual report of the department of agriculture, Federation of Malaya, for the year 1963, 6, 82 p

Johnston A (1964) Quarterly report for October–December, 1963, of the Plant Protection Committee for the South East Asia and Pacific region. FAO Publication, Bangkok, 20 p

Jorgensen J (1967) Studies on the cause of grey discoloration of seeds of white mustard (*Sinapis alba*). State seed control, 96th Annual Report, 1 July 1966–30 June 1967, Kobenhaun, 112 p

Jorstad L (1945) Parasitic fungi of cultivated and economic plants in Norway. I. Sack spore fungi (Ascomycetes) and conidial fungi (Fungi Imperfecti). Mediterranean Plant Pathology Institute, Oslo, 1, p 142

Juhans J (1934) Concerning seed-borne diseases. Mitteil Phytopathol Versuchsstat der Univ Tartu 19:13

Kadian AK, Saharan GS (1983) Symptomatology, host range and assessment of yield losses due to *Alternaria brassicae* infection in rapeseed and mustard. Indian J Mycol Plant Pathol 13:319–323

Kadow KJ, Anderson HW (1940) A study of horseradish diseases and their control. Illinois Agric Exp Stat Bull 469:531–583

Kanwar ZS, Khanna PK (1979) Mustard seed mycoflora in Central India-their effect on the health of seed, seedling and pod, and their control. Int Pest Control 21:83, 86–87, 92

Kaushik CD, Saharan GS, Kaushik JC (1984) Magnitude of losses in yield and management of Alternaria blight in rapeseed-mustard. Indian Phytopathol 37:398. (Abstr.)

Kear RW, Williams DJ, Stevens CC (1977) The effect of iprodione on the fungal deterioration of stored white cabbage. In: Proceedings of 1977 British Crop Protection Conference, Pests and Diseases, Brighton, pp 189–195
Khristov A (1979) Diseases of turnip in Bulgaria. Rastitelna Zashchita 27:27–28
Kidane A, Bekele E (1987) Major diseases of brassica and preliminary observations on yield loss due to Alternaria leaf spot. In: Proceedings of IDRC (Canada) 3rd Oil Crops Network Workshop held in Addis Ababa, Ethiopia, pp 53–57
Kikoina MR (1930) Note on the work of the laboratory for the investigation of storage of vegetables. Bull North Caucasian Plant Prot Stn Rostoff-on-Don 1930(6–7):287–288
Klemm M (1938) The most important diseases and pests of colza and rape. Dtsch Landw Pr, lxv 19:239; 20:251–252
Koleva-Sekutkovska M (1959) A contribution to the knowledge of parasitic flora in the People's Republic of Macedonia. Zasht Bilja (Plant Protection, Beograd) 51:107–117
Kolte SJ (1982) Diseases of rai and toria. Indian Farm 32:91–93
Kolte SJ (1985a) Disease management strategies for rapeseed-mustard crops in India. Agric Rev 6:81–88
Kolte SJ (1985b) Diseases of annual edible oilseed crops, vol II. CRC Press Inc, Boca Raton, 135 p
Kolte SJ, Awasthi RP, Vishwanath (1987) Assessment of yield losses due to Alternaria blight in rapeseed and mustard. Indian Phytopathol 40:209–211
Krishnia SK, Saharan GS, Singh D (2000) Genetic variation for multiple disease resistance in the families of inter-specific cross of *Brassica juncea x Brassica carinata*. Crucif Newslett 22:51–53
Kubota M, Abiko K (1998) Diseases occurred to plug seedlings of cabbage (in Japanese with English summary). Proc Kansai Plant Prot Soc 40:55–63
Kueh TK (1972) New plant disease records for Sarawak for 1972. Gardens' Bull Singap 27:123–129
Kumar A (1997) Assessment and economics of avoidable yield losses due to Alternaria blight in oilseed Brassicas. Plant Dis Res 12:152–156
Kumar A (2001) Assessment of yield losses at various levels of Alternaria blight infection in rapeseed-mustard. Himachal J Agric Res 25:24–26
Lapis DB, Ricaforte ET (1974) Alternaria leaf spot of crucifers in the Philippines. Philipp Agric 58:121–137
Leppik EE (1973) Diseases of crambe. Plant Dis Rep 57:704–708
Linnasalmi A (1952) Damping-off on herbaceous vegetables and ornamental plants grown under glass in Finland. Ann (bot-zool) Soc Zool-bot, Fenn Vanamo 26:1–120
Litzenberger SC, Stevenson JA (1957) A preliminary list of Nicaraguan plant diseases. Plant Dis Rep 243(Suppl):19
Louvet J (1958) The blackspot disease of colza, *Alternaria brassicae*. C R Acad Agric J France 44:694–701
Malencon G, Delecluse R (1937) Pathogenic fungi observed in Morocco. Bulletin de la Societe des Sciences Naturelles du Maroc 18:132–144
Marchionatto JB (1947) Parasitic fungi of plants, new or little known in Argentina. Publication misc. Ministry of Agriculture B. Aires, Ser A 3(37), 11 p
Mayee CD, Datar VV (1986) Phytopathometry. Marathwada Agricultural University, Parbhani, pp 110–111
McDonald WC (1959) Gray leaf spot of rape in Manitoba. Can J Plant Sci 39:409–416
McFarlane JS, Bardin R, Snyder WC (1954) An Alternaria leaf spot of the sugar beet. Proc Am Soc Sugar Beet Technol 8:241–246
McKay R (1956) Crucifer diseases in Ireland. At the sign of the three candles, Dublin, pp 52–54
McKinney HH (1923) Influence of soil temperature and moisture on infection of wheat seedlings by *Helminthosporium sativum*. J Agric Res 26:195–217
Mclean DM (1947) Alternaria blight and seed infection, a cause of low germination in certain radish seed crops. J Agric Res 75:71–79
Meah MB, Hau B, Siddiqua MK (2002) Relationships between disease parameters of Alternaria blight (*Alternaria brassicae*) and yield of mustard. Zeitschrift für Pflanzenkrankheiten und Pflanzenschutz – J Plant Dis Protect 109:243–251
Mehta PR, Garg DN, Mathur SC (1950) Important diseases of food crops, their distribution in India and Uttar Pradesh. Techn Bull Depart Agric UP 2, 13 p
Michereff SJ, Noronha MA, Filha MSX, Câmara MPS, Reis A (2012) Survey and prevalence of species causing Alternaria leaf spots on *Brassica* species in Pernambuco. Hortic Bras 30:345–348
Milbraith DG (1922) *Alternaria* from California. Bot Gazatte 74:320–324
Mondal KK, Bhattacharya RC, Koundal KR, Chatterjee SC (2007) Transgenic Indian mustard (*Brassica juncea*) expressing tomato glucanase leads to arrested growth of *Alternaria brassicae*. Plant Cell Rep 26:247–252
Moore WC (1944) Diseases of vegetables. In: Diseases of crop plants. A ten years' review (1933–1942). Bulletin No. 126. The Ministry of Agriculture and Fisheries, H.M.S.O., London, pp 40–57
Moore WC (1948) Diseases of crop plants (1943–1946). Bulletin No. 139. The Ministry of Agriculture and Fisheries, H.M.S.O., London, pp 28–29; 41–44; 53–54
Morton FJ (1964) Species of *Alternaria* on *Brassica* hosts in New Zealand. N Z J Bot 2:19–33
Myers CE (1942) The Penn State Ball head cabbage some problems encountered in its development. Bull Agric Exp Stn 430:52
Narain U, Singh J, Kaul AK (1982) Leaf spot of candy tuft caused by *Alternaria raphani*. Nat Acad Sci Lett 5:13
Neergaard P (1939) Annual report of the Phytopathological Laboratory of J. E. Ohlsen's widow from 1st April, 1938 to 31st March, 1939, 16 p

Neergaard P (1945) Danish species of *Alternaria* and *Stemphylium* taxonomy, parasitism, economical significance. Oxford University Press, London/Oxford, 559 p

Neergaard P (1948) Eleventh annual report from the J. E. Ohlsen's Phytopathological Laboratory, 1st August 1945 to 31st July 1947, 19 p

Negru A (1959) New and rare species of fungi from the Romanian People's Republic. Botaniceskie Materialy (Notul. Syst. Sect. Crypt. Inst. Bot. Acad. Sci. U.S.S.R.), 12:213–224

Nelen E (1966) Diseases of plants in the Soviet Far East. Zashch Rast Mosk 11:46–47

Nelen ES (1959) Alternariosis or blackspot of crucifer seed plants in the Primorskii Krai. Soobshch. dal'nevost. Fil. Sibir. Ot del. Akad. Nauk. S.S.S.R. 11:77–83

Nelson R (1926) Storage and transportation diseases of vegetables due to sub-oxidation. Mich Agric Exp Stn Techn Bull 81, 38p

NIAB (1985) Assessment key for oilseed rape diseases. Key no. 41, National Institute of Agricultural Botany, Cambridge

Nielsen O (1933) Experiments in the control of the siliqua fungus. Tidsskr Planteavl 39:437–452

Nijhawan HL, Hussain A (1964) Effect of Alternaria blight on the chemical composition of yellow sarson. Indian Oilseeds J 8:44–45

Nourani SL, Minassian V, Safaie N (2008) Identification, pathogenicity and distribution of *Alternaria* spp. of canola in Iran. Iran J Plant Pathol 44:33–36

Petrie GA (1975) Diseases of rapeseed and mustard: oilseed and pulse crops in Western Canada- a symposium, JT Harapiak (ed). Western Co-operative Fertilizer Ltd., Calgary, Alberta, Canada, pp 399–413

Petrie GA, Vanterpool TC (1966) Diseases of rape, mustard and cruciferous weeds in the Prairie Provinces. Can Plant Dis Surv 46:117–120

Pollacci G (1932) Report on the activity of the cryptogamic laboratory of Pavia (Phytopathological observatory for the Provinces of Cremona, Parma, Pavia and Piacenza, Italy) during the year 1931. Atti. 1st Bot. R. Univ. di Pavia ser 4:310–324

Ponce F, Mendoza C (1983) Fungal diseases and pests of rapeseed *Brassica napus* L. and *B. campestris* L. in the high valleys of Mexico. Phytopathology 73:124 (Abstr.)

Porter RH (1926) A preliminary report of surveys for plant diseases in East China. Plant Dis Rep Suppl 46:153–166

Pound GS (1946) Diseases of cabbage plants grown for seed in Western Washington. Washington Agric Exp Stn Bull 475:27

Pound GS, Cheo PC, Calvert OH, Raabe RD (1951) Extent of transmission of certain cabbage pathogens by seed grown in Western Washington. Phytopathology 41:820–828

Prasad R, Saxena D, Chandra S (2003) Yield losses by Alternaria blight in promising genotypes of Indian mustard. Indian Phytopathol 56:205–206

Prasada R, Khandelwal GL, Jain JP (1970) Morphology, physiology and control of *Alternaria brassicae* on taramira. Indian Phytopathol 23:105–110

Putnam LG, Petrie GA, McDonald H (1972) Alternaria black spot. In: Insect pests and diseases of rape and mustard. Rapeseed Association of Canada, Winnipeg, Publication No., 18, pp 25–26

Raabe A (1939) Investigations on parasitic fungal diseases of colza and rape. Zbl Bakt, Abt., 2e:1–3, 35–52

Rai JN, Sinha AK (1963) Effect of colchicine on the growth and sporulation of *Alternaria tenuis* Ness. ex Pers. Curr Sci 32:135–136

Ram RS, Chauhan VB (1998) Assessment of yield losses due to Alternaria leaf spot in various cultivars of mustard and rapeseed. J Mycopathol Res 36:109–111

Ramsey GB, Smith MA (1961) Market diseases of cabbage, cauliflower, turnips, cucumbers, melons and related crops. US Dep Agric Agric Handb 184:49

Rangel JF (1945) Two *Alternaria* diseases of cruciferous plants. Phytopathology 35:1002–1007

Rao BR (1977) Species of *Alternaria* on some crucifereae. Geobios 4:163–166

Redman CE, King EP, Brown JF (1967) Tables for converting Barratt-Horsfall Rating scores to estimate mean percentages. Eli Lilly Co., Indianapolis, 100 p

Reis A, Boiteux LS (2010) *Alternaria* species infecting Brassicaceae in the Brazilian neotropics: geographical distribution, host range and specificity. J Plant Pathol 92:661–668

Roger L, Mallamaire A (1937) Notes on African phytopathology. Ann Agric Afr Occup 1:187–206

Romero MF, Jimenez DRM (1979) Black spot: a disease of turnip rape recently recorded in Spain. Annales del Institute Nacional de investigaciones Agrarias, Proccion Vegetal No 9:11–31

Rosielle AA, Hamblin J (1981) Theoretical aspects of selection for yield in stress and non- stress environment. Crop Sci 21:943–946

Ryan EW, Staunton WP, Cassidy JC (1984) Diseases of vegetables. Research Report 1983. Horticulture. An Foras Taluntais, Dublin, pp 41–42

Saharan GS (1984) A review of research on rapeseed-mustard pathology in India. Paper presented in the Annual Rabi Oilseed Workshop held at Jaipur, India, 6–10 August 1984

Saharan GS (1991) Assessment of losses, epidemiology and management of black spot disease of rapeseed-mustard. In: Proceedings of GCIRC 8th international rapeseed congress, Saskatoon, Canada, , 9–11 July, p 84 (Abstr.); Volume 2, pp 465–470

Saharan GS (1992a) Management of rapeseed and mustard diseases. In: Kumar D, Rai M (eds) Advances in oil seed research, vol 1, 7. Scientific Publishers, Jodhpur, pp 152–188

Saharan GS (1992b) Disease resistance. In: Labana KS, Banga SS, Banga SK (eds) Breeding oilseed Brassicas, vol 12. Narosa Publishing House, New Delhi, pp 181–205

Saharan GS (1997) Disease resistance. In: Kalia HR, Gupta SK (eds) Recent advances in oilseed Brassicas, vol 12. Kalyani Publishers, Ludhiana, pp 233–259

Saharan GS, Chand JN (1988) Diseases of rapeseed and mustard. In: Diseases of oilseed crops (in Hindi), Chapter 3. Directorate of Publication, Haryana Agricultural University Press, Hisar, pp 84–91

Saharan GS, Kadian AK (1983) Analysis of components of horizontal resistance in rapeseed and mustard cultivars against *Alternaria brassicae*. Indian Phytopathol 36:503–507

Saharan GS, Kaushik JC, Kaushik CD (1982) Two new host records of *Alternaria brassicae*. Indian Phytopathol 35:172

Sangwan MS, Naresh M, Gandhi SK (2002) Some pathological studies on *Alternaria raphani* causing leaf and pod blight of radish. J Mycol Plant Pathol 32:125–126

Seem RC (1984) Disease incidence and severity relationships. Annu Rev Phytopathol 22:133–150

Seidle E, Rude S, Petrie A (1995) Final report: the effect of Alternaria black spot of canola on seed quality and seed yield and studies on disease control. Agriculture Development Fund Canada, Saskatoon, 41 p

Seow AL, Lim LG (1969) A list of leaf spot diseases in Singapore. Rev Mycol 34:79–82

Sharma AK (1985) An unrecorded leaf blight disease of taramira (*Eruca sativa* Mill.) from India caused by *Alternaria brassicicola* (Schew.) Wiltshire. Curr Sci 54:942–943

Sheir HM, Kassim MY, Abou-Heilah AN, Khan S (1981) Leaf spot diseases in Saudi Arabia. 1. Studics on lcaf spots of certain cruciferous plants. J Coll-Sci Univ Riyadh 12:101–111

Sherf AF, Macnab AA (1986) Diseases of crucifers. In: Vegetable diseases and their control, 2nd edn. The Ronald Press, New York, pp 251–306

Shrestha SK, Munk L, Mathur SB (2005) Role of weather on Alternaria leaf blight disease and its effect on yield and yield component of mustard. Nepal Agric Res J 6:62–72

Siddiqui MR (1963) Taxonomy and pathogenicity of the genus *Alternaria* with special reference to Indian species I. Alternaria tenuis Auct Journal of the Indian Botanical Society 42:260–272

Singh A, Bhowmik TP (1981) Persistence and efficacy of some fungicides against *Alternaria brassicae* the causal agent of leaf blight of rapeseed and mustard. 3rd international symposium on plant pathology, IARI, New Delhi, India, December 14–18, 1981, p 70

Singh R, Suhag LS (1983) Some pathological studies on *Alternaria alternata* causing leaf and pod blight of radish in Haryana. Indian Phytopathol 36:174–176

Singh RS (1987) Alternaria leaf spot or black spot of crucifers. In: Diseases of vegetable crops. Oxford and IBH Publication Company, New Delhi, pp 159–163

Smith IM, Dunez J, Phillips DH, Lelliott RA, Archer SA (1988) European handbook of plant diseases. Blackwell Scientific Publishers, Oxford/London, pp 366–368

Snyder WC, Baker KF (1943) Diseases of seed cabbage in California. Plant Dis Rep 27:394–398

Snyder WC, Baker KF (1945) Diseases of seed cauliflower in California. Plant Dis Rep 29:248–252

Stell F (1922) Some common diseases of kitchen garden crops. Proc Agric Soc Trinidad Tobago 22:779–785

Stoll K (1948) On the Alternaria blackening of Brassicas. NachrBl Dtsch PflSchDienst 2:174–178

Stoll K (1952) The origin, injurious effect and control of *Brassica* blackening. NachrBl Dtsch Pfjsch Dienst Berl NF 6:81–85

Su MT (1934) Report of the mycologist, Burma, Mandalay, for the year ending the 31st March, 1934. Rep Depart Agric Burma 1933–34:25–33

Subramanian CV (1971) Hyphomycetes. Indian Council of Agricultural Research, New Delhi, pp 801–820

Suhag LS, Singh R, Malik YS (1983) Assessment of losses caused by *Alternaria alternata* on radish seed crop and its control by chemicals. Indian Phytopathol 36:758–760

Taber RA, Vanterpool TC (1963). *Alternaria* species on rape in Western Canada. Proc Can Phytopathol Soc 30:19 (Abstr.)

Tahvonen R (1979) Seed-borne fungi on cruciferous cultivated plants in Finland and their importance in seedling raising. J Sci Agric Soc Finland 51:327–379

Tahvonen R (1981) Storage fungi of cabbage and their control. J Sci Agric Soc Finland 53:211–227

Tarr SAJ (1951) Plant pathology. Report Research Division, Ministry of Agriculture, Sudan Govt. 1948–1949, 47–65

Tarr SAJ (1954) Plant pathology. Report Research Division, Ministry of Agriculture, Sudan Govt. 1951–1952, 71–80

Teng SC (1940) Supplement to higher fungi of China. Sinensia 11:105–130

Tewari JP (1985) Diseases of canola caused by fungi in the Canadian prairies. Agric For Bull 8:13–20

Tewari JP (1991a) Current understanding of resistance to *Alternaria brassicae* in crucifers. Proc GCIRC 8th Int Rapeseed Congress, Saskatoon, Canada, vol, 2, 9–11 July, p 84 (Abstr.), pp 471–476

Tewari JP (1991b) Structural and biochemical bases of the black spot disease of crucifers. Adv Struct Biol 1:325–349

Tewari JP, Conn KL (1988) Incidence of the black spot of Canola caused by *Alternaria brassicae* (Berk.) Sacc. during 1987. Canadian Plant Dis Survey 68:103

Tewari JP, Conn KL (1993) Reactions of some wild crucifers to *Alternaria brassicae*. IOBC/WPRS Bulletin 16:53–58

Toro RA (1929) Plant disease notes from the Central Andes II. Phytopathology 19:969–974

Tosi L, Zazzerini A (1985) *Alternaria brassicae* (Berk.) Sacc. on rape (*Brassica napus* L. subsp. *oleifera* and turnip rape (*Brassica campestris* L. var. *oleifera* DC.). Informatore Fitopatol 35:43–44

Tripathi NN, Kaushik CD (1984) Studies on the survival of *Alternaria brassicae* the causal organism of leaf

spot of rapeseed and mustard. Madras Agric J 71:237–241
Tripathi NN, Saharan GS, Kaushik CD, Kaushik JC, Gupta PP (1987) Magnitude of losses in yield and management of Alternaria blight of rapeseed and mustard. Haryana Agric Univ J Res 17:14–18
Tupenevich SM, Shirko VN (1956) A study of cabbage seedling diseases. Bull Inst Appl Zool Phytopathol 4:147–154
Vakhurusheva TE (1983) Diseases of Swede rape grown in Leningrad province. Nauchno-tekhnicheskii Byulleten vsesoyuznogo ordena Lenina I ordena Druzhby Narodov Nauchno-issledovatel skogo istituta Rastenievodstva Imeni N I Vavilova 127:40–44
Van Schreven DA (1953) *Alternaria, Stemphylium* en *Botrytis* a antasting bij Koolzaad (*Brassica napus*). Tijdschr Plantenziekt 59:105–136
Vanterpool TC (1950) Rape-seedling blight and leaf spot. In: 29th annual report. Canadian Plant Disease Survey, 31 p
Vasudeva RS (1958) Diseases of rape and mustard. In: Singh DP (ed) Rape and mustard. Indian Central Oilseed Committee, Hyderabad, pp 77–86
Verma PR, Saharan GS (1994) Monograph on *Alternaria* diseases of crucifers. Saskatoon Research Centre, Technical Bulletin 1994–6E, Agriculture and Agri-Food Canada, Saskatoon, SK, Canada, 162 p
Verma PR, Saharan GS (1993) *Alternaria brassicae* (Berk.) Sacc., *A. brassicicola* (Schwein) Wiltsh. and *A. raphani* Groves and Skolko: Introduction, bibliography and subject index. Agriculture Canada Research Station, Saskatoon, Technical Bulletin, 81
Verona O (1932) On the so called pox of cauliflower. Boll R Inst Sup Agric di Pisa 8:291–300
Voisenat (1930) Examination of the state of health of seeds at the official seed testing station at Wageningen. Ann Sci Agron 47:744–754
Walker JC (1927) Diseases of cabbage and related plants. US Depart Agric Fanners' Bull 1439:30
Walker JC (1952) Diseases of crucifers, Chapter 6. In: Diseases of vegetable crops. McGraw Hill Book Company, New York/London, pp 150–152
Wallace GB (1952–1954) Annual reports of the Plant Pathologist, Lyamungu, Moshi. Rept. Dept. Agric. Tanganyika 1952, 55–60; Part 2: 71–77, 1954
Wallace GB, Wallace MM (1945) Tanganyika territory fungus list: recent records. VI. Mycology Circular Department of Agriculture Tanganyika, 15, 2
Ware WM (1936) Alternaria leaf-spot of Stock (Matthiola). Gardner's Chronical C 2596:236–237
Weber GF (1932) Some diseases of cabbage and other crucifers in Florida. Florida Agric Exp Stn Bull 256:62
Weber GF (1973) Bacterial and fungal diseases of plants in the tropics. University Florida Press, Gainesville, pp 86–88
Weimer JL (1924) Alternaria leaf spot and brown rot of cauliflower. J Agric Res 29:421–441
Weimer JL (1926) A leaf spot of cruciferous plants caused by *Alternaria herculea.* J Agric Res 33:645–650
Weiss F (1960) Index to plant diseases in the United States. Crops Research Division, Agricultural Research Service, U.S. Department of Agriculture, Handbook No., 165, pp 97–106
Wiltshire SP (1947) Species of *Alternaria* on Brassicae. CMI, England, Mycology Paper 20:1–15
Yoshii H (1933) On three species of *Alternaria* parasitic on cruciferous plants. Bull Sci Fakultato Terkultura, Kjusu Imper Univ 3:221–235
You MP, Simoneau P, Dongo A, Barbetti MJ, Li H, Sivasithamparam K (2005) First report of an Alternaria leaf spot caused by *Alternaria brassicae* on *Crambe abyssinicia* in Australia. Plant Dis 89:430

3 Pathogen

3.1 Introduction

Alternaria species pathogenic on cruciferous crops are *A. alternata*, *A. brassicae*, *A. brassicicola* and *A. raphani*. Among the four species of *Alternaria*, *A. brassicae* is more common on oil-yielding crops, while other three species are major pathogens of vegetable crops. Phylogeny, taxonomy, morphology, classification, infection process, pathogenesis, disease cycle and identification characteristics along with synonymous of all four species have been well documented. Identification, cloning and sequencing of virulence genes of *Alternaria* infecting crucifers have resolved some doubts about their relationship with cruciferous hosts. Pathogenicity factor and transcription factor *Amr 1* have been identified in *A. brassicicola*. A non-ribosomal peptide synthase gene (*AbNPs2*) is important for cell wall integrity, conidial viability and virulence of aged spores of *A. brassicicola*. More than 100 genes have been functionally analysed through various techniques like gene knockout and overexpressions making *A. brassicicola* the species of choice for functional genomics research. For minimum, maximum and optimum growth, and sporulation of *Alternaria* species pathogenic on crucifers, in vitro studies have been carried out. The suitable culture media, temperature, relative humidity, pH, nutrient sources and light and darkness conditions for growth, sporulation and spore germination have been determined. The pathogen survives through infected seed, crop residue, cruciferous hosts, weeds, microsclerotia and chlamydospores and all act as source of primary inoculum. Secondary infection during the crop season is through conidia produced on infected host plants (under high humidity), which spread through rain splashes and wind to complete the disease cycle. Major identification characteristics of *Alternaria* species infecting rapeseed–mustard crops are given in Table 3.1.

3.2 Historical

As early as 1836, Berkeley identified the causal fungus on plants belonging to the Brassicaceae as *Macrosporium brassicae* Berk., which was later renamed as *A. brassicae* (Berk.) Sacc. by Saccardo (1886). In 1922, 1926 and 1945, respectively, Milbraith, Weimer and Rangel also described the same fungus. Then in 1947, Wiltshire separated the small and big spore forms as *A. brassicicola* (Sch.) Wiltshire and *A. brassicae* (Berk.) Sacc., respectively.

Every modern attempt to define the phaeodictyosporic form genera *Alternaria* Nees ex Fries and *Stemphylium* Wallroth has involved the problem of treating numerous taxa which superficially resemble the type species of one or the other of these genera, but which are sufficiently different as to leave a measure of doubt. Chief among these anomalous species are *Stemphylium*

G.S. Saharan et al., *Alternaria Diseases of Crucifers: Biology, Ecology and Disease Management*,
DOI 10.1007/978-981-10-0021-8_3

Table 3.1 Identification characteristics of *Alternaria* species infecting rapeseed–mustard

Fungal structures	*A. brassicae*	*A. brassicicola*	*A. raphani*
Mycelium	Septate, brownish grey	Septate, olive grey to greyish black	Cottony whitish to greenish grey or dark grey
Conidiophore	Dark, septate, arise in fascicles, 14–74 μ × 4–8 μ	Olivaceous, septate, branched, 35–45 μ × 5–8 μ	Septate, olive brown, single or branched, 29–160 μ × 4–8 μ
Conidia	Brownish black, obclavate, muriform, produced singly or in chains or 2–3	Dark cylindrical to oblong, muriform, produced in chains of 8–10 spores	Olive brown to dark, obclavate, muriform, more or less pin pointed at each end, appears singly or in chains of up to 6 spores
Spore body (μ)	96–114 × 17–24	45–55 × 11–16	45–58 × 13–21
Spore beak length (μ)	45–65	None	1–25
Spore transverse septation	10–11	5–8	6–9
Longitudinal septation	0–6	0–4	3–6
Rate of growth and sporulation on media	Rudimentary slow growth	Black sooty colony with distinct zonations, fast growing with abundant sporulation	Cottony mycelial colony with less abundant sporulation
Infection	Penetrates leaf only through stomata	Penetrates leaf directly or through stomata	Direct penetration

lanuginosum Harz. and *Macrosporium consortiale* Thiimen, both of which, at one time or another, have been named *Alternaria*, *Stemphylium* or *Pseudostemphylium* (Simmons 1967).

Wiltshire (1933, 1938) pioneered the basic studies of this group of Hyphomycetes and published the results of his examination of the available type specimens. His descriptive literature was fundamental to the prevailing concepts of *Alternaria*, *Macrosporium* and *Stemphylium*. His major conclusions were that *Macrosporium* should be suppressed as a nomen ambiguum in favour of *Alternaria*, typified by *A. tenuis* Nees, the type specimen of which Wiltshire was unable to locate for examination, and that the limits of *Stemphylium* should be modified to include two sections.

Neergaard (1945), in his extensive treatment of species of *Alternaria* and *Stemphylium* occurring in Denmark, recognized the same taxonomic problems in handling species similar to *S. lanuginosum*. He also followed the lead of Wiltshire by retaining the two sections proposed for *Stemphylium*. Joly (1964) in his survey of *Alternaria* differentiated *Stemphylium* in its original sense (and in the sense of Wiltshire's section *Eustemphylium*) and transferred to *Alternaria* several of the taxa similar to *S. lanuginosum* which were considered controversial by earlier students of the group. Most taxonomists recognized that *Alternaria* and *Stemphylium* were inappropriate generic designations for species similar to *S. lanuginosum*. By their nomenclatural proposals or, more importantly, by their invariably excellent illustrations, conidiophores of members of the *S. lanuginosum* group bear no resemblance whatsoever to those of *S. botryosum*, and the conidial morphology of the group is fundamentally different from that of the type species of *Alternaria*.

There are two species of *Alternaria* infecting *Brassica* species which are probably the most controversial than any other species in this genus (Wiltshire 1947). The first is *A. brassicae* (Berk.) Sacc. described as *Macrosporium brassicae* Berk. in 1836, and the second is the fungus commonly known as *A. circinans* (Berk. & Curt.) Bolle. or *A. oleracea* Milbraith or incorrectly as *A. brassicae* (Berk.) Sacc. for which Wiltshire (1947) proposed the name *A. brassicicola* (Schwein.).

3.3 Phylogeny

For redefining the taxonomy of *Alternaria* and allied genera, 121 strains were included in the *Alternaria* complex alignment by Woudenberg et al. (2013). The alignment length and unique site patterns of the different genes and gene combinations were well defined. The original ITS alignment consisted of 577 characters of which the first 78 were excluded as this contained a non-alignable region. In the original *TEF1* alignment, 375 characters were coded as major inserts, which otherwise would negatively influence the phylogeny, resulting in a *TEF1* alignment of 269 characters. All the phylogenies, different phylogenetic methods and gene regions or gene combinations used on this dataset (data not shown; trees and alignment lodged in TreeBASE) show a weak support at the deeper nodes of the tree. The only well-supported node (Bayesian posterior probability of 1.0, RAxML Maximum Likelihood support value of 100) in all phylogenies separates *Embellisia annulata* CBS 302.84 and the *Ploeospora/Stemphylium* clade from the *Alternaria* complex. In the *Alternaria* clade, six monotypic lineages and 24 internal clades occur consistently in the individual, and combined phylogenies, although positions vary between the different gene regions or combinations used. The support values for the clades within *Alternaria* (called sections) were plotted in a heat map per gene and phylogenetic method used. The support values for the different phylogenetic methods vary, with the Bayesian posterior probabilities being higher than the RAxML bootstrap support values. The SSU, LSU and ITS phylogenies display a low resolution, which reflects in poor to no support of the sections. Therefore, Woudenberg et al. (2013) choose not to include them in the multigene alignments, except in the all-gene alignment. In the GAPDH phylogenies, sect. *Cheiranthus*, sect. *Nimbya* and sect. *Pseudoulocladium* are poorly supported, and *A. resedae* clusters separate from sect. *Cheiranthus*. In the *RPB2* phylogenies, the support values for sect. *Alternata*, sect. *Embellisioidos* and sect. *Euroka* are relatively low; *A. cumini* clusters in sect. *Embellisioidos* instead of sect. *Eureka* and *U. capsici* clusters separate from sect. *Pseudoulocladium*. The TEF1 phylogenies did not support sect. *Nimbya* and show relative low support for sect. *Cheiranthus*, sect. *Dianthicola*, sect. *Embellisioides*, sect. *Panax*, sect. *Phragmosporae* and sect. *Radicina* and *A. cumini* clusters outside sect. *Eureka*. In the two-region phylogenies, *U. capsici* clusters outside sect. *Pseudoulocladium* based on *GAPDH* and *RPB2*, *E. indefossa* clusters outside sect *Cheiranthus* based on *GAPDH* and *TEF1* and sect. *Eureka* are poorly supported based on *RPB2* and *TEF1*. The combined phylogeny based on the *GAPDH*, *RPB2* and *TEF1* sequences is displayed, as these are the genes with the best resolution.

The final *Pleosporineae* alignment included 74 strains, representing six families, and consisted of 2506 characters (SSU 935, LSU 796. RPB2 775) of which 700 were unique site patterns (SSU 111 LSU 145. RPB2 444). In the SSU alignment, a large insertion at position 446 in the isolates *Chaetosphaeronema hispidulum* CBS 216. 75, *Pleospora fallens* CBS 161.78, *Pleospora flavigena* CBS 314.80 and *Ophiosphaerella herpotrichia* CBS 620.86 was excluded from the phylogenetic analyses. A total of 43 202 trees were sampled after the burn-in. The type species of *Clathrospora*, *C. elynne*, forms a well-supported clade, located basal to the *Pleosporaceae* (Fig. 3.1) outside the *Alternaria* complex. The type species of *Comoclathris*, *C. lanata*, was not available for the study, but the two *Comoclathris compressa* strains cluster in a well-supported clade within the *Pleosporaceae* outside *Alternaria s. str.*

The *genus Alternariaster* with *Alternariaster helianthi* as type and only species also clusters outside the *Alternaria* complex and even outside *Pleosporaceae*; it belongs to the Leptosphaeriaceae instead (Fig. 3.1). *Embellisia annulata* is identical to *Dendryphiella salina* and forms a well-supported clade in the *Pleosporaceae* together with *Dendryphiella arenaria*. As the type species of *Dendryphiella*, *D. vinosa*, clusters outside the *Pleosporineae* (Cruz 2006; Jones et al. 2008), *Dendryphiella salina* and *D. arenaria* are placed in a new genus, *Paradendryphiella* (Woudenberg et al. 2013).

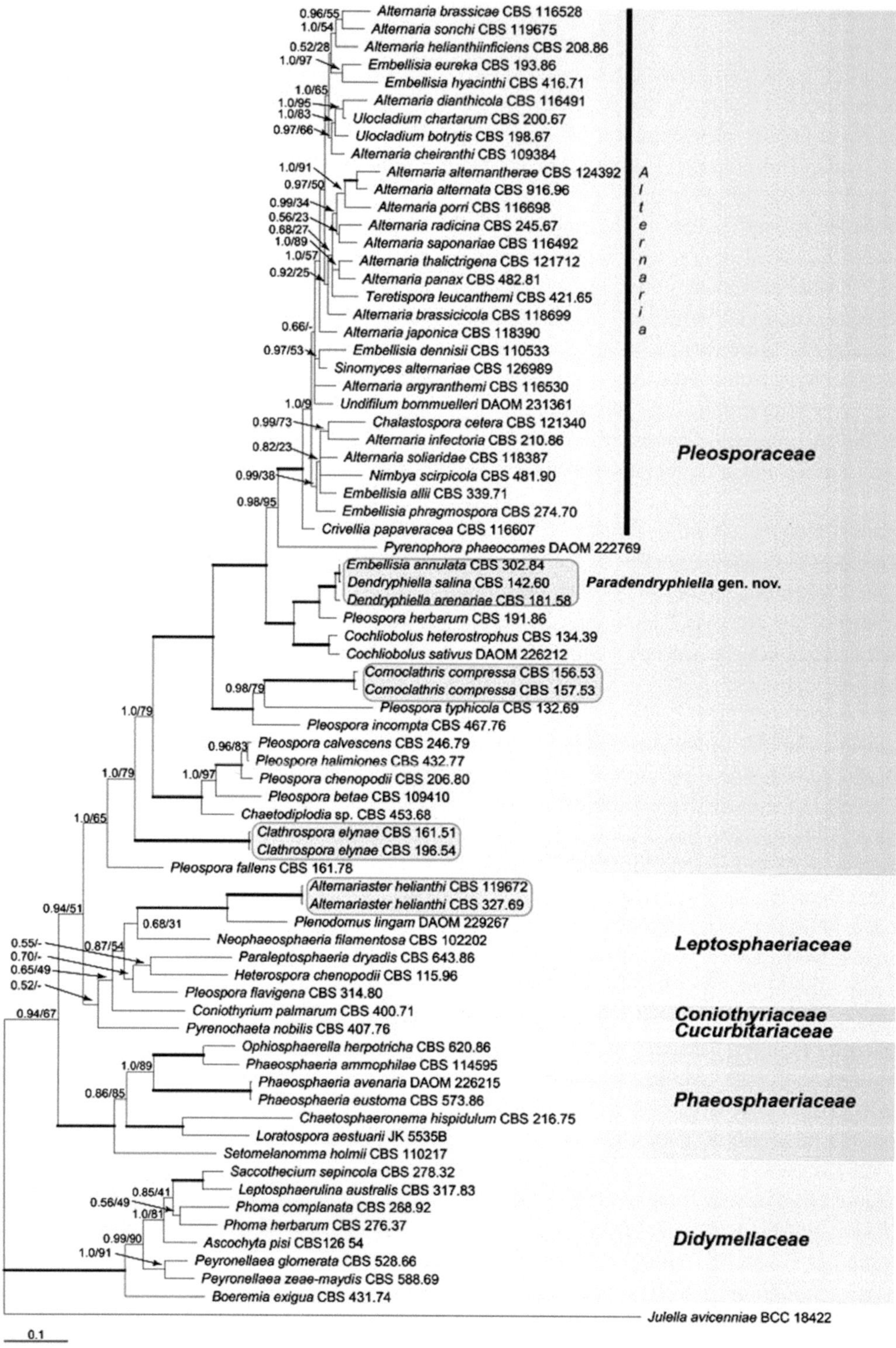

Fig. 3.1 Bayesian 50 % majority-rule consensus tree based on the SSU, LSU and RPB2 sequences of 74 strains representing the *Pleosporineae*. The Bayesian posterior probabilities (PP) and RAxML bootstrap support values (ML) are given at the nodes (PP/ML). Thickness lines a PP of 1.0 and ML of 100. The tree was rooted *to Julella avicenniae* (BCC 184220 (Woudenberg et al. 2013)

3.4 Taxonomy, Nomenclature and Morphology

Based on DNA sequence data in combination with a review of literature and morphology, the species within the *Alternaria* clade, 121 strains representing the *Alternaria* complex, were all recognized *Alternaria* by Woudenberg et al. (2013). This puts the genera *Allewia*, *Brachycladium*, *Chalastospora*, *Chmelia*, *Crivellia*, *Embellisia*, *Lewia*, *Nimbya*, *Sinomyces*, *Teretispora*, *Ulocladium*, *Undifilum* and *Ybotromyces* in synonymy with *Alternaria*, resulting in the proposal of 32 new combinations, 10 new names and the resurrection of 10 names. Species of *Alternaria* were assigned to 24 *Alternaria* sections, of which 16 are newly described and six monotypic lineages. The (emended) description of the genus *Alternaria*, the *Alternaria* sections and monotypic lineages with new *Alternaria* names and name combinations have been treated in alphabetical order. Finally, the description of the new genus *Paradendryphiella* is also provided (Woudenberg et al. 2013).

Alternaria Nees, Syst. Pilze (Würzburg) 72: 1816 [1816–1817]
- = *Elosia* Pers. Mycol. Eur. (Erlanga) 1: 12, 1822
- = *Macrosporium* Fr. Syst. Mycol. (Lundae) 3: 373, 1982
- = *Rhopalidium* Mont. Ann. Sci. Nat. Bot. Ser. 2, 6: 30: 1836
- = *Brachycladlum* Corda, Icon Fungorum hucusque Cogn (Prague) 2, 14: 1838
- = *Ulocladium* Preuss, Linnaea 24: 111, 1851
- = *Chmelia* Svob. Pol., Biologia (Bratislava) 21: 82, 1966
- = *Embellisia* E.G. Simmons, Mycologla 63: 380, 1971
- = *Trichoconiella* B. L. Jain, Kavaka 3: 39, 1976 [1975]
- = *Botryomyces* De Hoog & C. Rubio, Sabouraudia 20: 19, 1982 (nom illegit.)
- = *Lewia* M.E. Barr& E.G. Simmons. Mycotaxon 25: 289, 1986
- = *Ybotromyces* Rulamort, Bull. Soc. Bot. Centre-Ouest, Nouv. Ser. 17: 192, 1986
- = *Nimbya* E.G. Simmons, Sydowia 41: 316, 1989
- = *Allewia* E.G. Simmons, Mycotaxon 38: 260, 1990
- = *Crivellia Shoemaker* & Inderb, Canad. J. Bot. 84: 1308, 2006
- = *Chalastospora* E.G. Simmons, CBS Biodiversity Ser. (Utrecht) 6: 668, 2007
- = *Teretispora* E.G. Simmons, CBS Biodiversity Ser. (Utrecht) 6: 674, 2007
- = *Undifilum* B.M. Pryor, Creamer, Shoemaker, McLaln-Romero & Hambl., Botany 87: 190, 2009
- = *Sinomyces* Yong Wang bis & X.G. Zhang, Fungal Biol. 115: 192, 2011

Colonies effuse usually grey, dark blackish brown or black. Mycelium immersed or partly superficial; hyphae colourless, olivaceous brown or brown. Stroma rarely formed. Setae and hyphopodia absent. Conidiophores macronematous, mononematous, simple or irregularly and loosely branched, pale brown or brown, solitary or in fascicles. Conidiogenous cells integrated, terminal becoming intercalary, polytretic, sympodial or sometimes monotretic, cicatrized. Conidia catenate or solitary, dry, ovoid, obovoid, cylindrical, narrowly ellipsoid or obclavate, beaked or nonbeaked, pale or medium olivaceous brown to brown, smooth or verrucose, with transverse, and with or without oblique or longitudinal septa. Septa can be thick, dark and rigid, and an internal cell-like structure can be formed. Species with meristematic growth are known. Ascomata small, solitary to clustered, erumpent to (nearly) superficial at maturity, globose to ovoid, dark brown, smooth, apically papillate, ostiolate. Papilla short, blunt; peridium thin. Hamathecium of cellular pseudoparaphyses. Asci few to many per ascoma, (4–6) 8-spored, basal, bitunicate, fissitunicate, cylindrical to cylindro-clavate, straight or somewhat curved, with a short, furcate pedicel. Ascospores muriform, ellipsoid to fusoid, slightly constricted at septa, yellow-brown, without guttules, smooth, 3–7 transverse septa, 1–2 series of longitudinal septa through the two original central segments, end cells without septa, or with 1 longitudinal or oblique septum or with a

Y-shaped pair of septa (Ellis 1971; Holliday 1980; Woudenberg et al. 2013).

3.4.1 Type species: *Alternaria alternata* (Fr.) Keissl

The description of *Alternaria* s. str. by Woudenberg et al. (2013) is supported by (1) a well-supported phylogenetic node in multiple analyses; (2) high similarity of clades within *Alternaria* based on SSU, LSU and ITS data; and (3) variation in the order of clades between the different gene phylogenies, which is in congruence with low support values at these deeper nodes. They have followed the procedure introduced by Lawrence et al. (2013) to assign the taxonomic status of sections of Alternaria for the different clades found, thus allowing retaining the former generic names but associated with a different taxonomic status. For end users, this seems to be a more stable and understandable taxonomy and nomenclature (Woudenberg et al. 2013)

3.5 Classification

The genus *Alternaria* belongs to the kingdom Mycota, phylum Ascomycota, division Deuteromycota, subdivision Pezizomycotina, class Dothideomycetes, order Pleosporales and family Pleosporaceae. It has several species as saprophytes and parasites. Some species of *Alternaria* are the asexual anamorph of the ascomycetes *Pleospora*, while others are speculated to be anamorph of *Leptosphaeria*. The two major features of *Alternaria* species are the production of melanin especially in the spores and production of host-specific toxins in the case of pathogenic species. The genus *Alternaria* was established in 1817 with *A. alternata* (originally *A. tenuis*) as the type isolate. Because of the absence of an identified sexual stage for the vast majority of *Alternaria* species, this genus was classified into the division of mitosporic fungi or the phylum Fungi imperfecti. The key taxonomic feature of the genus *Alternaria* is the production of large, multicellular, dark-coloured (melanized) conidia with longitudinal as well as transverse septa (phaeodictyospores). These conidia are broadest near the base and gradually taper to an elongated beak, providing a club-like appearance. Conidia are produced in single or branched chains on short, erect conidiophores. *Alternaria* forms conidia that arise as protrusions of the protoplast through pores in the conidiophore cell wall. At the onset of conidial development, the apex of the conidiophore thickens, and a ring-shaped electron-transparent structure is deposited at the apical dome. At the central cavity of this electron-transparent structure, a pore is formed through the dissolving of the cell wall. Through this pore, cytoplasm only covered by the plasma membrane is pushed out. The turgor pressure required to push the cytoplasm through the pore is presumably provided by the well-developed, large vacuole that appear at this stage in the conidiophore cell. Subsequent to this, a nucleus migrates into the newborn conidium, and later on a cell wall is deposited (Honda et al. 1987, 1990). The melanin that is present in the conidia is concentrated in the outer region of the primary cell walls, which are derived from the original wall of the developing spore, and in the septa, which delimit individual spore cells in the multicellular conidium. After the cells have been delimited by septa, secondary cell walls are deposited, but these remain unmelanized, suggesting a developmental regulation of melanin deposition during conidiogenesis (Campbell 1969; Carzaniga et al. 2002; Kawamura et al. 1997). Melanin is probably actively involved in conidial development, since disruption of a melanin biosynthesis gene in *A. alternata* reduced conidial size as well as septal number (Kawamura et al. 1999).

A classification based on conidial characteristics is complicated by the existence of other fungal genera, such as *Stemphylium* and *Ulocladium*, which produce phaeodictyosporic conidia that resemble those of *Alternaria*. Based on the characteristics defined by Simmons (1995), *Stemphylium* and *Alternaria* species are discriminated by the appearance of the conidiophore apex and *Ulocladium* and *Alternaria* species by the appearance of the basal end of immature conidia. This differentiation is largely supported by a molecular analysis of ribosomal DNA sequences (Pryor and Gilbertson 2000).

Within the genus *Alternaria*, species are also primarily defined upon conidium characteristics. Over 100 species occurring worldwide have been described (Simmons 1992). However, errors in the taxonomy of *Alternaria* species have arisen due to the variability of its morphological characters, which are not only affected by intrinsic factors, but also by environmental conditions. As a result of this, it is feasible that single species have been accidentally divided into several (Rotem 1994). This is illustrated by the observation that species of *Alternaria* alone is capable of attacking over 100 hosts, and in addition, Groves and Skolko (1944) found that this species typically displays morphological variations. Obviously, this phenotypic variation does not justify assigning *A. alternata* like specimens to other species. However, it is this morphological variation that has probably resulted the description of certain *Alternaria* species that have never been verified by others (Rotem 1994).

Because of the large diversity of *Alternaria* species, a division into subgeneric groups has occasionally been proposed. However, to date there has not been one general classification of *Alternaria*. Neergaard (1945) proposed a classification based on catenulation (chain formation of conidia), while more recently an organization of the genus into species groups, each typified by a representative species, was proposed (Simmons 1992). Because of morphological similarity but pathological differences, strains of a certain species, especially *A. alternata*, *A. brassicae* and *A. brassicicola*, have been defined as *formae species* or 'pathotypes' (Nishimura and Kohmoto 1983; Verma and Saharan 1994; Thomma 2003).

3.6 Morphology of *Alternaria* species Pathogenic on Cruciferous Crops

The following species of this genus cause economically important diseases in Brassicaceae:

Alternaria alternata (Fr.) Keissler, Beih. Bot. Zbl., 29: 434,1912

= *Torula alternata* Fr., 1832, Syst. Mycol., 3: 500

= *A. tenuis* C.G. Nees, 1816/17, Syst. Pilze Schwamme: 72

Additional synonyms listed in Simmons (2007).

The reasons why the epithet al*ternata* should be used instead of the more commonly accepted one *tenuis* are clearly stated by Simmons (1967).

Colonies usually black or olivaceous black, sometimes grey. Conidiophores arising singly or in small groups, simple or branched, straight or flexuous, sometimes geniculate, pale to mild olivaceous or golden brown, smooth, up to 50-μ long, 3–6-μ thick with 1 or several conidial scars. Conidia formed in long, often branched chains; obclavate, obyriform, ovoid or ellipsoidal, often with a short conical or cylindrical beak; sometimes up to but not more than one-third the length of the conidium; pale to mid golden brown; smooth or verruculose with up to 8 transverse and usually several longitudinal or oblique septa; overall length 20–63 (37) μ, × 9–18 (13) μ; thick in the broadest part; beak pale, 2–5 μ thick (Fig. 3.2). An extremely common saprophyte found on many kinds of plants and other substrata including foodstuff, soil and textiles; cosmopolitan (Ellis 1971; Holliday 1980).

3.6.1 Type Species: *Alternaria brassicicola* (Schw.) Wiltshire

Alternaria brassicicola (Schw.) Wiltshire in Mycol. Pap. 20:8, 1947

= *Helminthosporium brassicicola* Schweinitz (as *Helminthosporium brassicola*) in Trans. AM. Phil. Soc. N.S., 4: 279, 1832

= *Macrosporium cheiranthi* Fr. var. *circinans* Berk. & Curt. in Grevilles, 3: 105, 1875

= *Alternaria circinans* (Berk. & Curt.) Bolle in Meded. Phytopath. Lab. Willie commelin Scholten, 7:26, 1924

= *Alternaria oleracea* Milbraith in Bot. Gaz., 74: 320, 1922 (Full synonymy given by Wiltshire in Mycol. Pap. 20: 1947)

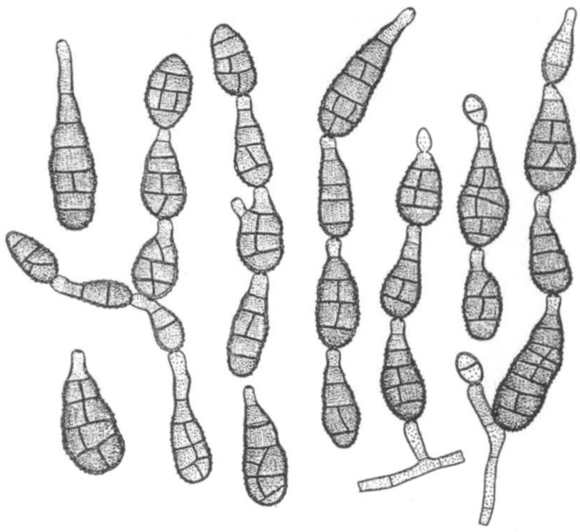

Fig. 3.2 *Alternaria alternata* (×650) Ellis 1971

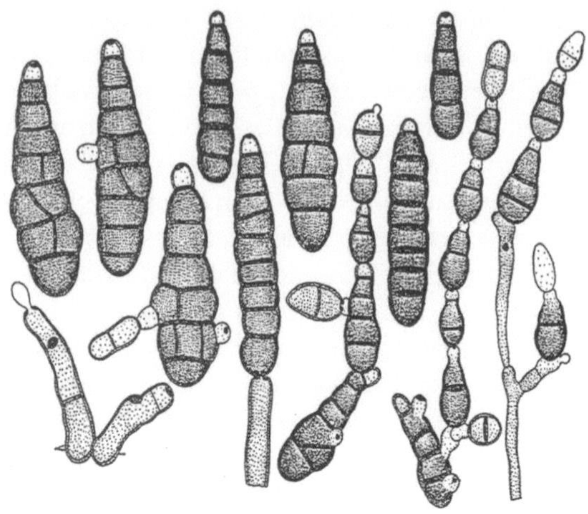

Fig. 3.3 *Alternaria brassicicola* (×650) Ellis 1971

Additional synonyms listed in Simmons (2007).

Colonies amphigenous, effused, dark olivaceous brown to dark blackish brown, velvety. Mycelium immersed; hyphae branched septate, hyaline at first, later brown or olivaceous brown, inter- and intracellular, smooth, 1.5–7.5-μ thick. Conidiophores arising singly or in groups of 2–12 or more; emerging through stomata; usually simple, erect or ascending; straight or curved; occasionally geniculate; more or less cylindrical, but often slightly swollen at the base; septate, pale to mild olivaceous brown; smooth; up to 70-μ long; 5–8-μ thick. Conidia mostly in chains of up to 20 or more, sometimes branched, acropleurogenous, arising through small pores in the conidiophore wall, straight, nearly cylindrical, usually tapering slightly towards the apex or obclavate. The basal cell rounded; the beak usually almost non-existent; the apical cell being more or less rectangular or resembling a truncated cone, occasionally better developed but then always short and thick, with 1–11, mostly less than 6, transverse septa, often constricted at the septa; pale to dark olivaceous brown; smooth or becoming slightly warted with age; 18–130-μ long; 8–30-μ thick in the broadest part, with the beak 1/6 the length of the conidium and 6–8-μ thick (Ellis 1968b, 1971; Holliday 1980). The fungus is confined to the Brassicaceae (Fig. 3.3).

3.6.2 Type Species: *Alternaria brassicae* (Berk.) Sacc.

Alternaria brassicae (Berk.) Sacc. in Michelia, 2:129 see also p. 172, 1880

= *Macrosporium brassicae* Berk. in Smith's Engl. Fl. 5, pt. 2 : 339, 1836. (Full synonymy given by Wiltshire in Mycol. Pap., 20, 1947)

Colonies amphigenous, effused, rather pale olive, hairy; the individual large conidia plainly seen under a x 20 binocular dissecting microscope. Mycelium immersed; hyphae branched, septate, hyaline, smooth, 4–8-μ thick. Conidiophores arising in groups of 2–10 or more from the hyphae; emerging through stomata; usually simple, erect or ascending; straight or flexuous; frequently geniculate; more or less cylindrical but often slightly swollen at the base; septate; mid-pale greyish olive; smooth up to 170-μ long; 6–11-μ thick; bearing one to several small but distinct conidial scars. Conidia solitary or occasionally in chains of up to 4, acropleurogenous, arising through small pores in the conidiophore wall; straight or slightly curved; obclavate; rostrate with 16–19 (usually 11–15) transverse septa and 0–8 (usually 0–3) longitudinal or oblique septa; pale or very pale olive or greyish olive; smooth or infrequently, very inconspicuously warted; 75–350-μ long and usually 20–30-μ

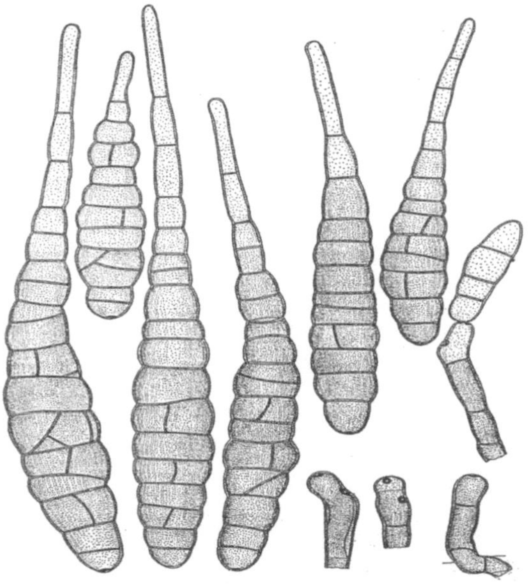

Fig. 3.4 *Alternaria brassicae* (×650) Ellis 1971

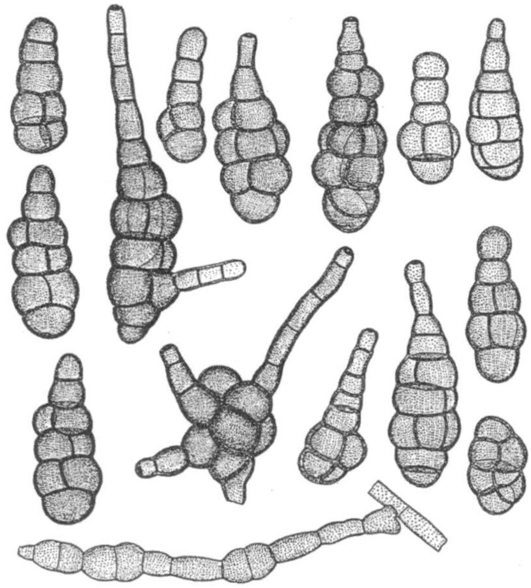

Fig. 3.5 *Alternaria raphani* (×650) Ellis 1971

(sometimes up to 40 μ) thick in the broadest part; the beak about 1/3 to 1/2 the length of the conidium and 5–9-μ thick (Fig. 3.4) (Ellis 1968a, 1971; Holliday 1980). This species produces chlamydospores. It is mostly confined to the Brassicaceae.

3.6.3 Type Species: *Alternaria raphani* Groves & Skolko

Alternaria raphani Groves and Skolko 1944, Can. J. Res., Sect. C, 22:227
= *Alternaria matthiolae* Neergaard 1945

Conidiophores simple or occasionally branched, septate, olivaceous brown, up to 150-μ long, 3–7-μ thick, sometimes swollen slightly at the tip and usually with a single conidial scar; conidia commonly in chains of 2–3, straight or slightly curved, obclavate or ellipsoidal, generally with a short beak, mid to dark golden brown or olivaceous brown, smooth or sometimes minutely verruculose, with 3–7 transverse and often a number of longitudinal or oblique septa, constricted at the septa, 50–130 (70)-μ long, 14–30 (22)-μ thick in the broadest part. Chlamydospores formed abundantly in culture, sometimes in chains at first 1 celled, round, finally many celled and irregular, brown. Conidiophores often develop from them (Ellis 1971; Holliday 1980). Unlike *A. alternata* and *A. brassicicola*, this species forms chlamydospores and the beak of the conidium is smaller than that of *A. brassicae* but longer than the almost non-existent beak of *A. brassicicola* (Fig. 3.5).

3.6.4 Type Species: *Alternaria cheiranthi* (Lib.) Bolle

Alternaria cheiranthi (Lib. Bolle 1924, as '(Fr.) Bolle', Meded. Phytopath. Lab. Willie Commelin Scholten, 7:55
= *Helminthosporium cheiranthi* Lib., 1827, apud Desm., Crypt. Fr. Exsicc., 213
= *Macrosporium cheiranthi* (Lib.) Fr., 1832, Syst. Mycol., 3: 374

Conidiophores arising singly or in groups, mostly simple but sometimes branched, straight or flexuous, septate, rather pale olive, often hyaline at the apex, smooth, up to 130-μ long, 5–8-μ thick with a single terminal scar at first but later with up to 4 scars, which may be borne close together without marked geniculation. Conidia mostly solitary, rarely in chains of 2, 3 or more; variously shaped, often pyriform, ovoid or elongate ovoid at first, later becoming irregular, mostly tapering to the apex, which may be drawn out into a beak; generally rounded at the base, with numerous transverse, longitudinal and oblique septa; light olive to golden brown; translucent, with the interior walls, which are dark; often plainly visible; smooth or with the wall pitted from the inside; 20–100-μ long; 13–32-μ thick in the broadest part (Fig. 3.6). Common on wallflowers and occasionally on other brassicaceous plants (Ellis 1971).

To establish the correct systematic designation of two common species of *Alternaria* on Brassicaceae, Wiltshire (1947) gave the following account of synonymy, which had appeared in the literature up to 1945. However, additional synonymous are listed by Simmons (2007).

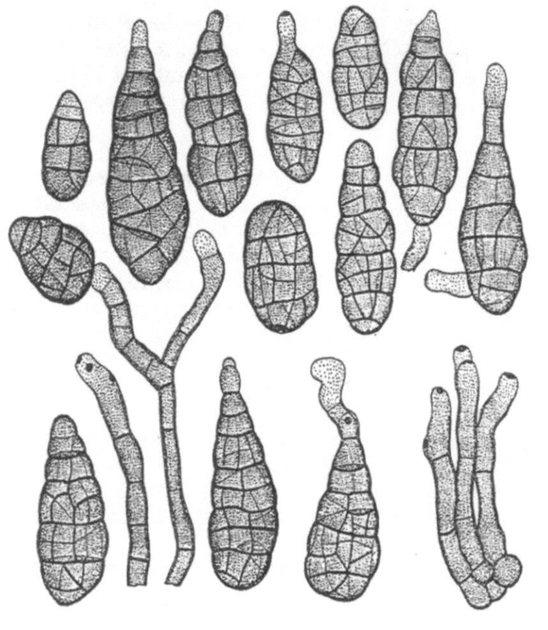

Fig. 3.6 *Alternaria cheiranthi* (×650) Ellis 1971

1. ***Alternaria brassicae*** (Berk.) Sacc.
   - 1836. *Macrosporium brassicae* Berk. In Smith's Engl. Flora. v, Part II P. 339
   - Hooker's British Flora, vol. ii, part II, 'Fungi' by M.J. Berkeley
   - 1836. *Puccinia* (?) *brassicae* Montagne. In Ann. Sci. Nat., Ser. II VI, p. 30
   - 1855. *Sporidesmium exitiosum* Kuhn. In Hedwigia, i p. 91
   - 1856. *Rhopalidium brassicae* Mont. & Fr. In Montagne's Syll. Crypt., P.297
   - 1859. *Polydesmus exitiosus* (Kuhn) Kuhn. In Kramkh. d. Kulturgew., P.165
   - 1880. *Alternaria brassicae* (Berk.) Sacc. var. minor Sacc. In Mich. ii, P. 172
   - 1882. *Cercospora bloxami* Berk. & Br. In Ann. & Mag. Nat. Hist. Ser. V, IX. P. 183, No.1882, 1879
   - 1882. *Macrosporium herculeum* Ellis & Martin, In Amer. Nat., XVI, p.1003; exsicc. N. Amer. Fungi, 1263
   - 1884. *Cercospora lepidii* Peck. In 35th Rept. N.Y. State Mus. 1881, p. 140.
   - 1886. *Alternaria brassicae* (Berk.) Sacc. var. *macrospora* Sacc. In Syll., IV, p.546
   - 1891. *Sporidesmium annii* Karst. In Symb. Myc. Fenn. XXX, P. 67
   - 1897. *Macrosporium brassicae* Berk. var. *macrospora* Eliason
   - 1901. *Sporidesmium brassicae* Massee. In Kew. Bull., 1901, P. 153
   - 1902. *Leptosphaeria exitiosa* (Kuhn) Rostrup. In Plantepatolagi Haandbog i Laeren om Plantesygdomme for Landbrugere, Havebrugere og Skovbrugere, P. 472
   - 1902. *Alternaria brassicae* (Berk.) Sacc. var. *exitiosa* (Kuhn) Ferraris. In Flora italica, P. 521
   - 1917. *Alternaria herculea* (Ell. & Mart.) Elliott. In Amer. Journ. Bot., IV, P. 472
   - 1924. *Alternaria brassicae* (Berk.) Bolle [nee. Sacc.]
   - 1944. *Alternaria macrospora* (Sacc.) Sawada. In descriptive catalogue of the Formosan fungi. Part V, Dept. Agric. Govt. Res. Inst. Formosa, Rept. 51, p. 123
   - 1945. *Alternaria exitiosa* (Kuhn) Jorstad. In Melding fra Statens. Plantepatogiske Institut. No.1, P. 594

It is extremely confusing that the name *A. brassicae* (Berk.) Sacc. has been applied, in accordance with Saccardo's usage in the Sylloge, to the species *A. brassicicola* by various authors, e.g. Voglino (1902), Ferraris (1912), Sawada (1931), Weber (1932), Yoshii (1933) and Fajardo and Palo (1934).

2. ***Alternaria brassicicola*** (Schwein) Wiltshire
   1832. *Helminthosporium brassicicola.* Schweinitz. In Syn. Fung. Amer. Bar. No. 2632, Trans. Amer. Phil. Soc., N.S., IV P. 279
   1855. *Sporidesmium exitiosum* Kuhn formae (B) *Alternarioides* and (Y) *luxuriosum* Kuhn. In Hedwigia, I, P. 91
   1859. *Polydesmum exitiosus* (Kuhn) Kuhn formae (B) *Alternarioides* and (Y) *luxuriosum* Kuhn) Kuhn. In Krankh. d. Kulturgew., p. 165
   1875. *Macrosporium cheiranthi* Fr. var. *circinans* Berk. & Curt. In Grevillea, III, P. 105
   1880. *Alternaria brassicae* (Berk.) Sacc. *minor* Sacc. In Mich., II, P. 172
   1886. *Macrosporium commune* Rabenh. var. *circinans* (Berk. & Curt.) Sacc. In Syll., iv, P. 524
   1886. *Alternaria brassicae* (Berk.) Sacc. In Syll., IV, p. 546
   1897. *Alternaria brassicae* (Berk.) Sacco var. *microspora* Brun. In Act. Soc. linn. Bordeaux, III, P. 149
   1902. *Helminthosporium brassicae* P. Henn. In Hedwigia, XLI, p. 117
   1922. *Alternaria oleracea* Milbraith. In Bot. Gaz., LXXIV, p. 320
   1924. *Alternaria circinans* (Berk. & Curt.) Bolle. In Meded. Phytopath. Lab. Willie Commelin Scholten; Baarn, VII, P. 26
   1933. *A. brassicae* (Berk.) 'Lindau' cited in error by Yoshii for *A. brassicae* (Berk.) Sacc.

Reviewing then the names that have been applied to *A. brassicicola*, it is evident that Schweinitz's name antedates the others by many years. There is no doubt that *A. oleracea* and *A. circinans* are identical with it and must rank as synonyms. The species is quite distinct from the true *A. brassicae* (Berk.) Sacc., and though *A. brassicicola* shows considerable variations on the host, in culture, it always produces the same characteristic, almost cylindrical, short beaked spores of 11.5–44 by 7–9 μ by which its identity can readily be established (Wiltshire 1947).

3.7 The Infection Process

In general, *Alternaria* species are foliar pathogens that cause a relatively slow destruction of host tissues through the reduction of photosynthetic potential. An infection leads to the formation of necrotic lesions, which sometimes have a target-like appearance due to growth interruptions caused by unfavourable conditions. The fungus resides in the centre of the lesion, which is surrounded by an un-invaded chlorotic halo, a symptom that is commonly observed for the infection process of necrotrophic pathogens. This zone is created by the diffusion of fungal metabolites like toxins (Agarwal et al. 1997; Tewari 1983). Members of the genus cause quiescent infections in which the fungus enters the tissue where it remains dormant until conditions become favourable for infection. *Alternaria* species generally do not affect water or nutrient transport throughout the plant, because they do not specifically target roots or vessels (Rotem 1994).

Alternaria has no known sexual stage or overwintering spores, but the fungus can survive as mycelium or spores on decaying plant debris for a considerable time or as a latent infection in seeds (Rotem 1994). If seed borne, the fungus can attack the seedling once the seed has germinated. In other cases, once the spores are produced, they are mainly spread by wind on to plant surfaces where infection can occur. Typically, weakened tissues, either due to stresses, senescence or wounding, are more susceptible to *Alternaria* infection than healthy tissues. The observation that saprobic *Alternaria* species can become parasitic when they meet a weakened host illustrates that the distinction between saprophytic and parasitic behaviour is not always evident.

Despite the taxonomic and pathogenic differences between *Alternaria* species, they cause similar infection patterns. Dormant spores have heavily melanized walls that, under favourable conditions, produce one or more germ tubes.

Subsequently, the germ tubes penetrate stomata, cuticle or wounds with or without the formation of small appressoria. In less virulent species, wounds and stomata are targeted, while more virulent species can also penetrate directly (Rotem 1991). Enzymatic processes in *Alternaria* infections are essentially similar to those in other diseases. The cuticle, which consists of a combination of cutin (a hydroxyl fatty acid polyester) and waxes, comprises the first line of defence to be overcome by directly penetrating fungal pathogen. For *A. brassicicola*, differential expression of cutinase genes was monitored between saprophytic and pathogenic stages of the fungus (Yao and Köller 1995). Furthermore, it was found that different cutinolytic enzymes are sequentially induced upon landing on, and penetration of, the cabbage leaf (Fan and Köller 1998). Constitutively produced cutinases are expressed during the initial contact of *A. brassicicola* with the cuticle. After reaching subcuticular layers, different cutinases that are active during saprophytic growth are induced (Trail and Köller 1993; Yao and Köller 1994). These cutinases are inducible by cutin monomers. This implies a switch between the parasitic and the saprophytic stage of the fungal pathogen so far. However, it has not been demonstrated that any extracellular hydrolase is crucially involved in fungal pathogenesis.

In addition to cutinases, lipases might also contribute to the establishment of infection. *A. brassicicola* found to produce a lipase that acts as a virulent factor (Berto et al. 1997). Anti- lipase antibodies were found to have an inhibitory effect on the in vivo infection of cauliflower leaves, as symptom development was inhibited in a dose-dependent manner despite normal germination of spores. However, the antibodies did not have an effect on fungal infection of dewaxed leaves, which suggests that this lipase has an early pathogenic activity during penetration (Berto et al. 1999).

About one-third of the total cell wall components in dicotyledonous plants are pectic polysaccharides. These components can be hydrolysed by fungal galacturonidases. *Alternaria citri* was found to be dependent on endopolygalacturonidase activity for establishing an infection, as a mutant lacking this activity was severely compromised in its pathogenic capacity. On the other hand, the tangerine pathotype of *A. alternata* did not show this dependency, possibly because this particular pathogen largely depends on toxin production for the colonization of its host (Isshiki et al. 2001).

For a specific *A. alternata* endoglucanase, it was demonstrated that its production is triggered by a pathogen-induced pH increase on the host (Eshel et al. 2002b). Correlation studies between enzyme production and symptom development suggest that endoglucanases and exoglucanases are involved in *A. alternata* pathogenicity (Eshel et al. 2000, 2002a, b).

3.8 Identification of *Alternaria* Genes

In order to identify candidate fungal pathogenicity genes and characterize a compatible host response, a suppression subtractive hybridization (SSH) cDNA library enriched for *A. brassicicola* and *Brassica oleracea* genes expressed during the interaction was created, along with a cDNA library representing genes expressed during nitrogen starvation (NS). A total of 3749 and 2352 expressed sequence tags (ESTs) were assembled into 2834 and 1264 unisequence sets for the SSH and NS libraries, respectively. Cramer et al. (2006) compared two methods to identify the origins (plant vs. fungal) of ESTs in the SSH library using different classification procedures, with and without the availability of a database representing the *A. brassicicola* whole genome sequence and Brassicaceae-specific genes. BLASTX analyses of the 2834 unisequence set using the GenBank nonredundant database identified 114 fungal genes. Further BLASTN analyses of the genes with unidentifiable origin using a database consisting of the 1264 fungal unisequence set from the nitrogen starved library identified 94 additional fungal genes. By contrast, BLASTN analyses of the same SS11 unisequence set using a partially assembled *A. brassicicola* whole genome draft sequence identified a total of 310 unisequences of fungal origin. Even a small number of organism-specific EST sequences can be very helpful to identify pathogen genes in a library derived from infected tissue, partially overcoming the limitation of the public databases for little studied organisms. However, using the whole genome draft sequence of *A. brassicicola*, one is able to identify approximately 30 % more fungal genes in the SS11 library than without utilizing this resource (Cramer et al. 2006).

The biosynthetic gene cluster of brassicicene C was identified in *A. brassicicola* strain ATCC 96836 from genome database search. In vivo and in vitro studies clearly reveal the function of *Orf 8* and *Orf 6* as a fusicoccadiene synthase and methyltransferase, respectively. The understanding towards the biosynthetic pathway promises construction of this type of diterpene compounds with genetic engineering (Minami et al. 2008).

3.9 Nuclear Ribosomal DNA Sequences

The sequences coding for the nuclear 18s rRNA, 5–8s rRNA and the internal transcribed spacers (*ITS1* and *ITS2*) were amplified by the polymerase chain reaction and sequenced for one isolate each of *Alternaria brassicae*, *A. brassicicola*, *A. raphani*, *A. alternata* and *Pleospora herbarum*. The 5–8s rDNA sequences from the four *Alternaria* species are identical and differed at only one base pair from that of *P. herbarum*. The internal transcribed spacer sequences especially *ITS1* are very variable in both base composition and length. The 18s rDNA sequence is highly conserved, but enough variability is present to distinguish genera clearly. Phylogenetic analysis of the sequence data sets by both parsimony and maximum likelihood methods clearly separated genera and species. All of the *Alternaria* species closely related to *Pleospora* also appeared to be more closely related to *Alternaria* than to *Leptosphaeria* (Jasalavich et al. 1995).

Currently taxonomists consider *Pleospora* and *Leptosphaeria* to be in separate Loculoascomycetes families, the Pleosporaceae and the Leptosphae-riaceae, respectively (Barr 1987; Eriksson and Hawksworth 1991), with the main differential criterion being the conidiomatal structure of the anamorphs. *Leptosphaeria* species have coelomycetous anamorphs, while genera in the Pleosporaceae have hyphomyceteous anamorphs (Barr 1987). The phylogenetic trees (Figs. 3.7 and 3.8) based on rDNA sequences support this taxonomic view. All the Loculoascomycetes included in the analysis, i.e. *Leptosphaeria* and *Pleospora* as well as *Alternaria*, formed a phyletic group. Within this group, there are two major lineages: the Leptosphaeriaceae composed of *Leptosphaeria* and the Pleosporaceae composed of *Pleospora* and *Alternaria*. The two *Ophiosphaerella* species, which are included only in the analysis based on the 5–8s rDNA and flanking internal transcribed spacers, fall within the *Leptosphaeria* group (Fig. 3.7). *Ophiosphaerella korrae* (Walker and Smith) Shoemaker and Babc has been placed in *Leptosphaeria* (Walker and Smith 1972), while *O. herpotricha* (Fr.; Fr) Walker was previously in *Phaeosphaeria* (Holm 1957), a segregate of *Leptosphaeria* (Holm 1957). For further details of the phylogeny of *Leptosphaeria*, see Morales et al. (1995).

Most species of *Alternaria* including the four species pathogenic to crucifers lack a known sexual stage. *Alternaria* and *Pleospora*, although distinct genera, are sister taxa (Fig. 3.8) based on the 18s rDNA sequence data. Simmons (1967) considered the *Alternaria*, *Stemphylium* and *Ulocladium* to be closely related on the basis of conidial development, as they all produce dictyoporospores. *Pleospora herbarum* has a *Stemphylium* anamorphs and provides the first molecular evidence that asexual species of *Alternaria* are indeed related to *Pleospora*. Therefore, the anamorphic *Alternaria* should be included in the Pleosporaceae.

Wiltshire (1947) considered *A. raphani* and *A. brassicae* to be closely related. It would appear, however, that *A. raphani* is actually more closely related to *A. brassicicola* than to *A. brassicae* based on rDNA sequence data. *Alternaria alternata*, *A. brassicae*, *A. brassicicola* and *A. raphani* formed a strong clade of very closely related sister taxa. The 18s rDNA resolved two subclades of species within *Alternaria* at a level of confidence of 95 % (Fig. 3.8), based on five informative sites contained within the 300 bp at 3′ end of sequence alignment. Complete resolution of the

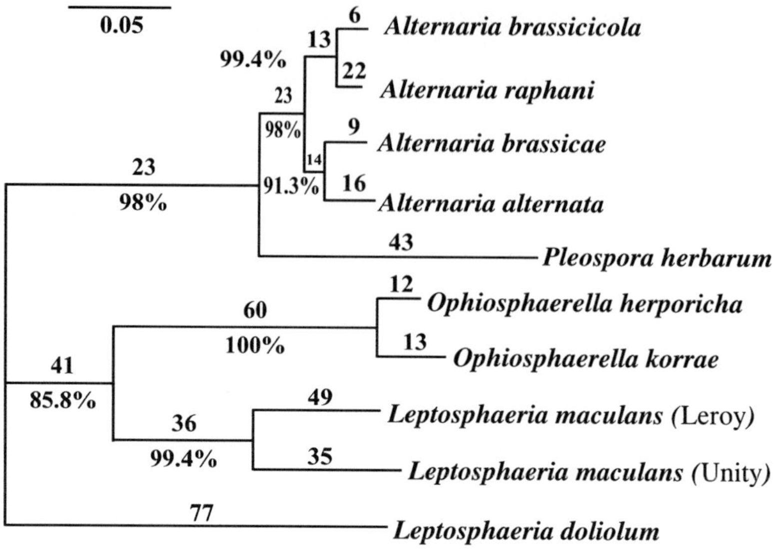

Fig. 3.7 The consensus tree generated by global parsimony bootstrap analysis of alignment of the 5–8s rDNA and flanking internal transcribed spaces (ITS1 and ITS 2) sequences. The percentages represent the proportion of 1000 bootstrap replications in which the taxa to the right of the node were placed together by the programme DNA PARS with randomization of the sequence input order. The other numbers represent the steps. Branch length (drawn in the horizontal dimension only) is the maximum likelihood estimates made by the programme DNAML. When the user tree was defined as the bootstrap consensus tree, the length of the vertical lines has no meaning and was adjusted arbitrarily for ease in labelling termini. *Leptosphaeria doliolum* was designated as the out-group (Jasalavich et al. 1995)

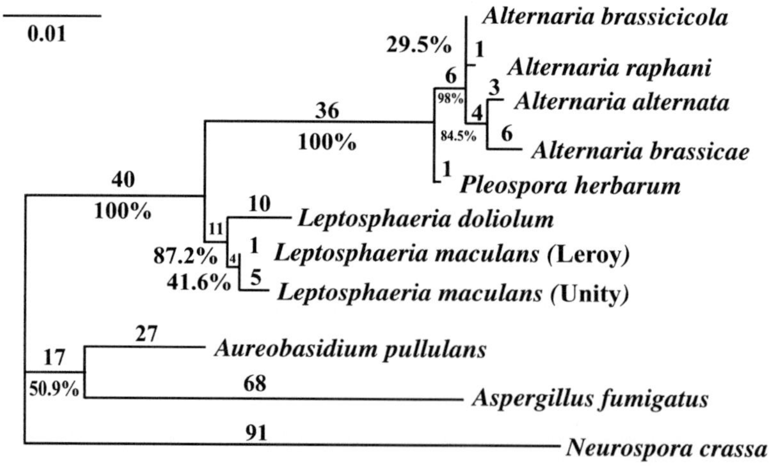

Fig. 3.8 The consensus tree generated by global parsimony bootstrap analysis of alignment of the 18s rDNA. The percentages represent the proportion of 1000 bootstrap replications in which the taxa to the right of the node were placed together by the programme DNA PARS with randomization of the sequence input order. The other numbers represent the steps. Branch length (drawn in the horizontal dimension only) is the maximum likelihood estimates made by the programme DNAML. When the user tree was defined as the bootstrap consensus tree, the length of the vertical lines has no meaning and was adjusted arbitrarily for ease in labelling termini. *Neurospora crassa* was designated as the out-group (Jasalavich et al. 1995)

species of *Alternaria* is achieved with the *ITS* sequence data (Fig. 3.7) which are much more variable and contained more phylogenetically informative sites than the 18 s rDNA. The *ITS* sequence data of *Leptosphaeria* are even more variable than those of *Alternaria*. Likewise, the range of genetic distance among the *Leptosphaeria* species is larger than those seen among the *Pleospora* and *Alternaria* species.

The genetic distances between *P. herbarum* and any of the four *Alternaria* species examined are on the same order of magnitude as that between the highly virulent and weakly virulent isolates of *L. maculans*. These differences suggest that genus *Alternaria* encompasses less genetic variability than does the genus *Leptosphaeria*. This may reflect the differences in size and reproductive strategies of the two genera. *Alternaria* comprises roughly 60 species (Rao 1969; Rossman et al. 1987), while *Leptosphaeria* includes a heterogeneous mixture of about 1689 taxa (Crane and Shearer 1991) and is very likely polyphyletic. Perhaps the four *Alternaria* species have separated from each other only recently in evolutionary time and so have not yet accumulated many changes in their rDNA sequences. Alternatively, perhaps plant host association may exert a pressure against further divergence. Along with phylogenetic relationship among the *Alternaria* species, portions of the rDNA sequence can have practical uses, e.g. production of diagnostic tools. The variability present in *ITS1* can be exploited to design species-specific probes and primers in order to identify rapidly species of *Alternaria* both in culture and in plant materials (Jasalavich et al. 1995).

3.10 Identification, Cloning and Sequencing of Virulence Genes

In many plant pathosystems, fungal secondary metabolites derived from non-ribosomal peptide synthetases (*NPSs*) are phytotoxic virulence factors or are antibiotics thought to be important for niche competition with other microorganisms. However, many of the functions of *NPS* genes and their products are largely unknown. Kim et al. (2007) investigated the function of one of the *A. brassicicola NPS* genes, *AbNPS2*. The predicted amino acid sequence of *AbNPS2* shows high sequence similarity with *A. brassicae AbrePsy1*, *Cochliobolus heterostrophus NPS4* and a *Stagonospora nodorum NPS*. The *AbNPS2* open reading frame was predicted to be 22 kb in length and encodes a large protein (7195 amino acids) showing typical *NPS* modular organization. Gene expression analysis of *AbNPS2* in wild-type fungus indicated that it is expressed almost exclusively in conidia and conidiophores, broadly in the reproductive developmental phase. *AbNPS2* gene disruption mutants show abnormal spore cell wall morphology and a decreased hydrophobicity phenotype. Conidium of *abnps2* mutants displays an aberrantly inflated cell wall and an increase in lipid bodies compared with wild type. Further, phenotypic analyses of *abnps2* mutants showed decreased spore germination rates both in vitro and in vivo and a marked reduction in sporulation in vivo compared with wild-type fungus. Moreover, virulence tests on *Brassica* with *abnps2* mutants reveal a significant reduction in lesion size compared with wild type but only when aged spores are used. Collectively, these results indicate that *AbNPS2* plays an important role in development and virulence (Kim et al. 2007).

3.11 Identification of Pathogenicity Factors

Functional redundancy of lipases with regard to *Alternaria* pathogenicity has been observed by many workers (Yao and Köller 1994, 1995; Berto et al. 1999; Cho et al. 2006). Interestingly, one of the factors responsible for the pathogenicity has been predicted to be secondary metabolite production. A non-ribosomal peptide synthase gene (*NPS6*) in *Cochliobolus heterostrophus* and *A. brassicicola* is found to direct the biosynthesis of a siderophores metabolite important for oxidative stress tolerance and pathogenicity (Oide et al. 2006). The secondary metabolite corresponding to or synthesized via *AbNPS2* has to be characterized. There is a need to further characterize secondary metabolite biosynthetic genes and their role in pathogenicity and fungal development.

Another important area of investigation in the *Alternaria*–Brassicaceae pathosystem is the fungal signal transduction. For example, disruption of the *Fus3/Kss1 MAP* kinase homologue (*Amk1*) in *A. brassicicola* resulted in a complete loss of pathogenicity as observed in other fungi (Cho et al. 2006, 2007). Interestingly, in the latter study, it was shown that addition of long polypeptide nutrients partially restored pathogenicity to the mutants. In addition, two novel virulence factors by Cho et al. (2009) were predicted to encode a transcription factor (*AbPro1*) and a two-component histidine kinase gene (*AbNIK1*). Both of these kinases are pathogenicity factors in phytopathogenic fungi. *Slt2* was found to be associated with cell wall integrity and *HOG* with oxidative stress tolerance (Xu 2000). The identification of virulence factors is the disruption of *Aso-1*, a gene required for hyphal fusion (anastomosis) which is also found to be required for pathogenicity in *Alternaria* species (Craven et al. 2008). Eventually over a hundred genes have been functionally analysed through various techniques like gene knockout and overexpression experiments making *A. brassicicola* the species of choice for functional genomic research to define conserved virulence mechanisms for this important genus of fungi (Oide et al. 2006; Cho et al. 2006, 2007; Kim et al. 2007; Cho et al. 2009).

With the objective of identification of *A. brassicicola*, an attempt was made to examine the role of cutinase genes in *A. brassicicola* pathogenesis (Yao and Köller 1994, 1995). In these studies, biolistic transformation was used to disrupt the *CUTAB1* gene. It affected the saprophytic growth since cutin was no longer able to be utilized as a sole carbon source, but this disruption had no significant effect on *A. brassicicola* pathogenicity. An extracellular lipase was found to be produced by *A. brassicicola* in vitro (Berto et al. 1999). In this study, anti-lipase antibodies were found to significantly decrease the ability of *A. brassicicola* to cause disease on cauliflower leaves. However, disruption of four predicted *A. brassicicola* lipase genes expressed during plant infection did not result in reduced virulence on cabbage (Cho et al. 2006).

To study regulatory mechanisms of pathogenesis, Cho et al. (2012) mined 421 genes in silico encoding putative transcription factors in a machine annotated draft genome sequence of *A. brassicicola*. Targeted gene disruption mutants for 117 of the transcription factor genes were produced and screened. Three of these genes are associated with pathogenesis. Disruption mutants of one gene (*AbPacC*) are non-pathogenic, and another gene (*AbVf8*) causes lesions less than half the diameter of wild-type lesions. Unexpectedly, mutants of the third gene, *Amr1*, cause lesions with twofold larger diameter than the wild-type and complementation mutants. *Amr1* is a homologue of *Cmr1*, a transcription factor that regulates melanin biosynthesis in several fungi. Cho et al. (2012) created gene deletion mutants of Δ*amr1* and characterized their phenotypes. The Δ *amr1* mutants used pectin as carbon source more efficiently than the wild type, were melanin deficient and are more sensitive to UV light and glucanase digestion. The *AMR1* protein is localized in the nuclei of hyphae and in highly melanized conidia during the late stage of plant pathogenesis. RNA *seq* analysis reveals that three genes in the melanin biosynthesis pathway, along with the deleted *Amr1* gene, are expressed at low levels in the mutants. In contrast, many hydrolytic enzyme-coding genes are expressed at higher levels in the mutants than in the wild type during pathogenesis. A gene important for survival in nature negatively affects virulence, probably by less efficient use of plant cell wall materials. The functions of the *Amr1* gene are important to be the success of *A. brassicicola* as competitive saprophyte and plant parasite.

3.12 Growth and Sporulation

3.12.1 Culture Media

A number of culture media have been reported suitable for growth and sporulation of *Alternaria* species pathogenic on brassicaceous plants (Table 3.2). According to Neergaard (1945), *A. brassicae* sporulates profusely on malt extract and standard nutrient agar, but very poorly on

Table 3.2 Culture media for growth and sporulation of *Alternaria* species pathogenic on Brassicaceae (Verma and Saharan 1994)

Liquid/Solid media	Reference
Alfalfa decoction	McDonald (1959)
Asthana and Hawker's	Gupta et al. (1969)
Brown's starch	Ansari et al. (1988); Gupta et al. (1969); Prasada et al. (1970)
Coconut	Lapis and Ricaforte (1974)
Corn meal	Ansari et al. (1988); Lapis and Ricaforte (1974); Singh (1980)
Czapek-Dox	Gupta et al. (1969); Lapis and Ricaforte (1974); Prasada et al. (1970)
Eruca sativa decoction	Prasada et al. (1970)
Glucose asparagines	Prasada et al. (1970)
Host decoction	Prasada et al. (1970)
Houston's	Singh (1980)
Kirchoff's	Ansari et al. (1988)
Leonian's agar	Lapis and Ricaforte (1974)
Malt extract	Ansari et al. (1988)
Mustard leaf extract	Ansari et al. (1988)
Oatmeal	Prasada et al. (1970); Singh (1980)
Pechay decoction	Lapis and Ricaforte (1974)
Potato dextrose	Ansari et al. (1988); Prasada et al. (1970)
Potato dextrose asparagine	Ansari et al. (1988)
Potato sucrose	McDonald (1959)
Rice meal	Singh (1980)
Richard's	Ansari et al. (1988); Gupta et al. (1969); Prasada et al. (1970)
Sabouraud's	Ansari et al. (1988); Prasada et al. (1970)
V-8	McDonald (1959)
Wheat meal	Singh (1980)

PDA. Atkinson (1950), on the other hand, found good sporulation on both PDA and malt agar. According to Ansari et al. (1988a), although *A. brassicae* grows and sporulates well on a wide range of media, PDA was found to be the best. Changsri (1960, 1961) found slightly poorer growth of *A. brassicae* in comparison to *A. brassicicola* and *A. raphani*, which grew well on a wide range of agar media. On PDA, Mukadam and Deshpande (1977) found that *A. brassicae* not only grew and sporulated poorly, but it also lost its ability to grow and sporulate with successive sub-culturing. Cultural characters of *A. brassicae* on different culture media are shown in Table 3.3 (Prasada et al. 1970).

Alternaria brassicae from *Eruca sativa* grows best on glucose-asparagine agar medium and Richard's liquid medium. Mycelial growth is maximum in Richard's medium and least in Brown's medium. It is capable of utilizing nitrate, nitrite, ammoniacal and organic forms of nitrogen (Khandelwal et al. 1970; Prasada et al. 1970).

Single-spore transfers of *A. brassicae* on 10 % alfalfa decoction agar produce the largest amount of spores (McDonald 1959). It grows well on pechay decoction agar, corn meal agar, Czapek-Dox agar, coconut agar, Leonian agar, PDA, V-8 juice agar, oat meal agar, nutrient agar and prune agar, while sporulation is abundant on all but corn and oat meal agar (Lapis and Ricaforte 1974). Billotte (1963) induced abundant sporulation of *A. brassicae* by the slow desiccation of culture in open Petri dishes preceded by removal of the aerial mycelium and washing in running water. According to Degenhardt (1973), V-8 juice agar with rose bengal plus streptomycin stimulates sporulation of both *A. brassicae* and *A. raphani.*

According to Gupta et al. (1969), calcium nitrate and, to a lesser extent, potassium nitrate support maximum growth of *A. brassicae* isolated from *B. oleracea* var. *botrytis.* Increased growth is correlated with the carbohydrate content of the medium. Buchwaldt et al. (1984) found that on Czapek-Dox medium, singrin (allylglucosinolate) does not influence growth rates of *A. brassicae*, but at increasing singrin concentrations, colonies become darker; on PDA, colonies are dark at all singrin concentrations. On PDA, singrin has no effect on growth rates in darkness, but in UV light, the high singrin concentration increases the mycelial growth slightly.

Alternaria brassicicola grows and sporulates well on a wide range of agar media (Changsri 1960, 1961), but shows marked selectivity in utilizing different carbon sources. Lactose is the best source of carbon for maximum growth followed by glycerol. The fungus prefers ammoniacal nitrogen over nitrate and nitrite. Growth of *A. brassicicola* is stimulated by DL-phenylalanine, DL-valine, asparagine, glycine and DL-serine (Jain 1974). The biosynthesis of lipids and phospholipids is best on Czapek's medium (Aizina et al. 1976).

Alternaria raphani grows well in culture but sporulation is rather poor (Changsri 1961). Spores of *A. raphani* are produced plentiful in most brassicaceous leaf decoction agar media, in which chlamydospores are produced abundantly. The density of the thallus and sporulation decreases as the amount of leaf decoction per litre is decreased from 400 to 6.5 g. Atkinson (1950) also reports abundant sporulation by wounding plate cultures and removing the lids of the culture plates. *Alternaria raphani* does not require an exogenous vitamin source for growth, although early growth is stimulated by the addition of vitamins (Taber et al. 1968). *Alternaria raphani* grows faster when glutamic acid is supplied as a nitrogen source. Taber et al. (1968) in their extensive nutritional studies report that *A. brassicae*, *A. brassicicola* and *A. raphani* grow well on most carbon sources; starch supports the most rapid growth, whereas methyl cellulose, filter paper strips and mannitol are very poor carbon sources.

Although the three species differ strikingly in their rates of utilization of mannitol, both *A. raphani* and *A. brassicicola* synthesize this polyol. Phenylalanine supports faster growth of *A. brassicicola* than other nitrogen sources, whereas glutamic acid and ammonium succinate stimulate more rapid growth of *A. raphani* and *A. brassicae* (Taber 1964).

3.12.2 Temperature and Relative Humidity

The optimum temperature for maximum sporulation of *A. brassicae* and *A. raphani* is between 23 and 25 °C (Changsri and Weber 1963; Degenhardt 1973; McDonald 1959; Neergaard 1945; Singh 1980; Taber 1964). *Alternaria brassicae* has more demanding growth requirements, with a distinct optimal growth peak at 22.5 °C (Gupta et al. 1972; Singh 1980). The temperature growth optimum of *A. brassicicola* is 25–27 °C (Sarkar and Sen Gupta 1978), but growth continues to the extremes of 6 and 37 °C. According to Changsri (1960, 1961) and Changsri and Weber (1960, 1963), the optimum temperatures for growth in culture of *A. brassicicola*, *A. brassicae* and *A. raphani* are 24–28 °C, 20–24 °C and 24–28 °C, respectively. However, according to Taber et al. (1968), *A. raphani* and *A. brassicae* grow better between 20 and 25 °C on malt agar, whereas *A. brassicicola* grows well over a wider temperature ranges. Lapis and Ricaforte (1974) report profuse mycelial growth and sporulation of *A. brassicae* at 16–24 °C, but according to Ansari et al. (1989a), the optimum is 23 °C.

Alternaria brassicae isolated from *Crambe abyssinica* grows well on malt agar and nutrient agar at 0.5–33 °C with optimum growth at 23 °C; sporulation is maximum at 20–30 °C. *Alternaria brassicicola* grows well between 0.5 and 38 °C with an optimum growth at 25^{2}C; sporulation is maximum at 17–30 °C. *Alternaria alternata* grows well from 0.5 to 40 °C with optimum growth and sporulation at 25–27 °C and 20–30 °C, respectively (Czyzewska 1970). According to Maude (1986), at least 12 h of continuous high relative humidity and temperature of more than 14 °C are required for abundant sporulation of *A. brassicae* and *A. brassicicola*. Below 14 °C, the sporulation is delayed under more humid conditions; at 8 °C, at least 30 h of high relative humidity is required for sporulation. Both fungi produce viable spores for 12 weeks on leaf and 20 weeks on stubble under field conditions. Ansari et al. (1989a) observed 95–100 % relative humidity optimum for mycelial growth and sporulation of *A. brassicae*.

3.12.3 Hydrogen Ion Concentrations (pH)

The optimum pH requirement for growth and sporulation of *A. brassicicola*, *A. brassicae* and *A. raphani* are 6.0–8.0, 7.1–8.0 and 7.1–8.0, respectively (Changsri 1961; Changsri and Weber 1963). According to Gupta et al. (1969), *A. brassicae* isolated from *B. oleracea* var. *botrytis* tolerates a wide pH range from 3.0 to 9.0, the optimum being 5.5. No sporulation occurs at pH levels below 3.0 and above 9.0. Very good sporulation occurs at pH levels between 5.0 and 6.5. According to Taber et al. (1968), *A. raphani* grows well over a pH range of 4.8–7.2. Mycelial growth occurs in pH range of 2.9–8.2 with an optimum at 6.5 (Ellis 1968a, b; Ansari et al. 1989a).

Biomass production and sporulation of *A. brassicae* and *A. brassicicola* at different pH have been assessed. The maximum biomass (495.00 mg) of *A. brassicae* is observed at pH 5.0 followed by 480.65 mg biomass at pH 6.0 and minimum biomass of 160.5 mg at pH 2. Abundant sporulation of *A. brassicae* is at the pH 6.0, moderate sporulation is at the pH 5.0 and 7.0, and no sporulation at pH 2.0 and 3.0. The pH range of 5.0 to 7.0 favours abundant sporulation (++++) in the case of *A. brassicicola*, whereas moderate sporulation is at pH 8.0, and no sporulation at pH 2.0 and 3.0.

In the case of *A. brassicicola*, the maximum biomass (590.3 mg) is observed at pH 6.0 followed by 530.4 mg biomass at pH 7.0 and the lowest biomass, 193.5 mg, at pH 2.0. The pH of the medium is also one of the major factors responsible for the growth and sporulation of the pathogen. The best suited pH range for the growth of *A. brassicae* and *A. brassicicola* are 5.0–6.0 and 5.0–7.0, respectively (Table 3.4). The pH range from 3.0 to 8.5 is most suitable for both mycelial development and conidial formation in *Alternaria* spp. including *A. brassicae* with optimum pH range from 5.0 to 6.5 (Kumar and Choudhary 2006).

3.12.4 Light and Darkness

Maximum growth and sporulation by *A. brassicae* occur with alternating light and darkness. Continuous light completely inhibits sporulation. In light, a definite zonation of spore production occurs (Ansari et al. 1989a; Changsri and Weber 1963; Gupta et al. 1972; Mridha 1986; Mukadam and Deshpande 1979a; Taber 1964). According to Sasaki et al. (1985), sporulation of *A. brassicae* proceeds normally in darkness, but is inhibited by monochromatic radiation from 350 to 520 nm and reduced by continuous irradiation of UV radiation shorter than 350 nm. Light intensity is deleterious for *A. brassicae* spore germination. With the increase in light intensity from 0 to 5000 lx, there is reduction in per cent spore germination, number and size of lesions per leaf (Table 3.5). Maximum spore germination with number and size of lesions per leaf is observed under dark conditions (Kadian and Saharan 1984).

Zone formation in cultures of *A. brassicicola* is affected by light. Three minutes of light exposure or 6 h of darkness alternating from the original source of light is required to induce zonation. Spore development is plentiful under combinations of intermittent light and darkness (Changsri 1961; Changsri and Weber 1963). Taber (1964) found that the longer the wave length, the greater is the production of conidia in relation to chlamydospores by *A. raphani*.

3.13 Perpetuation

It is believed that *Alternaria* species infecting brassicaceous plants survive and perpetuate through infected seeds, diseased plant debris, pathogen propagules in the soil and cultivated/weed hosts in a particular agroecosystem (Chupp 1925; Chupp and Sherf 1960; Dixon 1981; Ellis 1968a, b, 1971; Humpherson-Jones 1989; Kolte 1985a, b; Neergaard 1945; Putnam et al. 1972; Saharan 1992a, b; Saharan et al. 2005; Sherf and Macnab 1986; Tsuneda and Skoropad 1977; Vaartnou and Tewari 1972a, b; Verma and Saharan 1994; Weber 1973). Seed-borne *Alternaria* are capable of surviving as viable conidia and/or as internal mycelium for periods long enough for the seed to be harvested, stored, transported and finally sown. The proportions of *Brassica* seed carrying *Alternaria* spp. are very high. Richardson (1970) found 40 and 10 %, respectively, of *Brassica* seed infected with *A. brassicicola* and *A. brassicae*. Infection levels up to 50 % on *B. oleracea* var. *capitata* cv. Houston

Table 3.3 Cultural characters of *Alternaria brassicae* on different culture media (Prasada et al. 1970)

Medium	Average radial growth mm[a]		Colony character	Sporulation[b]
	6 days after	12 days after		
Potato dextrose agar	31	55	Colony circular, growth profuse with concentric rings cottony, olivaceous black	++
Glucose asparagine medium	48	69	Colony irregular, mycelium scanty, cottony without concentric rings, brownish in colour	++
Oat meal agar	31	65	Colony circular, cottony growth with numerous concentric rings buff olive in colour	++
Sabouraud's agar	33	59	Colony circular smooth fluffy without concentric rings, brownish in colour	++
Brown's agar	23	48	Colony circular, cottony growth without concentric rings, olivaceous in colour	+
Czapek-Dox agar	26	56	Colony circular dense with definite zonation, olivaceous in colour	++
Richard's agar	43	76	Colony circular dense centre raised surrounded by velvet light brownish in colour	+++
Host decoction	40	68	Colony circular rough cottony yellowish in colour	+++
CD @5 %	±2.68	±2.06		

[a]Colony diameter recorded as average of four replications
[b]+ = poor sporulation, ++ = good sporulation, +++ = excellent sporulation
Treatments – highly significant treatment No. 5 6 1 3 4 8 7 2 (after 6 days)

Evergreen and up to 90 % on turnip rape (*B. rapa*) are reported from Canada (Petrie 1974). In the UK, infected seed is the primary source of the *Brassica* dark leaf spot fungi *A. brassicae* and *A. brassicicola* in autumn sown oilseed rape and vegetable brassicas (Humpherson-Jones 1985; Gladders 1984). In oilseed rape, *A. brassicae* is the dominant species, but in vegetable brassica seeds, *A. brassicicola* predominates. Although seed of vegetable brassicas including Brussels sprouts may be infected with *A. brassicicola*, *A. brassicae* has not been detected in the seed of this crop (Humpherson-Jones and Maude 1982a, b; Humpherson-Jones 1985; Humpherson-Jones and Phelps 1989). Nevertheless, *A. brassicae* has recently assumed increasing dominance as a pathogen of Brussels sprouts. However, in tropical countries like India, *A. brassicae* gets eliminated from the seeds of oilseed brassicas during storage from April to September at 25–35 °C

Table 3.4 Biomass and sporulation index of *A. brassicae* and *A. brassicicola* at different pH levels (Kumar and Choudhary 2006)

	A. brassicae			*A. brassicicola*		
Initial pH of filtrate	Dry mycelial weight (mg)	Sporulation	Final pH of the filtrate	Dry mycelial weight (mg)	Sporulation	Final pH of the filtrate
2.0	160.0	–	3.5	193.5	–	3.4
3.0	244.6	–	5.2	270.7	–	5.6
4.0	390.4	++	6.3	396.4	++	6.0
5.0	495.0	+++	7.2	500.0	++++	7.5
6.0	480.6	++++	7.5	590.3	++++	7.0
7.0	370.2	+++	7.5	530.4	++++	7.5
8.0	280.5	++	7.9	386.6	+++	7.4
Mean	346.0			409.7		

	SEm±	CD (0.05 %)
pH of media	6.41	18.57
Pathogen	2.87	08.31
Interaction	9.07	26.27

Table 3.5 Influence of light intensity on *A. brassicae* infection (Kadian and Saharan 1984)

	Disease intensity[a]		
Light intensity	Spore germination after 24 h (%)	No. of lesions/ leaf	Size of lesions/leaf (mm)
0 lx	78	7.0	19.40
1000 lx	63	6.2	15.50
2000 lx	60	5.9	13.80
5000 lx	49	3.8	08.20

[a]Average of 20 leaves

Table 3.6 Survival of *Alternaria brassicae* in rapeseed–mustard seed during storage at Ludhiana and Hisar in 1979 (Saharan 1984)

	Per cent seed infection		
Months	RLM-198	Prakash	Yellow Sarson
April	30.0	11.8	15.5
May	24.0	1.7	1.9
June	16.0	0.0	0.0
July	10.0	0.3	0.0
August	0.0	0.0	0.0
September	0.0	0.0	0.0

(Table 3.6) temperature (Chahal 1981; Kolte 1985a; Saharan 1992a; Mehta et al. 2002). *Alternaria raphani* is internally seed borne in all parts of dormant radish seed (Atkinson 1950). According to Dixon (1981), infected seed is the main avenue of transmission of three important *Alternaria* spp. of *Brassica*. It may be both externally as conidia adhering to the testa and internally as mycelia present within the seed tissues. The latter may lead to complete destruction of the embryo. Four phases of seed transmission can be distinguished: (1) transmission from seed to developing seedlings, (2) transmission from seed to adjacent seed, (3) transmission from adult plant to seed by fungal growth through the green siliquae coat into the moist atmosphere with the seed pod and (4) transmission from adult plant to adult plant. In the Philippines, *A. brassicae* remains viable for 12–14 months in infected leaves (Lapis and Ricaforte 1974). According to Tripathi and Kaushik (1984), *A. brassicae* overwinters with plant debris buried in the field below the depth of 7.5 cm. According to Mehta et al. (2002), incorporation of thoroughly mixed plant debris in soil gives 100 % disease incidence (Table 3.7). The pathogen may survive in temperate condition during winter season and result in infection to the summer crop. The pathogen may spread to subtropical regions from temperate regions with the withdrawal of monsoon as well as change of the wind currents from western to

Table 3.7 Alternaria blight incidence (%) in rapeseed–mustard following inoculation with diseased debris[a] at different depth (Mehta et al., 2002)

Inoculum depth	Disease incidence (%)[b]
1" below soil	77.77 (81.98)
2" below soil	92.85 (80.83)
3" below soil	45.23 (42.94)
Thoroughly mixed in soil	100.00 (90.00)
CD at 5 %	–(28.87)

[a]Stored at −10 °C

[b]Figures in parenthesis are angular transformed values

eastern areas. It is evident from the observations that the disease may appear during October–November on *Brassica campestris* var. *toria* sown in the month of August–September and may further spread to main crop from toria in the subtropical areas of India (Mehta et al. 2002).

Ansari et al. (1989b) observed that in temperate and subtropical countries, *A. brassicae* survives and perpetuates through diseased plant debris. In the UK, infected debris of *B. oleracea* seed crop is shown to be a major source of *A. brassicicola* spores (Humpherson-Jones and Maude 1982a). Infected debris of *Brassica* crops, with *A. brassicae* and *A. brassicicola* remaining on the ground after harvest, may provide a source of dark leaf spot infection, which may be implicated in the spread of the disease within and between crops (Humpherson-Jones 1989). On leaves of oilseed rape and cabbage, the fungi produce viable spores for as long as the leaf tissue survives which is up to 6 weeks in winter oilseed rape and up to 8 weeks in summer. On cabbage, spore viability lasts up to 8 weeks in winter and up to 12 weeks in summer (Humpherson-Jones and Hocart 1983). Infected Crambe seed is the main source of primary inoculum of *A. circinans* (Holcomb and Newman 1970). The virulence of *A. brassicicola*, also from Crambe, can be retained for one year by culturing on filter paper and storage in Petri dishes at 5ºC (Kilpatrick 1975, 1976).

Alternaria raphani is reported to survive for 5 years in soil culture (Atkinson 1950, 1953). The fungus shows no change in cultural characteristics after its 5-year period in dry soil, nor any loss of virulence when plants of radish, stock and wallflower are inoculated (Atkinson 1953). It is also possible that *A. brassicae* and *A. raphani* survive through chlamydospores (Atkinson 1953; Tsuneda and Skoropad 1977; Vaartnou and Tewari 1972a, b). According to Vaartnou and Tewari (1972a, b), hyphal chlamydospores of *A. raphani* are produced on the stem, siliquae and seed of infected plants. Such spores have been found to remain viable even after prolonged deep freezing of the infected material. Tsuneda and Skoropad (1977) observed that conidia of *A. brassicae* are transformed into microsclerotia (Fig. 4.2; Plate 3.1). Such structures are round and darkly pigmented, resist desiccation and function in a similar manner to those of the sclerotia produced by other fungi. More microsclerotia are formed only on the previously affected and partially decayed plant tissues suggesting the possibility of survival of the fungus through such structures. These structures on germination have been found to produce numerous new conidiophores and conidia. Numerous brassicaceous weeds, forage brassicas and other weed hosts also serve as sources of primary inoculum of *Alternaria* spp. infecting several economically important brassicas (Table 2.2).

3.14 Spore Germination

3.14.1 Effect of Culture Media

The germinability of spores of *A. brassicae* and *A. brassicicola* is higher when cultivated on rich media than the spores produced on complex or poorly nutritive media. The presence of metabolic inhibitors in the growth media reduces spore germinability even in the presence of nutrients (Czapek-Dox agar) and causes abnormalities in their morphology. Spores of both the species show maximum germinability irrespective of their age (up to 20 days). Increase in the age of the spores (30 days onwards) either increases

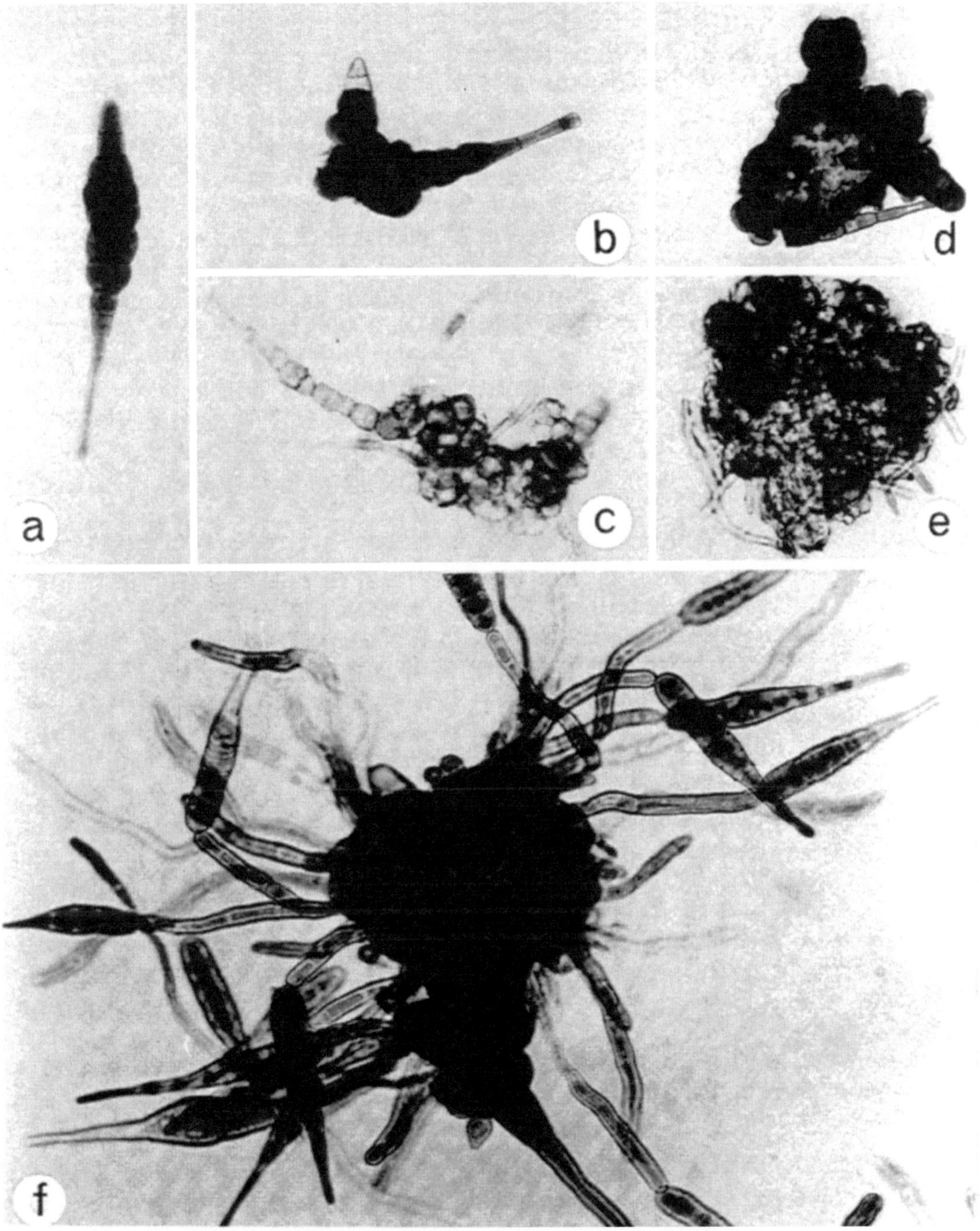

Plate 3.1 Light micrographs showing the development of microsclerotium from *Alternaria brassicae* conidium: (**a**) conidium, (**b**) initial stage in the formation of a microsclerotium, (**c**) half-developed microsclerotium with about 50 cells, (**d**) mature microsclerotium still showing head and beak of the original conidium ×400, (**e**) germination of a mature microsclerotium to form hyphae ×400 and (**f**) germination of a frozen-thawed microsclerotium showing many new conidia ×650 (Tsuneda and Skoropad 1977)

their latent period and/or reduces their germinability, even in the presence of nutrients (Gupta et al. 1969).

3.14.2 Effect of Temperature and Relative Humidity

For *Alternaria brassicae* causing brown rot of cauliflower, the optimum temperature for spore germination is 32–35 °C with the minimum of <1.5 °C and the maximum of 40–46 °C; as no temperature between 40.5 and 46 °C was tried, the exact maximum temperature was not determined. The optimum temperature for mycelial growth is 25–27 °C (Weimer 1925). Degenhardt et al. (1982) observed excellent germination of *A. brassicae* spores at 15 °C with 90 % relative humidity and 6–8 h of host leaf wetness period in contrast to very poor germination of *A. brassicicola* spores under similar conditions. Gupta et al. (1972) found 17–19 °C to be optimal for conidial germination of *A. brassicae*. According to Ansari et al. (1988b), *A. brassicae* spores germinate over a very wide range of 10–28 °C with an optimum of 23 °C. However, Kadian and Saharan (1984) recorded 25 °C as the optimum temperature for spore germination of *A. brassicae*. Spore germination commences after 4 h and reaches its peak after about 24 h; relative humidity of >90 % is essential for maximum spore germination (Fig. 3.9).

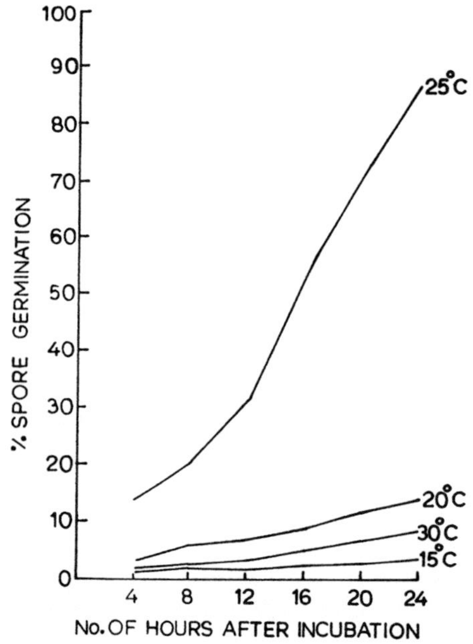

Fig. 3.9 Influence of temperature on spore germination of *A. brassicae* (Kadian and Saharan 1984)

Conidial germination of *A. brassicicola* is optimal at 30 °C (Gupta et al. 1972). Sarkar and Sen Gupta (1978) observed that conidia of *A. brassicicola* germinate best at 90–100 % relative humidity and a temperature of 22–32 °C. Self-inhibition of *A. brassicicola* occurs when the spore load in the germination medium is 0.2 optical density. The self-inhibiting compound is volatile at 35 °C, and denatures at 90 °C, and is thought to be a substance of low molecular weight (Mukadam 1982). Barton and Fine (1958) obtained 86–90 % spore germination of *A. brassicicola* in the presence of 200 ppm gibberellic acid. Conidia of *A. alternata* causing leaf and pod blight of radish germinate at 15–30 °C with an optimum at 20–25 °C (Singh and Suhag 1983).

3.14.3 Effect of Host Extract and Exudates

Leaf extracts of *Brassica* spp. are generally inhibitory to spore germination of *A. brassicae*. Leaf extracts and exudates of both susceptible and resistant cultivars are inhibitory, but inhibitory effects of resistant cultivars like Tower and

Table 3.8 Effect of leaf exudates and leaf extracts from different cultivars of rapeseed and mustard on spore germination of *A. brassicae* (Kadian and Saharan 1984)

	% Spore germination after 24 h of incubation[a]	
Cultivars	Leaf exudates	Leaf extracts
Tower	40	26
RC-781	49	29
YRT-3	66	47
CSR-448	61	44
CSR-741	58	45
CSR-142	62	46
RH-30 (Prakash)	71	49
Control (water)	88	91

[a]Average of 200 spores incubated at 25 °C

RC-781 are more pronounced; generally, leaf exudates are more inhibitory than leaf extracts (Table 3.8) (Kadian and Saharan 1984). The effect of leaf exudates of Yellow Sarson (*B. rapa*) and Taramira (*E. sativa*) is reported to vary with the host variety, age of host plant and maturity of the leaves (Sharma et al. 1985).

3.14.4 Effect of Light Intensity

Direct exposure to strong light intensity is deleterious to *A. brassicae* spores, inhibiting germination and reducing host infection (Kadian and Saharan 1984). Maximum sporulation of *A. brassicae* occurs with alternating light and darkness. Continuous light completely inhibits sporulation (Ansari et al. 1989a; Changsri and Weber 1963; Gupta et al. 1972; Mridha 1986; Mukadam and Deshpande 1979; Sasaki et al. 1985; Taber 1964).

3.15 Seed Infection

Alternaria blight has been reported to be a serious seed-borne disease of several seed-producing brassicaceous crops in many countries of the world (Crosier and Patrick 1940; Ellis 1968a, b; Green 1947; Groves and Skolko 1944; Holcomb and Newman 1970; Holtzhausen 1978; Holtzhausen and Knox-Davies 1974; Jouan et al. 1972; Kilpatrick 1975, 1976; Kolte 1985a, b; Kothanur et al. 1982; Nipoti 1978; Noble and Richardson 1968; Petrie 1974; Pound et al. 1951; Richardson 1970, 1979; Schimmer 1953; Van Schreven 1953; Sivapalan and Browning 1992; Verma and Saharan 1994; Wiltshire 1947). High recovery of *A. raphani* from radish seeds in Michigan (McLean 1947), and *A. brassicicola* from cabbage seeds in Washington (Pound et al. 1951), has been reported. In South Africa, all three species of *Alternaria*, i.e. *A. brassicae*, *A. brassicicola* and *A. raphani*, have been isolated from seeds of cabbage, cauliflower, broccoli and Brussels sprouts (Holtzhausen and Knox-Davies 1974; Knox-Davies 1980). In Egypt, testing for seed health in cabbage, cauliflower, turnip and radish showed the presence of *A. brassicicola* and *A. raphani* (Michail et al. 1979). In Finland, 91 % of white cabbage and red cabbage seed lots were found to be infected with *A. brassicicola. Alternaria brassicae* was found in 4 % of the cabbage and in 31 % of the rape seed lots. *Alternaria raphani* was found in 30 % of the radish and black radish seed lots (Tahvonen 1979).

In the UK, between 1981 and 1984, up to 25 % of *B. napus* seeds and 8.5 % of *B. rapa* (turnip) seeds yielded *A. brassicae*, and 55 % and 13 % seeds of *B. oleracea*, respectively, yielded *A. brassicicola* and *A. brassicae* (Humpherson-Jones 1985). *Alternaria brassicae* was detected at varying levels on seed of rape harvested from different climatic districts of New South Wales during 1983. The highest levels were observed on seed samples from the South West slopes (North) and Riverina (av. 22.8 and 14.2 %, respectively) with much lower levels of infection (av, <8.0 %) on samples from central and northern districts (Stovold et al. 1987). In Scotland, up to 89 % of seeds of *B. napus* were recorded as being infected with *A. brassicae* (Kothanur et al. 1982).

In Australia, samples of *Brassica oleracea* seed showed *A. brassicicola* infection between 24 and 37 % of seeds, with 4–8 % infection in the embryo tissues. Inoculation of seed with *A. brassicicola* results in loss of vigour in germinated seedlings, followed by death. The fungus retains its viability and pathogenicity on seed stored for up to 12 months. This investigation indicates that a high proportion of commercially available *Brassica* seed is contaminated with *A. brassicicola* and may therefore be a primary source of infection in *Brassica* crops in Australia (Sivapalan and Browning 1992).

In Lithuania, *Alternaria* seed infection was 10–100 % in winter rape and 16.0–93.6 % in spring rape seed samples (Brazauskiene and Petraitiene 2006). In Japan, *A. brassicicola* accounted for 57–95 % in cabbage seeds (Kubota et al. 2006). Some other fungi, viz. *A. alternata*, *Rhizopus oryzae*, *Fusarium avenaceum*, *Aspergillus niger*, *A. flavus*, *A. ostianus* and *Penicillium* spp., have also been recorded in cabbage seed from Japan.

In Canada, Petrie (1974) reported *A. brassicae* and *A. raphani* as important seed-borne pathogens of oleiferous brassicas. *Brassica rapa* (turnip rape) had higher levels of *A. brassicae* than *A. raphani*; *B. napus* (rape) seed contained considerably lower levels of both species. In heavily infested seed of *B. rapa*, 73 % of the *A. brassicae* and 90 % of the *A. raphani* occurred on the seed surface. Storage of infected seed for 6–8 months at 25 °C reduced the levels of infestation by more than 50 %. In rapeseed–mustard, the frequency of infection of seed by *A. brassicae* varies with the number of lesions on the siliquae. Seed infection diminished with the length of storage period at higher temperature in summer months (Chahal 1981). Mustard seeds showing 42 % infection of *A. brassicae* in April became completely free when stored at 45 and 35 °C for 2 and 3 months, respectively (Shivpuri and Siradhana 1989). According to Vishnuavat et al. (1985), *A. brassicae* seed infection in mustard was eliminated during storage from March to July at 11.4–39.2 °C, but survived at 5 °C in cold storage.

Alternaria species comprised 86 % of the fungal flora isolated from samples of Crambe seed from Illinois, Indiana, Maryland and Ohio; *A. tenuis* was isolated from 47 % and *A. brassicicola* from 39 % of the samples (Kilpatrick 1976). The high percentage of *A. brassicicola*-infected seed indicates that infection is internal and that the fungus is capable of systemic infection and direct penetration of pods. The recovery of *Alternaria* spp. is better through the blotter method than the agar plate method. Ashraf and Chaudhary (1989) recovered 20.8 % of *A. alternata* from rapeseed seed on blotter and 10.7 % on Agar plate; the recovery of *A. brassicae* was 10.7 % and 10.5 %, respectively. Jain et al. (1982) recovered 31.5–10.5 % *A. alternata* and 6.0–1.5 % *A. brassicae* from rapeseed seed through the blotter method but none through the agar plate method. In Bangladesh, Latif et al. (2006) isolated *Alternaria, Aspergillus, Chaetomium, Curvularia, Fusarium, Penicillium* and *Rhizopus* from mustard seeds. Domsch (1957) described the infection of rape and cabbage seeds by *A. brassicae* and *A. brassicicola.* The relationship between *A. brassicicola* and *Brassica* seeds was determined by Knox-Davies (1980). The fungus first colonizes the testas of inoculated seeds as they rupture and then attacks embryos after a further 24–30 h. Testas of naturally infected cabbage seeds are colonized prior to rupture and largely in the hilum area. Spores lodge in the cavities and folds of tissue depressions at the hilum and micropyle. Spore germination and hyphal development are frequently poor at the hilum, but in some cases mycelial development is extensive over the hilum. Development on the remainder of the testa is generally poor, the only extensive growth being present on damaged seeds (Plate 3.2). According to Vannacci and Pecchia (1988), *A. raphani* occurs in all seed parts of the radish, and conidia are present on the seed coat. Similar observations were made by Atkinson (1950).

3.15.1 Location of Seed-Borne Infection

Component plating reveals the infection of *A. brassicae* in the seeds of rapeseed and mustard mainly internally in the seed coat and rarely in the embryo (Table 3.9). The range of internal seed coat infection varies between 3 and 16 %, but embryo infection is only 1 %. In infected seeds, sporulation of *A. brassicae* is first observed in the hilum region of seeds after 48 h of incubation. In infected seeds, hilum infection is recorded after 96 h, and by this time, sporulation is also recorded on other sites of the seed coat. The fungus continues to develop on infected seeds up to 240 h (Table 3.9). The sporulation of the fungus around the hilum area is significantly higher than on the remaining parts of the seed coat (Shrestha et al. 2000).

3.15.2 Disease Transmission in the Field

When infected seeds are sown, symptoms on the cotyledonary leaves become visible in 2–3 weeks as small round or irregular brownish to black necrotic spots, particularly on the margins. Under a stereomicroscope, the spots are dark brown in the centre surrounded by concentric rings of light brown tissues followed by light-yellow haloes.

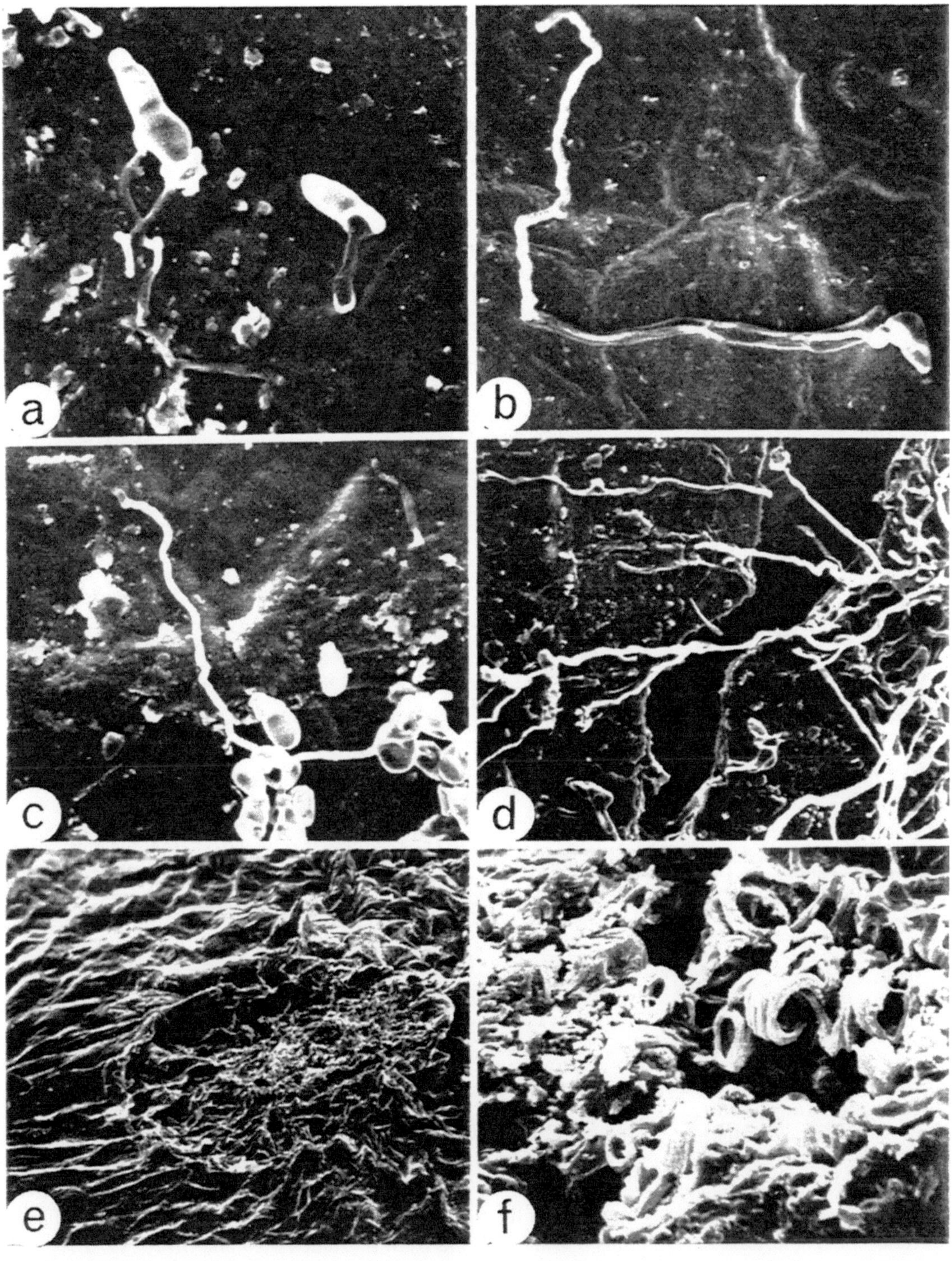

Plate 3.2 Scanning electron micrographs of the surfaces of cabbage seeds inoculated with *Alternaria brassicicola* and germinated on water agar for 24 h at 21 °C: (**a**) germinated spores with collapsed germ tube and initials of turgid aerial hyphae, (**b**) germinated spore with collapsed germ tube and turgid aerial hyphae, (**c**) turgid aerial germ tube with collapsed tip, (**d**) profuse development of hyphae on damaged seed, (**e**) roughened hilum area with broken vascular elements of the funiculus and (**f**) spiral thickening of xylem vessels of the broken vascular elements of the hilum shown in figure (Courtesy: Knox-Davis 1979)

Table 3.9 Location and sporulation of *Alternaria brassicae* in different parts of *Brassica* seeds examined after 2, 4, 6, 8 and 10 days of infection (Shrestha et al. 2000)

Sample	*Brassica* spp.	Cultivar	Per cent infection[a]		Per cent sporulation[b]		T test between location of sporulation
			Seed coat	Embryo	Hilum	Remainder of seed coat	
1	*B. campestris* var. *toria*	Local	3.0	0	3.7	1.5	3.8[a]
2	*B. campestris* var. *toria*	Local	5.0	0	4.5	2.7	3.6
3	*B. campestris* var. *toria*	Local	16.0	1	15.5	2.2	5.3
4	*B. juncea*	TL-15	8.0	0	6.0	4.0	5.7
5	*B. juncea*	Vaibhav	11.0	1	7.2	5.2	4.9
6	*B. juncea*	Kranti	12.0	0	8.2	3.5	4.6

[a]Mean of 100 seeds
[b]Mean of 400 seeds
*Significant $t(0.05) = 3.2$

Table 3.10 Transmission of *Alternaria brassicae* in the field by infected *Brassica* seeds in 1991 (Shrestha et al. 2000)

Sample	*Brassica* spp.	Cultivar	Per cent seed infection[a]	Mean number of emerged seedlings	Mean % of emerged seedlings having symptoms on cotyledonary leaves
1	*B. campestris* var. *toria*	Local	10.5	260.0	3.8
2	*B. campestris* var. *toria*	Local	17.2	262.2	7.2
3	*B. campestris* var. *toria*	Local	27.7	248.5	8.7
4	*B. juncea*	TL-15	20.7	221.5	8.2
5	*B. juncea*	Vaibhav	26.5	191.0	9.7
6	*B. juncea*	Kranti	39.0	199.0	11.1

[a]Mean of four replications; 400 seeds sown in each replicate

Table 3.11 Transmission of *Alternaria brassicae* in the field by infected *Brassica* seeds in 1992 (Shrestha et al. 2000)

Sample	*Brassica* spp.	Cultivar	Per cent seed infection	Mean number of emerged seedlings[a]	Mean % of emerged	
					Seedlings having symptoms on cotyledonary leaves	Seedlings having symptoms on cotyledonary leaves
1	*B. campestris* var. *toria*	Vikas	12.0	279.5	6.2	13.9
2	*B. juncea*[b]	Krishna	2.0	314.7	08	1.4
3	*B. juncea*[c]	Krishna	28.5	260.5	8.2	20.1
4	*B. juncea*	Pusabold	31.5	211.2	14.2	18.3
5	*B. juncea*	Varuna	54.0	189.7	16.2	28.2

[a]Mean of four replications; 400 seeds sown in each replicate
[b]Seeds collected from mid-hill
[c]seeds collected from terai (plain)

On the underside of cotyledons, opposite to the necrotic spots, the areas are slightly shrunken forming shallow cavities. Occasionally, olive brown conidia of *A. brassicae* are seen on sides of lesions, but mainly they are on the underside. The infected cotyledons soon start withering, become detached from the stem or remain hanging from it (Tables 3.10 and 3.11; Fig. 3.10a, b). Symptoms on the first true leaves are small brownish necrotic spots to distinct concentric lesions (1–5 mm diameter) surrounded by faint-yellow haloes (Shrestha et al. 2000).

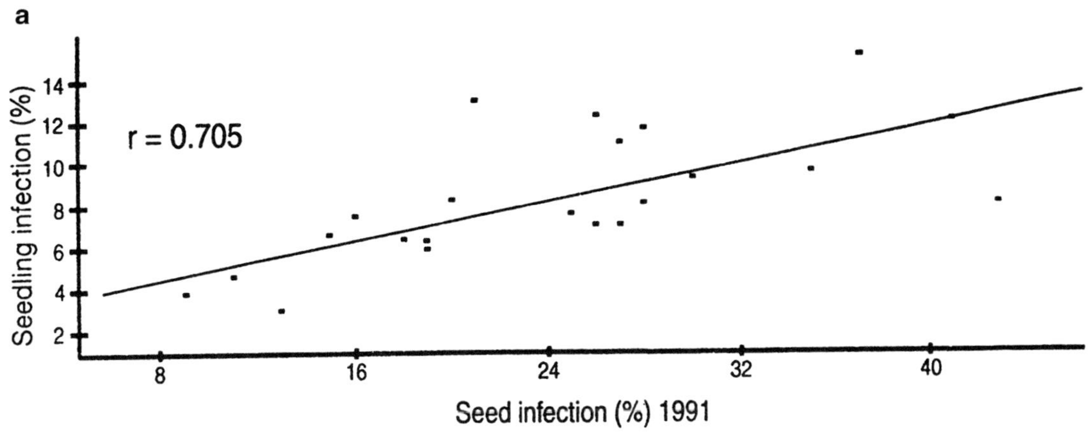

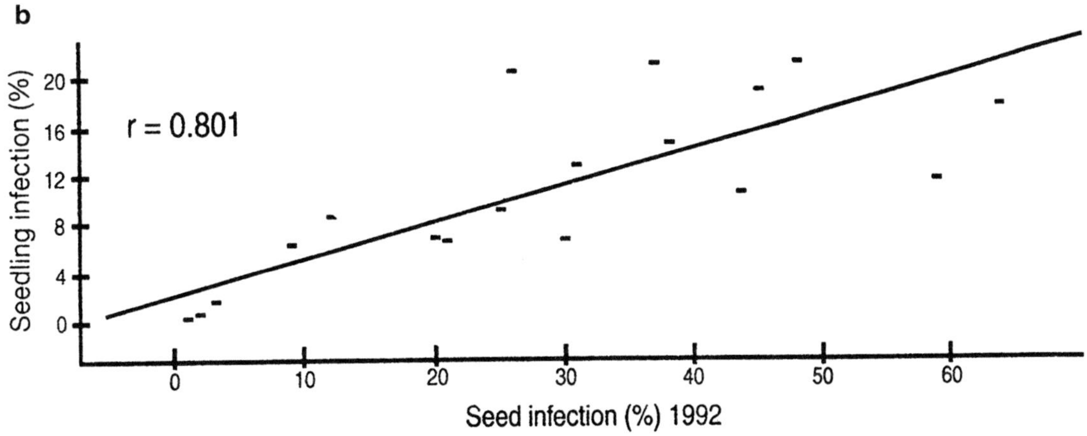

Fig. 3.10 (**a**) Relationship between seed infection and seedling infection due to *A. brassicae* in rapeseed and mustard in field. Correlation coefficient for both seasons was highly significant ($r = 0.0705$; $p<0.001$) in 1991. (**b**) Figures showing the relationship for the year 1992 ($r=0.801$; $P<0.001$). Similarly, the correlation between seedling infection and first true leaf infection was also significant ($r=0.562$; $P<0.01$) (Shrestha et al. 2000)

Table 3.12 Effect of seed treatment fungicides on seed-borne infection of *Alternaria brassicae* and on subsequent seedling infection in mustard (Shrestha et al. 2000)

Fungicides	Dose	Per cent seed infection[a]	Average count of normal seedlings[b]	Per cent infected seedlings[b]
Iprodione (Rovral)	2 g.a.i. /kg seed	0.5 (0.7)d	24.7 (5.0) a	0.5 (0.7) d
Thiram (Thiride)	2 g.a.i. /kg seed	2.5 (1.6)cd	22.7 (4.8) b	1.1 (1.0) c d
Mancozeb (Dithane M45)	2 g.a.i. /kg seed	3.0 (1.7)c	22.7 (4.8) b	1.5 (1.2) c
Captan (captan)	2 g.a.i. /kg seed	9.0 (2.9)b	19.7 (4.4) c	2.7 (1.6) b
Control (untreated)	–	20.2 (4.5)a	16.7 (4.0) d	7.5 (2.7) a
LSD (0.05)	–	(0.75)*	(0.18)*	(0.18)*
CV (%)	–	20.5	2.5	18.2

*Significant at ($P<0.001$)
Mean in a column with different letters are significantly different at ($P=0.05$)
[a]Mean of four replications, 100 seeds per replication
[b]Mean of four replicates, 25 seeds per replicate; value in parenthesis is square root transformation

3.15.3 Effect of Seed Treatment

Seed treatment with different fungicides eliminates or significantly reduces the inoculum of *A. brassicae* in infected seeds (Table 3.12). There are significant differences in efficacy of chemicals at 95 % confidence limit. Among the tested chemicals, iprodione is most effective followed by thiram, mancozeb and captan. All treated seeds have higher emergence and low per cent of seedling infection than untreated seeds. Seedling emergence from iprodione-treated seeds on wet blotters is 90 % normal followed by 91 % in thiram- and mancozeb-treated seeds (Table 3.12). The percentage of diseased seedlings in the seeds treated with iprodione, thiram and mancozeb is significantly lower ($P=0.05$) than in captan treatment and control (Shrestha et al. 2000).

References

Agarwal A, Garg GK, Devi S, Mishra DP, Singh US (1997) Ultrastructural changes in *Brassica* leaves caused by *Alternaria brassicae* and destruxin B. J Plant Biochem Biotechnol 6:25–28

Aizina AF, Zlatoust MA, Razumovskii PN, Yadovina VN (1976) Study on phospholipids of *Alternaria brassicicola* strain 13. Izv AN Mold SSR Biol ihim. N. 1:57–59

Ansari NA, Khan MW, Muheet A (1988) Effect of temperature and relative humidity on spore germination of *Alternaria brassicae* and development of Alternaria blight on oilseed crucifers. Rev Trop Plant Pathol 5:79–84

Ansari NA, Khan MW, Muheet A (1989a) Effect of some factors on growth and sporulation of *Alternaria brassicae* causing Alternaria blight of rapeseed and mustard. Acta Bot Ind 17:49–53

Ansari NA, Khan MW, Muheet A (1989b) Survival and perpetuation of *Alternaria brassicae* causing Alternaria blight of oilseed crucifers. Mycopathologia 105:67–70

Ashraf SS, Chaudhary KCB (1989) Mycoflora associated with rapeseed and their significance. Int J Trop Plant Dis 7:93–94

Atkinson RG (1950) Studies on the parasitism and variation of *Alternaria raphani*. Can J Res Sect C 28:288–317

Atkinson RG (1953) Survival and pathogenicity of *Alternaria raphani* after five years in dried soil cultures. Can J Bot 31:542–547

Barr ME (1987) Prodomus to class Loculoascomycetes. Hamilton I. Newell, Inc., Amherst

Barr ME, Simmons EG (1986) In Simmons. Mycotaxon 25:289

Barton LV, Fine JM (1958) Noninterference between effects of giberellic acid and fungicides. Contr Boyce Thompson Inst 19:291–294

Berkeley MJ (1836) Fungi. In: Smith JE (ed) The English flora (vol. V., II, 339)

Berto P, Belingheri L, Dehorter B (1997) Production and purification of a novel extracellular lipase from *Alternaria brassicicola*. Biotechnol Lett 19:533–536

Berto P, Comménil P, Belingheri L, Dehorter B (1999) Occurrence of a lipase in spores of *Alternaria brassicicola* with a crucial role in the infection of cauliflower leaves. FEMS Microbiol Lett 180:183–189

Billotte JM (1963) Une methode d'induction de la sporulation de *Alternaria brassicae* (Berk) Sacc. de colza en culture pure. Comptes Rendus Academie Agriculture de France 4:1056–1061

Bolle PC (1924) Die Durch Schwärzepilze (Phaeodictyae) Erzeugten Pflanzenkrankheiten. Meded. Phytopath. Lab. "*Willie Commelin Schölten*" Baarn (Holland) 7:77

Brazauskiene I, Petraitiene E (2006) The occurrence of *Alternaria* blight (*Alternaria* spp.) and Phoma stem canker (*Phoma lingam*) on oilseed rape in central Lithuania and pathogenic fungi on harvested seed. J Plant Protect Res 4:295–311

Buchwaldt L, Nielsen JK, Sorensen H (1984) Preliminary investigations of the effect of sinigrin on *in vitro* growth of three fungal pathogens of oilseed rape. Advances in the production and utilization of cruciferous crops. Proceedings of a Seminar CEC Programme Research Plant Protein Improvment, Copenhagen, pp 260–267

Campbell R (1969) An electron microscope study of spore structure and development in *Alternaria brassicicola*. J Microbiol Genet 54:381–392

Carzaniga R, Fiocco D, Bowye P, O'Connell RJ (2002) Localization of melanin in conidia of *Alternaria alternata* using phage display antibodies. Mol Plant-Microbe Interact 15:216–224

Chahal AS (1981) Seed-borne infection of *Alternaria brassicae* in Indian mustard and its elimination during storage. Curr Sci 50:621–623

Changsri W (1960) Studies of *Alternaria* spp. pathogenic on crucifereae. PhD thesis, University of Florida

Changsri W (1961) Studies of *Alternaria* spp. pathogenic on crucifereae. Disease Abstracts, 21, 1698

Changsri W, Weber GF (1960) Studies of *Alternaria* spp. pathogenic on Cruciferae. Phytopathology 50:631. (Abstr.)

Changsri W, Weber GF (1963) Three *Alternaria* species pathogenic on certain cultivated crucifers. Phytopathology 53:643–648

Cho Y, Davis JW, Kim KH, Wang J, Sun QH, Cramer RA Jr, Lawrence CB (2006) A high throughput targeted gene disruption method for *Alternaria brassicicola* functional genomics using linear minimal element (LME) constructs. Mol Plant Microbe Interact 19:7–15

Cho Y, Cramer R, Kwang-Hyung K, Davis J, Mitchell T, Figuli P, Pryor B, Lemasters E, Lawrence CB (2007) The *Fus3/Kss1* MAP kinase homolog *Amk1* regulates the expression of hydrolytic enzyme genes in the fungus *Alternaria brassicicola*. Fungal Genet Biol 44:543–553

Cho Y, Kim K-H, La Rota M, Scott D, Santopietro G, Callihan M, Mitchell TK (2009) Identification of virulence factors by high throughput targeted gene deletion of regulatory genes in Alternaria brassicicola. Mol Microbiol 72:1316–1333

Cho Y, Srivastava A, Ohm RA, Lawrence CB, Wang KH, Grigoriev IV, Marahatta SP (2012) Transcription Factor Amr1 induces melanin biosynthesis and suppresses virulence in *Alternaria brassicicola*. PLoS Pathog 8, e1002974. doi:10.1371/journal.ppat.1002974

Chupp C (1925) Alternaria leaf spot and blight of crucifers. In: Manual of vegetable garden diseases. The MacMillan Company, New York, pp 146–150

Chupp C, Sherf AF (1960) Crucifer diseases. In: Vegetable diseases and their control. Ronald Press Company, New York, Chapter 8, pp 237–288

Cramer RA, Rota CML, Cho Y, Thon M, Craven KD, Knudson DL, Mitchell TK, Lawrence CB (2006) Bioinformatic analysis of expressed sequence tags derived from a compatible *Alternaria brassicicola–Brassica oleracea* interaction. Mol Plant Pathol 7:113–124

Crane JL, Shearer CA (1991) A nomenclator of *Leptosphaeria,* V. Cesati, V., & Notaris, G. de. Illinois Nat Hist Survey Bull 34:1–355

Craven KD, Velez H, Cho Y, Lawrence CB, Mitchell TK (2008) Anastomosis is required for virulence of the fungal necrotroph *Alternaria brassicicola*. Eukaryote Cell 7:675–683

Crosier W, Patrick S (1940) Influence of chemical and thermal treatments on infection of cruciferous seedlings by *Alternaria* spp. and *Rhizopus nigricans*. Proc Assoc Seed Ann North Am 1939:116–120

Czyzewska S (1970) Effect of temperature on the growth and sporulation of *Alternaria* species isolated from *Crambe abyssinica*. Acta Mycol 6:261–276

Degenhardt KJ (1973). *Alternaria brassicae* and *A. raphani*: Sporulation in culture, and their effects on yield and quality of rapeseed. M. Sc. thesis, University of Alberta, Canada

Degenhardt KJ, Petrie GA, Morrall RAA (1982) Effects of temperature on spore germination and infection of rapeseed by *Alternaria brassicae*, *A. brassicicola* and *A. raphani*. Can J Plant Pathol 4:115–118

Cruz TEE dela (2006) Marine *Dendryphiella* species from different geographical locations: an integrated, polyphasic approach to its taxonomy and physioecology. Ph. D. dissertation. Fakültat für Lebenswissenschaften der Technischen Universität Carolo-Wilhelmina, Braunschweig, Germany

Dixon GR (1981) Pathogens of crucifer crops. In: Vegetable crop diseases. AVI Publishing Company, Inc., Westport, pp 119–123

Domsch KH (1957) The blackening of rape and cabbage siliquae. Z Pflanzenkr Pflanzen Schutz 64:65–79

Elliott JA (1917) Taxonomic characters of the genera *Alternaría* and *Macrosporium*. Am J Bot 4:439–476

Ellis MB (1968a) *Alternaria brassicae*. Commonwealth Mycological Institute. Descriptions of Pathogenic Fungi and Bacteria. No. 162. CMI, Kew

Ellis MB (1968b) *Alternaria brassicicola*. Commonwealth Mycological Institute. Descriptions of Pathogenic Fungi and Bacteria. No. 163. CMI, Kew

Ellis MB (1971) Dematiaceous hyphomycetes, vol 608. Commonwealth Mycological Institute, Kew, pp 464–497

Eriksson OE, Hawksworth DL (1991) Outline of the Ascomycetes. Syst Ascomycet 9:39–271

Eshel D, Ben-Arie R, Dinoor A, Prusky D (2000) Resistance of gibberellin-treated persimmon fruit to *Alternaria alternata* arises from the reduced ability of the fungus to produce endo-1,4-β-glucanase. Phytopathology 90:1256–1262

Eshel D, Lichter A, Dinoor A, Prusky D (2002a) Characterization of *Alternaria alternata* glucanase genes expressed during infection of resistant and susceptible persimmon fruits. Mol Plant Pathol 3:347–358

Eshel D, Miyara I, Ailing T, Dinoor A, Prusky D (2002b) pH regulates endoglucanase expression and virulence of *Alternaria alternata* in persimmon fruit. Mol Plant-Microbe Interact 15:774–779

Fajardo TG, Palo MA (1934) A serious leaf spot of Chinese celery cabbage, wongbok and other cruciferous plants in Trinidad Valley, Mountain Province Luzon. Phillip J Agric 5:143–153

Fan CY, Köller W (1998) Diversity of cutinases from plant pathogenic fungi, differential and sequential expression of cutinolytic esterases by *Alternaria brassicicola*. FEMS Microbiol Lett 158:33–38

Ferraris. (1912). Fl. Ital. Crypt., Fungi, 1(8):521

Gladders P (1984) Present and potential disease interaction between oilseed rape and vegetable Brassicas. In Proceedings of the 1984 British Crop Protection Conference, Pests and Diseases, 2, pp 791–798

Green DE (1947) Black spot disease of Brassica seed crops. Gardners' Chron Ser 3 122(3149):178–180

Groves JW, Skolko AJ (1944) Notes on seed-borne fungi II. Alternaria. Can J Res Sect C 12:217–234

Gupta RBL, Desai BG, Pathak VN (1969) Nutritional studies on *Alternaria brassicae* (Berk.) Sacc., influence of different nutrient media, pH, nitrogen and carbon/nitrogen ratio. Phyton 26:201–205

Gupta RBL, Desai BG, Pathak VN (1972) Effect of light on growth and sporulation of *Alternaria brassicae* (Berk.) Sacc. Phytopathol Mediterrania 11:61–62

Holcomb GE, Newman BE (1970) *Alternaria circinans* and other fungal pathogens on *Crambe abyssinica* in Louisiana. Plant Dis Rep 54:28

Holliday P (1980) Fungus diseases of tropical crops. Cambridge University Press, Cambridge, UK, pp 6–9, 19

Holm L (1957) Etudes taxonomiques sur les pleosporacees. Symbolae Botanicne Upsaliensis 14:1–188

Holtzhausen MA (1978) Seed-borne fungal pathogens and diseases of Japanese radish and their control in South Africa. Phytophylactica 10:107–113

Holtzhausen MA, Knox-Davies PS (1974) Pathogens of cruciferous vegetable crops in commercial seed samples in South Africa. Phytophylactica 6:289–294

Honda Y, Sota K, Nozu M (1987) An ultrastructural study of photo-induced conidiogenesis and dedifferentiation in *Alternaria solani*. Bull Fac Agric 21:141–154

Honda Y, Nanba N, Nozu M (1990) A further study on ultrastructural aspects of photo-induced conidiogenesis and dedifferentiation in *Alternaria solani*. Bull Fac Agric 24:39–46

Humpherson-Jones FM (1985) The incidence of *Alternaria* spp. and *Leptosphaeria maculans* in commercial *Brassica* seed in the United Kingdom. Plant Pathol 34:385–390

Humpherson-Jones FM (1989) Survival of *Alternaria brassicae* and *Alternaria brassicicola* on crop debris of oilseed rape and cabbage. Ann Appl Biol 115:45–50

Humpherson-Jones FM, Hocart MJ (1983) *Alternaria* diseases of *Brassica* seed crops. In 33rd annual report for 1982, National Vegetable Research Station, Wellesbourne, Warwick, UK, pp 63–64

Humpherson-Jones FM, Maude RB (1982a) Studies on the epidemiology of *Alternaria brassicicola* in *Brassica oleracea* seed production crops. Ann Appl Biol 100:61–71

Humpherson-Jones FM, Maude RB (1982b) Control of dark leaf spot (*Alternaria brassicicola*) of *Brassica oleracea* seed production crops with foliar sprays of iprodione. Ann Appl Biol 100:99–104

Humpherson-Jones FM, Phelps K (1989) Climatic factors influencing spore production in *Alternaria brassicae* and *Alternaria brassicicola*. Ann Appl Biol 114:449–459

Isshiki A, Akimitsu K, Yamamoto M, Yamamoto H (2001) Endopolygalacturonase is essential for citrus black rot caused by *Alternaria citri* but not brown spot caused by *Alternaria alternata*. Mol Plant Microbe Interact 14:749–757

Jain JP (1974) Carbon and nitrogen requirement of *Alternaria brassicicola*, the incitant of leaf blight disease of cauliflower (*Brassica oleracea* var. *Botrytis* L.). Beihefte Zur Nova Hedwigia 47:211–217

Jain SK, Singh PN, Singhal G (1982) Fungi associated with the seeds of *Brassica campestris* var. Sarson. Seed Res 10:87–90

Jasalavich CA, Morales VM, Pelcher LE, Seguin-Swartz G (1995) Comparison of nuclear ribosomal DNA sequences from *Alternaria* species pathogenic to crucifers. Mycol Res 99:604–614

Joly P (1964) Le Genre *Alternaria. Encyclopedic Mycologique*, vol 33. P. Lechevalier, Paris

Jones EBG, Klaysuban A, Pang KL (2008) Ribosomal DNA phylogeny of marine anamorphic fungi: *Cumulospora varia, Dendryphiella* species and *Orbimyces spectabilis*. Raffles Bull Zool 19(Suppl):11–18

Jouan B, Lemaire JM, Herve Y (1972) Study of cauliflower diseases (*Brassica oleracea* L. var. *Botrytis* D. C.) part 1. The parasites of winter cauliflower grown for seeds. Ann Phytopathol 4:133–155

Kadian AK, Saharan GS (1984) Studies on spore germination and infection of *Alternaria brassicae* of rapeseed and mustard. J Oilseeds Res 1:183–188

Kawamura C, Moriwaki J, Kimura N, Fujita Y, Fuji S, Hirano T, Koizumi S, Tsuge T (1997) The melanin biosynthesis genes of *Alternaria alternata* can restore pathogenicity of the melanin-deficient mutants of *Magnaporthe grisea*. Mol Plant-Microbe Interact 10:446–453

Kawamura C, Tsujimoto T, Tsuge T (1999) Targeted disruption of a melanin biosynthesis gene affects conidial development and UV tolerance in the Japanese pear pathotype of *Alternaria alternata*. Mol Plant-Microbe Interact 12:59–63

Khandelwal GL, Prasada R, Jain JP (1970) Studies on growth response of *Alternaria brassicae* isolated from *Eruca sativa* Lam. in different nitrogen media. Labdev J Sci Technol 8B:140–141

Kilpatrick RA (1975) Seed-borne fungi of Crambe and virulence of selected fungi. Proc Am Phytopathol Soc 2:62 (Abstr.)

Kilpatrick RA (1976) Fungal flora of Crambe seeds and virulence of *Alternaria brassicicola*. Phytopathology 66:945–948

Kim KH, Cho Y, Rota ML, Cramer RA Jr, Lawrence CB (2007) Functional analysis of the *Alternaria brassicicola* non-ribosomal peptide synthetase gene *AbNPS2* reveals a role in conidial cell wall construction. Mol Plant Pathol 8:23–39

Knox-Davies PS (1979) The nuclei of Alternaria brassicicola. Trans Br Mycol Soc 72:81–90

Knox-Davies PS (1980) Relationships between *Alternaria brassicicola* and *Brassica* seeds. Trans Br Mycol Soc 73:235–248

Kolte SJ (1985a) Disease management strategies for rapeseed-mustard crops in India. Agric Rev 6:81–88

Kolte SJ (1985b) Diseases of annual edible oilseed crops, vol II. CRC Press, Boca Raton, 135 p

Kothanur PR, Prassana, Lennard JH (1982) Incidence of Alternaria infection in oilseed rape (*Brassica napus* L.) crops in Scotland. Cruciferae Newslett 7:60–61

Kubota M, Abiko K, Yanagisawa Y, Nishi K (2006) Frequency of *Alternaria brassicicola* in commercial cabbage seeds in Japan. J Gen Plant Pathol 72:197–204

Kumar D, Choudhary U (2006) *In vitro* evaluation of role of hydrogen ion concentration on mycelial growth and sporulation of *Alternaria brassicae* and *Alternaria brassicicola*. J Res SKUAST-J 5:122–124

Lapis DB, Ricaforte ET (1974) Alternaria leaf spot of crucifers in the Philippines. Philip Agric 58:121–137

Latif MA, Saleh AKM, Khan MAI, Rahman H, Hossain MA (2006) Efficacy of some plant extracts in controlling seed-borne fungal infections of mustard. Banglad J Microbiol 23:168–170

Lawrence DP, Gannibal PB, Peever TL, Pryor BM (2013) The sections of *Alternaria*: formalizing species-groups concepts. Mycologia 105:530–546

Maude RB (1986) Seed treatment control of *Phoma* and *Alternaria* infections of *Brassica* seed. In British crop protection council monograph No. 33 Symposium on Healthy Planting Material, pp 209–210

McDonald WC (1959) Gray leaf spot of rape in Manitoba. Can J Plant Sci 39:409–416

McLean DM (1947) Alternaria blight and seed infection cause of low germination in certain radish seed crops. J Agric Res 75:71–79

Mehta N, Sangwan MS, Srivastava MP, Kumar R (2002) Survival of Alternaria brassicae causing Alternaria blight in rapeseed-mustard. J Mycol Plant Pathol 32:64–67

Michail SH, Al-Menoufi OA, Abo-Taleb EA (1979) Seed health testing, leaf spot and damping-off of certain crucifers in Egypt. Acta Phytopathol Acad Sci Hungary 14:41–48

Milbraith DG (1922) *Alternaria* from California. Botany Gazette 74:320–324

Minami H, Kim JS, Ikezawa N, Takemura T, Katayama T, Kumagai H, Sato F (2008) Microbial production of plant benzylisoquinoline alkaloids. Proc Nat Acad Sci USA 105:7393–7398

Morales VM, Jasalavich CA, Pelcher LE, Petrie GA, Taylor JL (1995) Phylogenetic relationship among several *Leptosphaeria* species based on their ribosomal DNA sequences. Mycol Res 99:593–603

Mridha MAU (1986) *In vitro* sporulation of *Alternaria brassicae*. Banglad J Bot 15:205–206

Mukadam DS (1982) Studies on self-inhibition in *Alternaria brassicicola* (Schw.) Wiltshire. Indian Bot Rep 1:37–39

Mukadam DS, Deshpande KB (1977) Effect of different substrates and age of the culture on spore germination in *Alternaria brassicae* and *Alternaria brassicicola*. Indian Phytopathol 30:374–378

Mukadam DS, Deshpande KB (1979) Role of light and temperature on growth, sporulation and subsequent spore germinability of *Alternaria brassicae* (Berk.) Sacc. Sci Cult 45:244–246

Neergaard P (1945) Danish species of *Alternaria* and *Stemphylium* taxonomy, parasitism, economical significance. Oxford University Press, London/Oxford, 559 p

Nees Von Esenbeck GG (1817) System der Plize Urid Schwamme. Wurzburg, 234 p

Nipoti P (1978) Investigations into the presence of *Alternaria brassicicola* on cabbage seeds. Infonnatore Fitopathol 28:21–23

Nishimura S, Kohmoto K (1983) Host-specific toxins and chemical structures from *Alternaria* species. Annu Rev Phytopathol 21:87–116

Noble M, Richardson MJ (1968) An annotated list of seed-borne diseases. Proc Int Seed Test Assoc 33:1–191

Oide S, Moeder W, Krasnoff S, Gibson D, Haas H, Yoshioka K, Turgeon BG (2006) *NPS6*, encoding a non-ribosomal peptide synthetase involved in siderophore-mediated iron metabolism, is a conserved virulence determinant of plant pathogenic ascomycetes. Plant Cell 18:2836–2853

Petrie GA (1974) Fungi associated with seeds of rape, turnip rape, flax, and safflower in Western Canada, 1968–73. Can Plant Dis Surv 54:155–165

Pound GS, Cheo PC, Calvert OH, Raabe RD (1951) Extent of transmission of certain cabbage pathogens by seed grown in Western Washington. Phytopathology 41:820–828

Prasada R, Khandelwal GL, Jain JP (1970) Morphology, physiology and control of *Alternaria brassicae* on Taramira. Indian Phytopathol 23:105–110

Pryor BM, Gilbertson RL (2000) Molecular phylogenetic relationships amongst *Alternaria* species and related fungi based upon analysis of nuclear ITS and mt SSU rDNA sequences. Mycol Res 104:1312–1321

Putnam LG, Petrie GA, McDonald H (1972) Alternaria blackspot. In: Insect pests and diseases of rape and mustard. Publication No. 18. Rapeseed Association of Canada, Winnipeg, pp 25–26

Rangel JF (1945) Two Alternaria diseases of cruciferous plants. Phytopathology 35:1002–1007

Rao VG (1969) The genus *Alternaria* from India. Nova Hedwigia 17:219–258

Richardson MJ (1970) Investigations on seed-borne pathogens of *Brassica* spp. Proc Int Seed Test Assoc 35:207–223

Richardson MJ (1979) An annotated list of seed-borne diseases. Phytopathological Paper No. 23. Commonwealth Mycological Institute Publication, 203 p

Rossman AY, Palm ME, Spielman LJ (1987) A literature guide for the identification of the plant pathogenic fungi. APS Press, St. Paul

Rotem J (1991) Assessment of production and dispersal of inoculum of *Alternaria macrospora* in various parts of the cotton canopy. Phytoparasitica 19:121–132

Rotem J (1994) The genus *Alternaria*: biology, epidemiology and pathogenicity. APS press, St. Paul

Saccardo PA (1886) Hyphomyceteae. In Sylloge fungorum omninum hucusque cognitorum. Pavia, Italy, 4, 807

Saharan GS (1984) A review of research on rapeseed-mustard pathology in India. Paper presented in the annual rabi oilseed workshop held at Jaipur, 6–10 August, 1984

Saharan GS (1992a) Management of rapeseed and mustard diseases. In: Kumar D, Rai M (eds) Advances in oil seed research. Scientific Publisher, Jodhpur, pp 151–188

Saharan GS (1992b) Disease resistance. In: Labana KS, Banga SS, Banga SK (eds) Breeding oilseed Brassicas. Narosa Publishing House, New Delhi, pp 181–200

Saharan GS, Mehta N, Sangwan MS (2005) Diseases of oilseed crops. Indus Publishing Company, New Delhi, 643 p

Sarkar B, Sen Gupta PK (1978) Studies on some aspects of the epidemiology of Alternaria leaf blight of mustard (*Brassica* sp.). Beitrage Zur Tropischen Landwirtschaft Und Veterinarmedizin 16:91–96

Sasaki T, Honda Y, Umekawa M, Nemoto M (1985) Control of certain diseases of greenhouse vegetables with ultraviolet-absorbing vinyl film. Plant Dis 69:530–533

Sawada (1931) Report of the Department of Agriculture, Government, Research Institute of Formosa, 51, 117 p.

Schimmer FC (1953) *Alternaria brassicicola* on summer cauliflower seed. Plant Pathol 2:16–17

Sharma AK, Maheshwari KK, Gupta JC (1985) Effect of leaf exudates of yellow sarson (*Brassica campestris* L. var. Sarson Prain) and Taramira (*Eruca sativa* Mill.) on conidial germination of *Alternaria brassicae* (Berk.) Sacc. Agric Sci Digest 5:131–132

Sherf AF, Macnab AA (1986) Diseases of crucifers. In: Vegetable diseases and their control, 2nd edn. The Ronald Press, New York, pp 251–306

Shivpuri A, Siradhana BS (1989) Seed-borne infection of *Alternaria brassicae* on Indian mustard and its elimination during storage. Indian Phytopathol 42: 275 (Abstr.)

Shrestha SK, Mathur SB, Munk L (2000) Transmission of *Alternaria brassicae* in seeds of rapeseed and mustard, its location in seeds and control. Seed Sci Technol J 8:75–84

Simmons EG (1967) Typification of *Alternaria, Stemphylium* and *Ulocladium*. Mycologia 59:67–92

Simmons EG (1992) *Alternaria* taxonomy: current status, viewpoint, challenge. In: Chelkowski J, Visconti A (eds) Alternaria biology, plant diseases & metabolites. Elsevier Science Publishers, Amsterdam, pp 1–35

Simmons EG (1995) *Alternaria* themes and variations (22–26). Mycotaxon 25:287–309

Simmons EG (2007) *Alternaria*: An identification manual. CBS Biodiversity Series No. 6, Utrecht, The Netherlands, 775

Singh DB (1980) Effect of culture media, pH and temperature on growth behaviour of *Alternaria brassicae* and *Drechslera graminea*. Proc Indian Nat Sci Acad B, 46:393–396

Singh R, Suhag LS (1983) Some pathological studies on *Alternaria alternata* causing leaf and pod blight of radish in Haryana. Indian Phytopathol 36:174–176

Sivapalan A, Browning JW (1992) Incidence *ofAlternaria brassicicola* (Schw.) Wiltsh. on *Brassica oleracea* seeds. Aust J Exp Agric 32:535–537

Stovold GE, Mailer RJ, Francis A (1987) Seed-borne levels, chemical seed treatment and effects on seed quality following a severe outbreak of *Alternaria brassicae* on rapeseed in New South Wales. Plant Protect Quarantine 2:128–131

Taber RA (1964) A study of pathogenicity and *in vitro* growth of *Alternaria* spp. isolated from *Brassica* spp. with special reference to *Alternaria raphani* from rape. M.Sc. thesis, University of Saskatchewan, Saskatoon, Canada

Taber RA, Vanterpool TC, Williard AT (1968) A comparative nutritional study of *Alternaria raphani, A. brassicae* and *A. brassicicola* with special reference to A. raphani. Phytopathology 58:609–616

Tahvonen R (1979) Seed-borne fungi on cruciferous cultivated plants in Finland and their importance in seedling raising. J Sci Agric Soc Finland 51:327–379

Tewari JP (1983) Cellular alterations in the blackspot of rapeseed caused by *Alternaria brassicae*. Phytopathology, 73:831. (Abstr.)

Thomma BPHJ (2003) Pathogen profile- *Alternaria* spp.: from general saprophyte to specific parasite. Mol Plant Pathol 4:225–236

Trail F, Köller W (1993) Diversity of cutinases from plant pathogenic fungi: purification and characterization of two cutinases from *Alternaria brassicicola*. Physiol Mol Plant Pathol 42:205–220

Tripathi NN, Kaushik CD (1984) Studies on the survival of *Alternaria brassicae* the causal organism of leaf spot of rapeseed and mustard. Madras Agric J 71:237–241

Tsuneda A, Skoropad WP (1977) The *Alternaria brassicae-Nectria inventa* host- parasite interfaces. Can J Bot 55:448–454

Vaartnou H, Tewari I (1972a) *Alternaria alternata* parasitic on rape in Alberta. Plant Dis Rep 56:676–677

Vaartnou H, Tewari I (1972b) *Alternaria* on polish-type rape in Alberta. Plant Dis Rep 56:633

Van Schreven DA (1953) *Alternaria, Stemphylium* en *Botrytis* a antasting bij Koolzaad (*Brassica napus*). Tijdschr Plantenziekt 59:105–136

Vannacci G, Pecchia S (1988) Location and transmission of seed-borne *Alternaria raphani* Groves and Skolko in *Raphanus sativus* L.: a case study. Archiv fur Phytopathologie und Pflanzenschutz 24:305–315

Verma PR, Saharan GS (1994) Monograph on *Alternaria* diseases of crucifers. Saskatoon Research Centre, Technical Bulletin 1994–6E, Agriculture and Agri-Food Canada, Saskatoon, SK, Canada, 162 p

Vishnuavat K, Agarwal VK, Singh RS (1985) Relationship of discoloration of mustard (*Brassica campestris* var. Sarson) seed infection by *Alternaria brassicae* and its longevity on seed. Seed Res 13:53–56

Voglino P (1902) Polydesmus exitiosus Kühn ed *Alternaria brassicae* (Berk.) Sacc. p.8

Walker J, Smith AM (1972) *Leptosphaeria narmari* and *L. korrae* spp. nov., two long spored pathogens of grasses in Australia. Trans Br Mycol Soc 58:459–466

Weber GF (1932) Some diseases of cabbage and other crucifers in Florida. Fla Agric Exp Stn Bull 256, 62p

Weber GF (1973) Bacterial and fungal diseases of plants in the tropics. University Florida Press, Gainesville, pp 86–88

Weimer JL (1925) Alternaría leaf spot and brown rot of Cauliflower. J Agric Res (1924) 29:421–441

Weimer JL (1926) A leaf spot of cruciferous plants caused by *Alternaria hercules*. J Agric Res 33:645–650

Wiltshire SP (1933) The foundation species of *Alternaria* and *Microsporium*. Trans Br Mycol Soc 18:135–160

Wiltshire SP (1938) The original and modern conceptions of *Stemphylium*. Trans Br Mycol Soc 21:211–239

Wiltshire SP (1947) Species of *Alternaria* on Brassicae. CMI, England. Mycol Paper 20:1–15

Woudenberg JHC, Groenewald JZ, Binder M, Crous PW (2013) *Alternaria* redefined. Stud Mycol 75:171–212

Xu JR (2000) Map kinases in fungal pathogens. Fungal Genet Biol 31:137–152

Yao C, Köller W (1994) Diversity of cutinases from plant pathogenic fungi: cloning and sequence analysis of a cutinase gene from *Alternaria brassicicola*. Physiol Mol Plant Pathol 44:81–92

Yao C, Köller W (1995) Diversity of cutinases from plant pathogenic fungi: different cutinases are expressed during saprophytic and pathogenic stages of *Alternaria brassicicola*. Mol Plant-Microbe Interact 8:122–130

Yoshii H (1933) On three species of *Alternaria* parasitic on cruciferous plants. *Bulletin Science, Fakultato. Terkultura*. Kjusu Imp Univ 3:221–235

4 Infection Process, Pathogenesis and Disease Cycle

4.1 Introduction

Depending upon the growing season, *Alternaria* species in crucifers cause primary infection through the windborne spores produced from over summered or overwintered primary inoculum in infected seeds, plant debris, collateral hosts and weeds. Under congenial environmental conditions, spores (conidia) germinate readily in the presence of moisture by producing a germ tube, which penetrates the hosts directly or through stomata. Further, the process of infection and pathogenesis is governed by a number of external and internal factors of a specific host–pathogen interaction. Induction of cell wall-degrading enzymes, secretion of toxins, enforcement of fungal cell wall architecture and detoxification of host defence molecules are essential functions for successful infection and pathogenesis of *Alternaria* species. Genes important for each of these functions have been identified in various fungal species. A transcription factor gene, *AbVF19*, has been discovered from *A. brassicicola*, which positively regulates hydrolytic enzyme-coding gene. Another transcription factor gene, *Amr1*, negatively regulates a subset of these genes during the late stage of pathogenesis and positively regulates melanin biosynthesis during conidiogenesis. It seems that the suppressive functions of *Amr 1* gene contribute to the specialized adaptation of *A. brassicicola* as an efficient and successful facultative parasites of cruciferous plants.

4.2 Infection and Pathogenesis

The primary infection results from the windborne spores produced on infected plant debris of the previous crop, weeds and other collateral hosts growing in the near vicinity. Infected seed as a source of primary inoculum is important in temperate regions. Chlamydospores and microsclerotia may initiate primary infection of *A. raphani* and *A. brassicae* (Fig. 4.2). Conidia readily germinate in the presence of moisture by giving rise to a germ tube, which emerges from any cell of the spore. Theoretically, only one spore can cause infection in the brassicaceous seedlings indicating a threshold infection value at a very low level (Czyzewska 1971). Germ tubes from germinated spores of *A. brassicae*, *A. brassicicola*, *A. raphani* and *A. alternata* generally penetrate the undamaged tissues of many brassicaceous hosts directly (Changsri and Weber 1963; Czyzewska 1971; Minkevicius 1932; Neergaard 1945; Schreven 1953; Weimer 1926), although indirect penetration through stomata has been reported in *A. brassicae* (Changsri 1961; Tsuneda and Skoropad 1978a).

Alternaria conidia are considered as 'dry conidia', which are wind dispersed with the

G.S. Saharan et al., *Alternaria Diseases of Crucifers: Biology, Ecology and Disease Management*, DOI 10.1007/978-981-10-0021-8_4 maximum release taking place in the mid-afternoon. Since there is no report on the numerical threshold for infection, lesions may consequently be initiated from a single germinating conidium. Maude and Humpherson-Jones (1977) found that infected marrow-stem kale *(B. oleracea* var. *acephala)* liberates 50 spores per cubic metre of air just prior to harvesting (190 spores per cubic metre of air at cutting) or 3200 spores per cubic metre of air when windrowed crops are harvested. Spores can be carried up to 1000-m downwind during harvesting. Sharma and Gupta (1978) trapped *A. brassicae* and *A. brassicicola* spores from the air during October to April on Brown Sarson crop fields under Indian conditions. Conidia of *A. brassicae* were intercepted at all hours of exposure, but conidia of *A. brassicicola* were seen only for a specific period during the day. Maximum conidial catch was observed in March, which coincides with the maximum manifestation of disease in the field. The main mode of perpetuation is on debris of infected plants. Mehta et al. (2002) have reported that diseased debris placed in deep freezer conditions (−10 °C) were able to cause 100 % infection in the coming season when mixed in the soil as compared to the debris placed in the field and laboratory conditions. It was further confirmed that pathogen survives in field for 2 years in the diseased debris collected from the temperate conditions (Canada) as compared to the subtropical conditions of India (Mehta and Tewari 2002). Local dispersal is through air, splashes and occasional driving rain drops. The primary infection appears on the cotyledonary leaves forming the source of secondary infection for the entire crop. For infection, a minimum period of 4 h of leaf wetness is required. Increase in period of leaf wetness at 25 °C increases infection and spread of the disease rapidly. Under favourable temperature conditions, and the presence of dew, the spores infect other parts of the plant as well. The infection occurs through the stomata, and under favourable climatic conditions, the new lesions arise within 4–6 days bearing spores (Verma and Saharan 1994).

Frequent infection occurs when water is present on the host surface due to dew formation. The infection processes of both *A. brassicae* and *A. brassicicola* are initiated if water is present for 5–8 h. For initiation of infection in the case of *A. brassicae* in cauliflower leaves and heads, 25–31 °C temperature is reported as optimum (Weimer 1924); the disease increases with the increase in humidity of the surrounding air. In rapeseed, *A. brassicae* establishes itself in less than 6 h at 22 °C (Louvet and Billotte 1964). For infection of cabbage plants, both *A. brassicae* and *A. brassicicola* require free water with an optimum temperature of 25 °C for *A. brassicicola* and 15 °C for *A. brassicae.* At these temperatures, a minimum of 16 h is necessary for the initiation of infection and 48–72 h for optimal disease development. Alternating wet and dry (70–80 % RH) periods of 16 h and 8 h, respectively, restrict infection by both species. The differential temperature requirements of the two species become most marked at 10 °C and 96 h of incubation after inoculation; *A. brassicicola* fails to produce significant infection, whereas infection by *A. brassicae* produces numerous lesions (Humpherson-Jones and Hocart 1983).

The initial infection process including conidial germination and mode of germ-tube penetration are similar in both susceptible (Varuna) and resistant (PAB 9511) cultivars, although considerably slower in PAB 9511 due to tolerance. Apparent differences between the two cultivars, however, are noticeable by 1 day after infection. At the plant surface, impeded fungal growth, poor proliferation of spores and active suppression of dense hyphal growth are the factors involved in the expression of tolerance against *A. brassicae* in cultivar PAB 9511 (Giri et al. 2013).

In *A. brassicicola,* cuticular penetration is most frequent and stomatal penetration occurs to a lesser extent. Penetration is preceded by the formation of appressoria from the tip of the germ tubes, which are either spherical or club shaped and 3–8 mm in diameter. Stomatal penetration is more common with *A. brassicae.* Once within the host, the epidermal cells are fully invaded, the mycelia ramify through and between the mesophyll and palisade cells, and the entire leaf is soon parasitized. Early in the post-penetration phase, the invaded epidermal cells become necrotic, and the parenchyma tissues ahead of the advancing hyphae often collapse (Dixon 1981).

On the leaves of *B. napus* cv. Midas and *B. rapa* cv. Torch, conidia of *A. brassicae* germinate at the rate of 12.1 % and 19.5 %, respectively, at 9 h after inoculation. Conidia usually germinate by producing either germ tubes or secondary conidia. Penetration of leaves by *A. brassicae* is abundant at 24 h and occurs either with or without the formation of appressoria. Penetration of the cultivar Torch leaves occurs either directly through epidermal cells or indirectly through stomata, while Midas leaves are penetrated almost exclusively through the stomata. Black spot lesions develop within 48 h after inoculation (Tsuneda and Skoropad 1978b). According to Tewari (1983, 1986), *A. brassicae* in rapeseed becomes subcuticular after direct penetration. This is followed by colonization of the epidermal and the mesophyll cells. In the leaves of rapeseed cultivars Candle and Altex, the pathogen heavily colonizes the necrotic center and is not present in the chlorotic area indicating that a diffusible metabolite may be directly or indirectly responsible for leaf chlorosis. The plasma membrane is the first target of the diffusible metabolite. Subsequently, the chloroplasts are either directly or indirectly affected leading to leaf chlorosis. Relative to chloroplasts, the effect on mitochondria is seen at a much later stage. The cells in the necrotic area are almost completely devoid of cellular organelles and reveal electron-dense lamellar deposits (Plates 4.1 and 4.2; Tewari 1983). *Alternaria* infection in the *C. abyssinica* leaves is preceded by toxic substances secreted by the fungus (Czyzewska 1971).

The same may be true for other *Brassica*. According to Suri and Mandahar (1984, 1985, 1986) and Suri et al. (1983), cytokinin-like substances appear to be actively involved in infection and pathogenesis of *A. brassicicola.* However, cytokinin, 6-benzyl amino purine, reduces disease symptoms and mycelial growth within plant tissues. It also inhibits in vitro growth of *A. brassicae* (Sharma et al. 2010). Poapst et al. (1979) suggested the role of the hormone ethylene in screening cabbage leaves infected with *A. brassicicola.* Production of cellulases and pectolytic enzymes by *A. brassicae* causing leaf blight disease of rapeseed–mustard has been observed (Nehemiah and Deshpande 1976; Suri and Mandahar 1982). The differential susceptibility of the brassicaceous hosts to *A. brassicae*, *A. brassicicola* and *A. raphani* has been correlated with differences in amino acid content (Changsri 1961). The population of the pathogens varies with the age, variety of the host, maturity of leaves and climatic conditions. The colonies increase when the temperature and relative humidity of the field are 21.5 °C and 85 %, respectively. Both *A. brassicae* and *A. brassicicola* occur in greater profusion on leaves of *B. rapa* var. Yellow Sarson than on leaves of *E. sativa*. The younger leaves have lower incidence of disease of either pathogen compared to the older leaves (Sharma et al. 1986). Species of *Alternaria* pathogenic to crucifers are prolific toxin producers, which facilitate their necrotic lifestyle. The role of toxins in the process of infection has been described in Chap. 10 in Sect. 10.5. Besides role of toxins in *Alternaria* pathogenesis, relatively few genes and/or gene products have been identified that contribute to or are required for pathogenicity.

4.3 Transcription Factors Associated with Pathogenesis

Out of 138 *A. brassicicola* transcription factor genes with zinc finger domains, and other regulatory genes analysed to date, only six genes including *AbSte12*^{C}, *AbPac*^{C}, *AbVf19*^{C}, *Amr1*, *AbVf8* and *AbPro1*^{C} are strongly associated with pathogenesis (Table 4.1). *AbSte12* is a homologue of the *Ste12* gene with mutants that are either non-pathogenic or reduced in virulence in several fungi (Wong and Dumas 2010). *Ste12* is regulated by a MAP kinase, *Fus3/Kss1*. Mutants of either homologous gene in *A. brassicicola* show defective conidial development and are non-pathogenic (Cho et al. 2007, 2009). The *AbPac*^{C} gene has a single copy in the genome sequence of *A. brassicicola* and shows a high sequence similarity ($E=0.0$) with the *Pac*^{C} gene in other fungi. *AbVf19* regulates suite of genes that are important for decomposing and utilizing plant materials during

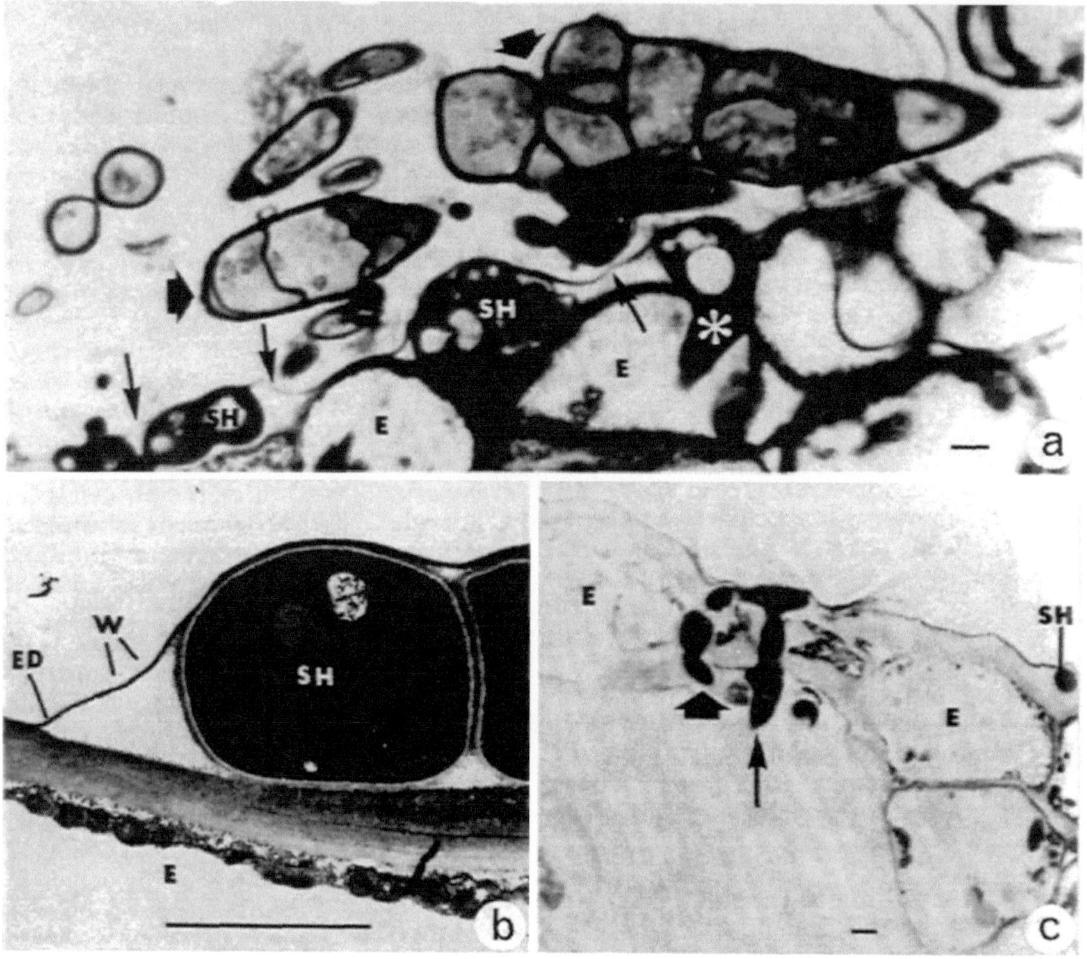

Plate 4.1 *Alternaria brassicae* on leaves of *Brassica rapa* cv. Candle. Bar = 5.0 μm: (**a**) light micrographs showing germinating conidia (*broad arrows*) and subcuticular hypha (SH) penetrating (*asterisk*) into epidermal cell (E) (*narrow arrows*) of cuticle ×1200; (**b**) transmission electron micrograph showing subcuticular hyphae (SH), epicuticular wax layer (W) and electron-dense layer (ED) of cuticle ×7000; and (**c**) light micrograph showing a hypha penetrating through a stroma (*narrow arrow*) and another one penetrating into a subepidermal position (*broad arrow*) without becoming subcuticular ×800 (Tewari 1986)

the late stage of plant infection. Regardless of their function, however, all mutants studied to date except *Amr* 1 are non-pathogenic or show a reduction in virulence (Cho et al. 2012).

4.3.1 Melanin Biosynthesis and Virulence in *A. brassicicola*

Melanin is an ubiquitous pigment that plays an important role in protecting fungi from the damaging effects of environmental stress. It accumulates in the cell walls of hyphae and conidia during the late stationary phase of mycelial growth and also forms under stressful conditions, including ultraviolet irradiation, a hyperosmotic environment, nutrient deprivation or an accumulation of toxic wastes during in vitro culture. Overexpression of a *Cmr1* ortholog in *Bipolaris oryzae* (*Bmr1*) causes continuous induction of the three downstream genes in the melanin synthesis pathway and increases melanization of colonies (Kihara et al. 2008). The loss of function of *Cmr1*, its homologues or their downstream genes in other fungi result in melanin deficiency. Cho et al. (2012) identified a transcription factor *Amr1*,

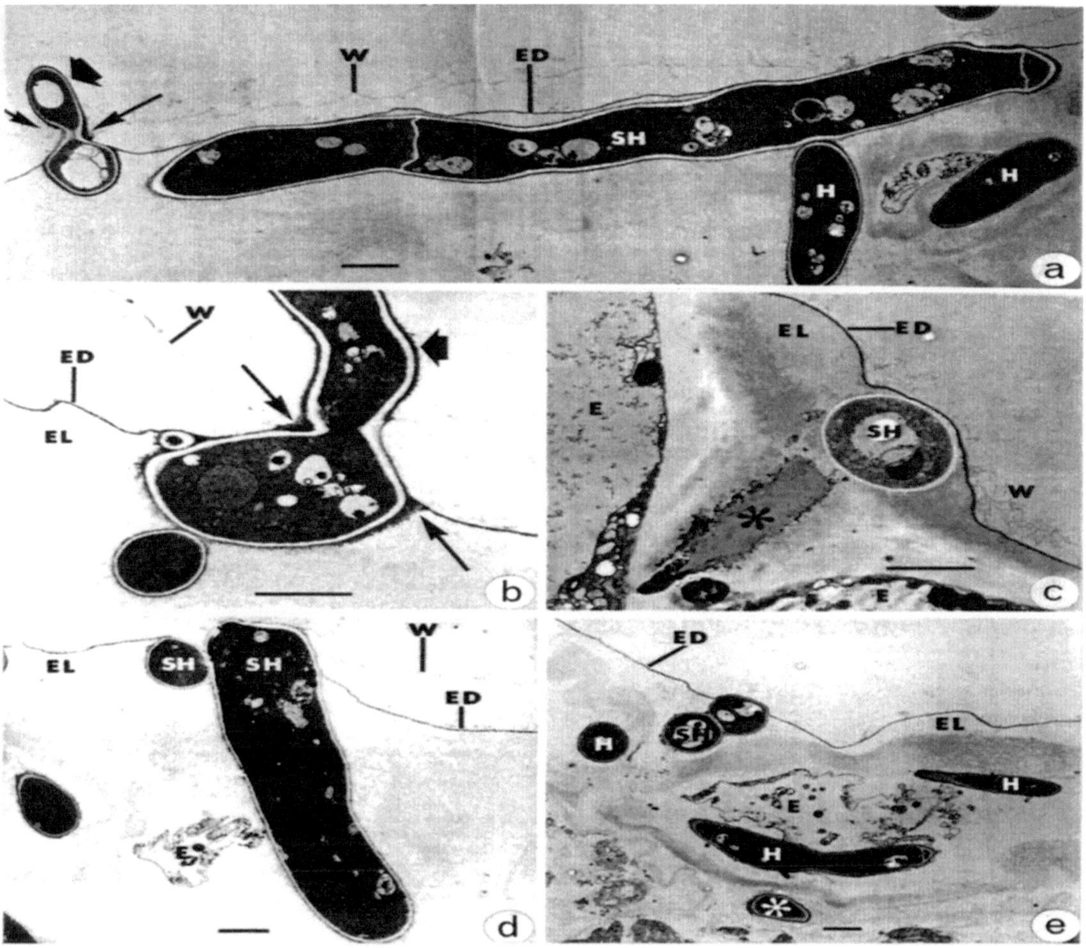

Plate 4.2 Transmission electron micrographs of ultrathin sections of *A. brassicae* on the leaves of *Brassica napus* cv. Altex. Bar = 2.0 μm: (**a**) a penetrating germ tube (*broad arrow*), a subcuticular hypha (SH) and two other hyphae (H), embedded in the cell wall of the host adaxial leaf epidermal cell. Note the electron-dense material at the point of penetration (*narrow arrows*), the layer of epicuticular was (W) detached from the electron-dense layer of the cuticle (ED) ×4200; (**b**) a penetrating germ tube (*broad arrow*). Note the electron-dense material at the point of penetration (*narrow arrows*), the detached epicuticular wax layer (W) and electron-translucent layer (EL) of the cuticle ×8050; (**c**) junctional region between the two epidermal cells (E) showing subcuticular hyphae (SH) and glancing section of a hypha passing through the cell wall (*asterisk*). The latter hypha is penetrating between the two cells ×7400; (**d**) section showing two subcuticular hyphae (SH), one of which is penetrating into the cell wall of the epidermal cell (E) ×4200; and (**e**) section showing subcuticular hyphae (SH) a few other hyphae (H) in the epidermal cells and one hypha (*asterisk*) below the epidermis close to the palisade mesophyll tissue ×3000 (Tewari 1986)

which regulates the melanin synthesis pathway in *A. brassicicola*. The melanin synthesis-associated structural genes *Bm1* and *Pks* are located together in a 30-kb region and their organization, and orientation is the same as in the closely related fungus *Cochliobolus heterostrophus*. Interestingly, melanin is not associated with virulence in *A. brassicicola* as has been reported in other fungi. However, there appears to be a linkage between melanin biosynthesis and reduction in virulence, perhaps due to a lifestyle switch from pathogenesis to reproduction (Cho et al. 2012).

Table 4.1 Summary of *Alternaria brassicicola* transcription factor domains based on Pfam scans (Cho et al. 2012)

Description	Gene	KO 1[a]	KO 2[b]	Pathogenesis-associated genes
Zn finger domain	221	4	117	*AbVf8*, *AbVf19*[C], *AbSte12*[C], *AbPac*[C], *AbPro1*[C], *Amr1*
Fungal-specific TF	72	1	0	
PHD finger	1	–	–	
Helix-turn-helix	29	1	–	
Regulatory receiver	23	5	–	
Leucine zipper	21	2	–	
High-mobility box group	15	3	–	
Helix-loop-helix	13	5	–	
DNA-binding myb	10	–	–	
Jumonji	7	–	–	
Homeobox	4	–	–	
β-Regulatory protein	3	–	–	
T regulatory	2	–	–	
Total	421	21	117	

[a]Genes of previously screened targeted deletion mutants (15)
[b]Genes of mutants created and screened in this study
[c]Genes whose functions were previously characterized (15, 20)

4.3.2 Mutation of the *Amr1* Gene Unexpectedly Causes Increased Virulence

Amr1 gene knockout mutants, both gene disruption (*amr1*) and gene deletion (Δ*amr1* and Δ*amr1*; *amr1p-GEP*), are easily recognized by the lack of melanin in their conidia, orange-coloured colonies and the secretion of a pink pigment. These phenotypes are similar to those found in mutants of its homologues in other fungi (Kihara et al. 2008). The Δ*amr1* mutants are more susceptible to UV light and enzyme digestion. Unexpectedly, the *Amr1* gene knockout mutants are more virulent than the wild type, producing larger lesions, and require fewer conidia for successful infection. Mutation of *Amr1* homologues in the phytopathogenic fungi, *Magnaporthe grisea* and *Colletotrichum lagenarium*, produces melanin-deficient conidia. The increased virulence of the Δ*amr1* mutants is unexpected and unique among melanin-deficient mutants of pathogenic fungi.

4.4 The Cause of Increased Virulence in Δ*amr1* Mutants

The pink pigment secreted by Δ*amr1* mutants is neither toxic to host plants nor beneficial for the wild-type *A. brassicicola* to infect host plants. Furthermore, melanin-deficient mutants of another gene, *abpks7*, show no changes in virulence. Neither the presence of pink pigment nor the lack of melanin is associated with the increased virulence of the Δ*amr1* mutants. Additionally, there is no difference in germination or vegetative growth between the Δ*amr1* mutants and the wild type in the presence of various chemicals. However, the mutants are more susceptible than wild type to UV light and the cell wall-degrading enzymes, glucanase. The mutant's ability to respond to stress is not associated with their increased virulence, since plant glucanases are known to be defence enzymes (Cho et al. 2012).

The ability of the mutants to outgrow the wild type when citrus pectin was the major carbon source suggested that the mutants were more efficient in utilizing the protein component of plant cell walls and middle lamellae. The expression profile comparisons identified 30 hydrolytic enzyme genes expressed more in the mutant than in the wild-type *A. brassicicola* during pathogenesis. However, they included only two pectate lyase genes and one pectin esterase gene. Other genes upregulated in the mutant compared to the wild type included gene coding for two cutinases, a lipolytic enzyme and 20 other various glycoside hydrolases (GH). Interestingly, 6 of the 22 genes in the GH61 gene family are expressed at higher levels in the Δ*amr1* mutant. The GH61 family might have a role in degrading cellulose, lignocelluloses, chitin or other polysaccharides

(Karkehabadi et al. 2008). The GH61 family may be involved in digesting pectin's side chains, which are structurally diverse in constituting sugars and glycosidic linkages. These may digest other cell wall components, such as cellulose and hemicelluloses along with other glycoside hydrolases.

Induction of the *Amr1* gene in *A. brassicicola* did not occur until 64-h post inoculation (hpi) (Fig. 4.1). Suppression of the *Amr1* gene might be important during pathogenesis when energy is allocated for colonization of host tissues and overcoming host defence mechanisms, rather than producing reproductive dispersal structures like conidia. In the wild type, the transcripts of hydrolytic enzyme genes were slightly increased compared to the actin transcripts, as previously reported (Karkehabadi et al. 2008). In addition, when conidiation was initiated, the induction of *Amr1* was accompanied by a sharp increase in transcripts of its downstream genes in the melanin synthesis pathway. There was also small increase in the transcripts of hydrolytic enzyme-coding genes in the wild type. However, many of these genes mostly expressing glycoside hydrolases, a few cutinases and pectate digestion enzymes were expressed at higher levels in the Δ*amr1-4* and Δ*amr1*; *Amr1p*-GFP mutants did not express *Amr1*. This suggests that *Amr1* suppressed genes predicted to be involved in the digestion of lipids and carbohydrates. This occurred during late pathogenesis (88 hpi) when they were still needed to some extent for either killing (Noda et al. 2010) or digesting host tissues. Whether the *Amr1* gene is involved in the regulation of most hydrolytic enzymes sequentially induced in *A. brassicicola* during pathogenesis has yet to be investigated. Other genes encoding a necrosis-inducing protein, two P450s, the aegerolysin peptidase homologue and other hypothetical proteins might have contributed to the increased virulence of the *amr1* mutant and warrant further investigation. Interestingly, 24 genes expressed at higher levels in Δ*amr1* were expressed at lower levels in the Δ*Abvf19* deletion mutants that displayed reduction in virulence, suggesting opposite roles of the two transcription factor genes in the regulation of pathogenesis-associated genes. This finding may aid in the selection of specific targets for future analysis of their functions in pathogenesis (Cho et al. 2012).

4.5 Evolution of Virulence in *A. brassicicola*

The melanin-deficient mutant phenotype and expression pattern of the *Amr1*, *Brn*, *Brn2* and *Scd1* genes in *A. brassicicola* suggest that the function of the *Amr1* gene is to regulate melanin synthesis during conidiogenesis. This function is highly conserved in other fungi. In addition to regulating the melanin biosynthesis pathway, *Amr1* negatively regulated virulence_ENREF_1. The evolution of virulence has been a subject of extensive theoretical analysis. The trade-off hypothesis produced a number of models showing a contested relationship between virulence and transmissibility. It challenges the hypothesis that strong virulence can seriously damage or kill host plants and threaten pathogen survival, so parasites will ultimately evolve to be less virulent. However, analysis of ecological and epidemiological factors related to virulence has produced inconsistent results. Therefore, a modified trade-off theory predicted that strong parasites became moderately virulent to their hosts over time but did not completely lose their virulence. Cho et al. (2012) identified a transcription factor suppressing virulence but enhancing the ability of a fungus to preserve itself in space, and time via melanin synthesis, primarily in conidia. This study provides an example of a transcription factor gene important for survival in nature that plays a critical role in negatively regulating virulence in the necrotrophic plant pathogens, *A. brassicicola*. The suppression of virulence observed is unique and might have been acquired by this fungus, as this phenotype has not been observed in loss-of-function mutants of *Amr1* homologues in other species of plant pathogenic fungi. An ancestral fungal strain of *A. brassicicola* without the ability to suppress virulence by the activity of *Amr1* could have been a more virulent pathogen. Negative regulation of virulence is rare but exists in other phytopathogenic fungi.

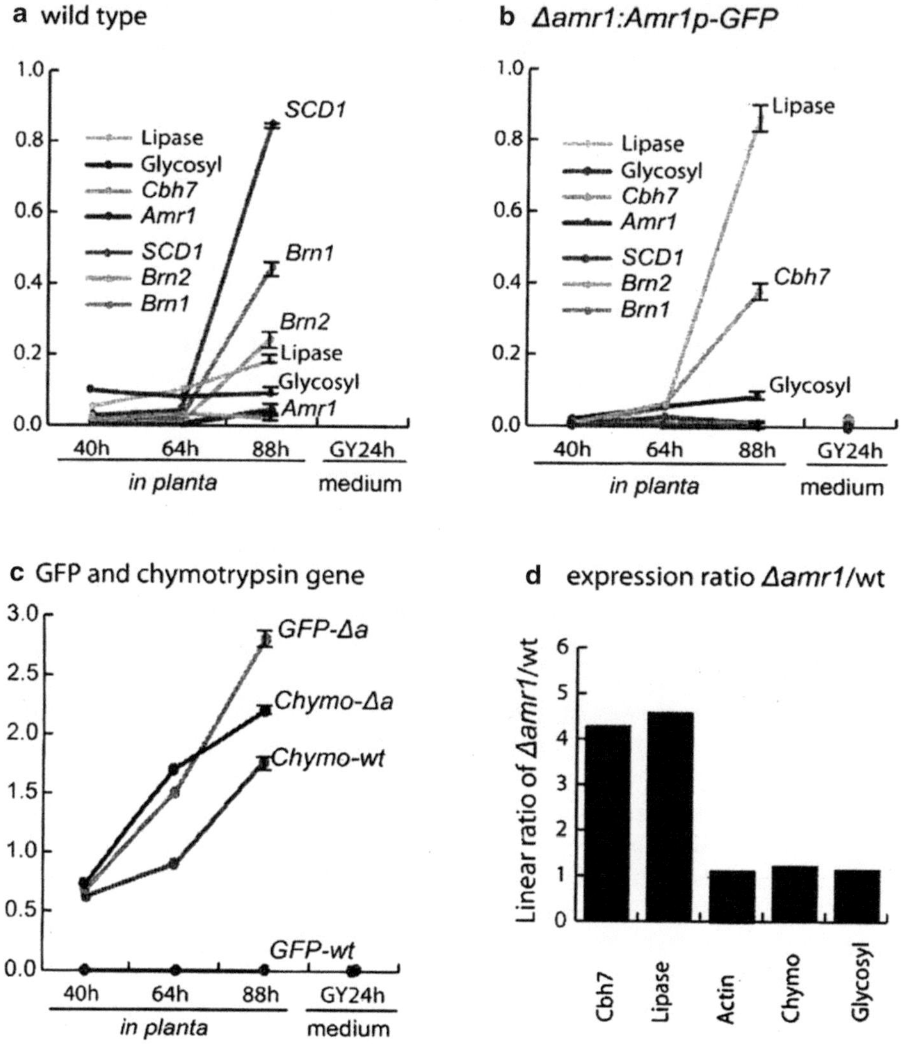

Fig. 4.1 Expression of melanin biosynthesis-associated genes and four hydrolytic enzyme-coding genes. (**a–c**) Relative transcription abundance of each gene was determined in comparison to actin gene transcripts in the same tissue. Y-axes show relative abundance of the transcript compared to the actin gene. (**d**) Expression ratio between the Δamr1 and wild type during the late stage of infection. A total of three biological replicates ($N=3$) were used for this study. Bars represent standard error. Wt = wild type, Δa = Δ*amr1*: *Amr1pGFP*, GY = glucose yeast extract broth. SCD1 = scytalone dehydratase, *Brn1*-13HN reductase, *Brn2*-14HN reductase, Cbh7 = cellobiohydrolase, *Amr1* = *Alternaria* melanin regulation, chymo = chymotrypsin (Cho et al. 2012)

Currently, *A. brassicicola* is a prolific saprophyte and necrotroph, efficiently colonizing susceptible, weakened or dead host plants. Its pathogenicity, however, is inhibited on more resistant host plants. Cho et al. (2012) speculated that the suppressive functions of *Amr1* contribute to the specialized adaptation of *A. brassicicola* as an efficient and successful facultative parasite.

4.6 Identification of Pathogenicity Factors

The functional redundancy of lipases has been revealed by Yao and Koller (1994, 1995), Berto et al. (1999) and Cho et al. (2006) with regard to pathogenicity. Interestingly, one of the factors responsible for the pathogenicity has been predicted to be secondary metabolic production. A non-ribosomal peptide synthase gene (*NPS6*) in *Cochliobolus heterostrophus* and *A. brassicicola* was found to direct the biosynthesis of a siderophore metabolite important for oxidative stress tolerance and pathogenicity (Oide et al. 2006). The secondary metabolic corresponding to or synthesized via *AbNPS2* has yet to be characterized. However, research is needed to further characterize secondary metabolic biosynthetic genes and their role in pathogenicity and fungal development, in the *Alternaria–Brassica* pathosystem fungal signal transduction, e.g. disruption of the *Fus3*/*Kss1*. MAP kinase homologue (*Amk1*) in *A. brassicicola* resulted in a complete loss of pathogenicity as observed in other fungi (Cho et al. 2006, 2007). Interestingly, in the later study, it was shown that addition of long polypeptide nutrients partially restored pathogenicity to the mutants. In addition, two novel virulent factors were predicted to encode a transcription factor (*AbPro1*) and a two-component histidine kinase gene (*AbNIK1*) (Cho et al. 2009). Both of these kinases are pathogenicity factors in phytopathogenic fungi. *Slt2* was found to be associated with cell wall integrity and HOG kinase homology with oxidative stress tolerance (Xu 2000). Another major work pertaining to the studies related to the identification of virulence factors was the disruption of *Aso-1*, a gene required for hyphal fusion (anastomosis), which was also found to be required for pathogenicity in *Alternaria* species (Craven et al. 2008). Eventually, over a hundred genes have been functionally analysed through various techniques like gene knockout and overexpression experiments making *A. brassicicola* the species of choice for functional genomics research to define conserved virulence mechanisms for this important genus of fungi (Oide et al. 2006; Cho et al. 2006, 2007, 2009; Kim et al. 2007).

For the identification of *A. brassicicola*, an attempt was made to examine the role of cutinase genes in *A. brassicicola* pathogenesis (Yao and Köller 1994, 1995). In these studies, biolistic transformation was used to disrupt the *CUTAB1* gene. Disruption of *CUTAB1* affected saprophytic growth since cutin was no longer able to be utilized as a sole carbon source, but this disruption had no significant effect on *A. brassicicola* pathogenicity. An extracellular lipase was found to be produced by *A. brassicicola* in vitro (Berto et al. 1999). In this study, anti-lipase antibodies were found to decrease the ability of *A. brassicicola* to cause disease on cauliflower leaves significantly. However, disruption of four predicted *A. brassicicola* lipase genes expressed during plant infection did not result in reduced virulence on cabbage (Cho et al. 2006). Results of a non-ribosomal peptide synthase gene (*NPS6*) in *Cochliobolus heterostrophus* and *A. brassicicola* suggest to direct the biosynthesis of a siderophore metabolite important for oxidative stress tolerance and pathogenicity (Oide et al. 2006). In another study, non-ribosomal peptide synthase gene (*AbNPS2*) was found to be important for cell wall integrity, conidial viability and virulence of aged spores of *A. brassicicola* (Kim et al. 2007). The secondary metabolite corresponding to or synthesized via *AbNPS2* has yet to be characterized.

4.7 Disease Cycle

It is believed that *Alternaria* survives through infected seed, plant debris, soil and weed hosts. It has been proved that *Alternaria* does not survive through seed to cause infection in the coming season due to high-temperature conditions during summer months in storage houses of North India. However, the possibility of its survival through seed on hills cannot be ruled out. *Alternaria* was not recovered from infected seeds of Yellow Sarson and raya stored under Hisar (Haryana) and Ludhiana (Punjab) conditions after June and July months. *Alternaria brassicae* survives with plant debris buried in the field at 7.5-cm depth in the form of microsclerotia and chlamydospore in the infected leaves. *Alternaria brassicae* is known to become seed borne although this mode of survival and dispersal is not very significant. The main mode of perpetuation is on debris of infected plants. Mehta et al. (2002) have reported that diseased debris placed in deep freezer conditions (−10 °C) were able to cause 100 % infection in the

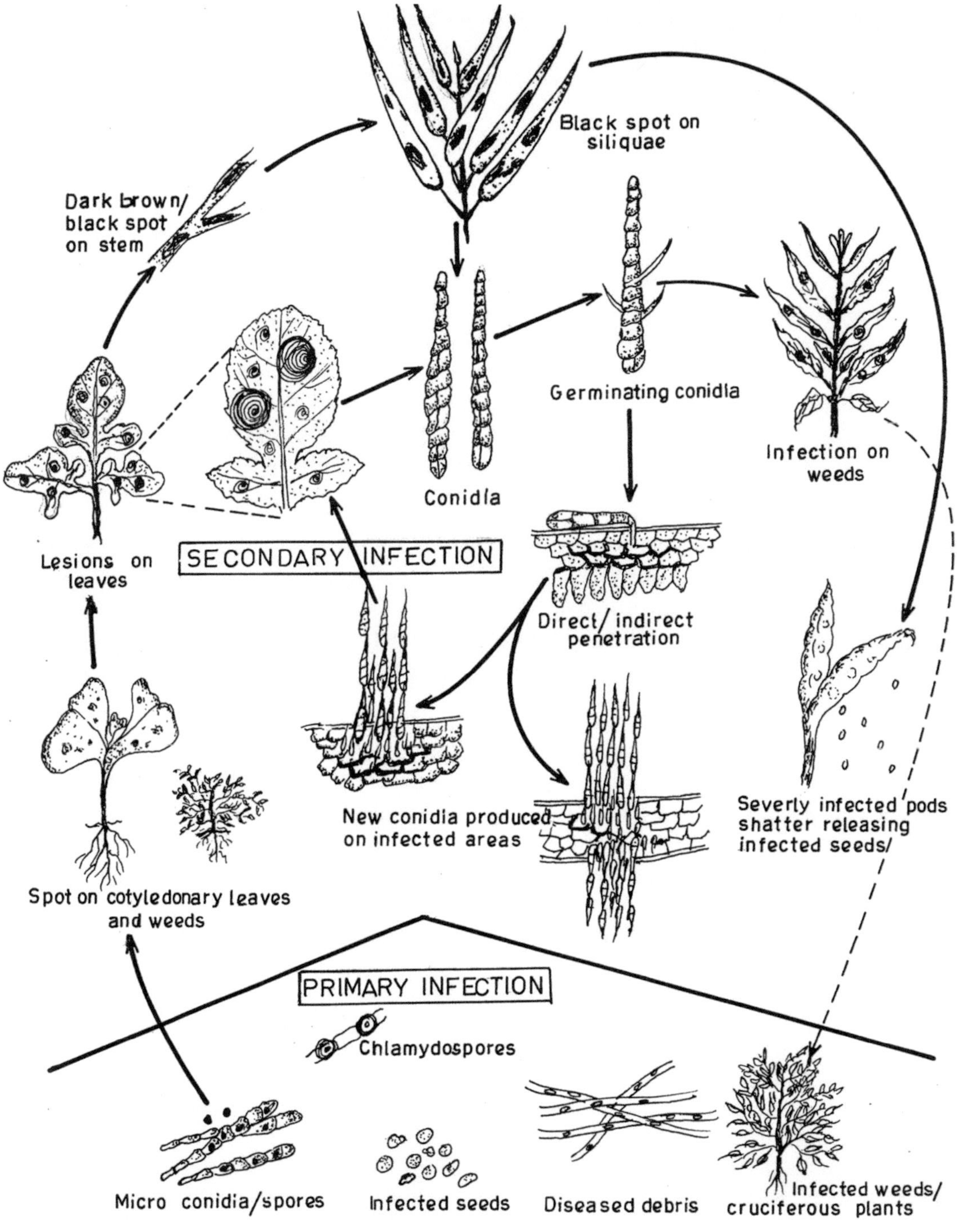

Fig. 4.2 Disease cycle of *Alternaria* on crucifers (Saharan et al. 2005)

coming season when mixed in the soil as compared to the debris placed in the field and laboratory conditions. It was further confirmed that pathogen survives in field for 2 years in the diseased debris collected from the temperate conditions (Canada) as compared to the subtropical conditions of India (Mehta and Tewari 2002). Local dispersal is through air, splashes and occasional driving raindrops. Spores are produced abundantly in wet weather. Spore release, which is passive, is stimulated by falling humidity and restricted by a high humidity so that air spore concentration in a diseased crop exhibits a distinct diurnal periodicity. The maximum concentration of spores is in the early afternoon and minimum in the early morning (Verma and Saharan 1994). The primary infection appears on the cotyledonary leaves forming the source of secondary infection for the entire crop. For infection, a minimum period of 4 h of leaf wetness is required. Increase in period of leaf wetness at 25 °C increases infection and spread of the disease rapidly. Under favourable temperature conditions, and the presence of dew, the spores infect other parts of the plant as well. The infection occurs through the stomata, and under favourable climatic conditions, the new lesions arise within 4–6 days bearing spores. The pathogen penetrates in the tissues of the pods and infects the seed. The disease cycle of the pathogen is presented in Fig. 4.2.

References

Berto P, Comménil P, Belingheri L, Dehorter B (1999) Occurrence of a lipase in spores of *Alternaria brassicicola* with a crucial role in the infection of cauliflower leaves. FEMS Microbiol Lett 180:183–189

Changsri W (1961) Studies of *Alternaria* spp. pathogenic on cruciferae. Disease Abstract 21:1698

Changsri W, Weber GF (1963) Three *Alternaria* species pathogenic on certain cultivated crucifers. Phytopathology 53:643–648

Cho Y, Davis JW, Kim KH, Wang J, Sun QH, Cramer RA Jr, Lawrence CB (2006) A high throughput targeted gene disruption method for *Alternaria brassicicola* functional genomics using linear minimal element (LME) constructs. Mol Plant Microb Interact 19:7–15

Cho Y, Cramer R, Kwang-Hyung K, Davis J, Mitchell T, Figuli P, Pryor B, Lemasters E, Lawrence CB (2007) The *Fus3/Kss1* MAP kinase homolog *Amk1* regulates the expression of hydrolytic enzyme genes in the fungus *Alternaria brassicicola*. Fungal Gen Biol 44:543–553

Cho Y, Kim KH, La Rota M, Scott D, Santopietro G, Callihan M, Mitchell TK (2009) Identification of virulence factors by high throughput targeted gene deletion of regulatory genes in *Alternaria brassicicola*. Mol Microbiol 72:1316–1333

Cho Y, Srivastava A, Ohm RA, Lawrence CB, Wang KH, Grigoriev IV, Marahatta SP (2012) Transcription factor *Amr1* induces melanin biosynthesis and suppresses virulence in *Alternaria brassicicola*. PLoS Pathog 8, e1002974. doi:10.1371/journal.ppat.1002974

Craven KD, Velez H, Cho Y, Lawrence CB, Mitchell TK (2008) Anastomosis is required for virulence of the fungal necrotroph *Alternaria brassicicola*. Eukaryote Cell 7:675–683

Czyzewska S (1971) The pathogenicity of *Alternaria* spp. isolated from *Crambe abyssinica*. Acta Mycol 7:171–240

Dixon GR (1981) Pathogens of crucifer crops. In: Vegetable Crop Diseases. AVI Publishers Company Inc., Westport, pp 119–123

Giri P, Taj G, Meena PD, Kumar A (2013) Microscopic study of *Alternaria brassicae* infection processes in *Brassica juncea* cultivars by drop plus agarose method. Afr J Microbiol Res 7:4284–4290

Humpherson-Jones FM, Hocart MJ (1983) *Alternaria* diseases of *Brassica* seed crops. In: 33rd annual report for 1982, National Vegetable Research Station, Wellesbourne, Warwick, UK, pp 63–64

Karkehabadi S, Hansson H, Kim S, Piens K, Mitchinson C, Sandgren M (2008) The first structure of a glycoside hydrolase family 61 member, Cel61B from *Hypocrea jecorina*, at 1.6 A resolution. J Mol Biol 383:144–154

Kihara J, Moriwaki A, Tanaka N, Tanaka C, Ueno M, Arase S (2008) Characterization of the *BMR1* gene encoding a transcription factor for melanin biosynthesis genes in the phytopathogenic fungus *Bipolaris oryzae*. FEMS Microbiol Lett 281:221–227

Kim KH, Cho Y, La Rota M, Cramer R, Lawrence C (2007) Functional analysis of the *Alternaria brassicicola* non-ribosomal peptide synthetase gene *AbNPS2* reveals a role in conidial cell wall construction. Mol Plant Pathol 8:23–39

Louvet J, Billotte JM (1964) Influence des facteurs climatiques sur les infections du colza par *l'Alternaria brassicae* et consequences pour la lutte. Ann Epipht 15:229–243

Maude RB, Humpherson-Jones FM (1977) Dark leaf spot of Brassica (*Alternaria brassicicola*). In: 27th annual report for 1976, National Vegetable Research Station, Wellesboume, Warwick, UK, pp 95–96

Mehta N, Tewari JP (2002) Perpetuation of *Alternaria brassicae* on diseased plant debris of rapeseed and mustard. J Mycol Plant Pathol 32:368 (Abstr.)

Mehta N, Sangwan MS, Srivastava MP (2002) Survival of *Alternaria brassicae* causing Alternaria blight in rapeseed-mustard. J Mycol Plant Pathol 32:64–67

Minkevicius A (1932) Investigations on the influences of narcosis on the susceptibility of plants to fungus infection. Phytopathol Z 2:99–152

Neergaard P (1945) Danish species of *Alternaria* and *Stemphylium* taxonomy, parasitism, economical significance. Oxford Univiversity Press, London/Oxford, 559 p

Nehemiah KMA, Deshpande KB (1976) Cellulase production and decomposition of cotton fabric and filter paper by *Alternaria brassicae*. Indian Phytopathol 29:55–57

Noda J, Brito N, Gonzalez C (2010) The *Botrytis cinerea* xylanase Xyn11A contributes to virulence with its necrotizing activity, not with its catalytic activity. BMC Plant Biol 10:38

Oide S, Moeder W, Krasnoff S, Gibson D, Haas H, Yoshioka K, Turgeon BG (2006) NPS6, encoding a non-ribosomal peptide synthetase involved in siderophore-mediated iron metabolism, is a conserved virulence determinant of plant pathogenic ascomycetes. Plant Cell 18:2836–2853

Poapst PA, Ramsoomair BA, Gaurley CO (1979) On the promotion of senescence in *Brassica oleracea* var. Capitata by *Alternaria brassicicola* and by *Botrytis cinerea*. Can J Bot 57:2378–2386

Saharan GS, Mehta N, Sangwan MS (2005) Diseases of oilseed crops. Indus Publishing Company, New Delhi, 643p

Schreven DAV (1953) *Alternaria, Stemphylium* en *Botrytis* a antasting bij Koolzaad (*Brassica napus*). Tijdschr Plantenziekt 59:105–136

Sharma SK, Gupta JS (1978) Seasonal and diurnal variation in the air-spore over a field of Brown Sarson. Indian Phytopathol 31:389–391

Sharma AK, Gupta JS, Dixit RB (1986) Incidence of pathogenic conidia on leaf surface of Yellow Sarson and Taramira. J Indian Bot Soc 65:550–552

Sharma N, Rahman MH, Liang Y, Kav NNV (2010) Cytokinin inhibits the growth of *Leptosphaeria maculans* and *Alternaria brassicae*. Can J Plant Pathol 32:306–314

Suri RA, Mandahar CL (1982) Production of pectolytic enzymes by *Alternaria brassicae*. Indian J Mycol Plant Pathol 12:351–352

Suri RA, Mandahar CL (1984) Secretion of cytokinins *in vivo* and *in vitro* by *Alternaria brassicicola* and their role in pathogenesis. Experientia Basel, Birkhauser 40:462–463

Suri RA, Mandahar CL (1985) Involvement of cytokinin-like substances in the pathogenesis of *Alternaria brassicae* (Berk.) Sacc. Plant Sci 41:105–109

Suri RA, Mandahar CL (1986) Cytokinin-like substances and accumulation of labelled metabolites at infection sites of *Alternaria brassicicola*. Mycopathologia 94:153–156

Suri RA, Mandahar CL, Gill PS (1983) Study of secretion of cytokinins by *Alternaria brassicicola* and their possible role in pathogenesis in mustard. Indian J Plant Pathol 1:117–121

Tewari JP (1983) Cellular alterations in the black spot of rapeseed caused by *Alternaria brassicae*. Phytopathology 73:831. (Abstr.)

Tewari JP (1986) Subcuticular growth of *Alternaria brassicae* in rapeseed. Can J Bot 64:1227–1231

Tsuneda A, Skoropad WP (1978a) Phylloplane fungal flora of rapeseed. Trans Br Mycol Soc 70:329–334

Tsuneda A, Skoropad WP (1978b) Behavior of *Alternaria brassicae* and its mycoparasite *Nectria inventa* on intact and excised leaves of rapeseed. Can J Bot 56:1333–1340

Verma PR, Saharan GS (1994) Monograph on *Alternaria* diseases of crucifers. Saskatoon Research Centre, Technical Bulletin 1994–6E, Agriculture and Agri-Food Canada, Saskatoon, SK, Canada, 162 p

Weimer JL (1924) Alternaria leaf spot and brown rot of cauliflower. J Agric Res 29:421–441

Weimer JL (1926) A leaf spot of cruciferous plants caused by *Alternaria herculea*. J Agric Res 33:645–650

Wong SHJ, Dumas B (2010) *Ste12* and *Ste12*-Like proteins, fungal transcription factors regulating development and pathogenicity. Eukaryote Cell 9:480–485

Xu JR (2000) MAP kinases in fungal pathogens. Fungal Genet Biol 31:137–152

Yao C, Köller W (1994) Diversity of cutinases from plant pathogenic fungi: cloning and sequence analysis of a cutinase gene from *Alternaria brassicicola*. Physiol Mol Plant Pathol 44:81–92

Yao C, Köller W (1995) Diversity of cutinases from plant pathogenic fungi: different cutinases are expressed during saprophytic and pathogenic stages of *Alternaria brassicicola*. Mol Plant-Microbe Interact 8:122–130

5 Epidemiology and Forecasting

5.1 Introduction

Alternaria blight of crucifers occurs and develops in the epidemic form under the combination of a number of factors including availability of susceptible host cultivars, source of virulent pathogen inoculum in the vicinity, and favourable environmental and cultural conditions. Temperature range of 15–25 °C, relative humidity >90 %, wind velocity of 2–5 km/h, and intermittent rains are the most abiotic conducive conditions for development of *Alternaria* blight under field conditions. In addition, closer spacing of 30 × 15 cm, high doses of nitrogen (>80 kg/ha), frequent irrigation and growing of susceptible cultivars further rapidly increase the disease severity in rapeseed–mustard. The disease progression under field conditions as influenced by environmental conditions and dates of sowing has been measured through the development of models. Disease forecasting models have been developed taking into account the leaf wetness period; number of rainy days; minimum, maximum and optimum temperatures and relative humidity; date of sowing; variety and species of *Brassica* crops grown under different agro-ecological conditions.

5.2 Disease Development in Relation to Environmental Conditions

Alternaria blight of Brassicaceae caused by *A. brassicae* is favoured by low temperature, high humidity and splashing rain (Table 5.1) (Dey 1948; Humpherson-Jones and Phelps 1989; Louvet 1958; Louvet and Billotte 1964; Jose et al. 2005). In Canada (McDonald 1959) and Holland (Van Schreven 1953), optimum temperature for the development of *A. brassicae* is reported to be between 20 and 24 °C. In India, a temperature range of 15–25 °C, relative humidity of 70–90 %, intermittent winter rains and wind velocity around 2–5 km/h (Fig. 5.1) have been reported to be most conducive to *Alternaria* blight development in mustard (Ansari et al. 1988; Saharan 1991; Saharan and Kadian 1984; Saharan et al. 1981). In France, Louvet (1958, 1963) observed severe development of *A. brassicae* at a temperature range of 15–22 °C, relative humidity of 80 % for a period of 36 h and under stormy and high wind velocity conditions. Similar conducive conditions have also been observed in India by Awasthi and Kolte (1989) and Chahal and Kang (1979). Domsch (1957)

G.S. Saharan et al., *Alternaria Diseases of Crucifers: Biology, Ecology and Disease Management*,
DOI 10.1007/978-981-10-0021-8_5

reported relative humidity between 95 and 100 % at least for 18 h and temperature of 21–27 °C for three successive days as essential for epidemic development of *Alternaria* disease of rape and cabbage. *Alternaria brassicae* produces abundant spores after a succession of wet and dry periods, which are probably spread over short distances by rain and wind and over greater distances by the action of wind alone during dry weather (Louvet and Billotte 1964). To a very limited extent, slugs can spread the disease by ingesting spores, which remain viable in the alimentary canal and can infect plants when excreted (Hasan and Vago 1966).

According to Awasthi and Kolte (1989), *Alternaria* blight of rapeseed–mustard develops best during rosette to flowering stages; relative humidity from 67 to 73 %, rainfall >70 mm, 5–7 h of sunshine/day and a minimum temperature range of 7–10 °C concomitant with the maximum temperature range of 20–23 °C have been positively correlated (r=0.511–0.805) with the severity of disease. Increase in age beyond 30 days is also positively correlated (r=0.9802) with the increase in disease severity. Indian mustard is susceptible to leaf infection phase during rosette to flowering stage, when the relative humidity 81–94 % is concomitant with total rainy days of 4–11. Low relative humidity with less frequent rain, irrespective of the other factors, does not result in high severity of the leaf infection phase. For *Alternaria* blight severity on pods, relative humidity shows 0.6 % variation, whereas the combined effect of relative humidity and maximum and minimum temperatures accounts for 70.3 % variation (Yadav and Brar

Table 5.1 Weather parameters congenial for *Alternaria brassicae* under (a) field and (b) laboratory conditions (Jose et al. 2005)

Authors	Maximum temperature (°C)	Minimum temperature (°C)	Relative humidity (%)	TFD (days)	Rainfall (mm)	Wind velocity (km/h)
(a) Field conditions						
Awasthi and Kolte (1994), Pantnagar, India		7–10	>67	>6	>70	
Chahal and Kang (1979), Punjab, India			>80			
Dang et al. (1995), Hisar, India	16.3–25.7	1.6–9.7	79.6–96.5			2.5–6.0
Kolte and Singh (1997), Pantnagar, India	20–23	7–10	67–73		>70	
Ansari et al. (1989), Aligarh, India	16.3–25.7	1.6–9.7	79.6–96.5			2.5–6.0
Saharan and Kadian (1984), Hisar, India	18.7–23.14	6.8–10.12	53.8–80.8			4.5–8.2

(b) Laboratory conditions			
Authors	**Optimum temperature**	**Relative humidity**	
Humpherson-Jones and Phelps (1989), Warwick, UK	18–24	91.5–100	
Kadian and Saharan (1984), Hisar, India	25	90	
Kennedy and Graham (1995), Warwick, UK	18–24	87	
Ansari et al. (1989), Aligarh, India	21–30	85–100	

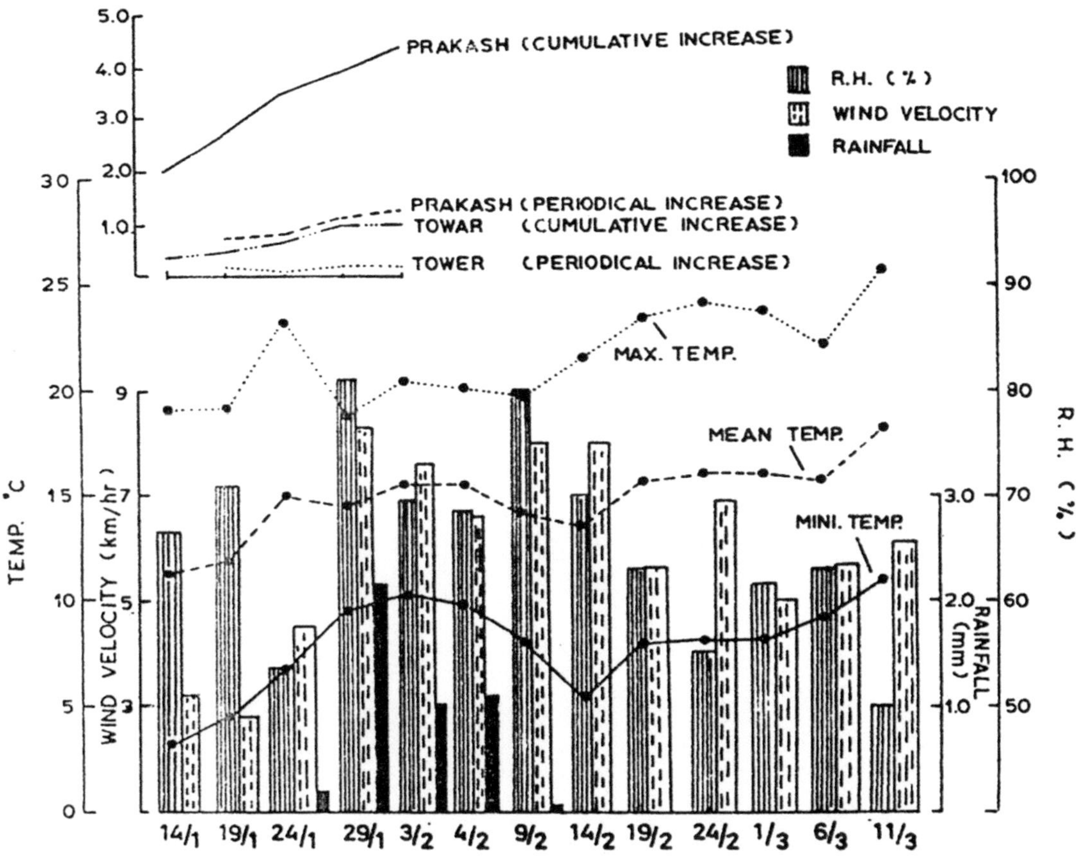

Fig. 5.1 Influence of weather variables on development of *Alternaria* blight (Saharan and Kadian 1984)

2003). Similarly, Chahal (1986) reported that the susceptibility of Brown Sarson to *A. brassicae* increases with the age of the host. Maximum susceptibility of rapeseed plants has been observed at 55–85-day-old plants (Sarkar and Sen Gupta 1978). For maximum infection and disease development in mustard, a minimum period of 4-h leaf wetness is essential. Longer periods of leaf wetness at 25 °C increase the infection frequency (Table 5.2) on the leaves (Kadian and Saharan 1984; Saharan and Kadian 1984; Saharan 1991). Reduced light intensity is more favourable for lesion development. Conversely, spores produced under high light intensities show reduced germination (Kadian 1982; Kadian and Saharan 1984; Mukadam and Deshpande 1979; Meena et al 2011). Awasthi and Kolte (1994) observed that the combined effect of rainfall (RF), relative humidity (RH) and minimum temperature influences blight severity on leaves, while total rainy days, RH and maximum and minimum temperatures have enormous effect on the disease on pods; RH (67 %), total rain duration (TRD) (>6), RF (>70 mm) and minimum temperature (7–10 °C) are more conducive for *Alternaria* blight severity on leaves, whereas the RH and TRD factors significantly influence spread of pod infection.

Dang et al. (1995) during 1988–1990 crop season at Hisar, India, studied the development and temporal progression of *Alternaria* leaf blight and correlated the disease severity with environmental conditions on different rapeseed–mustard cultivars. Results indicated that cumulative intensity increase is significant among the years, cultivars and observation intervals. The maximum increase in *Alternaria* blight is on Yellow Sarson (YSPb-24) and Brown Sarson (BSH-1). However, the progression is least on

Table 5.2 Influence of temperature and leaf wetness on infection of *A. brassicae* (Saharan and Kadian 1984)

[a]Disease intensity after hours of leaf wetness												
	4		8		12		16		20		24	
Temperature (°C)	No. of lesion/ leaf	Size (mm)	No. of lesion/ leaf	Size (mm)	No. of lesion/ leaf	Size (mm)	No. of lesion/ leaf	Size (mm)	No. of lesion/ leaf	Size (mm)	No. of lesion/ leaf	Size (mm)
15	0.1	0.1	0.3	0.2	0.5	0.6	0.6	0.5	0.7	0.7	0.7	0.7
20	0.9	1.0	1.2	1.4	1.6	1.7	1.5	1.9	1.6	1.6	1.7	1.5
25	3.6	5.0	4.2	8.1	6.0	13.9	6.8	14.3	7.0	14.6	7.4	15.2
30	0.2	1.5	0.4	1.3	0.8	1.9	1.5	1.5	0.7	1.7	0.8	1.6

[a]Average of ten leaves

mustard cultivar RH-8113. The variation can be explained up to 58–60 % with combined effect of temperature and wind velocity. When the cumulative increase is compared with the prevailing environmental conditions, the disease progresses fairly well with maximum temperature ranging between 16.3 and 25.7 °C, minimum temperature between 1.6 and 9.7 °C, maximum RH between 79.6 and 96.5 % and wind velocity between 2.5 and 6.0 km/h. The disease progression curves (Fig. 5.2) for different varieties showed linear relationship between disease intensity and crop growth stage (R^2=0.97 to 0.99). Hong and Fitt (1996) also reported a strong correlation between *Alternaria* blight development with leaf wetness period and temperature. Meena (2005) and Meena et al. (2011) reported that disease severity increases with delay in date of sowing. The value of AUDPC (area under disease progress curve) and 'r' value (apparent infection rate) are more in the variety 'Varuna' with the delayed sowing. The severity of *Alternaria* blight is significantly lower in October-sown crop (Kumar and Kumar 2006). The spread of the disease is more in the broadcasting method than in the line sowing (45 cm). The disease intensity also decreased with the addition of K (40 kg/ha) along with the recommended dose of fertilizers. Chattopadhyay et al. (2005) analysed the data for *Alternaria* blight progression and development from eight locations using cultivar 'Varuna' sown on ten dates at weekly intervals and revealed that the disease's first appearance on leaves occurred 42 and 139 days after sowing (DAS). The disease on pods appeared later between 67 and 142 DAS and being highest at 99 DAS. Severity of *Alternaria* blight on leaves was positively correlated to a daily temperature maximum (*T* max.) of 18–27 C, daily temperature minimum (*T*min) of 8–12 °C, daily temperature mean (*T* mean) of >10 °C, relative humidity morning (RH mor) >92 %, relative humidity evening (RH even.) >40 % and RH mean of >70 % in the preceding week. Disease severity on pods was favoured by a daily T max of 20–30 °C, daily T mean of >14 °C, RH mor >90 %, daily RH mean of >70 %, sunshine >9 h and leaf wetness >10 h. Progression of *Alternaria* on cultivar Varuna sown on different dates at Bharatpur (India) is presented in Fig. 5.3.

Alternaria brassicicola fails to produce significant infections at 10 °C after 96 h, while *A. brassicae* produces lesions on the host under similar conditions (Humpherson-Jones et al. 1983). *Alternaria brassicicola* lesions on overwintered leaf litter of *B. oleracea* grown for seed production produce high concentrations of spores in the spring and are able to initiate new infections on foliage and subsequently on inflorescence and siliquae. A vertical disease gradient develops in maturing crops, the lowest siliquae becoming infected first and infection spreading

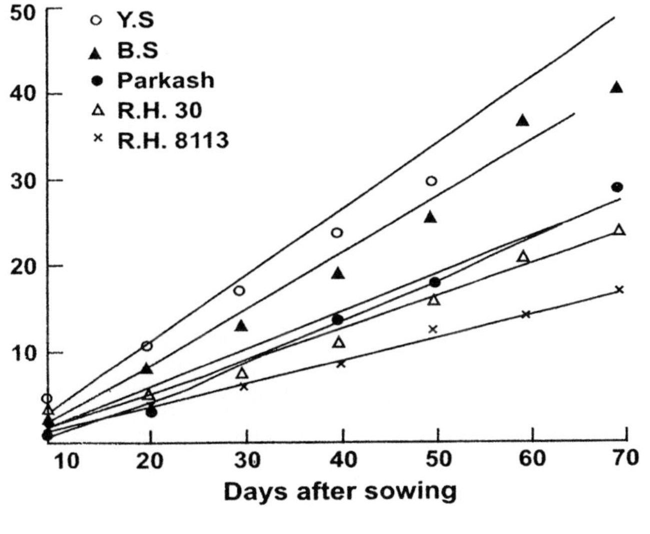

Fig. 5.2 Temporal progression of *Alternaria* leaf blight on five varieties of rapeseed–mustard during 1988 to 1990 crop seasons (pooled data) (Dang et al. 1995)

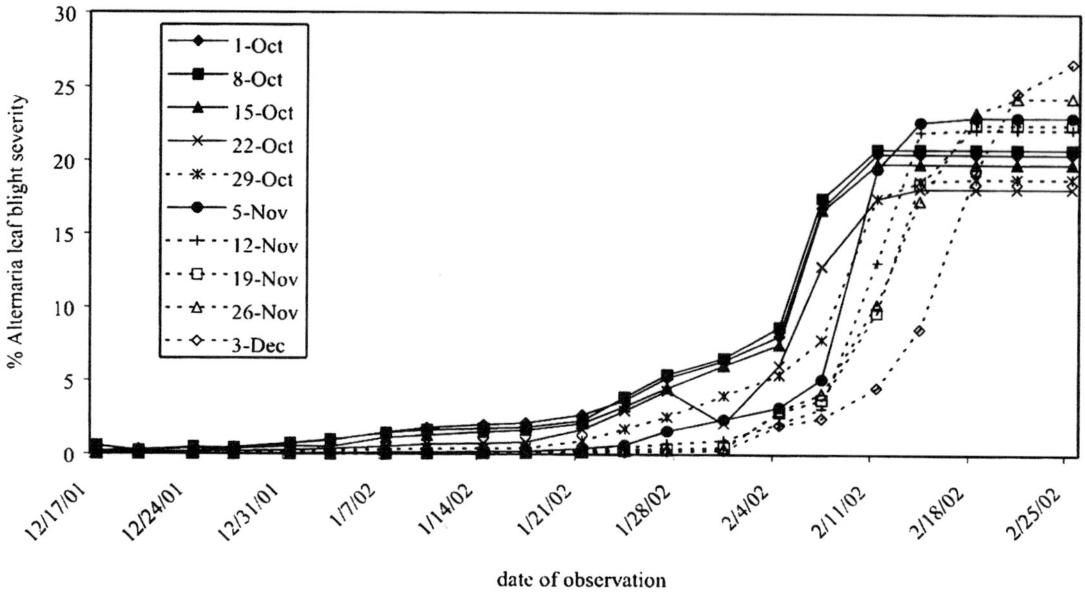

Fig. 5.3 Progress of *Alternaria* leaf blight severity on cultivar Varuna of Indian mustard in different dates of sowing at Bharatpur in 2001–2002 (Chattopadhyay et al. 2005)

slowly upwards. Spores are produced abundantly after 20 h leaf wetness at a mean temperature of 13 °C or more. Their release is stimulated by a reduction in RH but inhibited at a constant high RH resulting in a daily cycle in air spore concentrations with minimum occurring in the early morning and maximum in the early afternoon (Figs. 5.4 and 5.5); (Humpherson-Jones and Maude 1982). Peaks of high spore concentration are usually associated with dry days, shortly after rain, high temperature or high wind speed for conidial dispersal of *A. brassicicola* on Chinese cabbage at 10 h. The number of conidia trapped at a height of 25 cm above ground level is greater than that at 50, 75 and 100 cm (Fig. 5.6) (Chen et al. 2003). Sporulation in *A. brassicae,* and *A. brassicicola* on naturally infected leaf discs of oilseed rape, and cabbage requires RH>91.5 % and 87 %, respectively. The optimum temperature for sporulation is 18–24 °C for *A. brassicae* and 20–30 °C for *A. brassicicola* at which temperature both fungi produce spores in 12–14 h., and in temperature above 24 °C, sporulation in *A. brassicae* is inhibited. White light inhibits sporulation in *A. brassicae* with the degree of inhibition increasing with increasing light intensity (Humpherson-Jones and Phelps 1989).

The primary infection occurs on the cotyledonary leaves of mustard forming the source of secondary infection for the entire crop under field. A minimum of 4 h of leaf wetness is required for infection. Increased leaf wetness duration at 25 °C increases infection and spreads the disease rapidly. Under favourable temperature conditions, and presence of dew, the spores infect other parts of the plant. The infection occurs through the stomata, and under favourable climatic conditions, the new lesions arise within 4–6 days bearing spores. The pathogen penetrates the tissues of the pods and infects the seed (Saharan et al. 2005). The congenial factors for *Alternaria* spores germination have been reported as darkness or low light intensity (<1000 lx), 25 °C temperature and more than 90 % RH.

The favourable environmental factors for disease progression under field conditions have been reported at temperature 12–25 °C, RH>70 % with intermittent winter rains or irrigation, wind velocity between 2 and 5 km/h, closer plant spacing (30 × 15 cm) and high doses of nitrogen

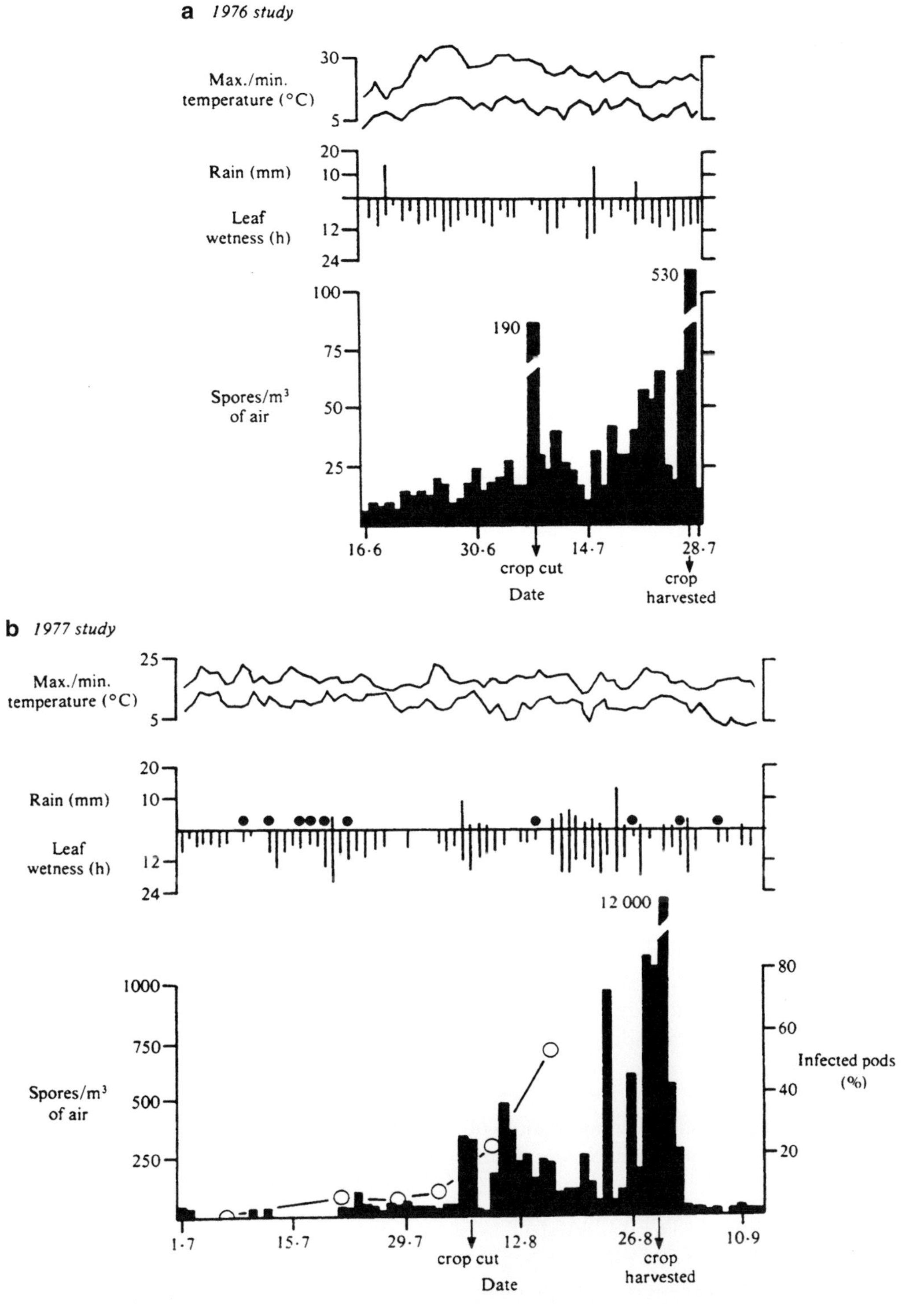

Fig. 5.4 The mean daily concentration of *Alternaria brassicicola* spores in the air within *Brassica oleracea* seed production crops: (**a**) 1976 and (**b**) 1977 study; ● >0.2 mm rain; 0–0 infected pods (Humpherson-Jones and Maude 1982)

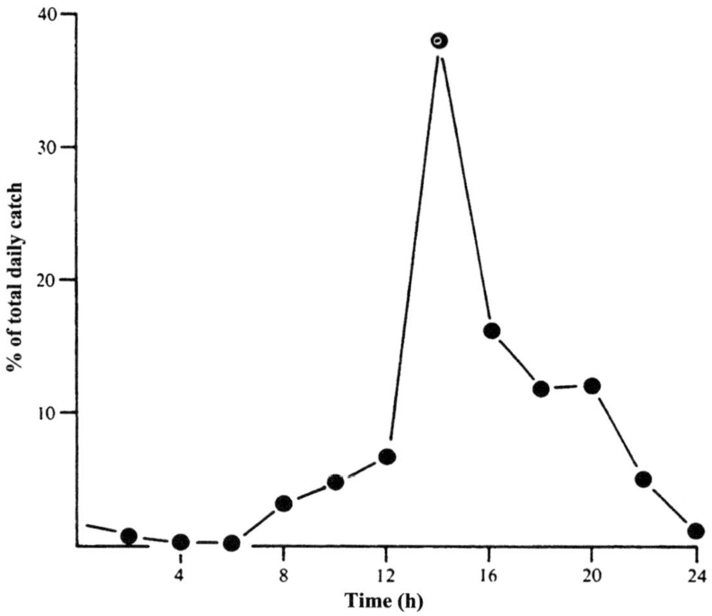

Fig. 5.5 The hourly concentration of *Alternaria brassicicola* spores in the air within a cabbage seed crop. Mean values of seven high spore count days (Humpherson-Jones and Maude 1982)

(80 kg /ha). The temperature maximum of 26–29 °C with RH Av >65 % favours the disease development (Sangeetha and Siddaramaiah 2007).

Under Haryana conditions, a raya crop sown in the last week of October recorded 52 % disease, while that sown in the third week of November had only 15.5 % disease (Saharan et al. 2005). Under Hisar conditions, *Alternaria* spores were trapped in a 7-day volumetric spore trap, 10–11 days before the disease appearance and their concentration increased and reached maximum in March. The spores were trapped maximum during 10.00 am to 2.00 pm (46 % of total spores) and minimum during 10.00 pm to 6.00 am. *Alternaria* spore concentrations declined sharply after 2.00 pm (Figs. 5.7 and 5.8) (Singh 2005). Based on the concentration of spores trapped, a prediction equation was developed as follows:

$$\begin{aligned}\mathrm{ALTn} = 21.50 + 1.164\mathrm{ALTc} + 0.128\mathrm{ALTp} \\ - 0.165(\mathrm{T\ max}) - 0.154(\mathrm{T\ min}) \\ - 0.203(\mathrm{RH\ mor}) - 0.095(\mathrm{RH\ even}),\end{aligned}$$

where

ALTn = Expected *Alternaria* blight;
ALTc = Disease intensity in current week;
ALTp = Disease intensity of previous week. The R^2 value recorded was 0.84.

Compared to the older plants, younger plants of radish are less susceptible to *A. raphani*. The disease intensities in 28 days old and younger plants were 32.5 % and 17 %, respectively. However, increase in age beyond 28 days did not have much influence on disease development. There is a direct correlation between the host age and disease intensity (Table 5.3). In case of *A. alternata*, 30–35-day-old radish plants have also been reported highly susceptible to *Alternaria* blight (Suhag et al. 1985). Disease intensities data in Table 5.4 reveal that 10-day-old cultures of *A. raphani* are more virulent than the older cultures; as the age of the culture increases, the disease intensity decreases with least in 30-day-old culture. This is also true with several other pathogens where young cultures are in more active metabolic

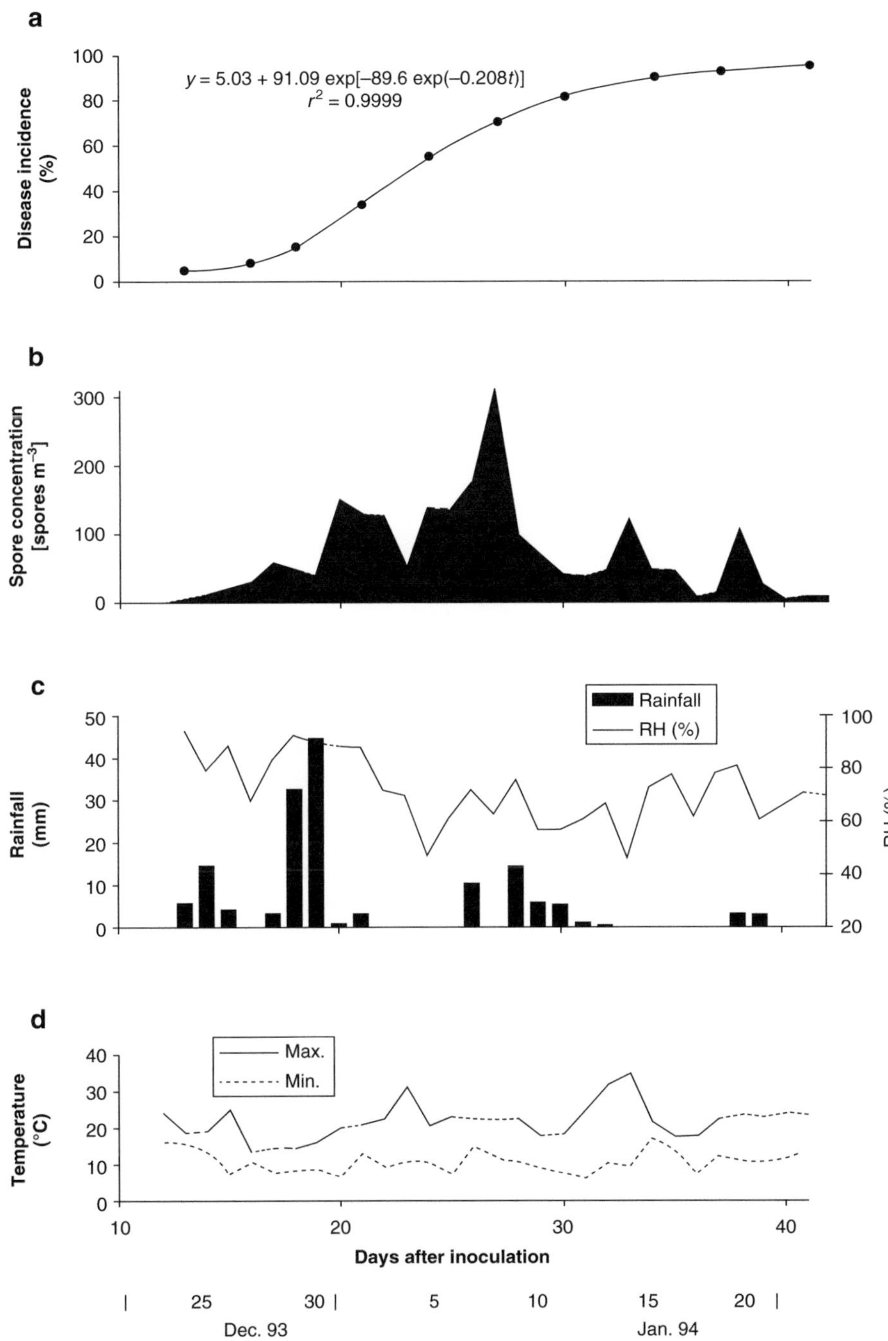

Fig. 5.6 (**a**) Disease incidence (%) of dark leaf spot on Chinese cabbage and the Gompertz growth curve and model; (**b**) spore concentration of *Alternaria brassicicola* in the air; (**c**, **d**) Meteorological data during the experimental period in summer (1993–1994) (Chen et al. 2003)

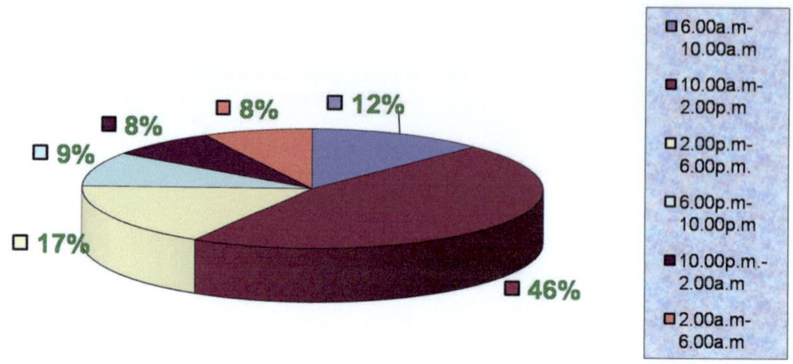

Fig. 5.7 Spore percentage (*A. brassicae*) trapped over crop canopy during different time interval of the day (Singh 2005)

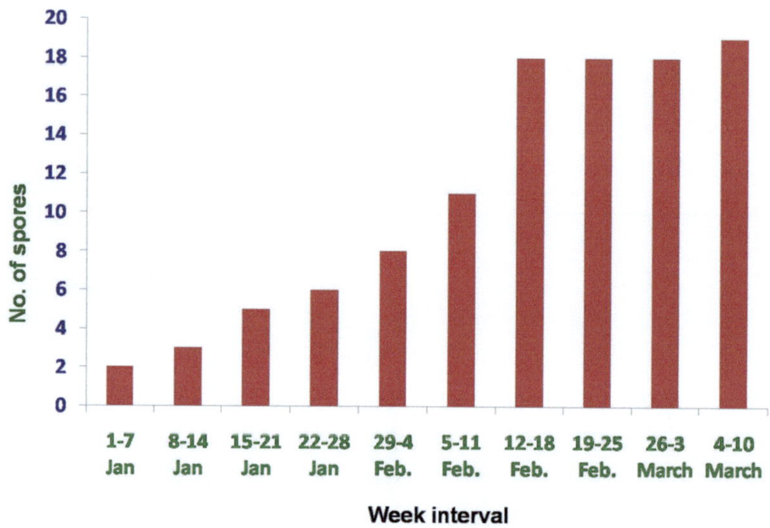

Fig. 5.8 Weekly average number of spores of *Alternaria brassicae* (Singh 2005)

stage than the older ones. Disease intensity is directly correlated with the inoculum load. The maximum disease intensity (25.1 %) is recorded with *A. raphani* inoculum load of 2.0×10^3 spores ml^{-1}; further increase in inoculum load did not cause significant increase in disease intensity (Table 5.5) (Sangwan et al. 2002).

The effects of humidity, seed infection, temperature and nutrient stress on development of *A. brassicicola* seedling blight of cabbage have been studied under controlled environment chambers by Bassey and Gabrielson (1983a). Cotyledon lesions are more severe with extended moist incubation following wound inoculation. The temperature of 30 °C is optimum for wire-stem disease from naturally infected seed. Natural wire stem does not occur below 15 °C even with a heavily infected seed lot, and little occurs below 20 °C with a moderately or lightly infected seed lot.

According to Singh (1988), the conidia of *Alternaria* appear in the air 8–13 days prior to the first onset of disease on the crop. The high percentage of disease incidence and disease intensity indices is closely related to the high percentage of *Alternaria* spore load in the air. Maximum incidence of mustard blight depends upon high RH (77–94 %) and high temperature of 28–30 °C. *Alternaria alternata* causing leaf and pod blight of radish develops fast when mean temperature is 20.6–23.8 °C and relative humidity from 57.9 to 67.4 % (Singh and Suhag 1983; Suhag et al. 1985). Under field conditions, *A. raphani* infection progresses rapidly at 22–26 °C. At high soil moisture content, infection is less at 18 °C (Atkinson 1950). *Alternaria* blight development on *Crambe* is dependent on warm and damp weather (Czyzewska 1971).

Table 5.3 Effect of age of radish plants on blight intensity caused by *Alternaria raphani* (Sangwan et al. 2002)

Age of plants at inoculation (days)	Disease intensity (%)	
7	7.3	(1.7)
14	14.4	(6.7)
21	16.7	(8.7)
28	34.5	(32.5)
35	32.1	(28.7)
42	29.8	(25.0)
CD (p=0.05)		(12.9)

Values in the parentheses are arc transformed

Table 5.4 Effect of age of culture (*Alternaria raphani*) on blight of radish (Sangwan et al. 2002)

Age of culture (days)	Disease intensity (%)	
10	28.2	(22.4)
15	26.6	(20.1)
20	21.4	(13.2)
25	19.6	(11.4)
30	19.4	(10.1)
CD (p=0.05)		(5.1)

Values in the parentheses are arc transformed

Table 5.5 Effect of inoculum (*Alternaria raphani*) load on blight intensity in radish (Sangwan et al. 2002)

Inoculum load (×10^3 conidia / ml^{-1})	Disease intensity (%)	
0.0	6.3	(1.2)
0.5	13.8	(5.7)
1.0	20.7	(12.5)
1.5	23.9	(16.5)
2.0	30.0	(25.1)
2.5	29.8	(24.8)
3.0	29.6	(24.5)
CD (p=0.05)		(4.9)

Values in the parentheses are arc transformed

5.3 Disease Development in Relation to Nutrition and Cultural Conditions

Closer spacing (30 × 15 cm), high doses of nitrogen (>80 kg N/ha) and frequent irrigations are known to rapidly increase severity of *Alternaria* blight of mustard (Kadian and Saharan 1988; Saharan 1991, 1992a; Saharan and Kadian 1984). In Prague (Czechoslovakia), higher doses of NPK increased *A. brassicae* infection in winter rape pods. Similarly, top dressing with nitrogen in spring significantly increased *Alternaria* intensity on rape pods (Stankova 1972). Planting time has a major influence on the incidence of disease in mustard crops. In Haryana (India), 52 % disease incidence has been recorded on mustard crop sown in the last week of October. However, when the crop is sown in the third week of November, infection fell to 15.5 % (Saharan 1984). Severity of *Alternaria* blight of mustard decreases with delay in sowing at Bharatpur (Meena et al. 2011). The *Alternaria* pod blight of radish seed crop is significantly less in the normal season transplanting of 15th December in the Punjab (India), whereas *Alternaria* twig blight is significantly less in the early transplanting of 15th November (Sandhu et al. 1984, 1985).

The dwarfing agent, Cycocel (CCC), sprayed on rapeseed crop tends to decrease infection of *A. brassicae,* whereas alar increases the infection (Strzelczyk and Rozej 1974). The fungicide Calixin (75 % w/w tridemorph) applied at 1.125 ml in 4-L water predisposes cabbage plants to *A. brassicae* and *A. brassicicola* infection (Munro 1984). Incidence of dark leaf blight (*A. brassicicola*) on Chinese cabbage is higher in plots with row orientation parallel to the average wind direction than in plots with row orientation at right angles to the average wind direction (Chen and Price 2002).

The role of NK (N90 kg ha^{-1}+K 40 kg ha^{-1}) in decreasing the severity of *Alternaria* blight is more pronounced than PK (P 40 kg ha^{-1}+K 40 kg ha^{-1}), NP and K (40 kg ha^{-1}) applications. The decrease in *Alternaria* blight severity due to K is probably due to increased production of phenolics in plants, which inhibit conidial germination and decrease sporulation of *A. brassicae*. The decrease in the severity of *Alternaria* blight due to NK application consistently increases seed yield more than 68 % than those of control and other treatments. The K-fertilized plants also show increased resistance to lodging, increased 1000 seed weight and decreased seed infection. Seeds obtained from K-fertilized plants show good seed germinability and vigorous seedling growth (Sharma and Kolte 1994).

5.4 Disease Development in Relation to Host Resistance

Alternaria blight progression on highly susceptible (*B. juncea* cv. Prakash) and moderately resistant (*B. napus* cv. Tower) cultivars of rapeseed–mustard is governed by reduction in infection frequencies coupled with reduction in number and size of lesion formation due to host resistance (Figs. 5.9 and 5.10). In highly susceptible cultivar Prakash, the onset of disease occurs earlier, with a greater abundance of large lesions. Such cultivars have a shorter latent period, abundant conidial production per lesion and higher infection rate (0.46) in comparison to moderately resistant cultivar Tower (Saharan 1991, 1992b; Saharan and Kadian 1983). On cultivar Tower, infection is delayed up to 25 days with the occurrence of very few lesions of restricted size (Table 5.6). Tower gets 15.6 % of leaf infection compared to 38.5 % in Prakash.

The rate of disease development is faster on varieties belonging to *B. juncea* (RH-30, RH-8113, RH-8695, RH-8546) and *B. campestris* (YSPb-24, BSH-1, Candle, Shiva) compared to *B. carinata* (HC-2, HC-9001), *B. napus* (GSH-1) and *B. alba* (Mehta et al. 2008). Kumar (2008) observed that three genotypes, viz. PR-8988, PR-9024 and Kranti, have partial resistance and show lowest blight severity. The size of the spots is positively and significantly correlated with disease severity, leaf defoliation and sporulation.

5.5 Disease Development in Relation to Flea Beetle

Transmission of *A. brassicicola* to cabbage by flea beetles (*Phyllotreta cruciferae*) has been reported by Dillard et al. (1998). *Alternaria* leaf spot develops on plants that are infested with the contaminated flea beetles. Faeces obtained from flea beetles that fed on cabbage infected with

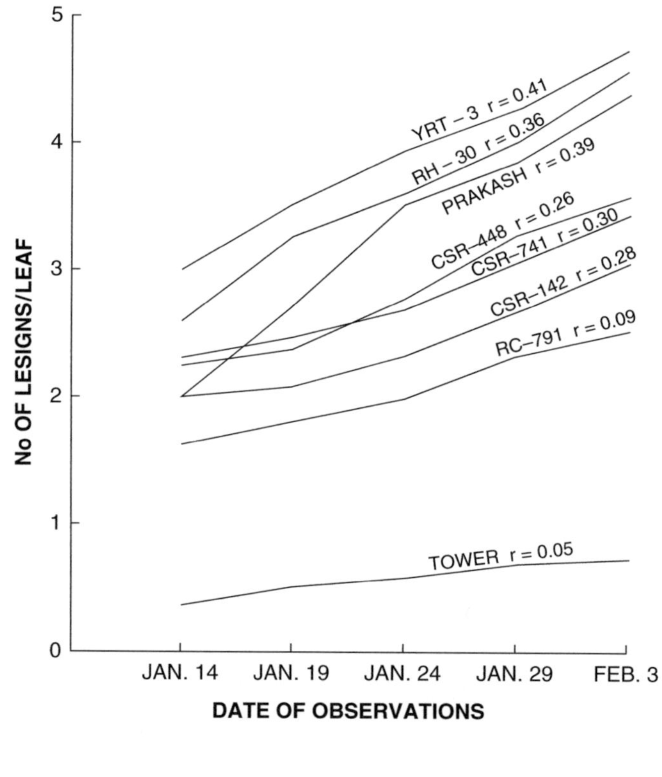

Fig. 5.9 Increase in number of *Alternaria* lesions on rapeseed–mustard cultivars (Saharan and Kadian 1983)

Fig. 5.10 Increase in size of *Alternaria* lesions on rapeseed–mustard cultivars (Saharan and Kadian 1983)

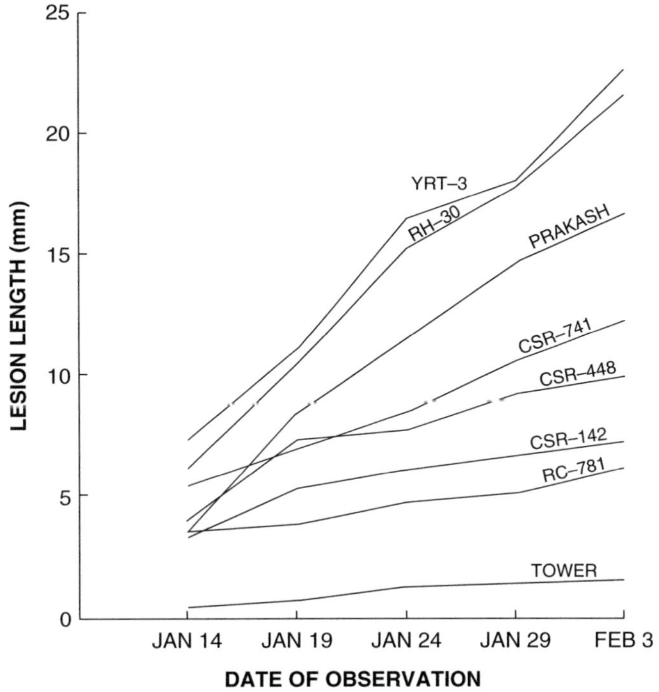

Table 5.6 Factors influencing resistance/susceptibility of different cultivars of rapeseed–mustard against *Alternaria brassicae* (Saharan and Kadian 1983)

Cultivars	Leaf surface	Stomata[-1] No. (per sq. cm)	Stomata[-1] Size (mm)	Lesions/leaflet[2] No.	Lesions/leaflet[2] Size (mm)	Sporulation[3] per lesion	Incubation [4] period (days)	Latent period (days)
Tower	Upper	145	15.7	1.8	1.9	80	10	16
	Lower	210	25.9					
RC-781	Upper	166	17.3	3.3	5.7	92	8	12
	Lower	245	26.7					
CSR-448	Upper	196	19.2	3.7	10.7	120	7	10
	Lower	303	27.9					
CSR-741	Upper	189	19.3	6.3	14.6	125	7	9
	Lower	292	20.0					
CSR-142	Upper	199	19.3	4.4	10.6	120	8	11
	Lower	306	28.0					
YRT-3	Upper	225	22.0	10.2	22.3	250	6	9
	Lower	403	30.8					
RH-30	Upper	221	22.0	8.1	19.1	240	6	9
	Lower	392	30.6					
Prakash	Upper	229	22.7	8.7	18.9	260	6	8
	Lower	413	31.2					

*1 and 3 average of 100 observations
**2 and 4 average of 10 leaves

A. brassicicola contain intact and broken conidia of *A. brassicicola* and undigested pieces of cabbage leaf. The conidia are viable after passing through the flea beetles, as evidenced by their germination on glass slide used for collecting the faeces. Conidia of *A. brassicicola* are observed by scanning electron microscopy on all parts of the flea beetle bodies including wings, mouthparts, antennae and legs.

5.6 Disease Development in Relation to Barrier Crops

When oats were used as barrier crops to control the secondary spread of *Alternaria* blight of rapeseed–mustard under Hisar, Indian conditions, the percent disease intensity decreased in subsequent strips of the crop. The percent reduction in disease intensity was 23.6 % in first strip, which increased to 66.2 % in the fifth strip over the inoculated strip (Table 5.7). The barrier crop (oat) was found superior over three sprays of Dithane M-45 for the control of disease (Singh 2005).

5.7 Models to Describe the Progress of the Disease

Several models have been proposed to describe the progress of compound interest diseases like *Alternaria* disease of crucifers (Fontem et al. 1991).

1. Gompertz equations: $Y = -\ln\left[-\ln\left(y\right)\right]$
2. Logisitic equations: $Y = -\ln\left[y/\left(1-y\right)\right]$

   These equations were compared by Berger (1981) to describe plant disease progress.
3. Weibull function: $Y = \left(\ln\left[1/\left(1-y\right)\right]\right\}\hat{}\,1/c$

   It has received much attention recently because of its flexibility and simplicity (Fontem 1986).
4. Gregory power function: $y = axb$, in which y is the proportion of disease at x units of distance from the source, a is the value of y at $x = I$ and b is the rate of change in y with change in x; i.e. b is an estimate of the slope of the gradient, and the value of b is usually negative. The transformation equation is $\log(y) = \log(a) + b \log(x)$. Since this model cannot be used to predict disease at the inoculum source (distance $x = 0$), Gregory's model was modified by Mundt and Leonard (1985) to include a translation factor (m) [$y = a(x+m)b$], which allows the estimation of disease at the source when $x = \text{o}$. Values of m are approximately the radii of the sources.
5. Kiyosawa and Shiyomi (1972) model: [$y = a \exp(-bx)$].

   This model has been used to describe spore dispersal in cultivars with multilines.
6. Lambert et al. (1980) model: [$y = a \exp(ax\hat{}n)$] with mixed power and exponential terms and a shape parameter n.
7. Maffia (1985) hollow-shaped curves: [$y = ax^b \exp(nx)$]. Sometimes $1n[y/(1-y)]$ or $-1n[-1n(y)]$ is plotted over $\log_{10}(x)$ or x to linearize hollow-shaped curves.

To illustrate some possibilities, Fontem et al. (1991) evaluated progress and spread of dark leaf spots in three cabbage cultivars during two seasons by fitting representative curves of eight gradient models (Fig. 5.11).

Three growth models (logistic, Gompertz and Weibull) are fitted to the severity values by nonlinear regression. The shape parameter (c) for the Weibull function averages 3.61. The average daily epidemic rates with the logistic model are $K = 0.06$ in the winter and $K = 0.11$ in the spring.

Table 5.7 Effect of barrier crop (oat) on the development of *Alternaria* blight of rapeseed–mustard (Singh 2005)

Disease intensity in inoculated strip (%)	Disease intensity in the subsequent strips intercepted by barrier crop (%)				
	1	2	3	4	5
71.6	54.7 (23.6)	46.5 (35.0)	38.9 (45.6)	31.7 (55.7)	24.2 (66.2)

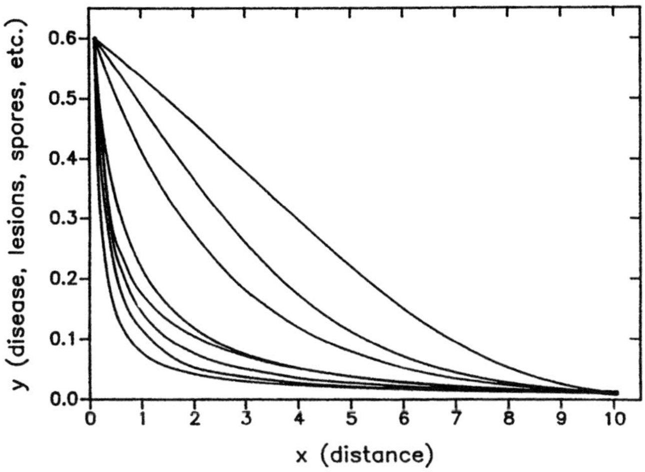

Fig. 5.11 Representation curves of eight gradient models. The curves from the bottom to the top of the figure at $x=1.0$ were generated by the following models: $y=ax^b$(Gregory); $\ln[y/(1-y)]$vs.log (x); $y=a\ \exp\ (bx^n)$, $n=0.2$ (Lambert et al.,); $Y=ax^b\ \exp(nx)$, $n=-0.2$ (Hoerl); $-\ln\ [-\ln(y)]$ vs. log (x); $y=a\ \exp\ (-bx)$ (Kiyosawa and Shiyomi); $\ln[y/(1-y)]$ vs. x; and $-\ln\ [-\ln(y)$ vs. x. All curves began with proportion $y=0.6$ at distance $x=0.1$ and decreased to $y=0.01$ at $x=10.0$. Different curve shapes are possible with the models of Lambert et al. and Hoerl by using other values for the shape parameter (n). The curve shown for the Gregory model typifies many of the steep gradients observed for *Alternaria brassicicola* on cabbage in which y=disease proportion and x=metres (Fontem et al. 1991; Berger 1981)

Final disease severity (yt) at the source averages 0.52 in the winter and 0.97 in the spring. The nearly flat gradients from 1 to 6.7 m are fitted satisfactorily by each of above-mentioned gradient models, but the total gradient is described adequately only by the modified model of Gregory and the Hoerl function.

The three models are modified as:

1. Logistic (Berger 1981): $Yt=1/\{1+\exp\ [a+k_1t]\}$
2. Gompertz (Berger 1981): $Yt=\exp[-a\exp(-Kgt)]$
3. Weibull (Pennypacker et al. 1980): $Yt=1-\exp\{-Kw(t-a)]^C\}$

In these models, 'Yt' is the disease proportion at time t, and k is the rate parameter. The term "a" in the three models is a parameter for initial disease; in the logistic model, $a=-\ln\ [\mathrm{Yo}/(1-\mathrm{Yo})]$, and in the Gompertz model, $a=-\ln\ (\mathrm{Yo})$. In the Weibull function, a positions the curve on the time axis, and c is the curve-shape parameter. In these analyses, 'Yo' is the amount of disease at first observations, and the maximum amount of possible disease (Y max) is assumed to be 1.0. The R^2 and the residual sum of squares are used to evaluate the magnitude of variation among the data that is explained by the model (goodness of fit) (Cornell and Berger 1987; Daniel and Wood 1980). The pattern of scatter when residuals (Y observed – Y predicted) are plotted vs. time is used to confirm the suitability of the model (Cornell and Berger 1987). The progress of disease among cultivars is compared with observed initial disease (Yo), epidemic rate (k), final disease severity (Yf) and the area under the disease progress curve (AUDPC) (Shaner and Finney 1977).

Similar statistical methods are used to evaluate models for the spread of disease. A non-linear curve-fitting programme (Berger 1988) is used to evaluate five gradient models:

1. Gregory (1968): $y=ax^b$
2. Modified Gregory (Mundt and Leonard 1985): $y=a(x+m)^b$
3. Kiyosawa and Shiyomi (1972): $y=a\exp(-bx)$
4. Lambert et al. (1980): $y=a\ \exp\ (-bx^n)$
5. Hoerl (Daniel and Wood 1980; Maffia 1985): $y=ax^b\ \exp(nx)$

In these models, 'Y' is the proportion of disease severity at x units of distance from the inoculum source and b is the slope parameter. In the model of Gregory (1968), a is the disease proportion at one unit of distance from the source, whereas in the modified Gregory (Mundt and Leonard 1985) model, a is the disease proportions at 1-m units of distance. In the models of Lambert et al. (1980) and Hoerl (Daniel and Wood 1980), n is a parameter that modifies the curve shape, and this parameter increases the flexibility and applicability of the models if curve shapes vary.

Three other gradient models were examined in their linearized form because the models contained logarithms both to base e ($1n$) and to the base 10 (log), and the non-linear equations were rather complex.

1. Logistic gradient model (Berger and Luke 1979):
   $\left[y/(1-y)\right]=\ln\left[a/(1-a)\right]-b\log(x)$
2. Gompertz gradient model (Danos et al. 1984):
   $-\ln\left[-\ln(y)\right]=\ln\left[-\ln(a)\right]-b\log(x).$
3. A variant of the logistic gradient model (Minogue and Fry 1983):
   $\ln\left[y/(1-y)\right]=\ln\left[a/(1-a)\right]-bx.$

In these regressions, 'a' is the average proportion of disease on the plants closest (0.3 m) to the inoculum source. These transformations are compared by the R^2 and for significance of the slope of the regression by 't' test.

The volume under the disease progress surface (VUDPS), as suggested by Maffia (1985), and the isopathic rate (rate of movement in space of a given level of disease severity) (Berger and Luke 1979) are calculated from the progress and spread characteristics of the disease. The VUDPS is calculated from the AUDPC at each point in space as:

$$\text{VUDPS}=\sum\left\{\left[\left(A_i+1+A_i\right)/2\right]\left[Xi+1-X_i\right]\right\}$$

where A_i=the AUDPC at Xi units of distance from the source. The isopathic rates are calculated from the source to 6.7 m for the y=0.1 isopath in the winter season and y=0.05 isopath in the spring season.

The above-mentioned models have been applied in modified form by several *Brassica* scientists to forecast *Alternaria* disease of crucifers taking into account numbers of environmental, host age, variety, species, date of planting, disease on leaf and pods, time of first appearance of disease and other variables, which influence the disease development under field conditions (Magarey et al. 2005; Chattopadhyay et al. 2005; Mahapatra and Das 2014; Kumar et al. 2013; Awasthi and Kolte 1994; Dang et al. 2006; Mehta et al. 2002; Mehta 2014).

5.8 Disease Forecasting

A simple generic infection model has been developed for predicting infection periods by fungal foliar pathogens. The model is designed primarily for use in forecasting pathogens that do not have extensive epidemiological data. Most existing infection models require a background epidemiological data set, usually laboratory estimates of infection at multiple temperature and wetness combinations. The model developed can use inputs based on subjective estimates of the cardinal temperatures and wetness duration requirement. These inputs are available for many pathogens or may be estimated from related pathogens. The model uses a temperature response function, which is scaled to the minimum, and optimum values of the surface wetness duration requirement. The minimum wetness duration requirement (W min) is the number of hours required to produce 20 % disease incidence or 5 % disease severity on inoculated plant parts at a given temperature. The model was validated with published data from 53 controlled laboratory studies, each with at least four combinations of temperature and wetness. Validation yielded an average correlation coefficient of 0.83 and a root mean square error of 4.9 h., but there was uncertainty about the value of the input parameters for some pathogens. The value of W min varied from 1–48 h. and was relatively uniform for species in the genera *Alternaria. Alternaria brassicae* requires a W min of =6–8 h, T opt 18 °C, W max 22 h, *T* min 26 °C and T max

35 °C. Operationally, infection models may use hourly or daily weather inputs. In case of the former, information also is required to estimate the critical dry period interruption value, defined as the duration of a dry period at relative humidities <95 % that will result in a 50 % reduction in disease compared with a continuous wetness period. Pathogens are classified into three groups on their critical dry period interruption value. The infection model is being used to create risk maps of exotic pests for the US Department of Agriculture's Animal and Plant Health Inspection Service (Magarey et al. 2005).

Experiments were laid out at eight locations, viz. Bharatpur, Mohanpur, Berhampur, Pantnagar, Dholi, Kangra, Faizabad and New Delhi, with *Brassica juncea* cultivar 'Varuna' and an important cultivar in the locality sown on ten dates at weekly intervals to determine epidemiology and forecasting of *Alternaria* blight of *Brassica* in India. First appearance of *Alternaria* blight disease (*Alternaria brassicae*) on leaves of mustard occurred between 42 and 139 days after sowing (DAS); 44–72, 42–61, 69–83, 45–60, 67–84 DAS having higher frequencies in 1999–2000, 2000–2001, 2001–2002, 2002–2003 and 2003–2004, respectively, being highest in respective years at 45, 46, 75, 45, 76 DAS. The disease first appeared on pods between 67 and 142 DAS, being highest at 99 DAS. Severity of *Alternaria* blight disease on leaves was positively correlated to a maximum daily temperature of 18–27 °C, minimum daily temperature of 8–12 °C, daily mean temperature >10 °C, morning relative humidity (RHmor) >92 %, >40 % afternoon RH and mean RH of >70 % in the preceding week. Disease severity on pods was favoured by a maximum of 20–30 °C, daily mean temperature of >14 °C, morning RH of >90 %, daily mean RH of >70 %, >9 h of sunshine and >10 h of leaf wetness. Temperature and RH conditions favourable to disease development noted in the field matched with the laboratory findings. Regional- and cultivar- specific models could predict the crop age at which *Alternaria* blight would first appear on leaves and pods, the highest blight severity on leaves and pods and the crop age at which blight severity would be highest on leaves and pods at least 1 week ahead of the first appearance of the disease on the crop. This will allow making decision for timely and effective fungicidal sprays (Chattopadhyay et al. 2005).

The stepwise multiple regression analysis (MRA) was carried out to determine the meteorological parameters influencing variation in disease severity of *Alternaria* leaf blight (*Alternaria brassicae* and *A. brassicicola*) of mustard. Disease severity estimates (Y) was considered as dependent variable, whereas other weather parameters like temperature maximum (T max) and temperature minimum (Tmin), RH maximum (RH max) and RH minimum (RH min), total rainfall (RT), wind velocity evening (WV evening) and morning (WV morning), vapour pressure noon (VP noon) and morning (VP morning) and bright sunshine hours (BSH) were used as independent variables. The weather variables were found to influence the disease severity differently at five different sowing dates in the two consecutive years. The Gompertz equation was best linearized with the disease progress data followed by the logistic and the untransformed data sets. The linear prediction equations are (1) $Y = 3.203 - 0.356(\text{T min}) + 0.015\ (\text{RH min})$ for 20th October sowing; (2) $Y = -1.929 + 1.634(\text{WV morning}) + 0.067(\text{RT})$ for 5th November sowing; (3) $Y = -121.91 + 1.57(\text{T min}) + 1.083(\text{RH max}) + 0.29(\text{RH min}) - 2.27\ (\text{VP noon}) - 0.61\ (\text{VP morning}) - 17.17(\text{WV evening}) + 17.83(\text{WV morning}) + 1.65(\text{BSH}) + 0.113(\text{RT})$ for 20th November sowing; (4) $Y = -5.131 + 0.25\ (\text{VP noon}) + 0.256\ (\text{BSH}) + 0.057(\text{RT})$ for 5th December sowing and (5) $Y = -3.19 + 0.235(\text{Tmin})$ for 20th December sowing. Gompertz transformation is the best for linearizing and prediction of disease severity on all dates of sowing, and for early sowing, T min, RH min and WV morning influenced the disease progression (Mahapatra and Das 2014) (Table 5.8).

Models were developed for forewarning crop age at first appearance of *Alternaria* blight (Y_1), crop age at peak severity (Y_2) and maximum severity of *Alternaria* blight (Y_3) for different locations. The extent of weather influence on disease development depends not only on the total magnitude but also on the distribution of weather

Table 5.8 Prediction equations of two different growth models and untransformed data and their comparable factors at five different dates of sowing (Mahapatra and Das 2014)

Prediction equations				
1st date of sowing (20th October)	R^2	Adj. R^2	SE (est.)	RSS
PDI(%) = 1.089–0.063 Temp (min)**	0.67	0.64	0.13	0.016
PDI (Logit) = 12.46–1.637 VP(noon)** + 0.047 RH (min)*	0.78	0.74	1.99	3.94
PDI (Gompit) = 3.203–0.356 Temp (min)** +0.015RH (min)*	0.84	0.82	0.42	0.18
2nd date of sowing (5th November)				
PDI(%) = −1.118 + 0.41WV (morning)** +0.019RT*	0.61	0.54	0.58	0.3
PDI (Logit) = −5.319 + 4.248WV (morning)*	0.37	0.32	1.91	3.64
PDI (Gompit) = −1.929 + 1.634WV (morning)** + 0.067RT*	0.64	0.58	0.14	0.02
3rd date of sowing (20th November)				
PDI(%) = −28.757 + 0.412 T(min) * + 0.250 RH(max)* + 0.075 RH(min)*-0.564 VP(noon)*-0.157VP (morning)**-3.754 WV(evening) + 3.963 WV (morning) +0.451 BSH** +0.030TR**	0.95	0.83	0.009	0.009
PDI (Logit) = −315.514 + 0.361 T(max) +3.487 T(min) + 2.833 RH(max)** + 0.707 RH(min)**- 5.83 VP(noon)**- 1.48VP (morning)*- 44.31WV (evening)* + 45.41WV (morning)* +3.602BSH** + 0.288TR**	0.98	0.91	0.55	0.31
PDI (Gompit) = −121.91 + 1.57 T(min)** +1.083 RH(max)** + 0.29 RH(min)**- 2.27 VP(noon)**- 0.61VP (morning)- 17.17WV (evening)* + 17.83WV (morning)* + 1.65BSH** + 0.113TR*	0.97	0.89	0.28	0.078
4th date of sowing (5th December)				
PDI(%) = −1.272 + 0.079 VP (noon)** + 0.084BSH** + 0.018TR*	0.83	0.78	0.11	0.012
PDI (Logit) = −5.698 + 0.402VP (noon)**	0.50	0.46	0.94	0.88
PDI (Gompit) = −5.131 + 0.25VP (noon)** + 0.256BSH ** + 0.057TR*	0.88	0.78	0.13	0.11
5th date of sowing (29th December)				
PDI(%) = −0.653 + 0.075 T (min)**	0.73	0.71	0.41	0.17
PDI (Logit) = −5.827 + 0.376 T (min)**	0.66	0.63	0.76	0.58
PDI (Gompit) = −3.19 + 0.235 T (min)**	0.73	0.71	0.13	0.017

PDI = percent disease index, R^2 = coefficient of determination, Adj. R^2 = adjusted coefficient of determination, SE (est.) = standard error (estimate), RSS = residual sum of square

variables over small time intervals. But, use of data in small time intervals will increase the number of variables in the model, and in turn, a large number of model parameters will have to be evaluated from the data. This will require a long series of data for precise estimation of the parameters, which may not be available in practice. Thus, a technique based on relatively smaller number of model parameters, and inclusion of entire weather distribution, was attempted. The simplest way to solve the problem is to take weighted accumulation of weather variables, giving weight according to their importance in different time periods. Two approaches were tried using weekly weather data. In the first approach, unfavourable/favourable/highly favourable ranges of weather variables were obtained based on experience of experts. This information was utilized to decide the weights (subjective) depending on values of the weather variables in different weeks pertaining to these ranges. However, the results of this approach were not found to be satisfactory. The other approach was based on objective weights where correlation coefficients between variable to forecast and weather variables were taken as the indicators of relative importance in respective weeks. In this approach, for each weather variable, two indices were developed, one as the simple total value of weather variables and the other as weighted total; weights being correlation coefficients between variable to forecast and weather variables in respective weeks. The first index represents the total amount of different weather variables received by the crop during the period under consideration, while the other one takes care of distribution of weather variables with special reference to its importance in different weeks in relation to the variable to forecast. Similarly, for joint effects of weather variables, weather indices were developed as weighted accumulations of product of weather variables (taken two at a time), weights being correlation coefficients between variable to forecast and product of weather variables considered in respective weeks. The form of model was (Agrawal et al. 1986; Agrawal and Mehta 2007):

$$Y = a_0 + \sum_{i=1}^{p}\sum_{j=0}^{p} aij\,Zij + \sum_{i=0}^{p}\sum_{j=0}^{1} biij\,Ziij + e$$

Where

$$Zij = \sum_{w=n1}^{n2} rjiwXiw$$

$$Ziij = \sum_{w=n1}^{n2} rjiiwXiwXiw$$

Y, variable to forecast; *Xiw*, value of *i*th weather variable in *w*th week; *riw*, correlation coefficient between *Y* and *i*th weather variable in *w*th week; $r_{ii}w$ correlation coefficient between the product of *Xi* and *Xi*' in wth week; *p*, number of weather variables used; n_1, initial week for which weather data were included in the model; n_2 final week for which weather data were included in the model; and *e*, error term.

Stepwise regression technique was used for selecting important variables to be included in the model. The forecasting performance of various models was judged by mean absolute percentage error (MAPE).

$$\text{MAPE} = 1/n\sum\left|\frac{(Yt - Ft)}{Yt}\right| \times 100$$

where *Yt* is actual observation, *Ft* is the forecast form models and *n* is the total number of tested data.

Models along with coefficient of determination (R^2) and mean absolute percentage error (MAPE) are given in Table 5.7.2, which indicate that the models fitted well because the coefficient of determination is highly significant in most of the cases. MAPE are in the affordable range. The MAPE is high for maximum severity which may be due to sample fluctuations. Forecasts from the models were very close to the observed values in subsequent years. Therefore, these models, based on weekly weather data starting from week of sowing up to 6 weeks of crop growth, can be used for reliable forewarning of *Alternaria* blight. The reliable forewarnings through this approach are possible well in advance. In general, the models fitted well with all the coefficients of determination are highly significant. Forecasts from the models were very close to the observed values in subsequent years. Therefore, these models, based on weekly weather data starting from week of sowing up to 6 weeks of crop growth, can be used for reliable forewarning of *Alternaria* blight. Therefore, reliable forewarning for crop age at first appearance of disease, crop age at peak severity of disease and maximum severity of diseases in two different mustard varieties for *Alternaria* blight is possible well in advance (Kumar et al. 2013) (Table 5.9).

Prediction models developed by Awasthi and Kolte (1994) showed multiple regression analysis (MRA) equations between weather parameters as independent factors and prediction of *Alternaria* blight (AB) severity as dependant factors on the most commonly grown susceptible rapeseed Yellow Sarson cultivar T-151 and mustard cultivar 'Varuna' based on the data obtained over 7 years crop periods (1980–1987). Total rainfall and minimum temperature show significant contribution in MRA equations, whereas combined effect of relative humidity, rainfall and minimum temperature accounted for more than 98 %

Table 5.9 Models to forecast different characters of *Alternaria* blight in mustard crop along with coefficient of determination and MAPE in different varieties (Kumar et al. 2013)

Location	Variety	Character	Model	R^2	MAPE
Bharatpur	Rohini (on leaf)	Y_1	$81.16–1.38\ Z_{20}+0.06\ Z_{121}+0.001\ Z_{341}$	0.75	8.9
		Y_2	$2.95+0.75\ Z_{11}$	0.82	26.2
		Y_3	$7.59+0.006\ Z_{231}$	0.40	196.3
	Rohini (on pod)	Y_1	$11.42+0.0012\ Z_{340}+0.026\ Z_{231}$	0.29	24.7
		Y_2	$87.10–00.01\ Z_{240}+0.01\ Z_{231}+0.08\ Z_{241}$	0.40	17.6
		Y_3	$26.94+0.06\ Z_{31}+0.001\ Z_{241}$	0.63	229.6
	Varuna (on leaf)	Y_1	$60.65+0.002\ Z_{340}+0.01\ Z_{121}+0.006\ Z_{341}$	0.78	14.7
		Y_2	$7.82+0.54\ Z_{11}+0.005\ Z_{231}$	0.83	22.3
		Y_3	$4.68+0.34\ Z_{11}$	0.58	150.1
	Varuna (on pod)	Y_1	$36.11+0.08\ Z_{31}$	0.60	14.2
		Y_2	$–222.95+4.43\ Z_{20}-0.14\ Z_{120}+3.61Z_{11}+2.38\ Z_{31}$	0.63	5.4
		Y_3	$68.41\ -0.02\ Z_{130}+0.66\ Z_{21}+0.04\ Z_{131}$	0.71	132.6
Dholi	Pusa Bold (on leaf)	Y_1	$28.79 – 0.01\ Z_{240}+0.04\ Z_{121}+0.002\ Z_{131}$	0.96	10.4
		Y_2	$24.29–0.001\ Z_{140}+0.01\ Z_{131}$	0.72	7.0
		Y_3	$87.90+17.56\ Z_{20}+31.64\ Z_{21}+2.32\ Z_{31}$	0.52	48.5
	Pusa Bold (on pod)	Y_1	$68.95+0.39\ Z_{41}+0.01\ Z_{121}+0.007\ Z_{131}$	0.83	6.9
		Y_2	$81.10–0.003\ Z_{240}+0.01\ Z_{121}+0.005\ Z_{141}$	0.87	3.3
		Y_3	$144.65+1.87\ Z_{31}+0.33\ Z_{41}+0.01\ Z_{131}$	0.77	91.8
	Varuna (on leaf)	Y_1	$72.89–2.11\ Z_{21}+3.35\ Z_{41}+0.07\ Z_{121}-0.02\ Z_{341}$	0.93	8.4
		Y_2	$–17.86–0.05\ Z_{240}+0.01\ Z_{131}-0.007\ Z\ _{231}+0.10\ Z_{241}$	0.84	9.3
		Y_3	$–59.10+1.72\ Z_{31}+0.015\ Z_{121}$	0.55	51.9
	Varuna (on pod)	Y_1	$–206.31+0.61\ Z_{30}+0.02\ Z_{121}+0.005\ Z_{341}$	0.79	3.9
		Y_2	$106.97–0.03\ Z_{230}+0.029\ Z_{121}+0.03\ Z_{231}$	0.87	3.0
		Y_3	$121.89+0.02\ Z_{130}+0.045\ Z_{131}–0.01\ Z_{141}$	0.87	88.9
Behrampur	Binoy (on leaf)	Y_1	$57.38–0.025\ Z_{141}+0.03\ Z_{241}$	0.54	12.4
		Y_2	$92.84+5.17\ Z_{21}+3.94\ Z_{31}-0.11\ Z_{121}$	0.61	9.7
		Y_3	$–93.70+4.91\ Z_{10}+0.06\ Z_{250}+0.31\ Z_{121}$	0.62	61.0
	Binoy (on pod)	Y_1	$–99.41+0.021\ Z_{131}$	0.62	12.6
		Y_2	$93.91–0.04\ Z_{240}+0.09\ Z_{241}-0.004\ Z_{341}$	0.61	2.9
		Y_3	$170.01+0.01\ Z_{131}+0.005\ Z_{231}$	0.50	182.7
	Varuna (on leaf)	Y_1	$59.21+0.56\ Z_{30}+5.67\ Z_{31}+0.03\ Z_{121}$	0.57	22.3
		Y_2	$213+0.01\ Z_{130}-0.003\ Z_{140}+0.09\ Z_{121}$	0.67	9.7
		Y_3	$125.08+9.19\ Z_{10}+27.51\ Z_{11}$	0.46	61.9
	Varuna (on pod)	Y_1	$188.16+1.55\ Z_{31}+1.15\ Z_{41}+0.03\ Z_{121}$	0.56	23.3
		Y_2	$148.25+1.66\ Z_{31}+0.02\ Z_{121}$	0.60	7.4
		Y_3	$–31.168–0.007\ Z_{240}+0.01\ Z_{341}$	0.53	68.1

contribution for prediction of AB severity on leaves of both Yellow Sarson and mustard. Total rainy days (TRDs), relative humidity, maximum temperature and minimum temperature show significant contribution in MRA equations in AB severity on pods of mustard, whereas only the TRD factor shows significant effect for spread of pod infection in Yellow Sarson. It is also found that the disease intensity beyond 30 days age on both Yellow Sarson and mustard and increase in the age are positively correlated (r^2=0.777 to 0.980) with increase in the susceptibility of the crop, the maximum leaf disease severity being at rosette to flowering.

The data on stepwise multiple regression analysis equations for prediction of AB severity on leaves (Y) and r^2 value indicated that a combined effect of relative humidity, rainfall and minimum temperature accounts for more than 98 % variation in AB severity on leaves of both Yellow Sarson and mustard. The equations are:

AB leaf infection severity on Yellow Sarson

$$Y_1 = -73.07 + (0.71x_3) + (7.68x_2) + (0.67x_4)\, r^2 = 0.99$$

where Y_1 = *Alternaria* blight leaf infection severity, X_2 = mean minimum temperature, X_3 = percent relative humidity, X_4 = total rainfall.

AB leaf infection severity on mustard

$$Y_1 = 6.98 + (9.09\,x_2) + (0.67x_4) + (-0.79\,x_3)\, r^2 = 0.98$$

where Y_1 = *Alternaria* blight leaf infection severity, X_2 = mean minimum temperature, X_3 = percent relative humidity, X_4 = total rainfall

Dang et al. (2006) developed prediction equation for the development of *Alternaria* blight using Gompertz model and two factors can be explained by the following equation:

$$\mathrm{DI} = \mathrm{Exp}^{[\log A \,\mathrm{Log}(-B^*\mathrm{Time})]} + C^*\,\mathrm{Sowing\,day} + D^*\,\mathrm{Factor\ 1}$$

where DI = disease intensity (A and B are the two parameters of the Gompertz model and C and D are the coefficient of the sowing day and factor 1). All the varieties are based on different genetic make-ups, which reacted differentially to the natural inoculum, and factor 1 may be interpreted as weather index and factor 2 as contrast between the heating factor and moisture factor. The best-fitted models worked out for each variety were as:

$$\mathrm{RH-30: Y = E}^{[\mathrm{Log}(0.620)^*\log(4.998^*\mathrm{Time})]} - 0.890^*\,\mathrm{sowing\,Day} + 0.238\,\mathrm{factor\,1}; \quad (R^2 = 0.389)$$

$$\mathrm{RH-8113: Y = E}^{[\mathrm{Log}(2.547)^*\log(1.072^*\mathrm{Time})]} - 0.561^*\,\mathrm{sowing\,Day} + 0.086\,\mathrm{factor\,1}; \quad (R^2 = 0.482)$$

$$\mathrm{YSPb-24: Y = E}^{[\mathrm{Log}(2.199)^*\log(4.827^*\mathrm{Time})]} - 1.868^*\,\mathrm{sowing\,Day} + 0.105\,\mathrm{factor\,1}; \quad (R^2 = 0.539)$$

$$\mathrm{BSH-1: Y = E}^{[\mathrm{Log}(3.008)^*\log(1.335^*\mathrm{Time})]} - 1.260^*\,\mathrm{sowing\,Day} + 0.166\,\mathrm{factor\,1}; \quad (R^2 = 0.482)$$

where Y = percent disease intensity

The above equations revealed that for a given cultivar, if the sowing time is known, and weather parameters at particular time are known, disease incidence can be predicted using the above models. Mehta et al. (2008) developed a prediction model for adopting better disease management practices where four varieties each of *B. juncea* (RH-30, RH-8113, RH-8695, RH-8546), *B. campestris* (YSPb-24, BSH-1, Candle, Shiva), two of *B. carinata* (HC-2, HC-9001), one each of *B. napus* (GSH-1) and *B. alba* (local) were monitored for the development and progression of *Alternaria* blight. Each cultivar was inoculated artificially with *A. brassicae* spores when the crop was about 2-month-old. Studies conducted revealed that temperature and RH had prominent role in disease development in addition to varietal behaviour. The prediction equations so developed were at par with the observed values. Regression equations for the development of *Alternaria* blight on different varieties of rapeseed–mustard are:

B. juncea (var. RH-30)

$$Y_1 = -4.3635 + 0.3488\,X_1 \quad (R^2 = 0.66)$$
$$Y_1 = 4.4927 + 0.2755\,X_1 - 0.0789\,X_3 \quad (R^2 = 0.71)$$

B. campestris (var. YSPb-24)

$$Y_1 = -4.2310 + 0.3420\,X_1 \quad (R^2 = 0.68)$$
$$Y_1 = 4.7131 + 0.2680\,X_1 - 0.0797\,X_3 \quad (R^2 = 0.73)$$

B. carinata (var. HC-1)

$$Y_1 = -4.8061 + 0.3584\,X_1 \quad (R^2 = 0.64)$$
$$Y_1 = 6.4244 + 0.2654\,X_1 - 0.1001\,X_3 \quad (R^2 = 0.71)$$

where $X_1 = T$ max; X_3 = RH mor

Sangwan et al. (2002) demonstrated that Gompertz model can be effectively used for prediction of *Alternaria* blight with two factors A and B drawn from the analysis of weather parameters and disease progress.

Factor A: 0.091 × T max + 0.887 × T min + 0.036 × RH mor + 0.808 × RH eve – 0.644 × Sunshine h.

Factor B: 0.317 × T max + 0.317 × T min + 0.933 × RH mor + 0.347 × RH eve – 0.618 × Sunshine h.

These two factors (A and B) explained 60.3 and 24.5 % of total variation, respectively, and together explained 85 % of total variation among the weather variables. The model for each group is as follows:

$$\text{Disease} = \text{Exp.}\left[\log(a)X - \log(b)\times\text{Time}\right] + cX\ \text{Factors}(1) + dX\ \text{Factors}(2)$$

where *a* and *b* are constants of Gompertz model.

The equation for each genotype of *Brassica* was developed as follows:

B. juncea

$$\text{Exp.}[\log(2.915)\times\log(-0.430\times\text{Time}] - 0.154\times\text{Factors}(1) + 0.204\times\text{Factors}(2)$$
$$R^2 = 0.931\ (\text{observed vs predicted})$$

B. campestris

$$\text{Exp.}[\log(2.722)\times\log(0.470\times\text{Time}] - 0.203\times\text{Factors}(1) + 0.167\times\text{Factors}(2)$$
$$R^2 = 0.974\ (\text{observed vs predicted})$$

B. carinata

$$\text{Exp.}[\log(2.717)\times\log(0.479\times\text{Time}] - 0.202\times\text{Factors}(1) + 0.168\times\text{Factors}(2)$$
$$R^2 = 0.982\ (\text{observed vs predicted})$$

B. napus

$$\text{Exp.}[\log(2.908)\times\log(0.424\times\text{Time}] - 0.185\times\text{Factors}(1) + 0.170\times\text{Factors}(2)$$
$$R^2 = 0.975\ (\text{observed vs predicted})$$

B. alba

$$\text{Exp.}[\log(4.312)\times\log(0.270\times\text{Time}] - 0.284\times\text{Factors}(1) + 0.212\times\text{Factors}(2)$$
$$R^2 = 0.970\ (\text{observed vs predicted})$$

B. oleracea

$$\text{Exp.}[\log(2.823)\times\log(0.455\times\text{Time}] - 0.196\times\text{Factors}(1) + 0.192\times\text{Factors}(2)$$
$$R^2 = 0.984\ (\text{observed vs predicted})$$

Experiments were conducted at two locations for development of prediction models for two popular varieties of rapeseed–mustard belonging to two different *Brassica* genotypes, viz. *B. juncea* (RH-30) and *B. campestris* var. Yellow Sarson (YSPb-24). These were monitored for the progression of *Alternaria* blight where disease progression was faster on YSPb-24 compared to RH-30. Maximum lesion size, 6.31 mm, was recorded on YSPb-24, whereas it was 4.21 mm on variety RH-30. The favourable weather conditions for the progression of the disease were observed to be at T max of 20 °C with RH >90 %. The stepwise regression analysis revealed that the T max and RH mor played significant and positive roles in disease progression. The R^2 value recorded was >0.9 in all the cases, which showed that weather variables played major role in disease progression in addition to the varietal factors. The prediction equations developed for leaves and pods for a variety for two locations were as follows (Mehta et al. 2008). Similar equations were developed for other varieties as well.

A. **On leaves**

$$\text{RH}-30(\text{location}-1)Y = 1.984 + .196X_1 + .951X_5 \quad (R^2 = 0.92)$$
$$\text{RH}-30(\text{location}-2)Y = 2.156 + .235X_1 - 5.49 + 10^{-2}X_4 \quad (R^2 = 0.87)$$

B. **On pods**

$$\text{RH}-30(\text{location}-1)Y = -0.471 + 5.802 + 10^{-2}X_1 \quad (R^2 = 0.84)$$
$$\text{RH}-30(\text{location}-2)Y = -0.317 + 1.543 + 10^{-2}X_4 \quad (R^2 = 0.97)$$

where X_1=T max; X_2=T min; X_4=RHeve.; X_5=wind speed.

Jha et al. (2013) recorded that T max positively correlated with disease index. *T* max of 23.2, RH max of 80 % and RH min of 66 % with correlation coefficient (r)=0.73 for T min and r=0.51 of RH min favoured the disease development.

The regression equation developed for leaves ($Y1$) and siliqua ($Y2$) are:

$$\text{Leaves}: Y1 = -47.388 + 5.114\,\text{T min} - 2.371\,\text{Tmax} + 1.492\,\text{RH min} \quad R^2 = 0.7376$$
$$\text{Siliqua}: Y2 = 31.524 + 4.225\,\text{T min} - 1.883\,\text{T max}. \quad R^2 = 0.69203.$$

References

Agrawal R, Mehta SC (2007) Weather based forecasting of crop yields, pests and diseases -IASRI models. J Indian Soc Agric Stat 61:255–263

Agrawal R, Jain RC, Jha MP (1986) Models for studying rice crop-weather relationship. Mausam 37:67–70

Ansari NA, Khan MW, Muheet A (1988) Effect of temperature and relative humidity on spore germination of *Alternaria brassicae* and development of Alternaria blight on oilseed crucifers. Rev Tropical Plant Pathol 5:79–84

Ansari NA, Khan MW, Muheet A (1989) Survival and perpetuation of *Alternaria brassicae* causing Alternaria blight of oilseed crucifers. Mycopathologia 105:67–70

Atkinson RG (1950) Studies on the parasitism and variation of *Alternaria raphani*. Can J Res Sect C 28:288–317

Awasthi RP, Kolte SJ (1989) Effect of some epidemiological factors on occurrence and severity of Alternaria blight of rapeseed and mustard. In: Proceedings of the IDRC (Canada) Oil Crops: meetings held at Pantnagar and Hyderabad, India, pp 49–55

Awasthi RP, Kolte SJ (1994) Epidemiological factors in relation to development and prediction of Alternaria blight of rapeseed and mustard. Indian Phytopathol 47:395–399

Bassey EO, Gabrielson RL (1983) The effects of humidity, seed infection level, temperature and nutrient stress on cabbage seedling disease caused by *Alternaria brassicicola*. Seed Sci Technol 11:403–410

Berger RD (1981) Comparison of the Gompertz and logistic equations to describe plant disease progress. Phytopathology 71:716–719

Berger RD (1988) The analysis of effects of control measures on the development of epidemics. In: Kranz J, Rotem J (eds) Experimental techniques in plant disease epidemiology. Springer, Berlin, pp 137–151

Berger RD, Luke HH (1979) Spatial and temporal spread of oat crown rust. Phytopathology 69:1199–1201

Chahal AS (1986) Relationship of Alternaria blight with the age of Brown Sarson. Indian J Mycol Plant Pathol 16:166–167

Chahal AS, Kang MS (1979) Influence of the meteorological factors on the development of Alternaria blight of rape and mustard in the Punjab. Indian Phytopathol 32:171

Chattopadhyay C, Agrawal R, Kumar A, Bhar LM, Meena PD, Meena RL, Khan SA, Chattopadhyay AK, Awasthi RP, Singh SN, Chakravarthy NVK, Kumar A, Singh RB, Bhunia CK (2005) Epidemiology and forecasting of Alternaria blight of oilseed *Brassica* in India – a case study. Zeitschrift für Pflanzenkrankheiten und Pflanzenschutz (Journal of Plant Disease Protection) 112:351–365

Chen LY, Price TV (2002) Dark leaf spot (*Alternaria brassicicola*) on Chinese cabbage: temporal spread and its influencing factors. Aust J Agric Res 53:1095–1103

Chen LY, Price TV, Park-Ng Z (2003) Conidial dispersal by *Alternaria brassicicola* on Chinese cabbage (*Brassica pekinensis*) in the field and under simulated conditions. Plant Pathol 52:536–545

Cornell JA, Berger RD (1987) Factors that influence the value of the coefficient of determination in simple linear and non-linear regression models. Phytopathology 77:63–70

Czyzewska S (1971) The pathogenicity of *Alternaria* spp. isolated from *Crambe abyssinica*. Acta Mycol 7:171–240

Dang JK, Kaushik CD, Sangwan MS (1995) Quantitative relationship between Alternaria leaf blight of rapeseed-mustard and weather variables. Indian J Mycol Plant Pathol 25:184–188

Dang JK, Sangwan MS, Mehta N, Sharma OP, Dhandapani A (2006) Development of prediction model for Alternaria blight caused by *Alternaria brassicae* of rapeseed and mustard. Plant Dis Res 21:199–201

Daniel C, Wood FS (1980) Fitting equations to data. Wiley, New York, 458 p

Danos E, Berger RD, Stall RE (1984) Temporal and spatial spread of citrus canker within groves. Phytopathology 74:904–908

Dey PK (1948) Plant pathology. Administrative report, Agriculture Department, UP, 1946–47: pp 39–42

Dillard HR, Cobb AC, Lamboy JS (1998) Transmission of *Alternaria brassicicola* to cabbage by flea beetles (*Phyllotreta cruciferae*). Plant Dis 82:153–157

Domsch KH (1957) The blackening of rape and cabbage siliquae. Zeitschrift für Pflanzenkrankheiten und Pflanzenschutz (Journal of Plant Disease Protection) 64:65–79

Fontem DA (1986) Disease progress of dark leaf spot in fungicide sprayed cabbage. M.Sc. thesis, University of Florida, Gainesville, USA

Fontem DA, Berger RD, Weingartner DP, Bartz JA (1991) Progress and spread of dark leaf spot in cabbage. Plant Dis 75:269–274

Gregory PH (1968) Interpreting plant disease dispersal gradients. Annu Rev Phytopathol 6:189–212

Hasan S, Vago C (1966) Transmission of *Alternaria brassicicola* by slugs. Plant Dis Rep 50:764–767

Hong CX, Fitt BDL (1996) Factors affecting the incubation period of dark leaf and pod spot (*Alternaria brassicae*) on oilseed rape (*Brassica napus*). Eur J Plant Pathol 102:545–553

Humpherson-Jones FM, Maude RB (1982) Studies on the epidemiology of *Alternaria brassicicola* in *Brassica oleracea* seed production crops. Ann Appl Biol 100:61–71

Humpherson-Jones FM, Ainsworh LF, Maude RB, Bambridge JM, Spencer A (1983) Seed studies. In: 33rd annual report for 1982, National Vegetable Research Station, Wellesbourne, Warwick, UK, pp 64–65

Humpherson-Jones FM, Phelps K (1989) Climatic factors influencing spore production in *Alternaria brassicae* and *Alternaria brassicicola*. Ann Appl Biol 114:449–459

Jha P, Kumar M, Meena PD, Lal HC (2013) Dynamics and management of Alternaria blight disease of Indian mustard (Brassica juncea) in relation to weather parameters. J Oilseed Brassica 4:66–74

Jose TC, Chakravarty NVK, Srivastava AK, Bhagawati G, Katiyar RK, Singh HB (2005) Alternaria blight and its role in yield reduction of mustard crop- a brief review. Brassica 7:15–31

Kadian AK (1982) Studies on the Alternaria blight of raya [*Brassica juncea* (L.) Czern. and Coss.]. M. Sc., thesis, Haryana Agricultural University, Hisar, India

Kadian AK, Saharan GS (1984) Studies on spore germination and infection of *Alternaria brassicae* of rapeseed and mustard. J Oilseeds Res 1:183–188

Kadian AK, Saharan GS (1988) Progress of Alternaria blight of mustard in relation to cultural practices. Oil Crops Newslett 5:32–33

Kennedy R, Graham M (1995) Infection on oil-seed rape by Alternaria brassicae under varying conditions of temperature and wetness. In: 9th International rapeseed congress, Cambridge 2:601–603

Kiyosawa S, Shiyomi M (1972) A theoretical evaluation of mixing resistant variety and susceptible variety for controlling plant diseases. Ann Phytopathol Soc Jpn 38:41–51

Kolte SJ, Singh MP (1997) Alternaria blight and white rust of mustard: potential threat for cultivation. In: Agnihotri VP, Sarbhoy AK, Singh DV (ed) Management of threatening plant diseases of national importance. Malhotra Publishing House, New Delhi, India, pp 87–95

Kumar B (2008) Assessment of slow blighting (*Alternaria brassicae*) resistance and yield in Indian mustard [*Brassica juncea* (L.) Czern. & Coss.]. Indian Phytopathol 61:171–183

Kumar N, Kumar A (2006) Effect of cultural practices on *Alternaria* blight in *Brassica juncea* and *B. napus*. Indian J Agric Sci 76:389–390

Kumar A, Agrawal R, Chattopadhyay C (2013) Weather based forecast models for diseases in mustard crop. Mausam 64:663–670

Lambert DH, Villareal RL, Mackenzie DR (1980) A general model for gradient analysis. Phytopathol Z 98:150–154

Louvet J (1958) The black spot disease of colza, *Alternaria brassicae*. C R Acad Agric J 44:694–701

Louvet J (1963) Alternaria disease of rape and disease causing concern on other oleaginous plants. C R Journess d' Inf., Oleagineux Metr. 1961, 9

Louvet J, Billotte JM (1964) Influence des facteurs climatiques sur les infections du colza par *l'Alternaria brassicae* et consequences pour la lutte. Annal Epipht 15:229–243

Maffia LA (1985) Progress and spread of bean rust. Ph. D dissertation, University of Florida, Gainesville, 131 p

Magarey RD, Sutton TB, Thayer CL (2005) A simple generic infection model for foliar fungal plant pathogens. Phytopathology 95:92–100

Mahapatra S, Das S (2014) Effect of meteorological factors on progression of Alternaria leaf blight of mustard and comparison of Logistic and Gompertz growth models in predicting disease severity. Indian Phytopathol 67:155–158

McDonald WC (1959) Gray leaf spot of rape in Manitoba. Can J Plant Sci 39:409–416

Meena PD (2005) Epidemiological studies and biocontrol of Alternaria blight disease of rapeseed-mustard. Rajasthan University, Jaipur, Rajasthan, Ph.D. thesis, 131 p

Meena PD, Chattopadhyay C, Meena SS, Kumar A (2011) Area under disease progress curve and apparent infection rate of Alternaria blight disease of Indian mustard (*Brassica juncea*) at different plant age. Arch Phytopathol Plant Protect 44:684–693

Mehta N, Sangwan MS, Srivastava MP, Kumar R (2002) Survival of *Alternaria brassicae* causing Alternaria blight in rapeseed-mustard. J Mycol Plant Pathol 32:64–67

Mehta N, Sangwan MS, Kumar R, Niwas R (2008) Progression of Alternaria blight on different varieties of rapeseed- mustard in relation to weather parameters. Plant Dis Res 23:28–33

Mehta N (2014) Epidemiology and forecasting for the management of rapeseed-mustard diseases. J Mycol Plant Pathol 44:131–147

Minogue KP, Fry WE (1983) Models for the spread of disease: some experimental results. Phytopathology 73:1173–1176

Mukadam DS, Deshpande KB (1979) Role of light and temperature on growth, sporulation and subsequent spore germinability of Alternaria brassicae (Berk.) Sacco. Sci Cult 45:244–246

Mundt CC, Leonard KJ (1985) A modification of Gregory's model for describing plant disease gradients. Phytopathology 75:930–935

Munro JM (1984) Alternaria infection of cabbage following spray applications of a commercial formulation of tridemorph. *Tests of Agrochemicals & Cultivars*, No. 5 (*Annals of Applied Biology*, 104, Supplement), 32–33

Pennypacker SP, Knoble HD, Antle CE, Madden LV (1980) A flexible model for studying plant disease progression. Phytopathology 70:232–235

Saharan GS (1984) A review of research on rapeseed-mustard pathology in India. Paper presented in the Annual Rabi Oilseed Workshop held at Jaipur, 6–10 August 1984

Saharan GS (1991) Assessment of losses, epidemiology and management of black spot disease of rapeseed-mustard. In: Proceedings of GCIRC 8th international rapeseed congress, Saskatoon, Canada, vol 2, pp 465–470

Saharan GS (1992a) Management of rapeseed and mustard diseases. In: Kumar D, Rai (eds) Advances in oilseed research, vol 1,7. Scientific Publishers, Jodhpur, pp 152–188

Saharan GS (1992b) Disease resistance. In: Labana KS, Banga SS, Banga SK (eds) Breeding oilseed Brassicas. Narosa Publishing House, New Delhi, pp 188–200

Saharan GS, Kadian AK (1983) Analysis of components of horizontal resistance in rapeseed and mustard cultivars against *Alternaria brassicae*. Indian Phytopathol 36:503–507

Saharan GS, Kadian AK (1984) Epidemiology of Alternaria blight of rapeseed and mustard. Cruciferae Newslett 9:84–86

Saharan GS, Kaushik JC, Kaushik CD (1981) Progress of Alternaria blight on rye cultivars in relation to environmental conditions. In: Proceedings of 3rd international symposium on plant pathology, IARI, New Delhi, India, 14–18 December 14–18, 236 (Abstr.)

Saharan GS, Mehta N, Sangwan MS (2005) Diseases of oilseed crops. Indus Publishing Company, New Delhi, 643 p

Sandhu KS, Singh H, Dhillon TS (1984) Effect of Alternaria blight on radish varieties sown on different dates. J Res Punjab Agric Univ 21:543–546

Sandhu KS, Singh H, Kumar R (1985) Effect of different nitrogen levels and dates of planting on Alternaria blight and downy mildew diseases of radish seed crop. J Res Punjab Agric Univ 22:285–290

Sangeetha CG, Siddaramaiah AL (2007) Epidemiological studies of white rust, downy mildew and Alternaria blight of Indian mustard [*Brassica juncea* (Linn.) Czern. and Coss.]. Afr J Agric Res 2:305–308

Sangwan MS, Mehta N, Gandhi SK (2002) Some pathological studies on *Alternaria raphani* causing leaf and pod blight of radish. J Mycol Plant Pathol 32:125–126

Sarkar B, Sen Gupta PK (1978) Studies on some aspects of the epidemiology of Alternaria leaf blight of mustard (*Brassica* sp.). Beitrage Zur Tropischen Landwirtschaft Und Veterinarmedizin 16:91–96

Van Schreven DA (1953) *Alternaria, Stemphylium* en *Botrytis* a antasting bij Koolzaad (*Brassica napus*). Tijdschr Plantenziekt 59:105–136

Shaner G, Finney RE (1977) The effect of nitrogen fertilization on the expression of slow-mildewing resistance in Knox wheat. Phytopathology 67:1051–1056

Sharma SR, Kolte SJ (1994) Effect of soil applied NPK fertilizers on severity of black spot disease (*Alternaria brassicae*). Plant Soil 167:313–320

Singh H (2005) Diurnal pattern of aerospora of *Sclerospora graminicola* and *Alternaria brassicae* in relation to disease development. PhD thesis, Department of Plant Pathology, CCS Haryana Agricultural University, Hisar, 111 p

Singh NI (1988) The fungal air-spora of Imphal, its seasonal fluctuations and relationship with occurrence and intensity of major crop diseases. Int Aerobiol Newslett 28:15–17

Singh R, Suhag LS (1983) Some pathological studies on *Alternaria alternata* causing leaf and pod blight of radish in Haryana. Indian Phytopathol 36:174–176

Stankova J (1972) Varietal variability of winter rape in its susceptibility to dark leaf spot and the factors influencing the development of the disease. Rostlinna Vyroba 18:625–630

Strzelczyk E, Rozej A (1974) CCC and Alar effect on infection of different winter rape varieties by *Alternaria brassicae* (Berk.) Sacc. Roczniki Nauk Rolniczych A 100:7–15

Suhag LS, Singh R, Malik YS (1985) Epidemiology of pod and leaf blight of radish seed crop caused by *Alternaria alternata*. Indian Phytopathol 38:148–149

Yadav MS, Brar KS (2003) Relationship between meteorological factors and incidence of Alternaria blight and white rust of Indian mustard in South-western Punjab. Plant Dis Res 18:80–82

6 Pathogenic Variability

6.1 Introduction

The most dynamic and significant aspect in host–pathogen interactions is that characteristics of individuals within a species are not 'fixed' in their morphology, physiology, biochemistry and pathogenicity. During reproduction, all individuals are expected to be different from each other and from their parents in a number of characteristics, although they retain most similarities with them and belong to the same species (Agrios 2005). When individuals are produced asexually, the frequency and degree of variability among the progeny are reduced greatly, but even then, certain individuals among the progeny will show different characteristics. Three categories of such populations are of direct interest to the Brassicalogists of the world:

1. Populations that differ in their ability to attack particular varieties of *Brassica* hosts
2. Populations differing in their physiological adaptations to specific environmental conditions
3. Populations differing in their ability to tolerate the effect of toxicants

In *Alternaria–Brassica* host pathosystem, the following categories of variability exist: although the genus *Alternaria* is known as an imperfect fungus, it shows genetic variability within a species, and this variability might be due to the existence of mutation, somatic hybridization, heterokaryoses, uniform host selection, extensive dispersal and/or a cryptic sexual stage.

6.2 Historical Developments

Initially, variations in cultural characteristics and pathogenesis of different isolates of three *Alternaria* species infecting Brassicaceae hosts were observed during 1952–1953 by Stoll (1952) in *A. brassicicola* (vegetables), by Van Schreven (1953) in *A. brassicae* (*Brassica*) and by Atkinson (1953) in *A. raphani* (radish host pathosystem). Therefore, it can be considered as the beginning of research development on pathogenic variability in *Alternaria*–crucifers host pathosystem.

In *A. alternata* strains showing differences in their physiological and pathological characteristics have been isolated from *Crambe* (*C. abyssinica*). Strain A occurs on leaves, stem and siliquae, whereas strains B and C are mainly found on siliquae and leaves, respectively. In pathogenic ability, strain B has been reported to be most virulent, strain A as moderately virulent and strain C as least virulent (Czyzewska 1969, 1971). These strains have different temperature optima for sporulation. Strain A sporulates abundantly at 17–35 °C, whereas B requires 20–30 °C and strain C sporulates best at 12 °C (Czyzewska 1970).

Alternaria brassicae is generally most virulent on all brassicaceous hosts. Preliminary reports on variability in this species were made

G.S. Saharan et al., *Alternaria Diseases of Crucifers: Biology, Ecology and Disease Management*,
DOI 10.1007/978-981-10-0021-8_6

from Holland (Van Schreven 1953) and the UK (Mridha 1983). Isolates of *A. brassicae* from rapeseed (colza) showed differences in cultural growth on cherry agar and differed in their pathogenesis on seedlings. Similarly, Kolte et al. (1989, 1991) and Awasthi and Kolte (1989) distinguished three *A. brassicae* isolates, viz. A, C and D, on the basis of their morphology, sporulation, growth and cultural characteristics. On *B. carinata*, these isolates produce distinct types of lesions. Among the three isolates, isolate C is the most sporulating and isolate A the least. Unlike isolates B and C, isolate A produces chlamydospores. In a serological study, Kolte et al. (1991) indicated that the Pantnagar isolates A, C and D resembled the Bihar isolates BHl and BH2 and the Kanpur isolate K, respectively. None of these workers, however, used different *Brassica* host differentials to distinguish *A. brassicae* isolates on the basis of their virulence. Saharan and Kadian (1983) used eight commonly cultivated *Brassica* species to distinguish isolates of *A. brassicae* in India. In a cross-infection study and differential interactions on different hosts, they distinguished three clearly separable isolates and designated them as RM1, RM2 and V3 races, which were found to be virulent on rapeseed–mustard group of crops. Race RM1 from rapeseed–mustard was avirulent on *B. oleracea* var. *capitata.* Race RM2 from *B. rapa* (*B. campestris)* var. Brown and Yellow Sarson and *Eruca sativa* was avirulent on both *B. oleracea* var. *capitata* and *B. oleracea* var. *botrytis.* Race V3 was virulent on all eight host species tested including radish, cabbage and cauliflower (Table 6.1). This study clearly indicated the existence of distinct pathotypes in *A. brassicae* infecting different *Brassica* species. According to Mridha (1983), 13 UK isolates of *A. brassicae* tested on selected cultivars of winter oilseed rape differed in their virulence.

Alternaria brassicicola is generally more common on vegetable crops than on oil-yielding brassicas. Stoll (1952) characterized three isolates of this species from siliquae of cauliflower seed crop, showing highly aggressive, less aggressive and non-pathogenic behaviour. However, highly aggressive isolates were less frequent (7.48 %) than the moderately aggressive isolates (56.86 %). Cultural and morphological variations in the isolates of this species show no distinction in pathogenic behaviour (Campbell 1970; Campbell et al. 1968; Changsri and Weber 1963). Spontaneous occurrence of albino mutants of this species has been observed (Campbell 1970; Campbell et al. 1968).

Alternaria raphani is the major pathogen of radish but also occurs on other brassicaceous hosts. Atkinson (1953) obtained 312 isolates of this species from different geographical areas in Canada, classified them as 'wild type' and 'variant type' and found the former as being less virulent than the latter. No differences were observed in their nutritional requirements for growth. In a later study, Changsri and Weber (1963) also did not find any variations in the *A. raphani* isolates from *B. nigra*, *B. napus* and *B. rapa* from different geographical areas of Canada.

The last decade of twentieth century and twenty-first century can be considered as boom period for *Alternaria*–crucifers pathogenic variability research. During the first two decades of present century, pathogenic variability in *Alternaria*–crucifers system has been determined on the aspect including pathological, symptomatological, morphological, cultural, biochemical, nutritional, thermal and fungicidal sensitivity, proteomic analysis, genetical and molecular (Saharan et al. 2015). However, no standard internationally acceptable parameter for the selection of host differentials (single gene lines,

Table 6.1 Physiological races of *Alternaria brassicae* (Saharan and Kadian 1983)

Host differential	Races/pathotypes		
	RM1	RM2	V3
Brassica juncea	S[a]	S	S
B. rapa var. Sarson	S	S	S
B. rapa var. *dichotoma*	S	S	S
B. rapa var. *toria*	S	S	S
Eruca sativa	S	S	S
Raphanus sativus	S	S	S
Brassica oleracea var. *capitata*	R[a]	R	S
Brassica oleracea var. *botrytis*	S	R	S

[a]*R* resistant, *S* susceptible

isogenic lines) and nomenclature of pathotypes has been maintained. Each researcher has used different sets of host differential and system of pathotype nomenclature. Gupta et al. (2004), however, attempted to use *B. juncea* varieties in their set of host differentials and gave nomenclature to pathotypes as Bj-4, Bj-5, Bj-6 and Bj-7. They have also tried to maintain parity in the order of the discovery of pathotypes of *A. brassicae* by Saharan and Kadian (1983) who reported three pathotypes, viz. RM-1, RM-2 and V-3.

6.3 Pathological Variations

Out of four species of *Alternaria* known to occur on crucifers, *Alternaria brassicae* (Berk.) Sacc. is more severe and variable (Verma and Saharan 1994). Preliminary reports on variability in *Alternaria* species from rapeseed (colza) by Van Schreven (1953) in Holland and by Mridha (1983) in the UK showed differences in cultural growth on cherry agar and in their pathogenesis on seedlings. Although pathogenic variability in *A. brassicae* has been observed by various workers (Verma and Saharan 1994), information on the existence of distinct pathotypes using standard host differentials is rather limited (Saharan 1992a, b; Mehta et al. 2005a). According to Mridha (1983), 13 isolates of *A. brassicae* tested on selected cultivars of winter rape differed in their virulence. Similarly, Kolte et al. (1989, 1991) and Awasthi and Kolte (1989) also reported variability in *A. brassicae*.

None of these workers used different *Brassica* species to distinguish *A. brassicae* isolates on the basis of their reaction on host differentials. Using eight commonly cultivated *Brassica* species as differentials, Saharan and Kadian (1983) distinguished three *A. brassicae* isolates and designated them as RM-1, RM-2 and V-3 races, which were found to be virulent on rapeseed and mustard group of *Brassicas* (Table 6.1). Mehta et al. (2003) collected ten isolates from different agroclimatic zones of India and cross-inoculated them on a set of 17 host differentials. Among the ten isolates, isolate DLK was the most virulent infecting 16 differentials followed by RSR-1 and GDP, which infected 15 host differentials, but isolates could not be differentiated into pathotypes.

Using 11 *B. juncea* genotypes as host differentials, Gupta et al. (2004) identified four distinct *A. brassicae* pathotypes, viz. Bj-4 (BWL), Bj-5 (HSR), Bj-6 (RTK) and Bj-7 (REW). Pathotypes Bj-4 was most virulent infecting all 11 host differentials, and Bj-5 was least virulent infecting only six host differentials (Table 6.2). Incubation and latent periods also exhibited greater variability for host genotype x isolate interactions of *A. brassicae*. Minimum incubation period of three days was required for pathotypes Bj-4 (BWL) and Bj-6 (RTK) on cultivars Varuna and RH-30 (Table 6.3). Vishwanath and Kolte (1997) also recorded differential interactions between *Brassica* species and *A. brassicae* isolates A and C and avirulent isolate D. *Alternaria brassicae* isolate A showed significantly higher disease scores than isolate C on *B. napus* genotype PPNS1, *B. juncea* cv. PR15, *B. campestris* var. *toria* cvs. PT 303 and PT 30, *B. campestris* var. Yellow Sarson cv. T-151; isolate C showed significantly higher disease scores on *B. campestris* var. Yellow Sarson cv. PYST-6, *B. campestris* ssp. *rapifera* cv. Turnip Red and *B. alba* in comparison to isolate A. *Alternaria brassicae* isolate A is a highly virulent pathotype and isolate C is a moderately virulent pathotype. The toxigenicity studies of three isolates on leaves of various hosts showed isolate A causing more severe symptoms than the isolates C and D at both 1:10 and 1:100 dilutions. Toxin from isolate D produced maximum symptom severity score on *E. sativa*, but failed to produce symptoms on leaves of other host cultivars. Isolate A toxin supports significantly less seed germination and minimum plumule and radical lengths as compared to isolates C and D at 1:10 and 1:100 dilutions. Some differences among different genotypes, however, were observed with respect to seed germination and seedling growth with respect to toxins produced by the three isolates.

Fifteen *A. brassicae* isolates from rapeseed and mustard collected from different locations in Haryana (India) showed pathogenic diversity on 17 host differentials under greenhouse conditions

Table 6.2 Reaction of different isolates of *A. brassicae* on *B. juncea* host differentials (Gupta et al. 2004)

Reaction of isolate collected from various locations					
Host differentials	Bawal (BWL)	Hisar (HSR)	Rohtak (RTK)	Rewari (REW)	Number of VI/HD
EC-129126-1	+	–	–	–	1
EC-322090	+	–	+	+	3
EC-322092	+	–	+	+	3
EC-322093	+	–	+	+	3
Varuna	+	+	+	+	4
EC-287711	+	–	+	–	2
ZEM-1	+	+	+	+	4
RC-781	+	+	+	+	4
RH-30	+	+	+	+	4
RH-8113	+	+	+	+	4
Rajat	+	+	+	+	4
Infectivity size	11	6	10	9	–

VI/HD = virulent isolates per host differential
+ = denotes compatible interaction; – = denotes incompatible interactions

Table 6.3 Incubation and latent period (in days) of *Alternaria brassicae* isolates on host differentials under controlled conditions (Gupta et al. 2004)

Host differentials	Bawal (BWL)		Hisar (HSR)		Rohtak (RTK)		Rewari (REW)	
	IP	LP	IP	LP	IP	LP	IP	LP
EC-129126-1	4	7	–	–	–	–	–	–
EC-322090	4	7	–	–	5	11	7	13
EC-322092	4	7	–	–	5	11	7	13
EC-322093	3	7	–	–	4	7	7	13
Varuna	3	6	4	6	3	9	5	11
EC-287711	4	7	–	–	5	11	–	–
ZEM-1	5	7	4	7	4	7	6	12
RC-781	4	7	5	9	5	11	6	12
RH-30	3	6	4	6	3	9	5	11
RH-8113	3	7	4	7	4	9	4	9
Rajat	3	7	4	6	4	9	5	11

IP = incubation period; LP = latent period; – = no infection

(Kumar et al. 2003a, b). Isolates CHR-I, CHR-II, JND-II, JHR and SRS were most virulent infecting all 17 host differentials, followed by REW, RTK and SPT causing infection on 16 host differentials. Isolates BWL, BHI and KTL-II infected all differentials except *E. sativa,* and *B. alba*, whereas isolate HSR is least virulent producing symptoms only on 13 host differentials. Out of 17 host differentials, six differentials of *B. juncea*, *B. carinata*, *B. nigra*, *B. oleracea* var. *botrytis*, *B. rapa* and *B. alba* differentiated 15 isolates into eight pathotypes/races. The eight pathotypes identified, CHR-I, CHR-II, JND-II, JHR and SRS were grouped in the first group; CHR-III, JND-I, and KTL-I in the second group, and BHI and KTL-II in third group. However, isolates BWL, HSR, REW, RTK and SPT formed as individual group of pathotypes (Kumar et al. 2003a, b) (Table 6.4).

On differentials, each isolate collected from different locations of India behaves differentially (Table 6.5). Isolate DLK is most virulent as it infects 16 host differentials; RSR-I and GDP on 15 host differentials; HSR-I, HSR-III and RSR-II on 14 host differentials; and isolates HSR-II, GNR, LDH and KNR on only 13 differentials. The comparative study reveals that all the differentials are susceptible to different isolates of *A. brassicae*. The genotype *B. alba* is infected by only four isolates, whereas genotype *B. oleracea* var. *botrytis* is infected by five isolates. The range of incubation period varies from 3 to 13 days in isolates, but in general, majority took 3–5 days to cause the infection (Mehta et al. 2003).

Sangwan and Mehta (2007) analysed virulence pattern of 24 *A. brassicae* isolates on a set of 17 host differentials consisting of *Brassica* species. The isolates BTD, BBK, DSA, GNR, HSR and PNT had wide virulence pattern infecting all the 17 host differentials. Isolates BHP,

Table 6.4 Differential reactions of different isolates of *A. brassicae* when inoculated on a set of host differentials (Kumar et al. 2003a, b)

	Alternaria brassicae isolates									
Differentials	BWL	BHI	CHR-III	HSR	JND-I	KTL-I	KTL-II	REW	RTK	SPT
Brassica campestris var. *toria*	+	+	+	+	+	+	+	+	+	+
B. campestris var. Yellow Sarson	+	+	+	+	+	+	+	+	+	+
B. campestris var. Brown Sarson	+	+	+	+	+	+	+	+	+	+
B. juncea	+	–	–	–	–	–	–	+	+	–
B. napus	+	+	+	+	+	+	+	+	+	+
B. carinata	+	+	–	–	+	+	+	+	+	+
B. tournefortii	+	+	+	+	+	+	+	+	+	+
B. nigra	+	+	+	–	+	+	+	+	+	+
B. chinensis	+	+	+	+	+	+	+	+	+	+
B. pekinensis	+	+	+	+	+	+	+	+	+	+
Raphanus sativus	+	+	+	+	+	+	+	+	+	+
B. oleracea var. *capitata*	+	+	+	+	+	+	+	+	+	+
B. oleracea var. *botrytis*	–	+	+	+	+	+	+	–	+	+
B. rapa	+	+	–	–	–	–	+	+	+	+
B. caulorapa	+	+	+	+	+	+	+	+	+	+
Eruca sativa	+	+	+	+	+	+	+	+	+	+
B. alba	–	–	–	–	–	–	–	+	–	+

BRT, GDP, HSRP, JPR, NGN, *B. alba* and Midas-1 are able to infect 16 host differentials. Virulence of isolates *B. chin* and VRN was confined to 13 host differentials. The host differential *B. alba* var. Local was the least susceptible as only 12 isolates are pathogenic on it. In terms of incubation period, majority of the isolates required 3–5 days to initiate the disease, whereas isolates ASM, BHP, FRD, JPR and RSR needed longer period to produce symptoms on cruciferous vegetables. Eight host differentials differentiated all the 24 isolates into 14 pathotypes/races (Fig. 6.1, Tables 6.5, 6.6 and 6.7).

Meena et al. (2012) measured aggressiveness of *A. brassicae* isolates on host differentials in the form of lesion size. The host differentials *B. alba*, *B. juncea* (PAB, EC-399299), *E. sativa*, *B. carinata* and *B. napus* expressed minimum lesion size.

Pathogenic variability of purified 98 isolates collected from *Brassica* species was analysed to identify virulent pathotype for screening and breed resistant genotypes of oilseed *Brassica* (Singh et al. 2013). Eight variants were grouped on the basis of resistant and susceptible reactions, incubation and latent period, lesion size and *Alternaria* blight severity. Isolates from Rewari and Fatehabad districts were able to infect all host differentials, followed by Bhiwani district isolates on ten differentials and Rohtak isolate with only seven differentials. The Rewari district isolate had the shortest incubation (IP) and latent period (LP) of 4–5 and 6–7 days, respectively, compared to 6–8 and 8–10 days, respectively, of Rohtak district isolates. Maximum *Alternaria* blight severity (24.6 %) and maximum lesion size were also produced by the Rewari district isolates (group 1) compared

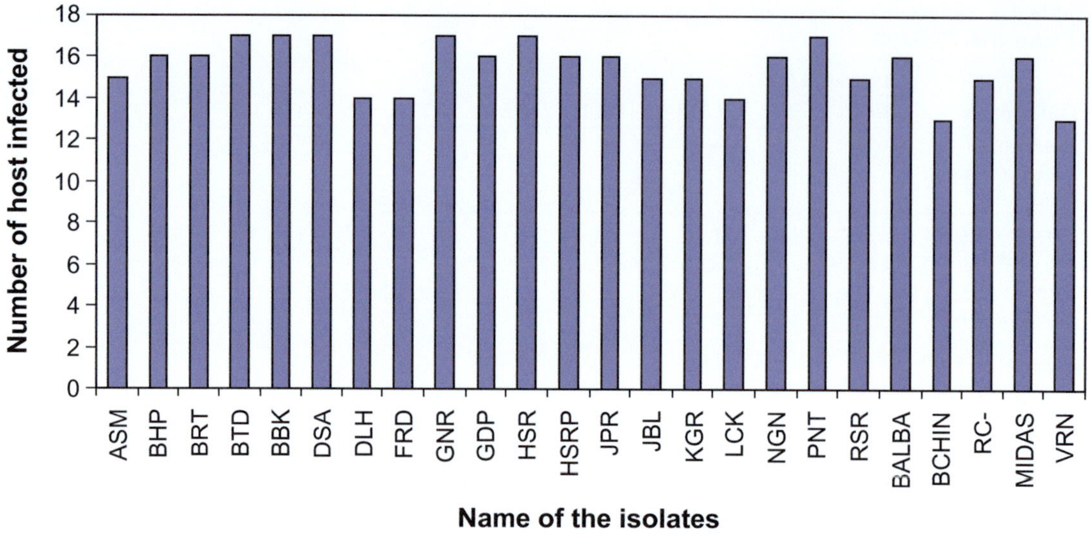

Fig. 6.1 Pathogenic reaction of different isolates of *Alternaria brassicae* on a set of host differential (Sangwan and Mehta 2007).

Table 6.5 Reaction of different isolates of *Alternaria brassicae* on *Brassica* differentials (Mehta et al. 2003)

	Reaction of *Alternaria brassicae* isolates									
Differentials	HSR-I	HSR-II	HSR-III	GNR	LDH	KNR	RSR-I	RSR-II	DLK	GDP
B. campestris var. *toria*	+(5)	–	+(5)	+(11)	+(9)	+(9)	+(9)	+(5)	+(11)	+(5)
B. campestris var. Yellow Sarson	+(5)	+(11)	+(5)	+(11)	+(5)	+(5)	+(5)	+(5)	+(5)	+(5)
B. campestris var. Brown Sarson	+(5)	+(5)	–	+(5)	+(5)	–	+(5)	+(5)	+(5)	+(3)
B. juncea	+(5)	+(5)	+(5)	+(5)	+(5)	–	+(5)	+(5)	+(5)	+(5)
B. carinata	+(13)	+(5)	+(5)	+(5)	+(11)	+(3)	+(9)	–	+(5)	+(5)
B. nigra	+(5)	+(9)	+(8)	+(11)	+(11)	+(9)	+(9)	+(9)	+(5)	+(8)
B. napus	–	+(5)	+(11)	–	+(11)	+(9)	+(11)	+(8)	+(6)	+(5)
B. chinensis	–	+(3)	+(5)	+(3)	+(5)	+(11)	+(3)	+(3)	+(5)	+(5)
B. pekinensis	+(11)	+(3)	+(5)	+(3)	+(3)	–	+(11)	+(5)	+(5)	+(5)
B. tournefortii	+(5)	+(3)	+(3)	+(3)	+(5)	+(3)	–	+(3)	+(3)	+(3)
B. alba	+(5)	–	+(5)	–	–	+(11)	–	+(5)	–	–
Eruca sativa	+(5)	+(9)	+(8)	+(3)	–	+(9)	+(9)	+(5)	+(5)	+(8)
B. oleracea var. *capitata*	+(9)	+(11)	+(3)	–	+(9)	+(9)	+(9)	+(5)	+(5)	+(3)
B. oleracea var. *botrytis*	–	–	+(5)	–	–	–	+(11)	+(5)	+(5)	+(5)
B. rapa	+(9)	+(9)	–	+(5)	+(9)	+(5)	+(9)	+(8)	+(5)	–
B. caulorapa	+(5)	+(5)	–	+(5)	+(9)	+(3)	+(9)	–	+(5)	+(3)
Raphanus sativus	+(5)	–	+(6)	+(9)	–	+(9)	+(11)	–	+(3)	+(5)

B Brassica, *HSR-I* Hisar-I, *HSR-II* Hisar II, *HSR-III* Hisar III, *GNR* Sri Ganganagar, *LDH* Ludhiana, *KNR* Kanpur, *RSR-I* R.S. Pura, Jammu, *DLK* Dhaula Kuan, *GDP* Gurdaspur, () incubation period, + infection, – no infection

Table 6.6 Pathogenic reaction of various isolates of *A. brassicae* from India on selected host differentials (Sangwan and Mehta 2007)

	Alternaria brassicae isolates																	
Differentials	ASM	BHP	BRT	DLH	FRD	GDP	HSRP	PR	JBL	KGR	LCK	NGN	RSR	*B alba*	*B chin*	RC-781	Midas-1	VRN
Brassica juncea	+	+	+	+	+	+	+	+	+	+	+	+	+	+	+	+	+	+
B. napus	+	+	−	+	+	+	+	+	+	+	+	+	+	+	+	+	+	+
B. chinensis	−	+	+	+	+	+	+	+	+	+	+	+	+	−	+	+	+	+
B. alba	+	+	+	−	+	−	−	+	−	−	−	−	−	+	−	−	−	−
B. oleracea var. *capitata*	+	−	+	+	−	+	+	+	+	+	+	+	+	+	+	+	+	+
Raphanus sativus	−	+	+	+	−	+	+	−	+	+	+	+	+	+	−	+	+	−
Eruca sativa	+	+	+	+	+	+	−	+	+	−	−	+	+	+	+	−	+	+
B. rapa	+	+	+	−	+	+	+	+	+	+	+	+	+	+	−	+	+	+
B. caulorapa	+	+	+	−	+	+	+	+	+	+	−	−	+	+	−	+	+	+

Table 6.7 Pathogenic behaviour of different isolates of *Alternaria brassicae* from India on a set of host differentials (Sangwan and Mehta 2007)

	Isolates of *Alternaria brassicae*																							
Differentials	ASM	BHP	BRT	BTD	BBK	DSA	DLH	FRD	GNR	GDP	HSR	HSRP	JPR	JBL	KGR	LCK	NGN	PNT	RSR	*B alba*	B chin	RC-781	Midas-1	VRN
B. campestris var. *toria*	+ (3)[b]	+ (3)	+ (4)	+ (7)	+ (3)	+ (3)	+ (4)	+ (3)	+ (3)	+ (5)	+ (3)	+ (5)	+ (6)	+ (3)	+ (4)	+ (5)	+ (3)	+ (3)	+ (9)	+ (3)	+ (3)	+ (4)	+ (5)	+ (3)
B. campestris var. Yellow Sarson	+ (4)	+ (3)	+ (3)	+ (6)	+ (3)	+ (3)	+ (4)	+ (3)	+ (3)	+ (5)	+ (5)	+ (5)	+ (3)	+ (3)	+ (4)	+ (8)	+ (4)	+ (4)	+ (5)	+ (3)	+ (3)	+ (4)	+ (5)	+ (3)
B. campestris var. Brown Sarson	+ (4)	+ (3)	+ (3)	+ (7)	+ (4)	+ (4)	+ (4)	+ (3)	+ (3)	+ (3)	+ (3)	+ (5)	+ (4)	+ (3)	+ (4)	+ (5)	+ (4)	+ (4)	+ (5)	+ (6)	+ (6)	+ (4)	+ (5)	+ (5)
B. juncea	+ (4)	+ (4)	+ (5)	+ (6)	+ (4)	+ (4)	+ (4)	+ (6)	+ (6)	+ (5)	+ (5)	+ (5)	+ (6)	+ (4)	+ (4)	+ (7)	+ (3)	+ (4)	+ (5)	+ (4)	+ (3)	+ (5)	+ (4)	+ (5)
B. napus	+ (4)	+ (4)	– (0)	+ (6)	+ (4)	+ (7)	+ (6)	+ (4)	+ (4)	+ (5)	+ (3)	+ (5)	+ (4)	+ (4)	+ (9)	+ (8)	+ (5)	+ (4)	+ (11)	+ (4)	+ (3)	+ (3)	+ (6)	+ (3)
B. carinata	+ (7)	+ (7)	+ (3)	+ (6)	+ (6)	+ (7)	+ (7)	– (0)	+ (4)	+ (5)	+ (4)	+ (5)	+ (7)	– (0)	+ (4)	+ (8)	+ (6)	+ (5)	+ (9)	+ (4)	+ (6)	+ (5)	+ (8)	– (0)
B. tournefortii	+ (4)	+ (4)	+ (4)	+ (7)	+ (4)	+ (4)	+ (4)	+ (3)	+ (4)	+ (3)	+ (4)	+ (5)	+ (4)	+ (4)	+ (4)	+ (6	+ (5)	+ (5)	– (0)	+ (4)	+ (5)	+ (3)	+ (7)	+ (5)
B. nigra	+ (4)	+ (4)	+ (3)	+ (3)	+ (3)	+ (4)	+ (6)	+ (4)	+ (4)	+ (8)	+ (3)	+ (8)	+ (7)	+ (4)	+ (4)	+ (5)	+ (5)	+ (5)	+ (9)	+ (4)	+ (5)	+ (5)	+ (7)	+ (3)
B. chinensis	– (0)	+ (7)	+ (6)	+ (3)	+ (6)	+ (7)	+ (5)	+ (7)	+ (4)	+ (5)	+ (5)	+ (6)	+ (7)	+ (7)	+ (5)	+ (5)	+ (5)	+ (6)	+ (3)	– (0)	+ (6)	+ (5)	+ (5)	+ (3)
B. pekinensis	+ (4)	+ (4)	+ (7)	+ (8)	+ (6)	+ (4)	+ (5)	+ (4)	+ (7)	+ (5)	+ (5)	+ (6)	+ (7)	+ (4)	+ (7)	+ (5)	+ (3)	+ (5)	+ (11)	+ (4)	+ (5)	+ (3)	+ (7)	+ (5)
B. alba	+ (4)	+ (7)	+ (7)	+ (6)	+ (7)	+ (7)	– (0)	+ (7)	+ (7)	– (0)	+ (6)	– (0)	+ (4)	– (0)	– (0)	– (0)	– (0)	+ (3)	– (0)	+ (3)	– (0)	– (0)	– (0)	– (0)
B. oleracea var. *capitata*	+ (4)	– (0)	+ (3)	+ (4)	+ (3)	+ (4)	+ (6)	– (0)	+ (4)	+ (3)	+ (3)	+ (5)	+ (9)	+ (4)	+ (4)	+ (8)	+ (5)	+ (5)	+ (9)	+ (4)	+ (6)	+ (6)	+ (5)	+ (8)
B. oleracea var. *botrytis*	+ (4)	+ (9)	+ (4)	+ (5)	+ (10)	+ (4)	+ (6)	+ (9)	+ (4)	+ (5)	+ (6)	+ (8)	+ (7)	+ (9)	+ (4)	+ (7)	+ (5)	+ (5)	+ (11)	+ (4)	+ (6)	+ (6)	+ (5)	+ (8)
B. rapa	+ (9)	+ (9)	+ (6)	+ (7)	+ (7)	+ (4)	– (0)	+ (4)	+ (4)	+ (5)	+ (3)	+ (6)	+ (7)	+ (9)	+ (4)	+ (5)	+ (5)	+ (5)	+ (9)	+ (7)	– (0)	+ (4)	+ (5)	+ (9)
B. caulorapa	+ (9)	+ (9)	+ (7)	+ (7)	+ (3)	+ (6)	– (0)	+ (9)	+ (4)	+ (3)	+ (5)	+ (9)	+ (9)	+ (9)	+ (6)	– (0)	– (0)	+ (5)	+ (9)	+ (4)	– (0)	+ (4)	+ (7)	+ (6)
Eruca sativa	+ (4)	+ (9)	+ (7)	+ (6)	+ (7)	+ (6)	+ (6)	+ (7)	+ (7)	+ (8)	+ (5)	– (0)	+ (9)	+ (4)	– (0)	– (0)	+ (5)	+ (6)	+ (9)	+ (7)	+ (6)	– (0)	+ (7)	+ (5)
Raphanus sativus	– (0)	+ (7)	+ (3)	+ (6)	+ (6)	+ (7)	+ (5)	– (0)	+ (4)	+ (5)	+ (3)	+ (9)	– (0)	+ (5)	+ (4)	+ (9)	+ (5)	+ (3)	+ (11)	+ (7)	– (0)	+ (5)	+ (5)	– (0)

[a]B. = *Brassica*

[b]Figure in parentheses indicates the incubation period

to Rohtak district isolates (group 6). *Brassica juncea* var. Varuna contracted the highest disease severity (24.6 %) and *B. alba* the least (2.9 %). Rewari district isolates can, therefore, be used for screening oilseed *Brassica* germplasm. *Alternaria* blight tolerance of *B. alba* genotypes can be harnessed as donor parent for breeding resistance/tolerant variety (Singh et al. 2013). Seven pathotypes (Abr 1 to Abr 7) of *A. brassicae* infecting rapeseed–mustard were identified on a set of six *Brassica* host differentials from Himachal Pradesh, India. Pathotype Abr 4 was most predominant and virulent infecting all the six test host differentials (Kumar et al. 2014).

6.4 Symptomatological Variations

The symptom variability exhibited by *A. brassicae* isolates on leaves of different *B. juncea* host differentials is generally in the form of medium-sized, circular, greyish brown spots, 6–8 mm in diameter, with three regular concentric raised rings of dark brown colour; no yellow halo (Plate 6.1, Fig. 2) is produced (BWL). Another type of symptoms produced by other isolates is in the form of large, circular lesions (8–10 mm), light green in the centre with concentric rings containing dark yellow halo (Plate 6.1, Fig. 1) around the spot (RTK). Third type of isolates produced large, irregular spots of 10 mm in diameter, dark brown in centre and greyish around the margins with concentric rings (Plate 6.1, Fig. 3) without yellow halo (HSR). In the fourth type, no yellow halo is produced, having medium-sized spots, light green in colour with only one concentric ring and very light pale ring (Plate 6.1, Fig. 4) around the spots (REW). The symptoms produced by four different isolates are stable and distinct on *B. juncea* host differentials; each distinct pathotype produces its own characteristic symptoms. This shows that symptom variability is a function of specific pathotype rather than host differentials' genetic variations (Gupta et al. 2004). These pathotypes are designated as Bj-4 (BWL), Bj-5 (HSR), Bj-6 (RTK) and Bj-7 (REW) in the order of their discovery as suggested by Saharan and Kadian (1983). Kolte et al. (1991) identified three pathotypes, A, C and D, on the basis of virulence and some spot characters, viz. spot colour, periphery colour, presence or absence of concentric rings and central region of the spot. Goyal et al. (2013) reported pathogenic variability among *A. brassicae* isolates on host genotypes, on the basis of many qualitative characters, including spot colour and periphery colour, central point and its colour, presence or absence of concentric rings and yellow halo region, and one quantitative character, viz. per cent disease severity. Three characters, i.e. central point colour, presence or absence of central point and yellow halo region of the spot, should be used to study variability among *A. brassicae* isolates. Pathogenic variability test revealed that all the isolates from rapeseed–mustard were pathogenic or aggressive at different rates on all 12 host differentials and produced different types of spots on different hosts (Figs. 6.2 and 6.3).

6.5 Morphological and Cultural Variations

Mehta et al. (2003) identified (Table 6.8) various isolates from different locations on the basis of the size of *A. brassicae* spores by designating them with the place of collection:

1. Hisar: HSR-I, *B. juncea* (var. RH-30); HSR-II, *B. campestris* var. Yellow Sarson (var. YSPb-24); and HSR-III, *B. tournefortii* (var. Local)
2. Sri Ganganagar: GNR, *B. juncea* (var. *Kranti*)
3. Ludhiana: LDH, *B. juncea* (var. RH-30)
4. Kanpur: KNR, *B. juncea* (var. RH-30)
5. Dhaula Kuan: DLK, *B. juncea* (var. RH-30)
6. Gurdaspur: GDP, *B. juncea* (var. RH-30)
7. R.S. Pura: RSR-I, *B. juncea* (var. RH-30), and RSR-II, *B. juncea* (var. RH-30)

The morphological characteristics of each isolate including size (length and breadth), number of septa, beak length, beak septa, etc., were recorded from 15-day-old culture.

Based on spore length, isolates were categorized into four groups, i.e. small (<100 μm),

Plate 6.1 Symptomatological variations of four isolates of *Alternaria brassicae* on *Brassica juncea*. Fig. 1, RTK; Fig. 2, BWL; Fig. 3, HSR; and Fig. 4, REW (Gupta et al. 2004)

medium (101–150 μm), long (151–200 μm) and very long (>200 μm). Group 1 includes GDP; group 2 includes HSR-I, HSR-III, GNR, KNR and RSR-I; group 3 contains LDH and DLK; and group 4 includes HSR-II and RSR-II isolates. The longest spore length was observed in the case of HSR-II and RSR-II (>200 μm), and the shortest in the case of GDP (94.45 μm). The breadth ranged from 13.5 to 36.0 μm with the maximum spore breadth in GNR and minimum in the case of HSR-II. The number of horizontal septa varied from 5 to 13, with maximum being in the case of HSR-II and minimum in the case of GDP (Table 6.8). The maximum beak length was observed in the case of HSR-II and minimum in the case of GNR (Mehta et al. 2003). The number of septa in beak varied from zero to six (Plate 6.2).

Conidial size variations in the *A. brassicae* isolates are due to nutrition rather than a characteristic pathological variation (Saharan and Kadian 1983). However, glaring differences in conidial size are noticed among the isolates even when the same medium is used for the growth of the isolates. It can be assumed that variation in the isolates may be inherent since isolates were collected from diverse agroclimatic zones. It is

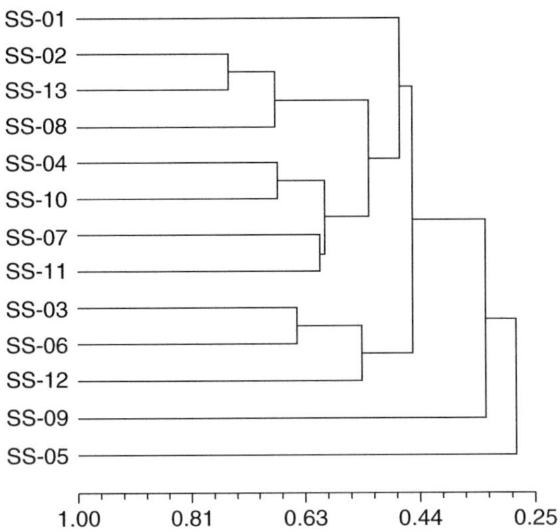

Fig. 6.2 Dendrogram showing pathogenic variability among 13 *A. brassicae* isolates in respect of five qualitative characters, i.e. spot colour, periphery colour, presence or absence of concentric rings, central point and yellow halo region, and one quantitative character, i.e. per cent disease severity (Goyal et al. 2013)

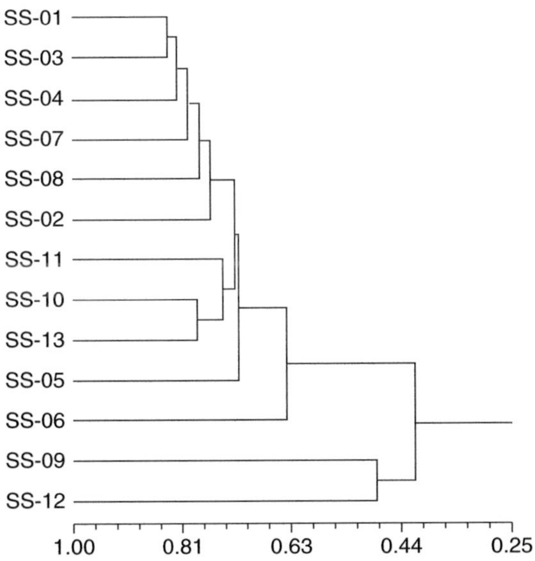

Fig. 6.3 Dendrogram showing molecular variability among 13 *A. brassicae* isolates based on RAPD fingerprints obtained from 100 RAPD primers (Goyal et al. 2013)

evident from the data that each isolate differed in their conidial size. Hence, these variations in the conidial size indicate the existence of variability in this pathogen in India (Mehta et al. 2003).

Conidial/spore measurement recorded on each isolate from Haryana (India) revealed that isolates differed in their conidial size. The average conidial length varied from 118.62 to 194.52 μm, being the maximum of isolate RTK and the minimum of isolate HSR. The range of conidial size varied from 81 to 300 μm. The average breadth of conidia varied from 14 to 23 μm, the thickest being of isolate JND-I and the thinnest of isolate CHR-III. The horizontal septations varied from 3 to 12 and vertical from 0 to 6 (Table 6.9). Some variations in beak size were also recorded. The average beak length varied from 39.99 to 119.07 μm. The longest beak was of isolate RTK and the smallest of JND-I. The average beak septations varied from 1.9 to 5.8 (Table 6.9). These observations revealed that variation in the conidial size existed in Haryana, India (Plate 6.3) (Kumar et al. 2003b).

Variations in morphology and cultural characteristics among 13 different geographical *A. brassicae* isolates in India were analysed by Goyal et al. (2011). All the isolates showed high level of variability in vitro in respect to conidial length, width, beak length and number of septa. Conidia of Nazirhat isolate (SS 04) are smallest with lowest number of septa. Substantial variations among the isolates were found in mycelial growth, and sporulation in different nutrient media, and artificial environmental conditions including temperature, relative humidity, light and hydrogen ion concentration. Different optimum temperature ranges were found for mycelial growth (25–30 °C) and sporulation (15–35 °C). All 13 isolates grew bested at 100 % relative humidity. However, they sporulate the most at different relative humidity (40–100 %). This reflects the adaptation of the respective isolates to the ambient conditions in the different cropping areas, which also may have induced the cultural variability. All the isolates did not grow and sporulate abundantly on the same nutrient medium. Asthana and Hawker's media were generally better for all the isolates. Variation in optimum pH and light conditions for mycelial growth and sporulation was also observed. Cluster analysis of data on cultural variability among 13 *A. brassicae* isolates found a close relationship

Table 6.8 Differences in morphological characters of different isolates of *Alternaria brassicae* from India (Mehta et al. 2003)

		Length (μm)		Breadth (μm)		Septation				Beak Length (μm)		Beak	
						Horizontal		Vertical					
Sr. No	Isolates	Av.	Range	Av.	Range	Av.	Range	Av.	Range	Av.	Range	Av.	Range
1	ASM	168.2	109.1–227.3	26.4	18.2–36.4	8.7	6–11	3.5	2–6	54.5	18.2–90.9	3.0	2–4
2	BHP	126.4	100.0–172.7	20.5	18.2–31.2	9.5	5–13	1.4	0–4	19.1	9.1–27.3	1.2	0–2
3	BRT	57.6	45.0–072.0	13.5	9.0–13.5	9.5	5–13	1.4	0–4	19.1	9.1–27.3	1.2	0–2
4	BTD	119.1	90.9–181.8	20.5	13.4–27.3	7.9	5–10	1.4	0–4	53.6	27.3–109.1	3.5	2–6
5	BBK	121.8	90.9–154.5	20.5	13.6–27.3	7.6	6–09	1.7	0–3	51.8	27.3–72.7	2.8	1–4
6	DSA	134.5	90.9–218.2	24.1	18.2–27.3	7.3	6–10	1.7	1–4	77.3	27.3–136.4	2.5	1–4
7	DLH	114.5	90.9–136.4	23.6	18.2–27.3	7.2	6–09	0.7	0–2	46.4	27.3–63.6	2.6	1–5
8	FRD	90.0	72.7–108.1	21.8	18.2–27.3	7.1	6–09	1.2	0–3	32.7	27.3–45.5	1.8	1–3
9	GNR	134.1	99.0–162.0	27.0	18.0–36.0	8.1	7–09	1.0	0–2	26.1	18.0–36.0	1.3	1–3
10	GDP	103.6	90.9–154.5	20.5	18.2–27.3	7.1	6–09	1.2	0–3	35.5	18.2–90.9	1.3	0–3
11	HSR	129.2	90.0–198.0	25.7	18.0–36.0	7.3	5–10	1.4	0–3	49.5	27.0–72.0	1.9	0–3
12	HSRP	194.5	172.7–245.4	20.9	18.2–27.3	10.4	8–14	1.6	0–4	75.5	63.6–109.1	2.8	2–13
13	JPR	130.0	90.9–145.4	20.5	13.6–27.3	6.6	5–08	0.7	0–2	55.4	36.4–127.3	1.5	0–6
14	JBL	128.7	72.7–181.8	23.2	18.2–27.3	6.6	5–08	2.9	1–4	49.1	27.3–136.4	2.3	1–6
15	KGR	134.5	109.1–172.7	13.5	18.2–27.3	10.4	8–14	1.6	0–4	75.5	63.6–109.1	2.8	2–13
16	LCK	129.4	90.9–136.4	20.5	18.2–27.3	7.6	6–09	1.4	0–3	49.5	27.0–72.0	2.8	2–13
17	NGN	72.9	54.0–103.5	15.3	18.0–13.5	9.0	8–11	4.2	2–7	19.8	9.0–45.0	1.4	0–2
18	PNT	172.3	127.3–243.4	22.7	18.2–27.3	9.8	8–12	3.5	2–7	54.2	9.1–136.4	3.4	1–6
19	RSR	190.9	136.4–218.2	25.5	18.2–36.4	10.2	6–13	3.8	0–7	54.5	36.8–81.8	3.4	2–6
20	*B alba*	100.9	72.8–163.6	16.8	13.6–18.1	7.5	5–09	2.3	1–4	47.7	36.4–90.9	2.5	1–9
21	*B chin*	104.5	90.9–181.8	16.8	13.6–27.3	7.9	6–09	2.3	1–4	41.8	27.3–90.9	1.7	1–3
22	RC-781	147.7	109.1–236.3	19.1	18.2–22.7	8.1	7–11	1.9	1–4	65.5	27.3–136.4	1.0	0–13
23	Midas-1	170.0	119.2–200.0	18.6	18.2–22.7	9.5	6–12	1.8	0–6	63.6	27.3–90.9	3.3	1–5
24	VRN	271.8	72.7–459.9	18.6	18.2–22.7	8.9	6–13	1.7	0–4	185.4	27.3–472.9	8.5	2.1

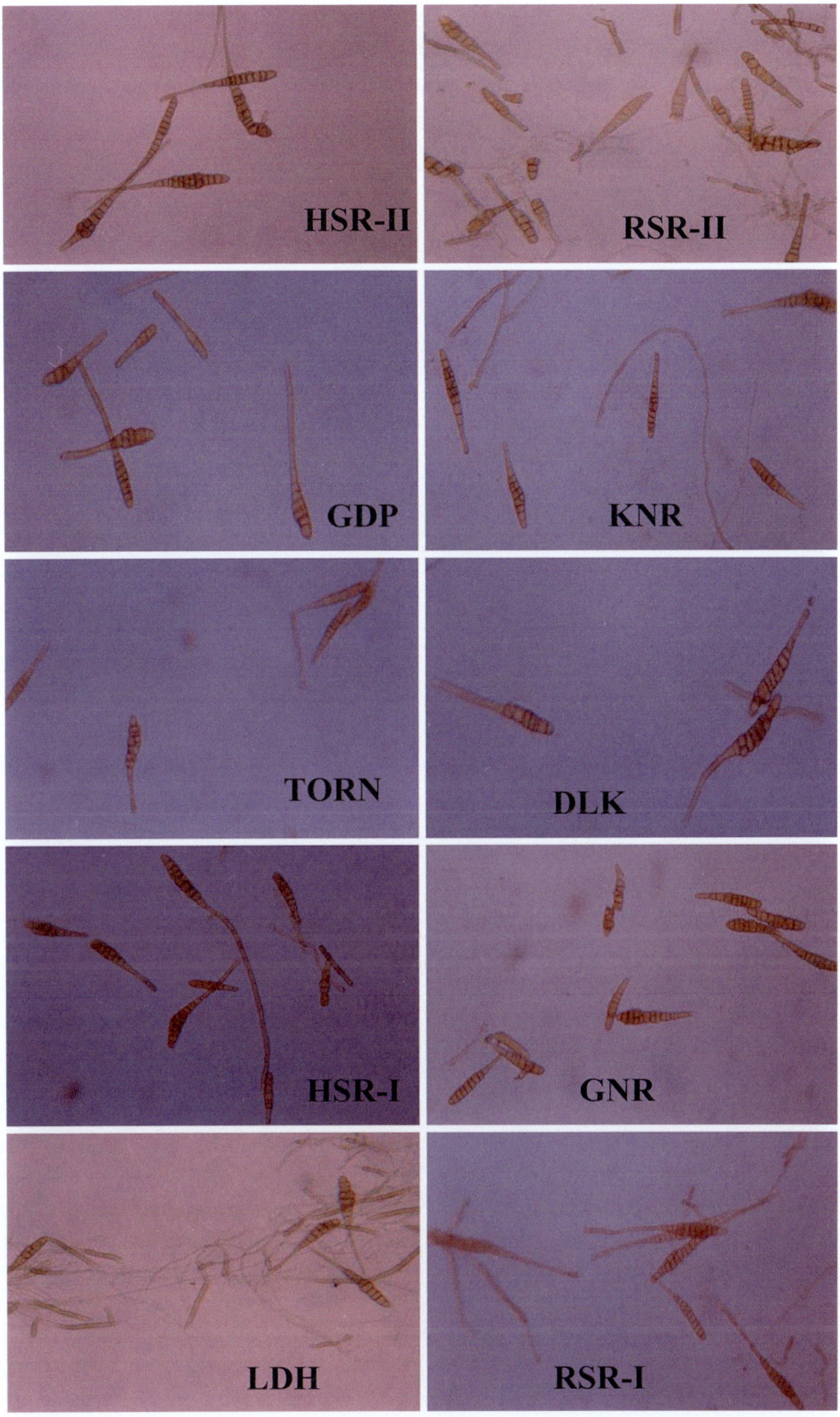

Plate 6.2 Morphological variations in conidia of *Alternaria brassicae* from India (Mehta et al. 2003)

Table 6.9 Conidial size of *Alternaria brassicae* from different locations (Mehta et al. 2003)

Isolates	Length (μm)		Breadth (μm)		Septation (no.) Horizontal		Septation (no.) Vertical		Beak length (μm)		Beak septation (no.)	
	Av.	Range	Av.	Range	Av.	Range	Av.	Range	Av.	Range	Av.	Range
HSR-I	129.1	90–198	25.6	18–36	7.3	5–10	1.4	0–3	49.5	27–72	1.9	0–3
HSR-II	231.3	189–270	20.2	13–27	11.6	11–13	1.7	0–3	105.3	81–135	3.6	2–6
HSR-III	140.8	108–180	22.0	18–27	8.4	7–10	1.0	0–2	52.6	36–72	2.3	1–4
GNR	134.1	99–162	27.0	18–36	8.1	7–9	1.0	0–2	26.1	18–36	1.3	1–3
LDH	153.9	135–171	25.2	18–27	7.3	5–10	1.6	0–3	75.6	63–99	3.1	2–4
KNR	144.9	117–180	25.6	18–27	8.9	7–10	1.6	1–2	53.5	27–72	2.6	1–4
RSR-I	109.8	90–135	22.9	18–36	6.8	6–9	1.3	0–2	33.7	18–45	1.7	0–2
RSR-II	218.7	153–270	23.4	13–36	11.0	9–13	2.1	0–4	52.2	36–90	2.9	2–4
DLK	163.3	121–225	23.4	18–27	9.5	7–12	1.1	0–2	30.6	18–40	1.8	2–4
GDP	94.4	90–117	20.7	13–36	7.2	6–9	1.0	0–2	30.6	18–40	1.8	1–2

Av. = Average

among isolates from Uttar Pradesh, Uttaranchal and Haryana, but distantly related to other states (Goyal et al. 2011).

Morphological characteristic of different *A. brassicae* isolates revealed variation in growth, shape and pigmentation of colony, conidial measurements and number of septa. Conidial length varied from 106.7 to 285.9 μm, width from 33.5 to 57 μm and beak length from 41.4 to 180.0 μm. The number of horizontal septa varied from 3.2 to 8.0 and vertical septa from 0.3 to 1.4. Different synthetic media showed profound variation in mycelial growth and sporulation indicating that the degree of sporulation in *A. brassicae* isolates is a function of nutrition (Tables 6.10, 6.11 and 6.12). Pathogens' aggressiveness demonstrated the existence of considerable variations in the level of tolerance of *Brassica* species to *A. brassicae* (Meena et al. 2012).

Variations in morphology and cultural characteristics were observed among 32 representative Indian geographical isolates of *A. brassicae* from cauliflower and rapeseed–mustard (Sharma et al. 2013). All the isolates showed high level of variability in vitro with respect to conidial length, width and number of septa. Conidia of isolates from Uttar Pradesh (CaABU4) were the smallest with lowest number of septa. Substantial variation among isolates was also observed in mycelial growth and sporulation on different nutrient media. All the isolates do not grow and sporulate abundantly on the same nutrient medium. However, potato dextrose agar, cauliflower (host) agar and carrot potato agar were suitable for all isolates. Cluster analysis of data on cultural variability among 32 *A. brassicae* isolates found a close relationship among isolates of both hosts cauliflower and mustard. Isolates from Uttar Pradesh, Delhi, Haryana and West Bengal are found to be similar to each other, whereas the Rajasthan isolates along with Tamil Nadu and Kerala isolate are distantly related to others. All the isolates are pathogenic in nature but directly related to the cultural and morphological characteristics. These isolates are further molecularly characterized by using internal transcribed spacer

Plate 6.3 Morphological variations in conidia of *Alternaria brassicae* from Haryana (India) (Mehta et al. 2003)

Table 6.10 Sporulation index of *Alternaria brassicae* isolates on different culture media (Meena et al. 2012)

Isolates	Brown's	Eilliot's	Asthana and Hawker's	Czapek's	Richard's
BAB-02	–	+	++	+	–
BAB-04	–	++	+	++	+++
BAB-06	–	+	+	+	–
BAB-08	–	+	++	+	–
BAB-18	–	++++	++	++	++
BAB-19	–	–	+	–	+
BAB-20	–	–	++++	+	++
BAB-23	–	–	++	+	+
BAB-28	–	++++	++++	++++	+++
BAB-29	–	+	+	+	–
BAB-30	–	–	++	+	++
BAB-39	–	–	++	–	–
BAB-40	–	–	++	+	+
BAB-41	–	++	+	+	–
BAB-42	–	–	+	+	+++
BAB-43	–	+	–	–	+
BAB-44	–	+	++	+	+
BAB-45	–	+	–	–	+
BAB-47	–	+	+	+	+
BAB-48	–	+	+	+	–
BAB-49	–	–	+++	+	–
BAB-50	–	++	+	++	+++

Sporulation index (number of spores per microscopic fields): – absent (nil); + trace (1–10); ++ mild (11–30); +++ moderate (31–50); ++++ abundant (more than 50)

region where all the isolates are found 56 % similar to each other and 99 % similar to the *A. brassicae* isolates present in NCBI database. *Alternaria brassicae* colonies varied in their cultural behaviour ranging from cottony, flurry to feathery, with smooth to rough margins, and white, off white to light brown in colour. The growth rate varied from slow, medium to fast, with fastest being in isolate KM and slowest in isolate JD. Significant morphological variations in conidial length, width (105 to 135 × 10 to 20 μm) and number of horizontal septa were observed. Isolates exhibited variations in disease severity, number and size of lesions. The dendrogram analysis based on molecular (DNA, RAPD) basis reveals two groups at 14 % similarity coefficient. Group 1 composed of seven isolates, namely VR, DV, P7, LM, P10, KR and ND, with 18 % similarity (82 % dissimilarity), while group 2 (Figs. 6.4 and 6.5) contained only three isolates including JD, KA and AS with only 24 % similarity (76 % dissimilarity) (Pramila et al. 2014).

6.6 Genetic Variability

Genetic variability in nucleotide sequence of ITS region of four *Alternaria* species (*A. brassicae*, *A. brassicicola*, *A. raphani*, *A. alternata*) infecting crucifers has not been recorded so far (Jasalavich et al. 1995). Cluster analysis of pathogenic variability data reveals a close relationship between Nazirhat (SS 04), Jaipur (SS 07), Sacha khera (SS-10) and Samalkha (SS-11) isolates. The use of 100 random amplified polymorphic DNA decamer primers indicates genetic variability among 13 *A. brassicae* isolates. Almost all the isolates show relationship according to their geographical origin except Sacha khera (SS-10) and Hatikhuti (SS-05) isolates.

Table 6.11 Conidial size of different geographical isolates of *A. brassicae* (Meena et al. 2012)

A. brassicae isolates	Length (μm)	Width (μm)	Beak length (μm)	No. of septa	
				Horizontal	Vertical
BAB-02	152.3	47.5	047.3	3.8	1.3
BAB-04	185.1	47.3	095.2	3.5	0.9
BAB-06	252.1	33.9	178.0	6.8	0.1
BAB-08	106.7	33.5	048.3	3.8	0.3
BAB-18	285.9	47.5	180.0	6.3	0.5
BAB-19	140.6	44.2	061.4	3.2	0.9
BAB-20	189.7	35.6	111.5	4.2	0.3
BAB-23	211.7	41.8	125.9	3.7	0.7
BAB-28	122.2	34.7	044.7	3.4	0.4
BAB-29	198.4	43.4	104.7	3.8	0.9
BAB-30	193.6	57.0	087.3	3.2	0.9
BAB-39	144.5	37.2	070.9	3.2	0.5
BAB-40	198.2	41.6	100.4	5.6	1.1
BAB-41	206.5	44.2	116.8	4.0	0.8
BAB-42	196.6	33.7	113.9	5.5	0.1
BAB-43	147.5	36.6	066.7	3.3	0.5
BAB-44	140.6	35.4	067.9	3.5	0.4
BAB-45	144.1	48.3	041.4	3.4	1.4
BAB-47	168.1	40.4	076.2	4.1	0.9
BAB-48	151.1	33.9	073.9	4.6	0.4
BAB-49	201.2	44.6	105.9	6.1	1.1
BAB-50	284.3	40.0	172.7	8.0	0.6
Mean	182.8	41.0	095.0	4.4	0.7
CV%	26.1	15.2	043.8	30.8	53.4

Pantnagar isolate (SS-09) was found closely related to Sacha khera (SS-10) isolate. No variability could be located among the *A. brassicae* isolates by internal transcribed spacer–amplified fragment length polymorphism molecular marker. Hence, pathogenic variability does exist among the isolates at the genomic level, but not in the highly conserved region of the genome of the pathogenic *A. brassicae* isolates (Goyal et al. 2013). However, internal transcript spacer analysis done by Sharma et al. (2013) shows that all isolates are 90–100 % similar to each other, indicating genetic similarity among different *A. brassicae* isolates that vary pathogenically. The analysis of 26 RAPD primers revealed a high level of genetic variability among ten isolates of *A. brassicae* from different *B. juncea* cultivars (Pramila et al. 2014). DNA fingerprinting using ISSR primers revealed wide genetic diversity among 32 isolates of *A. brassicae* from Himachal Pradesh. However, they were grouped into five major clusters independent to pathotypes (Kumar et al. 2014).

6.7 Molecular Techniques

RAPD analysis is easy, efficient, fast and reproducible than RFLP analysis in the detection of intraspecific variation in *A. brassicae*, *A. brassicicola* and *A. raphani* pathogenic to crucifers. Polymorphism within an *Alternaria* species by RAPD molecule marker has been described by many workers (Sharma and Tewari 1995, 1998; Kumar et al. 2008). Observing polymorphism among *A. brassicae* isolates from different

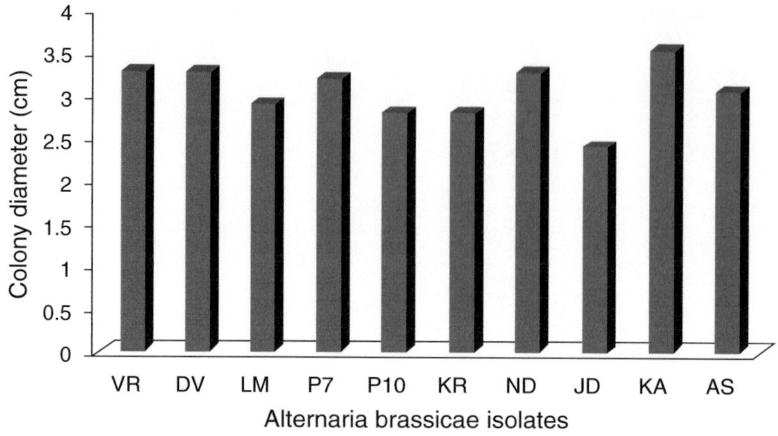

Growth of *A. brassicae* isolates on PDA

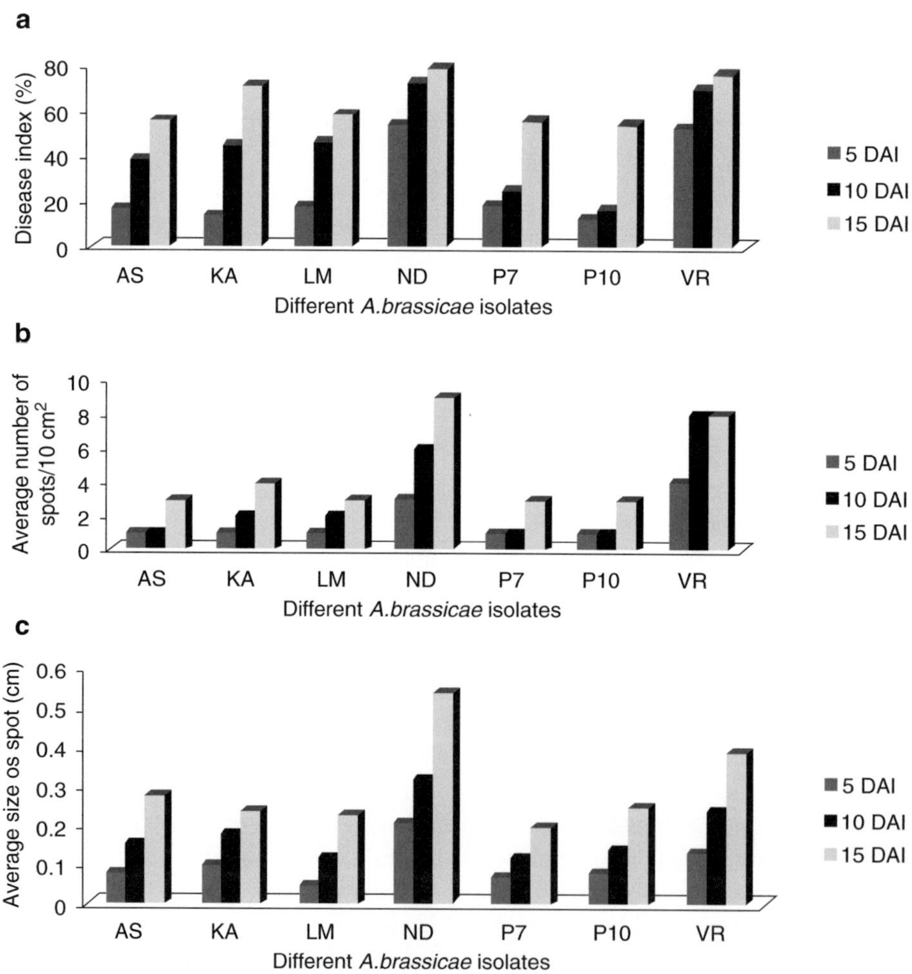

Fig. 6.4 (**a**) Disease index. (**b**) Average number of spores/10 cm^2. (**c**) Average size of spot (cm) on leaves of Divya inoculated by isolates grown on the PDA medium on 5, 10 and 15 days after inoculation (DAI) (Pramila et al. 2014)

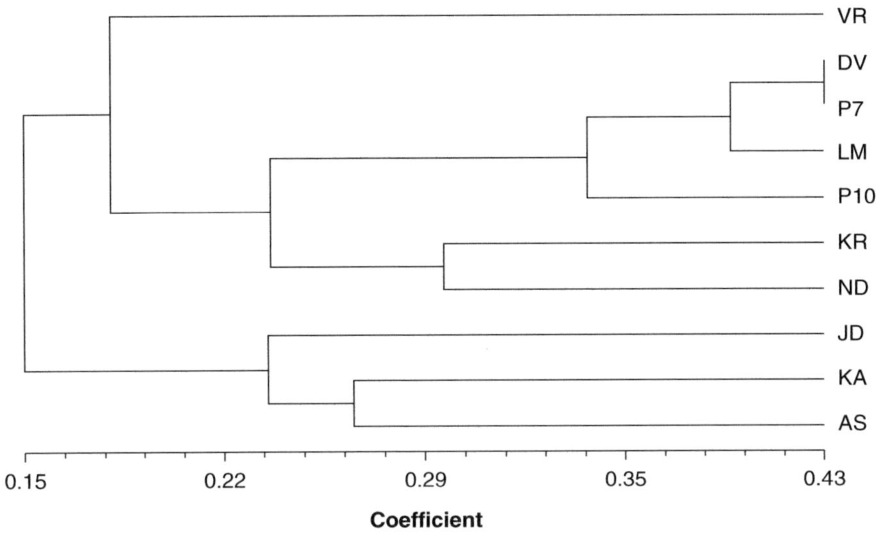

Fig. 6.5 Genetic divergence among ten isolates of *Alternaria brassicae* based on UPGMA cluster analysis (Pramila et al. 2014)

geographical regions of the world, Sharma and Tewari (1995, 1998), however, found low intra-regional variations among Indian and Canadian isolates with 75 % similarity. However, RAPD analysis of *A. brassicae* isolates from different geographical regions of India using more than one hundred primers suggested a high degree of polymorphism among isolates. The dendrograms from both pathogenic and molecular analyses seem to indicate that the Pant Nagar (SS-09) and Hatikhuti (SS-05) isolates are quite different from the others, and the two dendrograms follow the same trend (Goyal et al. 2013). BLAST analysis of the ITS of 32 *A. brassicae* isolates conducted by Sharma et al. (2013) showed high similarity among the isolates available at the NCBI database.

6.8 Proteome Analysis

Two isolates of *A. brassicae* with significant differences in virulence have been characterized at the proteome level. The morphological observations indicated the Ontario isolate to be more virulent by virtue of increased disease severity score as compared to the UAMH7476 isolate. This was further confirmed through histological observations that showed extensive colonization of the host tissue by the highly virulent isolate. Mycelial protein profiles of the two differentially virulent *A. brassicae* isolates were compared using two dimensional gel electrophoresis (2DE) and mass spectrometry (MS) in order to identify proteins that may be responsible for the differences. Several differences in the mycelial proteomes of the two isolates were recorded. The proteins that were significantly abundant in the more virulent isolate included a protein with conserved actin-related protein2/3 domain, enolase, malate dehydrogenase and serine protease. The differential protein expression pattern can be exploited to identify putative virulence and pathogenicity factors in *A. brassicae* (Sharma et al. 2012).

6.9 Nutritional Variability

Fourteen isolates of *Alternaria brassicae* causing *Alternaria* blight in rapeseed–mustard were characterized by their responses to various carbon and nitrogen sources, as well as to pH. All the isolates behaved differentially in growth and sporulation in relation to different carbon and nitrogen sources. Isolates KTL-I and BWL show

Table 6.12 Mycelial growth of *A. brassicae* under different temperature and relative humidity conditions (Meena et al. 2012)

Alternaria brassicae isolate	Radial growth (mm)[a]														
	Temperature (°C)[b]						Relative humidity (%)[c]								
	15	20	25	30	35	40	50	60	70	75	80	85	90	95	100
BAB-18	15.7	16.0	24.7	24.7	16.3	12.3	10.0	13.3	20.0	22.3	24.3	26.3	24.7	29.3	31.3
BAB-19	11.0	16.3	27.0	25.3	16.0	12.7	10.5	13.5	20.5	21.7	23.5	25.7	27.0	29.0	30.0
BAB-20	15.3	19.0	30.0	27.3	18.0	12.7	10.3	13.0	20.3	21.3	24.0	26.7	30.0	28.7	29.5
BAB-23	11.0	15.3	24.0	21.0	15.0	13.0	11.7	14.0	21.0	23.0	25.0	27.7	24.0	30.7	31.0
BAB-30	16.7	16.7	28.0	23.3	16.3	12.3	9.7	12.7	19.5	21.5	23.7	26.7	28.0	29.3	30.0
BAB-43	14.3	15.0	29.3	29.0	16.0	10.0	10.7	13.0	20.3	22.0	24.7	27.0	29.3	29.0	30.0
BAB-48	15.7	16.3	24.3	27.0	16.0	14.0	11.0	13.7	19.7	22.7	24.5	27.3	24.3	29.3	30.5

LSD temperature = 2.2, isolate = 1.8, relative humidity = 3.2, isolate = 2.1
($P < 0.05$): temperature x isolate = 2.1; relative humidity × isolate = 2.2

[a]Mean of three replications

[b]Temperature 25 °C

[c]Relative humidity 100 %

significantly higher growth on all the carbon sources, whereas isolate CHR-I yielded minimum growth and responded differentially to different carbon sources. Similarly, isolate REW showed more variation in sporulation than SPT and HSR isolates. Among the nitrogen sources evaluated, growth was maximum on sodium nitrate followed by potassium nitrate, ammonium nitrate and glycine. Irrespective of the nitrogen source, isolates KTL-I and CHR-II produced the maximum, whereas SRS and JD-II the minimum radial growth. Isolates BWL, CHR-I, CHR-II, RTK, KTL-I and SPT responded best on KNO_3-amended medium, whereas REW poorly on glycine; isolates BWL, CHR-II, JD-I, SRS and SPT sporulated best on KNO_3 and HSR very poorly on glycine. All isolates grew better at pH 7.5, but sporulated best at 5.0 after 21 days of incubation. On the basis of nutritional behaviour, all the isolates were placed into two major groups: isolates BHI, JD-I, JHR, REW, RTK, SRS, CHR-I, CHR-II, JD-II and HSR were placed in group 1, whereas isolates KTL-I, KTL-II and BWL formed the second group (Mehta et al. 2005b; Tables 6.13, 6.14, 6.15 and 6.16).

6.10 Biochemical Variability

Biochemical constituents of *A. brassicae* isolates differ significantly (Khurana et al. 2005b). The isolate BWL contains the maximum ortho-dihydric phenols and KTL-II the least (O.D. phenols); isolate REW contains the highest amount of total phenols (Table 6.17). The differences in the amount of total and ortho-dihydric phenols indicate the existence of variation in isolates. The isolate JD-II contains the maximum amount of both reducing and non-reducing sugars and RTK the minimum. The amount of reducing sugars also varied significantly between isolates. The isolates BWL, CHR-II, CHR-I and RTK have significantly lower reducing sugars than other isolates. Vishwanath and Kolte (1997) reported that *A. brassicae* isolate containing high amount of carbohydrate is an indicator of virulence. The estimation of total RNA content revealed significant differences among the isolates; isolate KTL-II contained the maximum amount of RNA, while SPT the minimum. On the basis of RNA contents, isolates can also be categorized into three groups. The protein content also differed significantly among isolates. Isolates CHR-I and JD-I contained the maximum amount of proteins and isolate KTL-I the least.

Similarly, there are significant differences among the isolates in their free amino acid content; isolates JD-I and SPT contained the highest and lowest amount of free amino acids, respectively. *Alternaria brassicae* isolates containing higher amount of proteins generally contain moderate amount of free amino acids and RNA, while those containing higher amount of RNA often contain moderate amount of proteins and free amino acids. On the basis of similarity in biochemical composition, all isolates were grouped into three categories: maximum, moderate and minimum. Isolates BHI, JHR, KTL-I, SRS, CHR-I and SPT formed the first group with maximum amount of biochemicals; isolates BWL, RTK, CHR-II, HSR, KTL-II and REW were in the second group with moderate amount, while isolates JD-I and JD-II were in the third group with the minimum amount (Khurana et al. 2005b).

6.11 Fungicidal and Plant Extracts Sensitivity

The variation in efficacy of fungicides in controlling *Alternaria* diseases of crucifers may be due to response of pathotypes prevalent in a region. According to Vishwanath and Kolte (1997), isolate A of *A. brassicae* showed more tolerance to ziram and Ridomil MZ-72 at 50 and 100 ppm as compared to isolate C. Similarly, isolate A exhibited maximum tolerance followed by isolates C and D against mancozeb and iprodione. Efficacy of several fungicides and neem products was evaluated against 15 *A. brassicae* isolates collected from different locations in Haryana (India) (Table 6.18). The fungicide Kitazin was highly effective against all isolates in inhibiting the spore germination, which was followed by Dithane M-45 and Ridomil MZ-72. Similarly,

Table 6.13 Effect of various carbon sources on the radial growth (cm) of various isolates of *Alternaria brassicae* from Haryana, India (Mehta et al. 2005b)

	Alternaria brassicae isolates														
Carbon sources	BHI	BWL	CHR-I	CHR-II	HSR	JD-I	JD-II	JHR	KTL-I	KTL-II	REW	RTK	SRS	SPT	Mean	SEM±
Dextrose	5.3	8.6	3.3	2.9	4.7	3.2	6.3	8.4	8.6	7.3	4.2	4.2	5.9	8.5	5.8	0.17
Fructose	5.2	8.9	1.9	5.4	2.7	4.6	5.4	3.5	8.9	7.1	4.1	3.3	3.2	8.5	5.2	0.19
Lactose	5.2	8.6	2.6	5.3	3.5	5.3	5.3	4.5	9.0	7.3	5.2	4.0	6.4	7.5	5.7	0.18
Mannitol	5.1	8.8	2.4	5.5	4.3	5.6	5.4	6.1	9.0	8.6	5.3	5.8	5.9	7.2	6.1	0.16
Sucrose	2.9	8.6	2.2	5.2	2.7	4.8	8.6	5.9	8.6	6.2	5.1	3.0	3.4	7.3	5.3	0.16
Control (without carbon)	3.1	6.4	1.6	2.1	2.6	5.8	6.3	3.6	6.8	6.6	2.9	4.6	1.9	6.3	4.3	0.17
Mean	4.4	8.4	2.3	4.4	3.4	4.9	6.2	5.3	8.5	7.2	4.5	4.1	4.5	7.6		

CD ($p = 0.05$ %): isolates (I) = 0.14; carbon source (C) = 0.09; I × C = 0.34

Table 6.14 Effect of various carbon sources on the sporulation of various isolates of *Alternaria brassicae* from Haryana, India (Mehta et al. 2005b)

Carbon sources	Spores/microscopic field of *Alternaria brassicae* isolates														
	BHI	BWL	CHR-I	CHR-II	HSR	JD-I	JD-II	JHR	KTL-I	KTL-II	REW	RTK	SRS	SPT	Mean	SEM±
Dextrose	6.5	3.8	4.3	6.3	6.0	5.3	6.3	4.5	3.3	5.8	7.3	7.0	5.3	4.5	5.4	0.9
Fructose	7.0	6.8	5.8	4.5	5.0	6.8	6.3	7.3	4.8	5.0	14.0	6.3	7.0	5.0	6.5	1.0
Lactose	2.8	7.3	10.5	2.8	4.8	7.3	5.0	2.5	4.8	3.0	3.5	4.0	6.0	4.8	4.9	0.9
Mannitol	6.5	6.5	6.8	6.8	6.3	5.0	7.3	4.8	6.3	6.5	16.0	6.8	7.0	4.5	6.9	1.1
Sucrose	5.0	5.8	5.3	4.0	4.8	4.0	2.8	6.5	6.3	4.8	14.3	9.8	4.8	5.5	5.9	0.9
Control (without carbon)	2.5	0.3	1.0	1.3	0.8	0.5	1.0	0.8	0.5	0.3	2.5	2.0	1.5	0.8	1.0	0.5
Mean	4.9	5.0	5.6	4.3	4.6	4.8	4.8	4.4	4.3	4.2	9.6	6.0	5.3	4.2		

CD ($p=0.05$ %); isolates (I) = 0.73; carbon sources (C) = 0.47; I × C = 1

Table 6.15 Effect of various nitrogen sources on the radial growth of different isolates of *Alternaria brassicae* from Haryana, India (Mehta et al. 2005b)

Nitrogen sources	Radial growth (cm) of *Alternaria brassicae* isolates														
	BHI	BWL	CHR-I	CHR-II	HSR	JD-I	JD-II	JHR	KTL-I	KTL-II	REW	RTK	SRS	SPT	MEAN	SEM±
Glycine	2.5	3.1	2.5	2.9	4.0	3.1	2.8	2.6	3.4	3.2	1.9	2.5	2.9	3.3	2.9	0.07
KNO_2	0.0	0.0	0.0	0.0	0.0	0.0	0.0	0.0	0.0	0.0	0.0	0.0	0.0	0.0	0.0	0.00
KNO_3	5.1	7.2	7.1	7.1	5.5	5.6	4.6	6.4	7.2	6.6	4.5	7.0	4.4	7.1	6.1	0.10
$NaNO_3$	6.1	7.5	7.4	7.4	7.5	6.2	6.0	6.4	7.3	6.6	5.0	6.5	4.0	6.5	6.5	0.09
NH_4NO_3	3.6	3.8	4.0	3.8	4.0	3.7	2.5	3.7	3.4	3.6	3.1	2.5	3.0	3.6	3.5	0.11
Control (without nitrogen)	3.5	3.8	4.5	4.5	4.3	4.1	3.3	4.3	4.3	3.4	5.5	3.9	3.7	4.1	4.1	0.08
Mean	3.6	4.3	4.3	4.4	4.3	3.9	3.3	4.0	4.4	4.0	3.4	3.8	3.1	4.2		

CD (p=0.05 %); isolates (I)=0.07; nitrogen sources (N)=0.04; I×N=0.17

Table 6.16 Effect of various nitrogen sources on the sporulation of different isolates of *Alternaria brassicae* from Haryana, India (Mehta et al. 2005b)

Nitrogen sources	Spores per microscopic field of *Alternaria brassicae* isolates															
	BHI	BWL	CHR-I	CHR-II	HSR	JD-I	JD-II	JHR	KTL-I	KTL-II	REW	RTK	SRS	SPT	Mean	SEM±
Glycine	1.5	2.2	2.5	2.7	1.0	2.7	2.0	3.5	2.2	1.7	2.2	2.2	3.0	1.7	2.2	0.6
KNO_2	0.0	0.0	0.0	0.0	0.0	0.0	0.0	0.0	0.0	0.0	0.0	0.0	0.00	0.0	0.0	0.0
KNO_3	8.7	10.0	8.5	10.0	9.7	11.5	8.2	9.2	9.7	9.5	7.5	9.2	11.2	10.5	9.5	1.2
$NaNO_3$	7.7	6.7	7.2	8.2	7.5	8.5	7.0	7.5	8.2	8.2	7.5	7.0	8.2	7.2	7.6	0.9
NH_4NO_3	5.2	5.5	6.2	4.0	4.2	4.2	4.7	3.7	5.5	4.5	3.7	4.0	5.2	4.2	4.7	0.7
Control (without nitrogen)	1.2	2.2	1.7	1.2	1.5	1.7	0.0	1.5	1.2	0.0	2.5	0.0	0.0	1.0	1.1	0.6
Mean	4.1	4.4	4.4	4.4	4.2	4.8	3.7	4.2	4.5	4.0	3.9	3.7	4.8	4.1		

CD ($P=0.05$): isolates (I) = 0.62; nitrogen sources (N) = 0.40; I × N = 1.53

Table 6.17 Biochemical constituents (mg/g)[a] in isolates of *Alternaria brassicae* (Khurana et al. 2005a, b)

Isolates	Biochemical constituents							
	Ortho-dihydric phenols	Total phenols	Non-reducing sugar	Reducing sugars	Total sugars	RNA	Proteins	Free amino acids
BHI	1.6	9.5	2.5	2.0	4.5	6.8	66.8	49.2
BWL	3.5	10.4	3.7	0.7	4.5	9.4	85.7	56.5
CHR-I	2.0	5.5	2.8	1.0	3.9	6.4	103.0	37.5
CHR-II	2.2	4.9	1.7	1.0	2.7	8.8	74.2	45.7
HSR	1.7	10.5	6.2	3.5	9.7	8.6	59.9	57.7
JD-I	2.1	6.1	6.1	1.4	7.5	9.6	100.4	63.9
JD-II	2.8	13.3	10.5	5.3	15.8	11.8	76.9	38.9
JHR	1.5	9.6	4.4	1.8	6.1	5.8	67.5	40.9
KTL-I	3.4	13.4	4.6	1.5	6.1	7.5	49.2	51.8
KTL-II	1.3	12.0	4.0	1.5	5.5	16.4	81.9	52.8
REW	2.2	13.5	2.2	1.6	3.8	6.9	85.2	50.2
RTK	1.4	9.7	1.5	1.1	2.6	11.6	89.9	41.3
SRS	2.2	7.9	4.8	3.2	8.0	6.5	68.7	40.3
SPT	2.4	8.2	2.8	2.2	5.0	4.5	70.4	25.6
C.D.(5 %)	0.13	0.33	0.49	0.16	0.28	0.72	3.63	8.11
SEM±	0.06	0.15	0.23	0.07	0.13	0.33	1.69	3.78

[a]Based on dry mycelial weight in 1000 mg

among four neem products, Achook and Bioneem were quite effective compared to Furpume and Nimbecidine. Variations were also observed among isolates in their sensitivity against fungicides. Isolates BHI, CHR-I and CHR-III were sensitive to all the fungicides, whereas JHR was sensitive only to Dithane M-45, Kitazin and Bavistin. In the case of neem products, HSR isolate was not sensitive to Achook, whereas Bioneem proved to be effective against CHR-I, CHR-III, HSR, KTL-I, KTL-II, REW and SPT isolates. Based on their sensitivity against fungicides and neem products, all isolates fell more or less in the same group (Tables 6.18 and 6.19). Isolates were categorized into eight groups on the basis of their differential sensitivity to fungicides and neem products. These groups were designated into eight pathotypes/races present under Haryana, Indian, conditions infecting rapeseed–mustard. Variations in the sensitivity of 14 *A. brassicae* isolates to extracts of bougainvillaea, garlic, Lawsonia, neem, mint and eucalyptus (Table 6.20) have also been observed (Kumar et al. 2004; Khurana et al. 2005a). Results revealed that garlic and neem leaves extracts proved to be quite effective against *Alternaria brassicae* in vitro and followed by mint, bougainvillaea, Lawsonia and eucalyptus. Among isolates, REW and RTK were quite sensitive to all the plant extract, whereas isolate SRS was sensitive to only neem and garlic extract. Isolates KTL-II, BHI and JHR were not sensitive to neem extract as compared to other isolates. Similarly in the case of garlic, isolates BHI, HSR, JD-II, KTL-I and SRS did not respond effectively as compared to other isolates. The eucalyptus and Lawsonia leaves extract which were ineffective, in general, appeared quite effective against REW and RTK isolates, clearly indicating the variation in the isolates of *Alternaria brassicae* (Table 6.20).

Evaluation of efficacy of ten fungicides against *A. brassicae* isolates from various parts of India showed that Emisan-6, in general, proved most effective in inhibiting spore germination followed by Ridomil MZ-72, while Sulfex and Blitox proved least effective (Sangwan and Mehta 2006). Isolates PNT, BHP, CAUL and *B.*

Table 6.18 Differential behaviour of various isolates of *Alternaria brassicae* against fungicides (Kumar et al. 2004)

	Alternaria brassicae isolates														
Fungicides	BWL	BHI	CHR-I	CHR-II	CHR-III	HSR	JND-I	JND-II	JHR	KTL-I	KTL-II	REW	RTK	SRS	SPT
Emisan-6	+	+	+	+	+	+	+	+	–	–	+	+	+	+	+
Wettable sulphur	–	+	+	+	+	–	–	–	–	+	–	–	–	+	–
Ridomil MZ-72	+	+	+	+	+	+	+	+	–	+	+	+	+	+	+
Blitox-50	+	+	+	+	+	+	+	+	–	+	+	+	+	–	+
Dithane M-45	+	+	+	+	+	+	+	+	+	+	+	+	+	+	+
Kitazin	+	+	+	+	+	+	+	+	+	–	+	+	+	+	+
Bavistin	–	+	+	–	+	+	+	+	+	–	–	+	+	+	–
Baynate	–	+	+	+	+	–	+	+	–	+	–	+	+	–	–

+ = sensitive to fungicide (less than 25 % spore germination); – = not sensitive to fungicide (more than 25 % spore germination)

Table 6.19 Differential behaviour of various isolates of *Alternaria brassicae* against neem products (Kumar et al. 2004)

Neem products	*Alternaria brassicae* isolates														
	BWL	BHI	CHR-I	CHR-II	CHR-III	HSR	JND-I	JND-II	JHR	KTL-I	KTL-II	REW	RTK	SRS	SPT
Furpume	–	–	–	–	–	+	–	–	–	–	–	–	–	–	–
Bio-neem	–	–	+	–	+	+	–	–	–	+	+	+	–	–	+
Nimbecidine	–	–	–	–	–	+	+	+	+	–	+	+	–	–	–
Achook	–	–	+	+	–	–	+	+	+	–	–	+	+	+	–

+ = sensitive to neem products (less than 25 % spore germination); – = not sensitive to neem products (more than 25 % spore germination)

Table 6.20 Sensitivity of different isolates of *Alternaria brassicae* collected from Haryana, India, to various plant extracts (Kumar et al. 2004)

Plant extracts	Per cent spore germination of *Alternaria brassicae* isolates														
	BHI	BWL	CHR-I	CHR-II	HSR	JD-I	JD-II	JHR	KTL-I	KTL-II	REW	RTK	SRS	SPT	Mean	SEM±
Bougainvillaea	31.41	29.00	38.73	29.7	41.7	40.4	34.6	48.8	20.1	36.8	11.2	13.7	60.2	35.8	33.73	1.36
	(34.0)	(32.3)	(38.5)	(33.0)	(40.2)	(39.4)	(36.0)	(44.3)	(26.6)	(37.5)	(19.5)	(21.6)	(50.9)	(36.8)	(35.0)	
Garlic[a]	23.8	15.2	19.3	13.3	24.9	23.1	22.0	16.7	22.7	16.6	9.2	7.4	24.2	18.3	18.3	1.97
	(29.1)	(22.9)	(25.9)	(21.3)	(29.9)	(28.6)	(29.7)	(23.9)	(28.3)	(24.0)	(17.5)	(15.7)	(29.4)	(25.2)	(25.1)	
Lawsonia	30.1	66.3	30.9	30.1	22.4	38.6	62.6	64.6	46.7	32.1	10.00	8.6	58.8	34.0	38.3	1.40
	(33.5)	(54.6)	(33.8)	(33.3)	(28.2)	(38.4)	(52.3)	(53.2)	(43.1)	(34.5)	(18.3)	(17.0)	(50.1)	(35.7)	(37.6)	
Neem	41.0	12.1	13.8	18.6	15.0	20.4	12.1	30.1	12.5	42.4	12.3	8.3	23.5	21.5	20.3	1.59
	(39.9)	(20.3)	(21.7)	(25.5)	(22.8)	(26.8)	(20.2)	(33.2)	(20.6)	(40.6)	(20.5)	(15.7)	(28.9)	(27.6)	(26.1)	
Mint	46.4	17.3	17.8	25.2	24.0	34.9	28.8	36.6	23.1	14.4	18.3	8.2	36.2	23.3	25.3	1.53
	(42.9)	(24.6)	(25.0)	(30.1)	(29.3)	(36.2)	(32.4)	(37.2)	(41.2)	(22.4)	(26.1)	(15.6)	(37.0)	(28.8)	(30.7)	
Eucalyptus	59.2	31.7	31.8	27.7	35.8	33.0	51.0	35.6	48.5	67.1	20.4	14.0	52.2	30.8	38.5	1.66
	(47.6)	(34.2)	(34.3)	(31.8)	(36.8)	(35.1)	(45.6)	(36.7)	(44.1)	(55.0)	(26.9)	(21.8)	(46.3)	(33.7)	(37.8)	
Control (water only)	75.5	75.7	76.9	77.2	77.1	81.5	76.5	72.5	71.0	83.2	78.7	79.3	82.6	74.6	77.3	1.75
	(60.3)	(60.5)	(61.2)	(61.8)	(61.4)	(64.6)	(61.0)	(58.4)	(57.4)	(65.8)	(62.9)	(63.2)	(65.4)	(59.7)	(61.7)	
Mean	43.9	35.3	32.7	31.7	34.4	38.8	41.1	43.6	34.9	41.8	22.9	19.9	48.2	34.0		
	(41.1)	(35.6)	(34.3)	(33.8)	(35.5)	(38.4)	(39.5)	(41.0)	(37.3)	(40.0)	(27.4)	(24.7)	(44.0)	(35.4)		

CD ($p = 0.05$ %): isolate (I) = 1.20; plant extract (PE) = 0.85; I × PE = 3.18; figures in the parentheses are arc sine values

[a]Cloves were used

alba were highly sensitive, whereas isolates FRD, *B. chin* and ASM were the least sensitive (Table 6.21) . Spore germination among isolates varied from 17.1 to 43.59 % (Sangwan and Mehta 2006). Differential behaviour of various isolates also indicated that isolates BHP and *B. alba* were more sensitive followed by CAUL, TRN, PNT and HSR. The isolates RC-781, FRD, B. chin, ASM, GRN and GDP responded similarly (Table 6.22).

6.12 Thermal Sensitivity

Differences among *A. brassicae* isolates in relation to their sensitivity to different temperatures have been reported. In general, there are no significant differences in spore germination in the temperature range of 20–30 °C. As the temperature rise, the viability of the spores declines, and in most isolates, except KTL-I and KTL-II, spores lose their viability at 55 °C. In isolates KTL-I and KTL-II, 5 % spores remain viable even at 55 °C. Based on spore germination, isolates BWL, CHR-II and JND-II are more resistant to high temperatures. In most isolates, only 10 % spores germinated at 45–50 °C as compared to 35–43 % in KTL-II and SPT isolates. The drastic reduction in spore germination at 45 °C indicates that the pathogen cannot survive during summer months in northern India (Table 6.23). Spores of only two isolates, KTL-I and KTL-II, germinate at 55 °C indicating their capability to withstand high temperature, which can have a significant implication on their survival. The isolates, which withstand highest temperature probably, have genetic resistance to high temperature. On the basis of their thermal sensitivity, the isolates were grouped into three categories by Kumar et al. (2003a). Isolates BWL, CHR-II and JND-I formed the first group since they lost 90 % spore viability at 45 °C; isolates BHI, CHR-I, CHR-III, HSR, JND-I, JHR, REW, RTK, SRS and SPT formed the second group as they lost 90 % spore viability at 50 °C; and isolates KTL-I and KTL-II formed the third group, where only 5 % spores survived at 55 °C.

6.13 Identification and Nomenclature of Pathotypes

Physiological races or pathotypes of plant pathogens are identified on the basis of infection types produced by them on specific set of cultivars called Differentials. The procedures and problems involved in the collection of diseased samples, isolation and purification of cultures, maintenance of specific isolates, techniques of inoculation and scoring of infection types have been described (Verma and Saharan 1994). In biotrophs, host pathosystem (*Puccinia*, wheat; *Melampsora*, flax) norm and standards of selection of host differentials acceptable at international level have been followed, but it has not been met in the studies conducted in *Alternaria*–crucifers system. Selection of standard host differentials consists of a set of host varieties termed Differentials, supplemented differentials (additional host varieties), single gene lines and near-isogenic lines. The use of such a set of host differentials can clear the picture of presence and identification of pathotypes in *Alternaria* spp. infecting crucifers.

Nomenclature of a race or pathotype has been done earlier as:

1. Arbitrary numbers: races are generally designated as number or letters in an arbitrary manner, generally in the order of their discovery, e.g. cereal rusts
2. Black's nomenclature: a race is designated on the basis of its virulence on a host resistance gene, e.g. *Phytophthora*–potato system, races R1 and R2, races R1 and R4, etc.
3. Virulence formulae: it is based on a race virulent and avirulent of particular gene for

Table 6.21 Sensitivity of different isolates of *Alternaria brassicae* from India to various fungicides (Sangwan and Mehta 2006)

	Per cent spore germination in various isolates														
Fungicides	RC-781	KGR	FRD	*B chin*	Varuna	ASM	*B alba*	BHP	PNT	CAUL	GRN	HSR	GDP	TRN	Mean
Dithane M-45	29.0	28.4	38.5	33.6	24.6	43.6	27.2	22.0	27.1	40.5	36.1	17.1	35.2	32.7	31.7
Sulfex	56.6	46.5	72.3	50.6	48.4	59.7	23.8	39.2	39.3	25.4	56.4	37.8	44.6	38.2	45.4
Blitox	48.2	33.4	63.3	53.7	40.9	47.8	37.4	50.0	33.5	33.3	44.2	25.6	56.5	45.5	43.8
Captan	30.9	32.8	33.7	43.4	33.1	45.2	25.0	21.6	26.3	20.9	51.9	41.2	40.5	24.8	33.6
Kitazin	36.6	26.8	48.1	50.7	35.8	54.4	33.6	22.9	24.2	27.4	35.2	23.4	40.6	29.7	35.0
Bavistin	38.4	23.3	51.4	43.8	34.8	50.5	21.2	30.9	18.4	21.4	41.6	26.9	34.5	25.0	33.0
Vitavax	29.6	30.4	40.3	38.3	72.7	48.4	23.6	22.0	14.7	25.5	35.8	27.0	44.9	22.1	31.1
Raxil	42.6	64.2	49.0	45.6	36.7	60.9	26.1	23.9	26.0	21.3	54.1	61.3	47.7	42.5	43.0
Emisan-6	06.9	05.7	16.5	16.9	09.8	12.4	12.0	11.9	06.8	9.0	14.5	16.3	17.2	12.7	12.0
Ridomil MZ-72	11.9	11.7	20.5	22.0	16.3	11.7	06.7	11.1	10.1	16.6	06.4	22.3	14.5	15.8	14.1
Control	87.6	85.4	80.2	84.5	82.8	85.3	88.2	85.2	85.1	84.7	83.7	74.7	78.3	78.6	83.6
Mean	39.8	36.2	46.7	44.8	36.6	47.3	30.4	31.0	28.3	29.6	42.7	34.0	41.3	35.2	-

CD (5 %): isolates (I), 2.46; fungicides (F), 1.27 I × F: 3.58

Table 6.22 Differential behaviour of various isolates of *A. brassicae* from India against different fungicides (Sangwan and Mehta 2006)

	A. brassicae isolates										
Fungicides	RC-781	KGR	Varuna	ASM	*B. alba*	BHP	PNT	CAUL	HSR	GDP	TRN
Dithane M-45	+	+	−	+	+	−	+	+	−	+	+
Sulfex	+	+	+	+	−	+	+	+	+	+	+
Blitox	+	+	+	+	+	+	+	+	+	+	+
Captan	+	+	+	+	−	−	+	−	+	+	−
Kitazin	+	+	+	+	+	−	+	+	−	+	+
Bavistin	+	−	+	+	−	+	−	−	+	+	−
Vitavax	+	+	+	+	−	−	−	+	+	+	−
Raxil	+	+	+	+	+	−	+	−	+	+	+
Emisan-6	−	−	−	−	−	−	−	−	−	−	−
Ridomil MZ-72	−	−	−	−	−	−	−	−	−	−	−

resistance, e.g. the formulae 6, 7, 10/5, 8, 9a and 11 for a race of *Puccinia* virulent on Sr 6, Sr7 and Sr10, but avirulent on Sr5, Sr8, Sr9a and Sr11

4. Habgood nomenclature
5. Virulence analysis

However, out of these criteria of nomenclature, *Brassica* researchers have adopted the first method in isolation (not taken into account earlier reports) giving their own arbitrary numbers not keeping parity with others and order of discovery. Even researchers had hesitation in designating pathotypes of *Alternaria* species except Saharan and Kadian (1983) and Gupta et al. (2004). The procedure and method of *Alternaria* species pathotypes designation adopted by Gupta et al. (2004) seem to be logical since it is based on the interaction of pathogen isolates with one specific genotype of a host species. Apparently, it meets gene for gene hypothesis in the absence of standard monogenic/isogenic host differential sets. The determinant attributes used by different workers for the identification of pathotypes of *Alternaria* species infecting crucifers are given in Table 6.24.

The utility and advantages of race/pathotype identification in *Alternaria* can boost *Brassica* production through the (1) development of resistant cultivars, (2) identification of new genes for resistance, (3) development of multigene-resistant cultivars and (4) identification of favourable gene combinations (Singh and Chand 1983).

Table 6.23 Variation in thermal sensitivity of different isolates of *Alternaria brassicae* from Haryana, India (Kumar et al. 2003a, b)

Temperature (°C)	Per cent spore germination of *Alternaria brassicae* isolates*															Mean
	BWL	BHI	CHR-I	CHR-II	CHR-III	HSR	JND-I	JND-II	JHR	KTL-I	KTL-II	REW	RTK	SRS	SPT	
30	90.3 (72.2)	93.5 (77.1)	96.4 (82.3)	91.5 (75.5)	100.0 (90.0)	94.7 (78.4)	94.1 (77.8)	91.1 (71.7)	95.9 (81.7)	93.3 (77.1)	98.9 (86.9)	96.8 (82.6)	92.2 (76.0)	94.0 (77.7)	91.9 (75.8)	94.3 (78.9)
35	90.2 (68.7)	87.8 (69.6)	88.1 (69.9)	80.1 (63.5)	85.6 (68.7)	83.8 (70.7)	87.3 (69.2)	85.1 (66.3)	88.5 (70.4)	87.0 (69.2)	91.0 (74.9)	90.8 (71.3)	88.1 (69.9)	90.1 (71.6)	84.6 (67.2)	87.5 (69.4)
40	80.4 (63.7)	79.3 (62.9)	73.6 (59.1)	80.3 (63.9)	81.7 (65.0)	79.8 (63.4)	81.8 (60.4)	76.9 (61.3)	76.8 (61.3)	78.3 (62.3)	80.2 (63.7)	79.3 (63.1)	83.6 (66.1)	75.6 (60.4)	69.4 (56.6)	78.4 (52.2)
45	8.3 (17.9)	23.4 (28.8)	17.3 (24.5)	9.2 (17.7)	15.2 (22.9)	12.8 (21.0)	15.6 (23.3)	9.6 (17.7)	17.0 (24.3)	43.8 (39.6)	36.0 (38.3)	17.7 (24.7)	10.5 (18.8)	18.2 (25.2)	35.4 (36.5)	19.3 (25.4)
50	3.7 (15.6)	15.9 (23.4)	14.5 (22.4)	6.7 (17.8)	14.0 (23.9)	6.6 (13.2)	2.4 (11.8)	4.0 (11.4)	9.2 (17.4)	11.7 (19.8)	11.7 (20.0)	9.2 (17.3)	8.5 (16.8)	10.3 (18.7)	12.4 (20.3)	9.4 (18.0)
55	0.0 (0.0)	0.0 (0.0)	0.0 (0.0)	0.0 (0.0)	0.0 (0.0)	0.0 (0.0)	0.0 (0.0)	0.0 (0.0)	0.0 (0.0)	5.6 (13.5)	5.4 (13.2)	0.0 (0.0)	0.0 (0.0)	0.0 (0.0)	0.0 (0.0)	0.7 (1.8)
Control (22 ± 1°C)	87.4 (68.4)	98.0 (83.1)	91.6 (73.0)	99.3 (87.6)	91.5 (73.2)	100.0 (90.0)	90.5 (72.1)	90.7 (72.2)	85.4 (67.6)	97.7 (85.6)	98.3 (84.8)	87.6 (69.5)	98.7 (85.5)	100.0 (90.0)	97.0 (80.2)	94.2 (78.9)
Mean	51.5 (43.8)	56.8 (49.3)	54.5 (47.3)	52.4 (46.6)	55.4 (49.1)	54.7 (48.1)	53.1 (44.9)	51.1 (43.0)	53.3 (46.1)	59.6 (52.4)	60.2 (54.5)	54.5 (46.9)	54.5 (47.6)	55.4 (49.1)	55.8 (48.1)	

CD (P = 0.05): temperature (T) = 1.5, isolates (I) = 2.2, T × I = 5.8

*figures in parentheses are angular transformed values

Table 6.24 Determinants of variability in *Alternaria* infecting crucifers (Saharan et al. 2015)

Determinant attributes	*Alternaria species*	Host	Pathotypes/races	Reference
Pathological	*A. alternata*	*Crambe*	A, B, C	Czyzewska, (1969, 1971)
	A. brassicae	Crucifers	RM-1, RM-2, V-3	Saharan and Kadian (1983)
	A. brassicae	Rape	13	Mridha (1983)
	A. brassicicola	Cauliflower	3	Stoll (1952)
	A. raphani	Radish	Wild Variants	Atkinson (1953)
	A. brassicae	*B. juncea*	Bj-4, Bj-5,Bj-6, Bj-7	Gupta et al. (2004)
	A. brassicae	*B. juncea*	DLK, RSR-1, GDP	Mehta et al. (2003)
	A. brassicae	*B. juncea*	A,C,D	Vishwanath and Kolte (1997)
	A. brassicae	*B. juncea*	8	Kumar et al. (2003a, b)
	A. brassicae	*Brassica* spp.	10	Mehta et al. (2003)
	A. brassicae	*Brassica* spp.	14	Sangwan and Mehta (2007)
	A. brassicae	*Brassica* spp.	8	Singh et al. (2008)
	A. brassicae	*Brassica* spp.	Abr 1 to Abr7	Kumar et al. (2014)
Symptomatological	*A. brassicae*	*Brassica* spp.	Bj-4, Bj-5, Bj-6, Bj-7	Gupta et al. (2004)
	A. brassicae	*Brassica* spp.	A,C,D	Kolte et al. (1991)
	A. brassicae	*Brassica* spp.	12	Goyal et al. (2013)
Morphological, cultural and nutritional	*A. brassicae*	*B. carinata*	A,B,C,D	Kolte et al. (1989, 1991)
	A. brassicae	*Brassica* spp.	4	Mehta et al. (2003)
	A. brassicae	*Brassica* spp.	4	Goyal et al. (2011)
	A. brassicae	*Brassica* spp.	5	Meena et al. (2012)
	A. brassicae	Cauliflower and rapeseed–mustard	2	Sharma et al. (2013)
	A. brassicae	*B. juncea*	2	Pramila et al. (2014)
	A. brassicae	Colza	-	Van Schreven (1953)
	A. brassicae	Rapeseed–mustard	2	Mehta et al. (2005b)
Biochemical	*A. brassicae*	Rapeseed–mustard	3	Khurana et al. (2005b)
	A. brassicae	Rapeseed–mustard	3	Vishwanath and Kolte (1997)

(continued)

Table 6.24 (continued)

Determinant attributes	*Alternaria species*	Host	Pathotypes/races	Reference
Genetical	*A. species (4)*	Crucifers	Genetical similarity in ITS region	Jasalavich et al. (1995), Goyal et al. (2013)
	A. brassicae	*B. juncea*	Vary pathogenically	Sharma et al. (2013)
	A. brassicae	*B. juncea*	Isolates genetically variable	Pramila et al. (2014)
	A. brassicae	Rapeseed–mustard	Isolates genetically variable	Kumar et al. (2014)
Molecular	*A. brassicae*	Crucifers	Polymorphism in isolate by RAPD analysis	Sharma and Tewari (1995, 1998)
	A. brassicicola	Crucifers	-do-	Goyal et al. (2013)
	A. raphani	Crucifers	-do-	Kumar et al. (2008)
Proteome level	*A. brassicae*	Crucifers	Variation in protein level of virulent and avirulent isolates	Sharma et al. (2012)
Thermo-sensitivity	*A. alternata*	*Crambe*	A, B, C	Czyzewska (1970)
	A. brassicae	*B. juncea*	3	Kumar et al. (2003a, b)
Fungicidal sensitivity	*A. brassicae*	Crucifers	A, C, D	Vishwanath and Kolte (1997), Sangwan and Mehta (2006)
			6	
	A. brassicae	Crucifers	8	Kumar et al. (2004), Khurana et al. (2005a, b)

References

Agrios GN (2005) Plant pathology, 5th edn. Elsevier Academic Press, New York, 922 p

Atkinson RG (1953) Survival and pathogenicity of *Alternaria raphani* after five years in dried soil cultures. Can J Bot 31:542–547

Awasthi RP, Kolte SJ (1989) Variability in *Alternaria brassicae* affecting rapeseed and mustard. Indian Phytopathol 42:275 (Abstr.)

Campbell R (1970) An electron microscope study of exogenously dormant spores, spore germination, hyphae and conidiophores of *Alternaria brassicicola*. New Phytol 69:287–293

Campbell R, Lamer RW, Madelin MF (1968) Notes on an albino mutant of *Alternaria brassicicola*. Mycologia 60:1122–1125

Changsri W, Weber GF (1963) Three *Alternaria* species pathogenic on certain cultivated crucifers. Phytopathology 53:643–648

Czyzewska S (1969) Alternaria blight of *Crambe abyssinica*. Acta Mycol 5:175–211

Czyzewska S (1970) Effect of temperature on the growth and sporulation of *Alternaria* species isolated from *Crambe abyssinica*. Acta Mycol 6:261–276

Czyzewska S (1971) The pathogenicity of *Alternaria* spp. isolated from *Crambe abyssinica*. Acta Mycol 7:171–240

Goyal P, Chahar M, Mathur AP, Kumar A, Chattopadhyay C (2011) Morphological and cultural variation in different oilseed *Brassica* isolates of *Alternaria brassicae* from different geographical regions of India. Indian J Agric Sci 81:1052–1058

Goyal P, Chattopadhyay C, Mathur AP, Kumar A, Meena PD, Datta S, Iquebal MA (2013) Pathogenic and molecular variability among Brassica isolates of *Alternaria brassicae* from India. Ann Protect Sci 21:349–359

Gupta K, Saharan GS, Mehta N, Sangwan MS (2004) Identification of pathotypes of *Alternaria brassicae* from Indian mustard [*Brassica juncea* (L.) Czern. and Coss]. J Mycol Plant Pathol 34:15–19

Jasalavich CA, Morales VM, Pelcher LE, Seguin-Swartz G (1995) Comparison of nuclear ribosomal DNA sequences from *Alternaria* species pathogenic to crucifers. Mycol Res 99:604–614

Khurana AK, Mehta N, Sangwan MS (2005a) Variability in the sensitivity of *Alternaria brassicae* isolates to plant extracts. J Mycol Plant Pathol 35:76–77

Khurana AK, Mehta N, Sangwan MS (2005b) Variation in Bio-chemical contents of *Alternaria brassicae* (Berk.) Sacc. isolates. J Mycol Plant Pathol 35:343–345

Kolte SJ, Awasthi RP, Vishwanath K (1989) Disease problems in Brassicas and research activities at Pantnagar. IDRC (Canada) Oil Crops. In: Proceedings of the three meetings held at Pantnagar and Hyderabad, India, pp 43–48

Kolte SJ, Bardoloi DK, Awasthi RP (1991) The search for resistance to major diseases of rapeseed mustard in India. In: Proceedings of the 8th international rapeseed congress, 9–11 July 1991, Saskatoon, Canada, p 60 (Abstr), vol 1, pp 219–225

Kumar S, Sangwan MS, Mehta N (2003a) Variability in *Alternaria brassicae* on the basis of thermal sensitivity. Plant Dis Res 18:96–98

Kumar S, Sangwan MS, Mehta N, Kumar R (2003b) Pathogenic diversity in isolates of *Alternaria brassicae* infecting rapeseed and mustard. J Mycol Plant Pathol 33:59–64

Kumar S, Sangwan MS, Mehta N, Kumar R (2004) Relative sensitivity of various isolates of *Alternaria brassicae* (Berk.) Sacc. to fungicides. J Mycol Plant Pathol 34:28–32

Kumar V, Haldar S, Pandey KK, Singh RP, Singh AK, Singh PC (2008) Cultural, morphological, pathogenic and molecular variability amongst tomato isolates of *Alternaria solani* in India. World J Microbiol Biotechnol 24:1003–1009

Kumar A, Katoch A, Sharma PN, Kumar V, Kumar A (2014) Pathogenic and genetic variability in *Alternaria brassicae* infecting rapeseed-mustard and evaluation of resistance sources. Indian Phytopathol 67:257–262

Meena PD, Rani A, Meena R, Sharma P, Gupta R, Chowdappa P (2012) Aggressiveness, diversity and distribution of *Alternaria brassicae* isolates infecting oilseed *Brassica* in India. Afr J Microbial Res 6:5249–5258

Mehta N, Sangwan MS, Srivastava MP (2003) Morphological and pathological variations in rapeseed and mustard isolates of *Alternaria brassicae*. Indian Phytopathol 56:188–190

Mehta N, Sangwan MS, Saharan GS (2005a) Fungal diseases of rapeseed mustard. In: Saharan GS, Mehta N, Sangwan MS (eds) Diseases of oilseed crops. Indus Publishing Company, New Delhi, pp 15–86

Mehta N, Khurana AK, Sangwan MS (2005b) Characterization of *Alternaria brassicae* isolates from Haryana on the basis of nutritional behaviour. In: Sharma RC, Sharma JN (eds) Integrated plant disease management. Scientific Publishers, Jodhpur, pp 295–303

Mridha MAU (1983) Virulence of different isolates of *Alternaria brassicae* on winter oilseed rape cultivars. In: Proceedings of 6th international rapeseed congress, Paris, France, pp 1025–1029

Pramila PG, Tasleem M, Taj G, Mal R, Kumar A (2014) Morphological, cultural, pathogenic and molecular variability amongst Indian mustard isolates of *Alternaria brassicae* in Uttarakhand. Afr J Biotechnol 13:441–448

Saharan GS (1992a) Disease resistance. In: Labana KS, Banga SS, Banga SK (eds) Breeding oilseed Brassicas. Narosa Publishing House, New Delhi, pp 181–200

Saharan GS (1992b) Management of rapeseed and mustard diseases. In: Kumar D, Rai M (eds) Advances in oilseed research, vol I. Scientific Publishers, Jodhpur, pp 152–188

Saharan GS, Kadian AK (1983) Physiologic specialization in *Alternaria brassicae*. Cruciferae Newslett 8:32–33

Saharan GS, Mehta N, Verma PR, Meena PD (2015) Analysis of variability determinants in *Alternaria-Brassica-* interactions. J Oilseed Brassica 6(1):126–141

Sangwan MS, Mehta N (2006) Categorization of *Alternaria brassicae* isolates on the basis of differential sensitivity to fungicides. Plant Dis Res 21:114–117

Sangwan MS, Mehta N (2007) Pathogenic variability in isolates of *Alternaria brassicae* (Berk.) Sacc. from different agro-climatic zones of India. Plant Dis Res 22:101–107

Sharma TR, Tewari JP (1995) Detection of genetic variation in *Alternaria brassicae* by RAPD fingerprints. J Plant Biochem Biotechnol 4:105–107

Sharma TR, Tewari JP (1998) RAPD analysis of three *Alternaria* species pathogenic to crucifers. Mycol Res 102:807–814

Sharma N, Rahman MH, Kav NNV (2012) A possible proteome-level explanation for differences in virulence of two isolates of a fungal pathogen *Alternaria brassicae*. J Plant Biochem Biotechnol 19:161–170

Sharma M, Deep S, Bhati DS, Chowdappa P, Selvamani R, Sharma P (2013) Morphological, cultural, pathogenic and molecular studies of *Alternaria brassicae* infecting cauliflower and mustard in India. Afr J Microb Res 7:3351–3363

Singh BM, Chand JN (1983) Identification of physiologic races-general consideration. In: Chand JN, Saharan GS (eds) Phytopathological techniques. HAU Press, Hisar, pp 109–115

Singh S, Singh RP, Singh HK, Kumar K (2008) Screening of *Brassica* genotypes for quality traits and reaction to Alternaria blight and white rust. Cruciferae Newslett 27:35–36

Singh R, Singh D, Singh H (2013) Variability in *Alternaria brassicae* incitant of Alternaria blight of oilseed *Brassica*. Res Crops 14:1082–1088

Stoll K (1952) The organ, injurious effect and control of *Brassica* blackening. NachrBl Dtsch Pfjsch Dienst Berl N F 6:81–85

Van Schreven DA (1953) *Alternaria, Stemphylium* en *Botrytis* a antasting bij Koolzaad (*Brassica napus*). Tijdschr Plantenziekt 59:105–136

Verma PR, Saharan GS (1994) Monograph on *Alternaria* diseases of crucifers. Saskatoon Research Centre Technical Bulletin 1994–6E, Agriculture and Agri-Food Canada, Saskatoon, SK, Canada, 162 p

Vishwanath K, Kolte SJ (1997) Variability in *Alternaria brassicae:* Response to host genotypes, toxin production and fungicides. Indian Phytopathol 50:373–381

7 Fine Structures and Electron Microscopy

7.1 Introduction

Among the four *Alternaria* species infecting cruciferous hosts, information on fine structures through electron microscopy has been generated on *A. brassicicola*. The sequence of events in the spore production and maturation has been described. Changes in the internal organelles of hypha, conidium and conidiophore have been observed. Conidiophores have a similar structure to mature hyphae, except that after spore production, they have pore in the tip and an annulus. There are variations in the number of nuclei in the cells of the vegetative hyphae, conidiophores and conidia with six chromosomes in dividing nuclei of vegetative hyphae (Campbell 1970a, b, 1972; Knox- Davis 1979).

7.2 Fine Structures and Electron Microscopy

The sequence of events in the spore production and maturation of *A. brassicicola* was described by Campbell (1964, 1969a, b, 1970a, b, 1972). Each cell of the mature spore of *A. brassicicola* has a two-layered wall, the layer distant from the protoplast being melanized. The mature septa are five layered, having two layers of secondary wall on either side of the septal partition, which itself is three layered. Each septum has one simple pore. New spores are produced by an outgrowth, through a pore, of the inner wall layer of the mother cell. Young spores have many small mitochondria, and much vesicular endoplasmic reticulum, and as they mature, lipid bodies and an unknown polyglucoside are produced. Mature spores have glycogen but very little, if any, lipid. The vesicular endoplasmic reticulum, multivesicular bodies and lomasomes are involved in wall formation. The plasmalemma has rectangular grooves in its outer surface and corresponding ridges on the inner one; both surfaces bear particles of two distinct sizes. The endoplasmic reticulum may be lamellated or vesicular and is involved in wall formation. The vesicles produced by the endoplasmic reticulum fuse with the plasmalemma. However, dormant spores (Campbell 1970b) have very thick, heavily pigmented, melanized walls with plugged septal pores. The small amounts of endoplasmic reticulum and the few mitochondria lie near the plasmalemma. The germ-tube walls arise from the inner layers of the spore walls. Lomasomes and endoplasmic reticulum vesicles are probably involved with this wall formation. During germination, mitochondria and ribosomes increase in numbers, first in the germinating cell and then in the germ tube. As the hyphae age, they produce lipid droplets and vacuoles, the latter finally fill most of the cell as the cytoplasm degenerates. Conidiophores have structure similar to mature hyphae, except that after spore production, they have a pore in the tip and an annulus. In an albino strain of *A. brassicicola*

G.S. Saharan et al., *Alternaria Diseases of Crucifers: Biology, Ecology and Disease Management*,
DOI 10.1007/978-981-10-0021-8_7

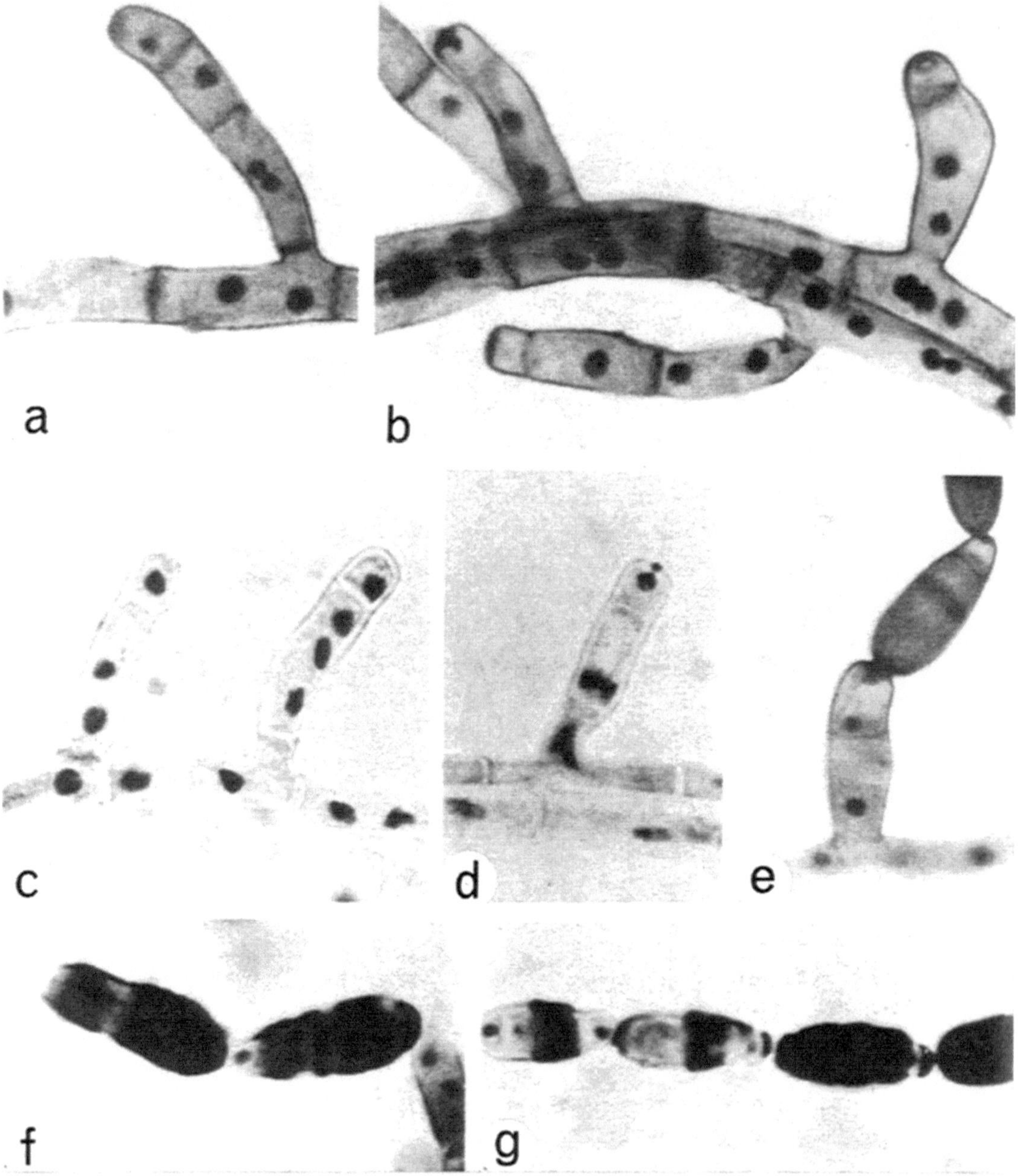

Plate 7.1 Conidiophores of *Alternaria brassicicola* showing (**a**) uni- and binucleate cells and apical cell with pore and annulus; (**b**) typical complement of nuclei in basal cells, anucleate terminal cells (*right and bottom*), pore and annulus (*right*) and nuclear material and cytoplasm wedged in pore in terminal cell (*left*); (**c**) typical complement of nuclei in subtending, basal and terminal cells; (**d**) nucleus wedged in pore of terminal of terminal cell; (**e**) developing conidia (only one shown) with nucleus wedged in pore between terminal cell of conidiophore and basal cell of developing conidium; and (**f, g**) conidia with thick, roughened wall material which stained intensely with Giemsa (Knox-Davies 1979)

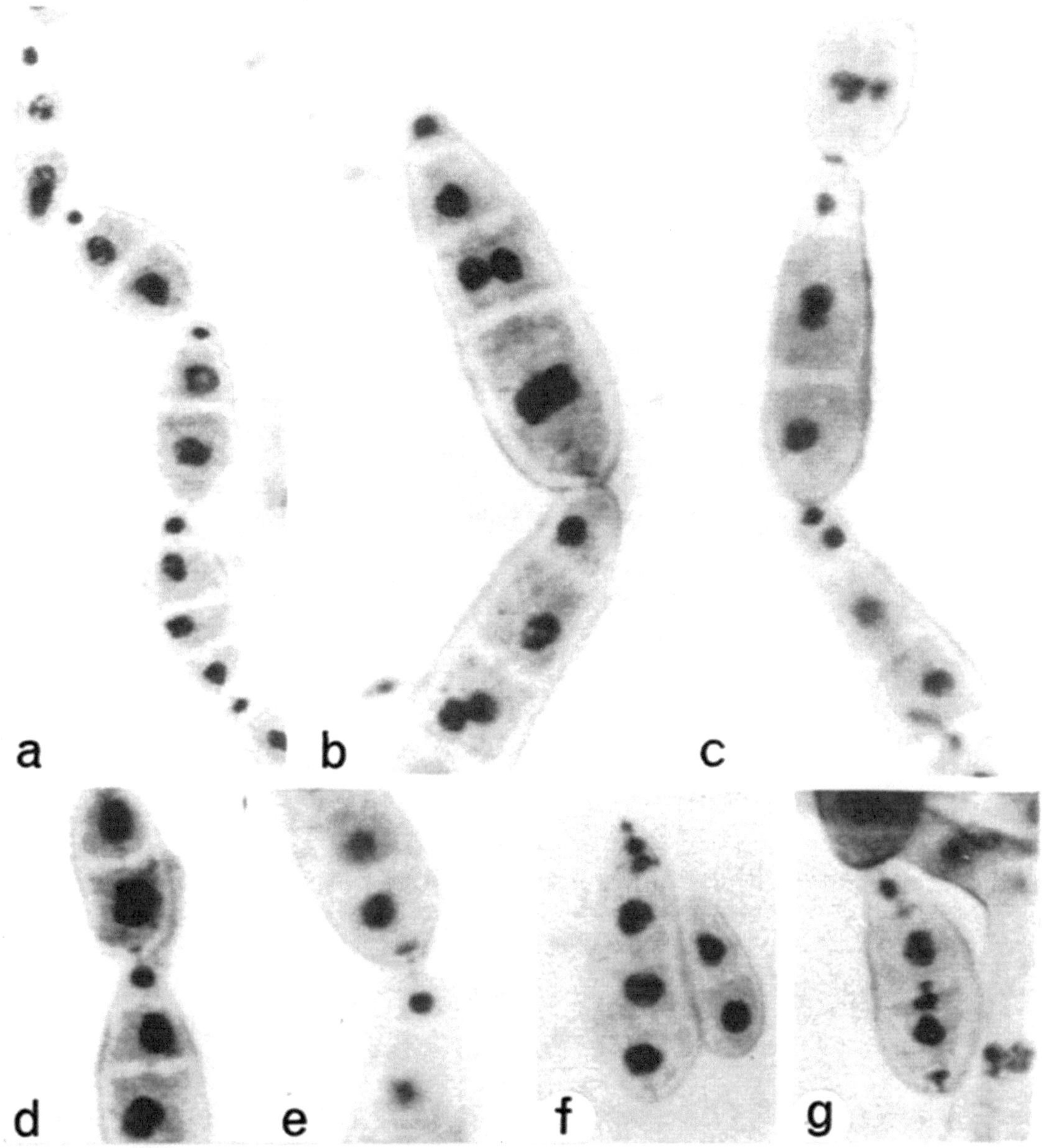

Plate 7.2 Conidia of *Alternaria brassicicola* showing (**a**) small, intensely staining nuclei of the terminal cells and wide pores of the basal cells; (**b**) septal pores, typical distribution of nuclei and small, intensely staining nucleus of the terminal cell; (**c**) Giemsa-stained material wedged in septal pores between adjacent conidia and in a septal pore within a conidium; (**d**, **e**) Giemsa-stained material wedged in the septal pores between adjacent conidia in a conidial chain; (**f**) Giemsa-stained material wedged in the terminal pore of a conidium; and (**g**) Giemsa-stained material wedged in basal and interstitial pores of a conidium (Knox–Davies 1979)

(Campbell et al. 1968), when compared with the wild type (Campbell 1969a, b, 1970a), there was little difference in the structure of the hyphae and the conidiophores, except that there were no electron dense deposits of melanin in the wall and the pore in the conidiophore had no annulus (Campbell 1970b).

Giemsa-stained preparations of *A. brassicicola* show variations in the number of nuclei in the cells of the vegetative hyphae from one to many, with the hyphal tip cells having up to 27 nuclei and older cells up to 33. There are fewer nuclei (0–3) in the mature conidiophore cells, and one or two nuclei in most cells of the conidia in 3-day-old cultures are grown on potato carrot agar. Nuclei are also seen wedged in the connections between conidia and in the basal apical pores of conidia after separation. They are less frequently wedged in septal pores in the conidia. Six chromosomes are present in dividing nuclei in the vegetative hyphae (Plates 7.1 and 7.2; Knox-Davies 1979).

References

Campbell CJ (1964) Studies on the leaf spot disease of *Brassica* spp. caused by *Alternaria brassicicola* (Schw.) Wilts. M.Sc. thesis, University of Exeter, UK

Campbell R (1969a) An electron microscope study of spore structure and development in *Alternaria brassicicola*. J Microbiol Genet 54:381–392

Campbell R (1969b) Further electron microscope studies of the conidium of *Alternaria brassicicola*. Arch Mikrobiol 69:60–68

Campbell R (1970a) An electron microscope study of exogenously dormant spores, spore germination, hyphae and conidiophores of *Alternaria brassicicola*. New Phytol 69:287–293

Campbell R (1970b) Ultra structure of an albino strain of *Alternaria brassicicola*. Trans Br Mycol Soc 54:309–313

Campbell R (1972) Changes in volume of reproductive structures during spore production by *Alternaria brassicicola*. Trans Br Mycol Soc 59:153–156

Campbell R, Lamer RW, Madelin MF (1968) Notes on an albino mutant of *Alternaria brassicicola*. Mycologia 60:1122–1125

Knox-Davies PS (1979) The nuclei of *Alternaria brassicicola*. Trans Br Mycol Soc 72:81–90

8 Biochemistry of Host–Pathogen Interaction

8.1 Introduction

A number of biochemical changes take place during the pathogenesis of *Alternaria* in the host and the pathogen, which produce various kinds of primary and secondary metabolites, which have a great role in host defence and virulence of pathogen.

8.2 Biochemical Changes in the Host

The effect of *A. brassicicola* infection on the changes of nucleic acids of *B. oleracea* var. *botrytis* has been studied by Maitra and Samajpati (1985). An increase in total nucleic acids (97 %), DNA (44 %) and RNA (150 %) in infected leaves is recorded. The increase in RNA content is due to an increase of each of the four nucleotides. However, guanylic acid increases significantly more than the other three nucleotides. The increase in G/C ratio is also due to an increase in guanylic acid.

Losses of soluble carbohydrates are high from dead or damaged seeds; moderate from immature seeds, or seeds which germinate slowly; and low from mature seeds with dark, leathery testas, which germinate normally. Loss of electrolytes is higher in dead, damaged, immature seeds and seeds that germinate slowly than in mature seeds that germinate normally (Knox-Davies 1980).

Atwal et al. (2004) have determined biochemical changes in relation to *Alternaria* leaf blight in Indian mustard. Healthy leaves of *B. juncea* possess significantly higher amount of total soluble sugars than those of different infected leaf parts at all growth stages. The depletion of sugars is highest in lesion areas of infected leaves (necrotic followed by chlorotic and green areas) as compared to healthy leaves at the initial stages of infection. However, the accumulation of sugars is noticed in the lesion as well as lesion-free areas of the infected leaves with progress of infection (Table 8.1). Sugar level is higher in chlorotic and green areas as compared to necrotic parts of the infected leaves. Both healthy and infected leaves have maximum sugar and starch content at 100 DAS, which declines thereafter. Higher levels of total sugars in chlorotic and green areas of infected leaves suggest that sugars are probably one of the factors responsible for resistance in *B. juncea*. They may contribute towards maintaining healthy flora of saprophytic microbes that help to inhibit pathogen growth (Juniper and Jeffree 1983). The lower accumulation in necrotic area can be due to utilization of sugars by the fungus for its own growth and establishment.

The chlorotic area of infected leaves possesses significantly higher levels of total and orthodihydroxy phenols, as compared to necrotic area, whereas no significant differences are observed in the green area of infected leaves as compared

G.S. Saharan et al., *Alternaria Diseases of Crucifers: Biology, Ecology and Disease Management*, DOI 10.1007/978-981-10-0021-8_8

Table 8.1 Carbohydrates, phenols and chlorophyll content (mg/g dry weight) in leaves of Indian mustard as influenced by *Alternaria* blight (Atwal et al. 2004)

Status of leaves/DAS		Starch	Total soluble sugars	Reducing sugars	Total phenols	Orthodihydroxy phenols	Flavonols	Total chlorophyll	Chlorophyll a	Chlorophyll b
Healthy	80	34.2 ± 2.2	72.1 ± 5.2	35.1 ± 3.0	2.6 ± 1.1	0.6 ± 0.3	2.5 ± 0.1	0.27 ± 1.2	0.11 ± 0.01	0.16 ± 2.3
	100	59.4 ± 0.9	134.8 ± 3.8	63.3 ± 1.3	4.9 ± 1.4	1.6 ± 0.1	2.3 ± 0.1	16.28 ± 2.0	5.78 ± 0.04	11.04 ± 0.9
	120	58.6 ± 0.6	104.7 ± 3.2	56.4 ± 2.6	7.0 ± 0.9	2.3 ± 0.1	1.5 ± 0.1	13.62 ± 1.4	4.69 ± 0.06	8.94 ± 0.6
Necrotic	80	11.2 ± 0.1	19.6 ± 2.1	11.2 ± 0.3	2.2 ± 0.1	0.5 ± 0.2	0.6 ± 0.03	–	–	–
	100	22.3 ± 0.1	36.7 ± 2.0	22.4 ± 0.9	3.0 ± 0.1	1.1 ± 0.1	0.8 ± 0.01	–	–	–
	120	18.4 ± 0.1	29.3 ± 2.4	18.3 ± 1.3	3.5 ± 0.1	2.3 ± 0.1	0.8 ± 0.02	–	–	–
Chlorotic	80	18.7 ± 1.1	36.7 ± 1.1	20.1 ± 1.7	3.5 ± 2.1	1.7 ± 0.3	1.9 ± 0.13	0.04 ± 0.3	0.01 ± 0.04	0.03 ± 0.04
	100	49.4 ± 2.3	78.9 ± 0.9	39.4 ± 2.9	6.7 ± 1.1	2.8 ± 0.2	2.2 ± 0.11	2.67 ± 0.3	1.04 ± 0.03	1.63 ± 0.07
	120	32.8 ± 1.1	66.6 ± 2.3	37.2 ± 1.3	8.6 ± 0.9	3.2 ± 0.3	1.5 ± 0.14	0.75 ± 0.7	0.11 ± 0.1	0.64 ± 0.12
Green	80	28.6 ± 0.1	48.2 ± 2.3	26.1 ± 4.9	2.5 ± 0.6	1.1 ± 0.4	2.2 ± 0.9	0.25 ± 4.3	0.10 ± 2.3	0.17 ± 0.01
	100	53.4 ± 2.6	99.4 ± 2.1	53.6 ± 1.6	4.7 ± 0.4	1.9 ± 0.1	2.0 ± 0.06	6.87 ± 1.9	6.47 ± 0.01	0.40 ± 0.03
	120	49.2 ± 0.2	78.3 ± 2.1	42.3 ± 2.0	6.9 ± 0.4	2.1 ± 0.1	1.1 ± 0.40	2.62 ± 1.5	2.14 ± 0.6	0.48 ± 0.04
CD *DAS H × I		2.91	2.81	2.21	0.10	0.20	0.060	0.10	1.21	0.46

DAS days after sowing * CD ($P < 0.05$); *H*-healthy leaves, *I*-infected leaves

to healthy leaves (Table 8.1). Total and orthodihydroxy phenols increase significantly with the increase in infection, and age of the plant, and are maximum at 120 DAS in all the cases. The presence of maximum levels of total and orthodihydroxy phenols in chlorotic area, the region next to the lesion area (necrotic), signifies their protective role to restrict the growth, spread of disease and invasion of the pathogen by formation of lignin and lignin-like substances, which are more toxic to fungi. Singh and Singh (1989) also found higher levels of total and orthodihydroxy phenols in leaves of chilli pepper after inoculation with cucumber mosaic virus. The content of flavonols is significantly lower in different infected parts of leaves as compared to the healthy leaves, and this decrease is more prominent in necrotic area followed by chlorotic and green, indicating less significant role of flavonols in the defence mechanism as compared to total and orthodihydroxy phenols.

The chlorophyll is found to be completely absent in necrotic areas and significantly low in chlorotic as compared to green areas of both infected and healthy leaves. Similar trend is observed in chlorophyll a and b content. However, the chlorophyll content of the green areas of the infected leaves is comparable to that of the healthy leaves at 80 DAS. The decline in chlorophyll content due to various biotic stresses has also been reported in other crops. Maiti et al. (2000) attributed the decline in chlorophyll content due to chloroplast structural modification by the fungus such as dilation of the whole chloroplast, separation of grana and accumulation of starch granules, which have a direct bearing on the photosynthetic capacity of chloroplast.

Brassica juncea cv. Varuna, *B. juncea* cv. PAB-9534 and *B. alba* showed susceptible, moderately resistant and tolerant disease reactions, respectively, to *A. brassicae* at different growth stages. In all stages of pathogen infection, disease severity and characteristic symptoms are more prominent in susceptible than the other two genotypes. The biochemical analysis of leaves of different mustard varieties revealed that total phenol, o-dihydroxy phenol, total sugar, reducing sugar, chlorophyll content and flavanol contents are more in resistant (*B. alba*) than in others. With progress of infection, total phenol, o-dihydroxy phenol and protein content increased in all three genotypes, while the chlorophyll, total sugar, reducing sugar and flavanol content decreased (Mathpal et al. 2011).

Green islands were observed around infected spots of *A. brassicicola* and *A. brassicae* on mustard leaves by Mandahar and Suri (1983, 1987). These green islands have higher content of cytokinin-like substances, compared to the surrounding yellowed and healthy tissues. The presence of starch in green islands is correlated with the formation of metabolic sinks because of their higher content of cytokinin-like substances (Suri et al. 1983). Cytokinin-like substances appear to be actively involved in infection and pathogenesis of *A. brassicicola* (Suri and Mandahar 1984, 1985, 1986). Production of pectolytic and cellulolytic enzymes by *A. brassicae* causing leaf spot of *B. rapa* has been observed (Srivastava and Srivastava 1982; Suri and Mandahar 1982). *Alternaria brassicae* produces invertase and amylase on cabbage decoctions (Weimer 1924), while infection in cauliflower leaves brings changes in polyphenol oxidase and peroxidase (Maitra and Samajpati 1982). The presence of large number of enzymes, viz. amylase, invertase, pectin methyl esterase, phosphorylase, aldolase, amidases, ribonuclease, alkaline phosphorylase, nucleophosphatases, catalase, dehydrogenases, deaminases, nucleodeaminases, glycerophosphatase, cellulose and β-glucosidase, in *A. alternata*, *A. brassicicola* and *A. raphani* infecting crucifers has been demonstrated (Dasgupta and Verma 1961, 1962; Verma 1964, 1971).

Alternaria brassicicola stimulates ethylene production in closed culture with floating leaf discs from cabbage. Production is increased by pre-culturing the fungus on media containing cabbage components, but which contains little or no methionine. It suggests that the nature of parasitism of this pathogen on cabbage is characterized by a latent capability to cause the production of the plant-senescing hormone ethylene (Poapst et al. 1979).

8.3 Biochemical Changes in the Pathogen

Alternaria brassicae and *A. brassicicola* produce antibiotics, which are active against bacteria, fungi and algae (Lindenfelser and Ciegler 1969). *Alternaria brassicicola* strain 13 is reported to form extracellular lipids like phospholipids, mono- and diglycerides, sterines, free fatty acids (predominant) and triglycerides (Aizina 1977). The biosynthesis of lipids and phospholipids is highest on Czapek-Dox medium (Aizina et al. 1976).

A melanoid pigment insoluble in strong alkali has been isolated from *A. brassicae* mycelium. Melanins are important for survival and longevity of fungal propagules. Abscisic acid (ABA), a naturally occurring plant growth regulator involved in the control of various plant processes, also has been characterized in the mycelia of the black spot fungus *A. brassicae* (Dahiya 1988; Dahiya et al. 1988). The germinating spores of *A. brassicae* and *A. brassicicola* isolated from mustard are able to hydrolyse the various carbohydrates, which are used in germination media. *Alternaria brassicae* has a greater ability to hydrolyze pectin and carboxymethyl cellulose than *A. brassicicola*, whereas the reverse is true for the hydrolysis of sodium polypectate (Mukadam and Deshpande 1979).

A cytokinin, 6-benzyl amino purine, is able to significantly reduce *Alternaria* blight disease symptoms and mycelial growth within plant tissues. This cytokinin is also able to inhibit the in vitro growth of *Alternaria brassicae* (Sharma et al. 2010).

The composition of amino acids extracted from *A. brassicicola* and *A. brassicae is* only slightly different from those of *A. raphani*. This similarity of amino acids in the first two-pathogen extracts is correlated with the severe to moderate susceptibility of the varieties of crucifers, and the difference of amino acids contained in *A. raphani* extract may have accounted for the severe susceptibility only on radish. One amino acid of Rf 0.66 of *A. raphani* extract corresponds to that of the radish extract at about the same level (Changsri 1961).

The conidia of *A. brassicicola* are remarkably resistant to phytoalexins including phaseollins, phaseollidin and phaseollinisoflavan, often germinating in assays using 100 μg/ml (Skipp and Bailey 1977). Similarly, 3-day-old spores were able to grow in relatively high levels of these phytoalexins. Dahiya and Tewari (1991) identified three plant growth factors from *A. brassicae* and treatment of seedlings of canola/rapeseed with two of them, N-methyl-2, 5-dimethyl-N-cinnamoylpiperazine and 3-carboxy-2-methylene-4-pentenyl-4-butenolide, reduced their growth, whereas 2-hydroxy-1′-methylzeatin increased the growth of seedlings.

8.4 Glucosinolates

Glucosinolates occur as glucosides throughout the Brassicaceae. These are present in the seed and other parts of oilseed rape. The hydrolysis products of glucosinolates include nitriles, isothiocyanates, oxazolidinethiones and thiocyanates, which can be unpalatable and toxic to non-ruminant animals. Thus, the presence of glucosinolates in rape meal has limited its use in animal feed, necessitating the development of the so-called 'double-low' lines. These produce seed low in erucic acid and glucosinolates. A decreased content of glucosinolates in oilseed rape tissues may have negative consequences for pest and disease incidence on the crop. The hydrolyzed products of glucosinolates are fungitoxic to rapeseed pathogens in vitro (Mithen et al. 1986), and there is also much indirect evidence to suggest that their presence contributes to resistance to a range of oilseed rape pathogens (Greenhalgh and Mitchell 1976; Mithen et al. 1987; Rawlinson et al. 1985). Following inoculation with *A. brassicae*, Doughty et al. (1991) examined the changes in the glucosinolate content of leaves in cultivars Bienvenu (single low) and Cobra (double low) and found that glucosinolate contents increased markedly, but the response depended on leaf age and cultivar. Glucosinolates accumulated more in the sixth leaf of Bienvenu. When *B. rapa* seedlings are inoculated with *A. brassicae*, 3-butenyl and

4-pentenyl isothiocyanates are reduced, together with dimethyl disulphide, dimethyl trisulphide, 4-oxoisophoron and a number of sesquiterpenes. The release of isothiocyanates is an evidence for the catabolism of glucosinolates during infection, which is a prerequisite for their involvement in resistance (Doughty et al. 1996).

8.5 Metabolites Produced

Alternaria brassicae when grown in liquid still culture produces host-specific phytotoxins, which have been shown to be cyclodepsipeptides. It also produces several other metabolites including the new drimane sesquiterpenes deoxyuvidin B, albrassitriol and isoalbrassitriol, as well as brassicadiol, a C_{15} prenylated pentaketide. These compounds show no phytotoxicity on canola (Ayer et al. 1987; Ayer and Pena-Rodriguez 1987a, b). The earliest reports on phytotoxin production by *A. brassicae* or *A. raphani* are those of Degenhardt (1973, 1978) and Husain and Thakur (1966). Cytotoxicity tests with extracts of culture filtrates of *A. brassicae* and *A. brassicicola* show that both species produce compounds, which are toxic to human epithelial (HEP2) cells, but there is no clear relationship between *Alternaria* infection of seed and its cytotoxicity. Metabolites produced by the two fungi are also phytotoxic, giving rise to necrosis of leaf discs and inhibition of seedling root growth, but extracts of the less specialized parasite, *A. alternata*, cause more severe phytotoxic effects on crucifer test plants. Extracts of culture filtrates from *A. brassicae* and *A. brassicicola* are found to show non-specific phytotoxicity (McKenzie et al. 1988). Several other metabolites produced during *Alternaria*–crucifers–host interaction and their role in pathogenesis as well as host defence have been covered in Chap. 10.

Most fungi produce secondary metabolites such as mycotoxins or antibiotics. These compounds are synthesized from precursors derived from primary metabolism and can be valuable as antifungal, antibacterial or antitumor agents. *Alternaria brassicicola* is known to produce a number of interesting compounds including antitumoric depudecin (Matsumoto et al. 1992), antibiotic complex brassicicolin (Ciegler and Lindenfelser 1969) and phytotoxic brassicicenes (Mac Kinnon et al. 1999). Depudecin is C_{11} compound containing two epoxide groups and shows antitumor activity (Matsumoto et al. 1992). Phytotoxic brassicicenes, fusicoccane-like diterpenoids, have been isolated from the culture filtrate of canola pathogen *A. brassicicola* (Mac Kinnon et al. 1999).

The fungal strain Sw-3 showing antimicrobial activity and isolated from soil of a Chinese cabbage patch in Korea was identified as *A. brassicicola* by its morphological characteristics and confirmed by the analysis of the 18S gene and ITS region of rDNA. The fungus showed a similarity of 99 % with *A. brassicicola* in the 18S rDNA sequence analysis. *Alternaria brassicicola* has been reported to produce an antitumor compound called depudecin. Strain SW-3 produces antimicrobial metabolites, in addition to depudecin, during sporulation under different growth conditions. The metabolite of the isolated fungus has strong antifungal activity against *Microsporum canis* and *Trichophyton rubrum* and antibacterial activity against *Staphylococcus aureus* and *Pseudomonas aeruginosa*. The amount and kind of metabolites produced by the isolates are affected by growth conditions such as nutrients and growth period (Jung et al. 2002). Metabolites and phytotoxins produced by *Alternaria* species infecting crucifers have been discussed in greater detail in Chap. 10.

References

Aizina AF (1977) The formation of extracellular lipids of *Alternaria brassicicola* strain 13. Izv AN Mold SSR Biol i khim N 3:87–88

Aizina A F, Zlatoust MA, Razumovskii PN, Yadovina VN (1976) Study on phospholipids of *Alternaria brassicicola* strain 13. Izv AN Mold SSR Biol i khim N 1:57–59

Atwal AK, Ramandeep Munshi SK, Mann APS (2004) Biochemical changes in relation to Alternaria leaf blight in Indian mustard. Plant Dis Res 19:57–59

Ayer WA, Pena-Rodriguez LM (1987a) Metabolites produced by *Alternaria brassicae*, the black spot pathogen of Canola, Part 2, Sesquiterpenoid metabolites. J Nat Prod 50:408–417

Ayer WA, Pena-Rodriguez LM (1987b) Metabolites produced by *Alternaria brassicae*, the black spot pathogen of canola. J Nat Prod 50:400–407

Ayer WA, Bains PS, Pena-Rodriguez LM, Tewari JP (1987) Production of a host specific phytotoxin by *Alternaria brassicae*. In: Proceedings of 7th International Rapeseed Congress, Poznan, Poland, pp 257 (Abstr.), pp 1256–1261

Changsri W (1961) Studies of *Alternaria* spp. pathogenic on Cruciferae. Dis Abstr 21:1698

Ciegler A, Lindenfelser LA (1969) An antibiotic complex from *Alternaria brassicicola*. Experientia 25:719–720

Dahiya JS (1988) Melanin biosynthesis in *Alternaria brassicae*. Indian J Microbiol 28:42–49

Dahiya JS, Tewari JP (1991) Plant growth factors produced by the fungus *Alternaria brassicae*. Phytochemistry 30:2825–2828

Dahiya JS, Tewari JP, Woods DL (1988) Abscisic acid from *Alternaria brassicae*. Phytochemistry 27:2983–2984

Dasgupta SN, Verma JP (1961) Studies in the enzyme make-up of *Alternaria* I. Qualitative demonstration of enzymes. Curr Sci 30:450–452

Dasgupta SN, Verma JP (1962) Studies in the enzyme make-up of *Alternaria* II. Quantitative distribution of enzymes. Proc Nat Inst Sci India 28 B:201–205

Degenhardt KJ (1973) *Alternaria brassicae* and *A. raphani*: sporulation in culture, and their effects on yield and quality of rapeseed. M. Sc. thesis, University of Alberta, Canada

Degenhardt KJ (1978) Alternaria black spot of rapeseed and mustard: phytotoxins and other aspects of the host-parasite interaction. Ph.D. thesis, University of Saskatchewan, Saskatoon, Canada

Doughty KJ, Porter AJR, Morton AM, Kiddie G, Bock CH, Wallsgrove R (1991) Variation in the glucosinolate content of oilseed rape (*Brassica napus* L.) leaves. II. Response to infection by *Alternaria brassicae* (Berk.) Sacc. Ann Appl Biol 118:469–477

Doughty KJ, Blight MM, Bock CH, Fieldsend JK, Pickett JA (1996) Release of alkenyl isothiocyanates and other volatiles from *Brassica rapa* seedlings during infection by *Alternaria brassicae*. Phytochemistry 43:371–374

Greenhalgh JR, Mitchell ND (1976) The involvement of flavor volatiles in the resistance to downy mildew of wild and cultivated forms of *Brassica oleracea*. New Phytol 77:391–398

Husain A, Thakur RN (1966) Production of a toxin by *Alternaria brassicae* (Berk.) Sacc. *in vitro*. Labdev J Sci Technol 4:144–145

Jung DS, Na YJ, Ryu KH (2002) Phylogenic analysis of *Alternaria brassicicola* producing bioactive metabolites. J Microbiol 40:289–294

Juniper BE, Jefree CE (1983) Plant surfaces. Edward Arnold, London

Knox-Davies PS (1980) Relationships between *Alternaria brassicicola* and *Brassica* seeds. Trans Br Mycol Soc 73:235–248

Lindenfelser LA, Ciegler A (1969) Production of antibiotics by *Alternaria* species. Dev Indian Microbiol 10:271–278

Mac Kinnon SL, Keifer P, William AA (1999) Components from the phytotoxic extract of *Alternaria brassicicola*, a black spot pathogen of canola. Phytochemistry 51:215–221

Maiti RK, Moreno-Limon S, Wesche-Ebeling P (2000) Responses of some crops to various abiotic stress factors and its physiological and biochemical basis of resistances. Agric Rev 21:155–167

Maitra S, Samajpati N (1982) Changes in polyphenoloxidase and peroxidase in the cauliflower (*Brassica oleracea* L.) leaves infected by *Alternaria brassicae* (Berk.) Sacc. In: Chattopadhyay SB, Samajpati N (eds) Advances in mycology and plant pathology. Oxford & IBH Publishing Company, New Delhi, pp 216–224

Maitra S, Samajpati N (1985) Nucleic acid content in leaf blight disease of *Brassica oleracea* var. Botrytis caused by *Alternaria brassicicola* (Schw.) Wiltshire. Phytopathol Z 114:69–72

Mandahar CL, Suri RA (1983) Secretion of cytokinins *in vivo* and *in vitro* by *Alternaria brassicae* and their role in pathogenesis. Trop Plant Sci Res 1:285–288

Mandahar CL, Suri RA (1987) Cytokinin-like activity in green islands formed around infection spots of *Alternaria brassicicola* on mustard leaves in nature and their role in pathogenesis. Biol Plant 29:76–79

Mathpal P, Punetha H, Tewari AK, Agrawal S (2011) Biochemical defense mechanism in rapeseed-mustard genotypes against Alternaria blight disease. J Oilseed Brass 2:87–94

Matsumoto M, Matsutani S, Sugita K, Yoshida H, Hayashi F, Terui Y, Nakai H, Uotani N, Kawamura Y, Matsumoto K (1992) Depudecin: a novel compound inducing the flat phenotype of NIH3T3 cells doubly transformed by ras- and src-oncogene, produced by *Alternaria brassicicola*. J Antibiot 45:879–885

McKenzie KJ, Robb J, Lennard JH (1988) Toxin production by *Alternaria* pathogens of oilseed rape (*Brassica napus*). Crop Res (Hort Res) 28:67–81

Mithen RF, Lewis BC, Fenwick GR (1986) *In vitro* activity of glucosinolates and their products against *Leptosphaeria maculans*. Trans Br Mycol Soc 87:433–440

Mithen RF, Lewis BC, Heaney RK, Fenwick GR (1987) Resistance of leaves of *Brassica* species to *Leptosphaeria maculans*. Trans Br Mycol Soc 88:525–531

Mukadam DS, Deshpande KB (1979) Role of light and temperature on growth, sporulation and subsequent spore germinability of *Alternaria brassicae* (Berk.) Sacc. Sci Cult 45:244–246

Poapst PA, Ramsoomair BA, Gaurley CO (1979) On the promotion of senescence in *Brassica oleracea* var. Capitata by *Alternaria brassicicola* and by *Botrytis cinerea*. Can J Bot 57:2378–2386

Rawlinson CJ, Muthyalu G, Poole VA, Cayley GR, Hulme PJ, Pickett JA (1985) Mustard oils, fungicides and disease. Rothamsted Exp Station Rep 1984:124–125

Sharma N, Rahman MH, Liang Y, Kav NNV (2010) Cytokinin inhibits the growth of *Leptosphaeria maculans* and *Alternaria brassicae*. Can J Plant Pathol 32:306–314

Singh MJ, Singh J (1989) Mechanisms of resistance to cucumber mosaic virus in chilli pepper (*Capsicum annuum* L.). In: Role of phenols and phenologies. Eucarpia VII meeting on genetics and breeding on capsicum and eggplant, Yugoslavia, pp 193–203

Skipp RA, Bailey JA (1977) The fungitoxicity of isoflavanoid phytoalexins measured using different types of bioassay. Physiol Plant Pathol 11:101–112

Srivastava SK, Srivastava AK (1982) Polygalacturonase and cellulase production by *Alternaria brassicae* (Berk.) Sacc. var. Sarson. New Bot 9:9–11

Suri RA, Mandahar CL (1982) Production of pectolytic enzymes by *Alternaria brassicae*. Indian J Mycol Plant Pathol 12:351–352

Suri RA, Mandahar CL (1984) Secretion of cytokinins *in vivo* and *in vitro* by *Alternaria brassicicola* and their role in pathogenesis. Experientia Basel Birkhauser 40:462–463

Suri RA, Mandahar CL (1985) Involvement of cytokinin-like substances in the pathogenesis of *Alternaria brassicae* (Berk.) Sacc. Plant Sci 41:105–109

Suri RA, Mandahar CL (1986) Cytokinin-like substances and accumulation of labelled metabolites at infection sites of *Alternaria brassicicola*. Mycopathologia 94:153–156

Suri RA, Mandahar CL, Gill PS (1983) Study of secretion of cytokinins by *Alternaria brassicicola* and their possible role in pathogenesis in mustard. Indian J Plant Pathol 1:117–121

Verma JP (1964) Studies in the enzyme make-up of *Alternaria*. V. Aldolase activity. Mycologia 56:909–918

Verma JP (1971) Studies in the enzyme make-up of *Alternaria*. IX. Nucleao-phosphatase activity. Indian Phytopathol 24:43–49

Weimer JL (1924) Alternaria leaf spot and brown rot of cauliflower. J Agric Res 29:421–441

9 Resistance

9.1 Introduction

Attempts have been made to transfer resistance from wild Cruciferae to cultivated *Brassica*. While transferring the resistance into cultivated *Brassica*, the knowledge of pathogenic variability, sources of resistance and methods to determine nature, mechanism and stability of resistance is essential. Various techniques to incorporate the resistance in the cultivated *Brassica* as well as to evaluate the resistance in the field, in screen house and in the laboratory conditions have been developed. The pathogenic variability, especially in *A. brassicae*, has been identified and requires special attention. The mechanisms of resistance in rapeseed and mustard due to epicuticular wax, as well as for pre-existing rapid accumulation of biochemical components, have been studied. The role of phytoalexins in the resistance has been confirmed. The importance of phytotoxic compounds including destruxin B, destruxin B2, homodestruxin B and desmethyldestruxin B in the pathogenesis of *A.* brassicae has been determined. The fact that the resistant *Brassica* varieties produce other metabolites including sesquiterpenes deoxyuvidin, β-albrassitriol, isoalbrassitriol and brassicadiol has also been confirmed. Scanning electron microscopy (SEM) has also revealed the sequestration of calcium during pathogenesis by *A. brassicae* (Saharan 1992, 1997; Saharan et al. 2003, 2005).

Genetic amelioration for disease resistance is considered to be the most effective, stable and economical method of protecting crops for ensuring their full productivity. But before the research on breeding for disease resistance is initiated, the researchers must fully understand regarding the method and rate of growth and multiplication and transmission of the pathogen. In nature, plants employ a wide range of mechanisms to restrict the growth of the pathogen. First mechanism operates before the parasitic contact, while the other mechanisms operate after the pathogen has made contact with the host. Selection for resistance implies measurement of plant's resistance which can be done by measuring the growth and development of the pathogen; the higher the reduction, the more resistant the host plant. This is generally not possible because in most cases, the pathogen is either invisible or partially visible. However, one can evaluate the pathogen's effects on the host in the form of symptoms. The information generated so far regarding the existence of many components, its multilayered nature and the mechanism of resistance in the *Alternaria–Brassica* pathosystem are presented in this chapter.

G.S. Saharan et al., *Alternaria Diseases of Crucifers: Biology, Ecology and Disease Management*,
DOI 10.1007/978-981-10-0021-8_9

9.2 Genetics of Host–Parasite Interaction

In *Brassica juncea* cv. RC781, resistance to *A. brassicae* is reported to be governed by a single dominant gene (Tripathi et al. 1980). However, Brun et al. (1987a, b, 1989) reported a high degree of resistance in *Sinapis alba* cv. Emergo against *A. brassicae.* Intraspecific crosses between *B. napus* and *B. juncea* have higher contents of epicuticular wax (Singh et al. 1999). Saharan and Kadian (1983) analysed different components of horizontal resistance (HR) in *Brassica* genotypes against *A. brassicae* and found large differences in the number of lesions, size of lesions, latent period, sporulation capability and infection rate in *B. napus* cv. Tower and in *B.* juncea cv. RC-781. In other *B. juncea* genotypes like CSR I42 and CSR 448, moderate level of HR has been recorded. In Tower, infection is delayed up to 25 days with very few lesions (0.95) of restricted size. Both Tower and RC-781 cultivars have longer latent periods of 18 and 12 days, respectively, compared to 3 days in cv. Prakash. The amount of conidial production in Prakash was higher (265 conidia per lesion) than in Tower (92 conidia per lesion) (Saharan 1991, 1992). Kolte (1987) suggested that the size of lesions and amount of sporulation may also be considered important criteria in determining the degree of resistance against *Alternaria* blight in *Brassica* species.

9.2.1 Inheritance of Resistance

Genetics of *Alternaria* blight resistance in inter- and intraspecific crosses of *B. juncea* and *B. carinata* have been determined by Krishnia et al. (2000). Population of six generations, viz. P_1, P_2, F_1, F_2, BC_1 and BC_2, of six crosses, viz. Pusa Basant × EC-322092, Kranti × EC-322092, Varuna × EC-322092, RH 30 × EC-322093, RH 30 × HC-1 and Varuna × PCC-2, were evaluated for the inheritance of resistance to *A. brassicae.* The F_1 hybrids showed intermediate disease intensity compared to their parents. The F_2 populations of all six crosses exhibited more susceptibility to *A. brassicae* than their respective F_1s. Performance of backcross progenies tends to deviate towards their respective recurrent parents. The X^2 values in joint scaling test were highly significant in all the crosses indicating simple additive dominance model to be inadequate for all the crosses. The six-parameter model revealed significant additive gene effects in all six crosses, whereas dominance gene effects only in the three crosses, viz. Varuna × EC-322092, RH 30 × EC-322092 and Varuna × PCC-2. However, there was preponderance of additive gene than dominance gene effects in all the six crosses. Non-allelic interaction and additive × additive (i) were significant in all the crosses except Pusa Basant × EC-322092 and Kranti × EC-322092, whereas additive × dominance epistasis was significant only in Pusa Basant × EC-322092, Kranti × EC-322092 and RH 30 × HC-1 crosses. The crosses Kranti × EC-322092, Varuna × EC-322092 and RH 30 × EC-322093 exhibited significant dominance × dominance (I) type epistasis. Based upon these results, it is advocated that such breeding methods should be employed which exploit both kinds of gene effects. Therefore, reciprocal recurrent selection or diallel selective matings can be successfully used to develop genotypes resistant to *Alternaria* blight (Tables 9.1, 9.2 and 9.3) (Saharan and Krishania 2001).

The per cent disease intensity (PDI) of *Alternaria* blight at leaf stage is less in the progenies of R × R cross families as compared to R self and R open families in each cross indicating that inter-mating between resistant plants helps in increasing the level of resistance. A similar improvement has also been reported for multiple disease resistance in other crops where the frequency of favourable genes increases in the population enhancing the probability of obtaining multiple disease resistance (Saharan and Krishania 2001). The utility of biparental techniques can be more pronounced if the additive or additive × additive types of genetic variations are coupled with repulsion phase linkage between genes (Singh et al. 1986). The selection of resistance to *Alternaria* blight in the inter-mated populations rather than in F_2 and F_3 populations had

Table 9.1 Per cent disease intensity (PDI) of *Alternaria* leaf blight (*A. brassicae*) on different generations of *Brassica* crosses (Krishnia et al. 2000; Saharan and Krishania 2001)

Cross	P_1	P_2	F_1	F_2	BC_1	BC_2
B. juncea* × *B. juncea						
Pusa Basant × EC-322092	43.42	5.75	23.55	26.65	34.35	17.52
	(41.21)	(13.81)	(29.06)	(31.05)	(35.91)	(24.73)
	±0.314	±0.184	±0.132	±0.088	±0.372	±0.190
Kranti × EC-322902	36.35	4.54	19.68	22.85	30.15	15.12
	(37.11)	(12.11)	(26.35)	(28.52)	(33.27)	(22.87)
	±0.290	±0.121	±0.143	±0.128	±0.218	±0.160
Varuna × EC-322092	46.95	6.32	28.35	33.33	38.85	16.98
	(43.72)	(14.54)	(32.20)	(35.24)	(38.59)	(24.27)
	±0.234	±0.211	±0.122	±0.086	±0.226	±0.134
RH 30 × EC-322093	38.25	5.58	20.75	22.26	34.12	16.45
	(38.23)	(13.69)	(27.06)	(28.39)	(35.73)	(24.04)
	±0.340	±0.290	±0.208	±0.090	±0.386	±0.342
B. juncea* × *B. carinata						
RH 30 × HC-1	41.41	6.58	27.35	30.45	36.35	20.75
	(40.05)	(15.00)	(31.56)	(33.52)	(37.11)	(27.06)
	±0.284	±0.202	±0.192	±0.100	±0.252	±0.222
Varuna × PCC-1	45.52	7.42	31.12	34.88	40.92	19.76
	(42.48)	(15.79)	(33.90)	(36.15)	(39.76)	(26.42)
	±0.384	±0.282	±0.204	±0.114	±0.306	±0.238

Figures in parentheses are angular transformed values

Table 9.2 Estimates of components of generation means on three-parameter model for *A. brassicae* on different crosses of oilseed *Brassica* (Krishnia et al. 2000)

Cross	m	d	h	X^2
B. juncea* × *B. juncea				
Pusa Basant × EC-322092	28.79	13.42**	1.81**	71.93**
	±0.31	±0.32	±0.50	
Kranti × EC-322902	25.68	12.39**	2.53**	86.42**
	±0.28	±0.28	±0.49	
Varuna × EC-322092	30.60	14.68**	3.21**	154.46**
	±0.28	±0.29	±0.48	
RH 30 × EC-322093	27.19	12.21**	1.51*	64.35**
	±0.34	±0.36	±1.51	
B. juncea* × *B. carinata				
RH 30 × HC-1	28.99	12.21**	5.04**	105.87**
	±0.31	±0.31	±0.55	
Varuna × PCC-1	30.87	13.57**	5.16**	105.26**
	±0.35	±0.36	±0.61	

*,**Significant at $P=0.05$ and $P=0.01$ level, respectively
m general mean, *d* sum of additive effects of genes, *h* sum of dominance effects of genes

been advocated by Singh and Singh (1989), since the generations of inter-varietal hybrids would obviate the harmful effects of linkages and linkage disequilibrium and shuffle the desirable genes in one recombinant (Comstock and Robinson 1952; Matzinger and Cockerham 1963; Gates et al. 1957).

The reduction in *Alternaria* blight at leaf stage is less in four crosses, viz. PCR 3 × Shiva, Pusa Basant × Shiva, Rajat × Shiva and Pusa

Table 9.3 Estimates of gene effects under six-generation mean analysis for *A. brassicae* in different crosses of oilseed *Brassica* (Krishnia et al. 2000; Saharan and Krishania 2001)

Cross	m	d	h	i	j	k
B. juncea* × *B. juncea						
Pusa Basant × EC-322092	31.05	11.18**	−1.37	−2.92	−2.52**	−5.22
	±0.30	±0.75	±1.97	±1.91	±0.83	±3.38
Kranti × EC-322902	28.52	10.40**	−0.06	−1.80	−2.10**	−8.56**
	±0.36	±0.61	±1.95	±1.69	±0.69	±3.01
Varuna × EC-322092	35.24	14.32**	−12.17**	−15.24**	−0.27	12.18**
	±0.29	±0.60	±1.75	±1.68	±0.69	±2.84
RH 30 × EC-322093	28.39	11.69**	7.08**	5.98**	−0.58	−19.48**
	±0.30	±0.85	±2.17	±2.09	±0.94	±3.81
B. juncea* × *B. carinata						
RH 30 × HC-1	33.52	10.05**	1.71	−5.74**	−2.48**	−4.43
	±0.32	±0.69	±1.95	±1.87	±0.77	±3.23
Varuna × PCC-1	36.15	13.34**	−7.48**	−12.24**	−.0.01	5.95
	±0.34	±0.74	±2.09	±2.00	±0.84	±3.47

*Significant at $P = 0.05$, and **Significant at $P = 0.01$ level, *m* general mean, *d* sum of additive effects of genes, *h* sum of dominance effects of genes, *i* sum of additive × additive epistatic effects of genes, *j* sum of additive × dominance epistatic effects of genes, *k* sum of dominance × dominance epistatic effects of genes

Bahar × Domo, as compared to others indicating that the parents involved in these crosses have similar degree of susceptibility; only slight improvement is made through selection in segregating generations. In general, the progenies of R × S families record greater variation in disease score. The numerical reduction in *Alternaria* blight (leaf phase) disease is observed in the progenies of S × S families in comparison to S-self plant progenies in each cross indicating that even susceptible plants possess resistant genes at different loci with minor effects.

The significant 'j'-type epistasis is observed in crosses Pusa Basant × EC-322092, Kranti × EC-322092 and RH 30 × HC-1, whereas 'I'-type epistasis is significant in crosses Kranti × EC-322092, Varuna × EC-322092 and RH 30 × EC-322093. The selection of resistant plants should be, therefore, done in advanced segregating generation rather than F_2. Therefore, the crosses possessing predominantly significant additive gene effect with additive × additive kinds of epistasis must be exploited to improve through simple selections. The highest GCV, PCV, h^2 and GG are recorded in progenies of R × S cross families in most of these crosses. The values of heritability and genetic gain are low in susceptible plant progenies as compared to resistant progenies. Almost a similar trend in all the crosses for *Alternaria* blight PDI at siliquae phase indicates a high-order correlation in *Alternaria* blight resistance at both leaf and siliquae phases (Saharan and Krishania 2001).

9.2.2 Disease Tolerance

Disease stress tolerance index (DSTI) was found an effective selection criterion for assessing the Indian mustard genotypes for their disease stress tolerance and yield potential. The genotypes Rajat, Kranti and RH-781, under normal sown, and Rajat, RL-1359 and Kranti, under late sown conditions, performed with uniform superiority under both non-disease stress and disease stress environments. Potential yield under controlled environment (Yp) was significantly and positively correlated with yield under disease stress environment (Ys). Potential yield, mean productivity (MP), disease tolerance (TOL), geometric mean productivity (GMP) and disease stress tolerance index (DSTI) under normal sown and non-disease stress conditions (Yp) had significant positive association with yield under disease stress environment (Ys), under late sown conditions (Gupta et al. 2002) (Tables 9.4 and 9.5).

Table 9.4 Estimates of disease stress tolerance attributes from the potential yield and yield under disease stress environment (DSI = 0.195) in Indian mustard (*B. juncea*) under normal date of sowing (Gupta et al. 2002)

Genotypes	Yp	Ys	MP	TOL	DSSI	GMP	DSTI
EC-129126-1	2508	2036	2272	472	0.96	2260	0.87
RH-781	2635	2063	2349	572	1.11	2332	0.93
PHR-1	2034	1753	1894	281	0.71	1888	0.61
Varuna	2586	2087	2337	499	0.99	2323	0.92
Kranti	2738	2217	2478	521	0.98	2464	1.04
RH 8113	2318	1886	2102	432	0.96	2091	0.75
RL 1359	2680	2025	2353	655	1.25	2330	0.93
Rajat	2807	2234	2524	573	1.05	2504	1.07
RH 30	2769	1897	2333	872	1.61	2292	0.90
Pusa Bold	2463	1856	2160	607	1.26	2138	0.78
PR-8805	2107	1826	1967	281	0.68	1961	0.66
RC-781	1838	1781	1810	57	0.16	1757	0.56
ZEM-1	1973	1564	1769	409	1.06	1757	0.53
Shiva	2413	2027	2220	386	0.82	2212	0.84
Mean	2419	1947	2183	473	1.00	2169	1.00
CD ($P = 0.05$)	183.07	106.87	139.06	111.92	0.19	136.04	0.09

Significant at $P = 0.05$; *Yp* yield under disease-controlled environment (kg/ha), *Ys* yield under artificially pathogen-inoculated environment, *DSI* disease stress intensity, *MP* mean productivity, *TOL* disease tolerance, *DSSI* disease stress susceptibility index, *GMP* geometric mean productivity, *DSTI* disease stress tolerance index

Table 9.5 Estimates of disease stress tolerance attributes from the potential yield and yield under disease stress environment (DSI = 0.209) in Indian mustard (*B. juncea*) under late date of sowing (Gupta et al. 2002)

Genotypes	Yp	Ys	MP	TOL	DSSI	GMP	DSTI
EC-129126-1	1691	1149	1420	542	1.53	1393	0.70
RH-781	1656	1281	1469	375	1.08	1456	0.76
PHR-1	1576	1294	1435	282	0.86	1428	0.73
Varuna	1865	1483	1674	382	0.98	1663	0.99
Kranti	1768	1463	1616	305	0.83	1608	0.93
RH 8113	1625	1493	1559	132	0.39	1557	0.87
RL 1359	1865	1481	1673	384	0.98	1662	0.99
Rajat	1849	1160	1727	244	0.63	1723	1.07
RH 30	1796	1452	1624	344	0.92	1615	0.94
Pusa Bold	1699	1358	1529	341	0.96	1519	0.83
PR-8805	1598	1005	1302	593	1.77	1267	0.58
RC-781	1575	1201	1388	374	1.14	1375	0.68
ZEM-1	1440	1125	1283	315	1.05	1272	0.58
Shiva	1357	1094	1226	263	0.93	1218	0.53
Mean	1669	1320	1494	348	1.00	1482	0.80
CD ($P = 0.05$)	88.97	105.34	91.62	66.40	0.19	93.72	0.10

Significant at $P = 0.05$, *Yp* yield under disease-controlled environment (kg/ha), *Ys* yield under artificially pathogen-inoculated environment, *DSI* disease stress intensity, *MP* mean productivity, *TOL* disease tolerance, *DSSI* disease stress susceptibility index, *GMP* geometric mean productivity, *DSTI* disease stress tolerance index

9.2.3 Components of Horizontal Resistance

While analysing different components of horizontal resistance (HR) in *Brassica* genotypes against *A. brassicae*, Saharan and Kadian (1983) reported large differences in the number and size of lesions, latent period, rate of sporulation and infection rate in *B. napus* cv. Tower and *B. juncea* cv. RC-781; *B. juncea* genotypes CSR 142 and CSR 448 showed moderate level of HR.

The latent periods of Tower and RC-781 were 18 and 22 days, respectively, compared to only 3 days in Prakash. The amount of conidial production was also considerably higher in Prakash (265 conidia per lesion) compared to Tower (92 conidia per lesion). Kolte (1987) suggested that the size of lesions and amount of sporulation are also important factors in evaluating resistance against *A. brassicae*. In genotypes PR-8988 and PR-9024, *A. brassicae* produced significantly reduced number of spots (4.36–15.89), smaller size of spots (2.12–6.17 mm), lower sporulation (0.30–1.84×10^3conidia), lower disease index (36.51–42.2 %), reduced apparent infection rate ($r = 0.047$–0.080) and lesser values of AUDPC (45.35–126.70) on leaf and pod, along with reduced leaf defoliation (38.40–44.40 %) in comparison to on national susceptible genotype Varuna (Tables 9.6 and 9.7). Therefore, these genotypes showed higher degree of partial resistance or slow blighting. The development and progression of *Alternaria* blight was slower in these two genotypes in comparison to the others. The yield (q ha^{-1}) of the genotype Kranti was highest (14.01) followed by genotype PR-9024 (11.14); yield day^{-1} was highest in genotype Divya (11.12 kg ha^{-1} day^{-1}) in comparison to the others (Kumar and Kolte 2001; Kumar and Singh 2006). Genotypes PR-8988 and PR- 9024 exhibit slow blighting and have the lowest number of lesions/10 cm^2 leaf area, reduced spot size, least number of conidia per spot, and the lowest disease index on leaf and pod (Kumar and Kolte 2006).

9.3 Morphological Resistance

The variations in the number and size of stomata of resistant and susceptible cultivars have been recorded (Saharan and Kadian 1983); numbers of stomata were maximum in susceptible cultivar Prakash and minimum in resistant cultivar Tower. Stomatal aperture was also narrower in resistant cultivars Tower and RC-781 compared to susceptible cultivars Prakash, RH-30 and YRT-3 (Table 9.8.). Significantly the lower number of stomata per unit area and the smaller aperture of the stomata are important morphological resistance factors in reducing infection of *Brassica* genotypes by *A. brassicae*.

9.4 Epicuticular Wax

The structure of wax in *Brassica* spp. and its role in host resistance has been studied by a number of workers (Conn and Tewari 1989a, b; Holloway et al. 1977; Saharan 1992; Skoropad and Tewari 1977; Tewari 1991a; Tewari and Skoropad 1976). In the canola-type cultivars Candle, Tobin and Altex, the wax is organized into an amorphous layer on which is situated a crystalline layer. The crystalline layer consists of platelike crystals and a layer of erect filamentous and rod-like crystals (Plates 9.1, 9.2 and 9.3; Conn and Tewari 1985, 1989a, b). The wax in *Brassica* spp. is complex both structurally and chemically (Conn 1986; Conn et al. 1984; Holloway et al. 1977). In the canola-type cultivars, the wax has the same nine major constituents (alkanes, esters, ketones, aldehydes, secondary alcohols, ketols, primary alcohols, triterpenols and fatty acids) but in varying proportions (Conn 1986; Conn et al. 1984; Tewari 1991a, b). In the *B. napus* ssp. *oleifera* cv. Altex, the major constituents of wax include C_{29} alkanes, C_{29} ketone, C_{29} secondary alcohol and C_{40}–C_{48} esters (Conn 1986; Tewari 1991a, b).

In providing resistance to the host against *Alternaria* spp., wax appears to present only a

Table 9.6 Components of *Alternaria* blight disease resistance and yield of mustard (*B. juncea*) genotypes (Kumar and Kolte 2001)

Genotypes	No. of spots per 10 cm²		Size of spot (mm)		Sporulation/spots (no. of conidia × 10³)		Disease index (%)		Apparent infection rate (r)		AUDPC (mm²)		Leaf defoliation (%)	Yield (q/ha)	Yield (kg ha⁻¹ day⁻¹)
	Leaf	Pod	Leaf	Pod	Leaf	Pod	Leaf	Pod	Leaf	Pod	Leaf	Pod			
Kranti	4.48	17.78	6.25	2.38	2.14	0.33	46.4	38.7	0.054	0.065	124.8	52.5	38.8	14.01	10.51
Varuna	5.10	25.50	7.71	5.76	3.04	0.63	53.9	52.5	0.063	0.115	153.0	58.5	57.6	10.05	8.21
PR-9650	5.56	23.47	7.74	3.27	3.48	0.63	56.1	53.8	0.065	0.140	155.5	65.0	64.8	7.73	6.64
Krishna	4.75	19.56	6.93	2.46	2.74	0.40	48.2	47.0	0.055	0.090	139.1	55.6	50.6	10.18	8.39
PR-8988	4.42	15.89	6.11	2.12	1.94	0.30	44.4	36.6	0.047	0.075	122.3	45.3	44.4	10.07	8.47
PR-9024	4.36	14.38	6.17	2.13	1.81	0.36	45.2	39.2	0.057	0.080	126.7	45.7	38.4	11.14	8.70
PR-8943	4.60	17.82	6.34	2.38	2.40	0.33	46.7	45.0	0.056	0.090	128.4	52.7	40.0	10.78	8.32
Divya	4.50	21.13	6.60	2.70	2.59	0.36	47.3	49.0	0.059	0.100	131.1	60.1	55.3	10.12	11.12
PR-9301	5.97	29.91	8.58	3.24	3.75	0.66	59.5	57.0	0.073	0.140	172.7	70.5	68.6	7.70	6.72
CD at 5 %	0.284	1.685	0.506	0.223	0.404	0.001	2.934	1.839	NS	NS	NS	NS	3.354	1.710	1.270

Table 9.7 Correlation coefficients (R) among different components of *Alternaria* blight disease resistance and yield of mustard (*B. juncea*) genotypes (Kumar and Kolte 2001)

Disease components and yield	Size of spot	Disease index	Leaf defoliation	Sporulation	AUDPC	Infection rate	Yield
No. of spots	0.883**	0.897**	0.812**	0.923**	0.956**	0.864**	–0.788*
Size of spots	–	0.985**	0.934**	0.982**	0.956**	0.893**	–0.668*
Disease index	–	–	0.905**	0.974**	0.949**	0.876**	–0.684*
Leaf defoliation	–	–	–	0.952**	0.884**	0.743*	–0.625 NS
Sporulation	–	–	–	–	0.973**	0.851**	–0.679*
AUDPC	–	–	–	–	–	0.903**	–0.790*
Infection rate	–	–	–	–	–	–	–0.770*

*Significant at 5 %; **significant at 1 %; *NS* non-significant

Table 9.8 Factors influencing resistance/susceptibility of different cultivars of rapeseed–mustard against *Alternaria brassicae* (Saharan and Kadian 1983)

Cultivars	Leaf surface	Stomata[1]		Lesions/leaflet[2]			Incubation period (days)[4]	Latent period (days)[4]
		No. (per sq. cm)	Size (mm)	No.	Size (mm)	Sporulation[3]		
Tower	Upper	145	15.7	1.8	1.9	80	10	16
	Lower	210	25.9					
RC-781	Upper	166	17.3	3.3	5.7	92	8	12
	Lower	245	26.7					
CSR-448	Upper	196	19.2	3.7	10.7	120	7	10
	Lower	303	27.9					
CSR-741	Upper	189	19.3	6.3	14.6	125	7	9
	Lower	292	20.0					
CSR-142	Upper	199	19.3	4.4	10.6	120	8	11
	Lower	306	28.0					
YRT-3	Upper	225	22.0	10.2	22.3	250	6	9
	Lower	403	30.8					
RH-30	Upper	221	22.0	8.1	19.1	240	6	9
	Lower	392	30.6					
Prakash	Upper	229	22.7	8.7	18.9	260	6	8
	Lower	413	31.2					

[1] and [3] average of 100 observations; [2] and [4] average of 10 leaves

physical barrier without a direct chemical effect (Conn 1986; Conn and Tewari 1989a, b; Skoropad and Tewari 1977; Tewari and Skoropad 1976). The wax forms a hydrophobic coating and reduces the deposition of water-borne inoculum. Wax also reduces the rate of conidial germination and the number of germ tubes produced by each conidium. The crystalline wax layer is made fluffy by enclosed air pockets, which may be responsible for the above two effects by impeding the movement of plant exudates. Plants of *B. napus* ssp. *oleifera* are very waxy compared to those of *B. rapa* ssp. *oleifera* which is more susceptible to *A. brassicae.* The leaves of cultivars Midas and Tower (*B. napus*), resistant to *Alternaria*, have appreciable amounts of epicuticular wax (Skoropad and Tewari 1977). According to Gupta et al. (1987a, b, c), *Alternaria*-resistant genotypes, viz. Tower, HNS-3 (*B. napus*),

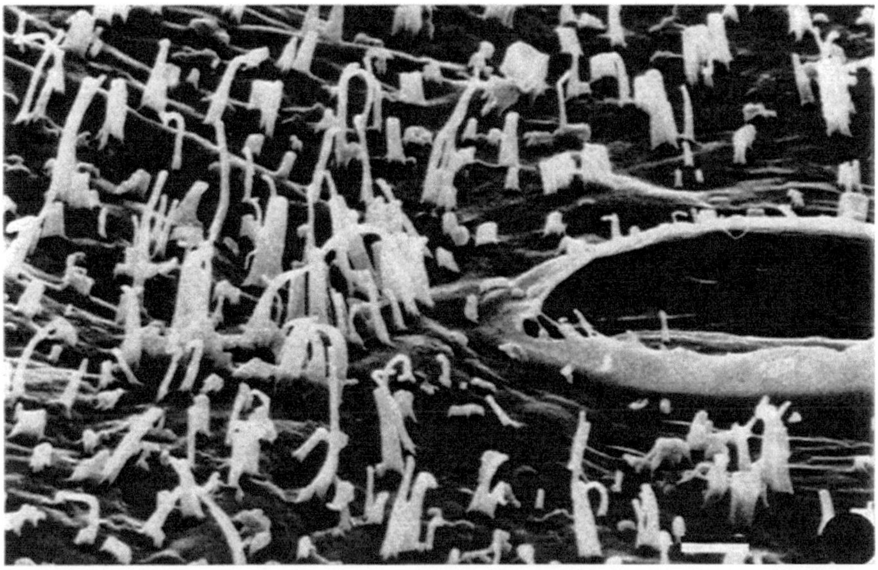

Plate 9.1 Scanning electron micrograph of air-dried, osmium vapour-fixed and gold-coated middle leaves of *Brassica rapa* cv. Candle showing wax crystals. Bar = 2 μm (Tewari 1991a, b)

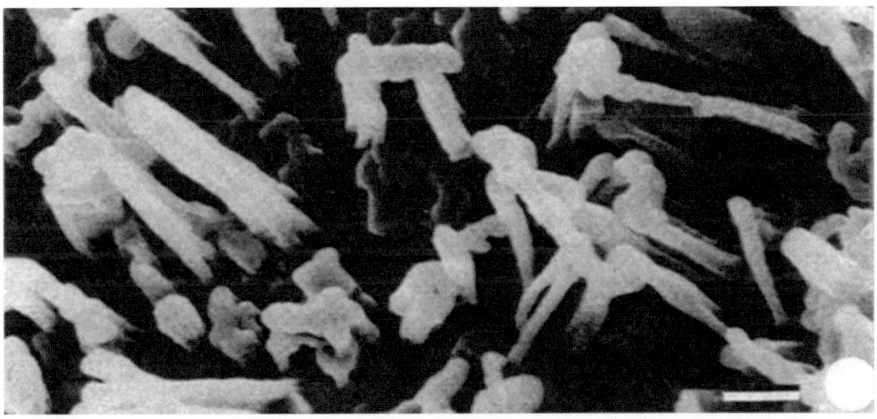

Plate 9.2 Scanning electron micrograph of air-dried, osmium vapour-fixed and gold-coated stem of *Brassica rapa* cv. Tobin showing flat and erect wax crystals. Bar = 1 μm (Tewari 1991a, b)

HC-2 (*B. carinata*) and *B. alba*, have higher amounts of wax on their leaves at different growth stages as compared to susceptible genotypes, BSH-1, YSPb-24 (*B. rapa*) and RH-30 (*B. juncea*). Gomez-Campo and Prakash (1999) identified three different epicuticular wax columns in *Brassica* species with a chromosome number (n) = 9, long columns (LC), short columns (SC) and netted columns (NC). The *B. napus* and *B. carinata* seemed to inherit the NC type of wax present in *B. oleracea*. Singh et al. (1999) observed that intraspecific crosses between *B. napus* and *B juncea* had high content of epicuticular wax.

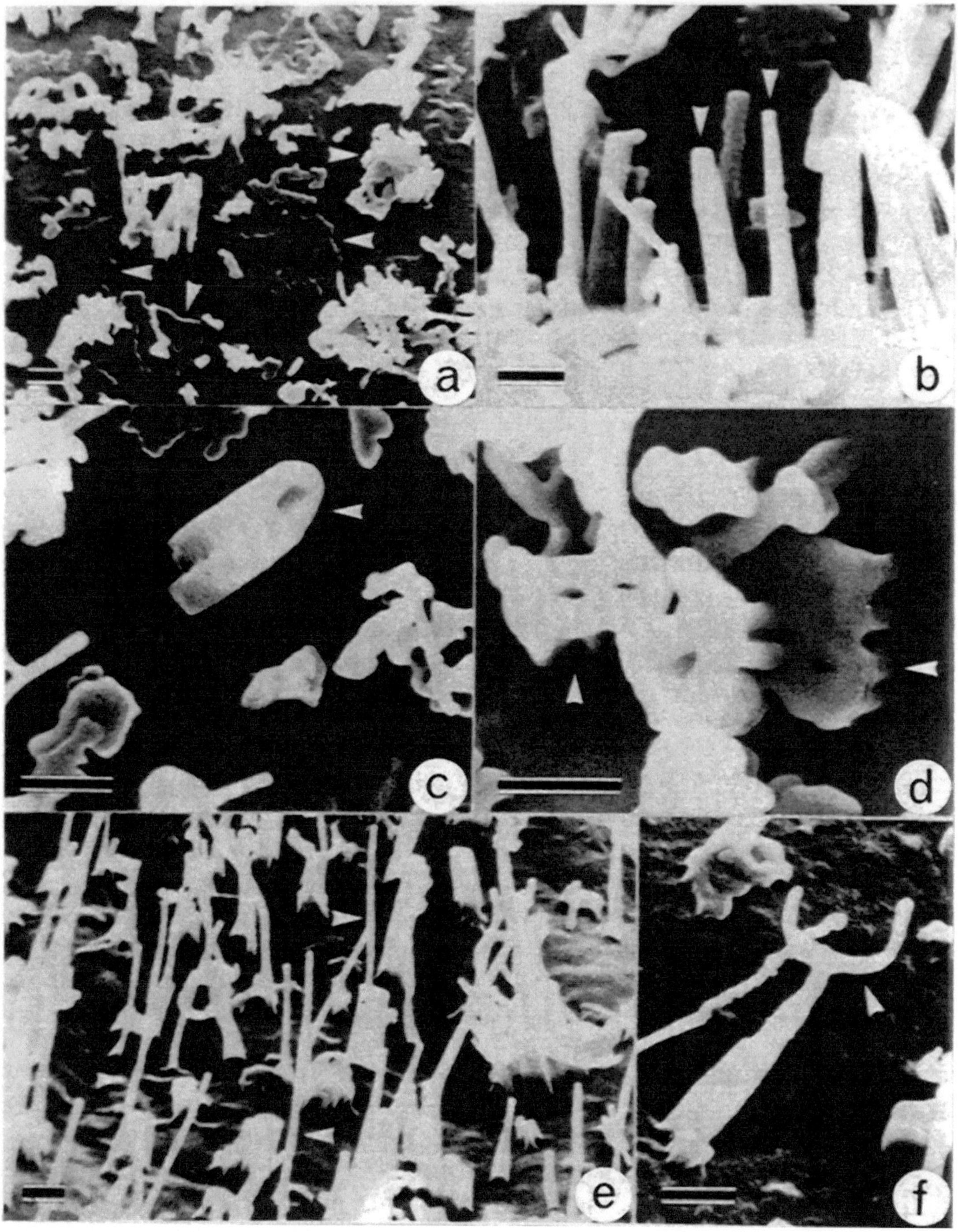

Plate 9.3 (**a**) Adaxial surface of an upper leaf of *Brassica napus* cv. Altex showing platelike wax crystals (*arrows*); (**b**) stem surface of cultivar Altex from the middle of a plant showing rods (*arrows*); (**c**) adaxial surface of an upper leaf of *B. rapa* cv. Tobin showing a fused rod (*arrow*); (**d**) adaxial surface of an *upper leaf* of *B. napus* cv. Westar showing fused rods (*arrows*) and growth rings in wax crystals; (**e**) adaxial surface of an *upper leaf* cv. Tobin showing filamentous wax crystals (*arrows*); and (**f**) adaxial surface of a *middle leaf* cv. Tobin showing a branched filamentous wax crystal (*arrow*). The plant surfaces depicted in a-f were prepared for SEM by air-drying method. Bar = 1 μm (Conn and Tewari 1989a, b)

9.5 Biochemical Basis of Resistance

The *Alternaria*-resistant *B. juncea* genotype RC-781 is characterized by relatively high concentrations of phenols and low concentrations of sugars and nitrogen compared to a susceptible genotype Prakash (Gupta et al. 1984). The leaf surface constituents including wax, total phenols, soluble nitrogen, total soluble and reducing sugars of susceptible genotypes BSH-1, YSPB-24 (*B. rapa*) and RH-30 (*B. juncea*) and resistant genotypes Tower, HNS-3 (*B. napus*), AC-2 (*B. carinata*) and local (*B. alba*) have been determined by Gupta et al. (1987a, b, c), at 30, 50, 70 and 90 days of plant growth. The concentration of the phenolic compounds is significantly higher in resistant compared to susceptible species at all plant growth stages. Total soluble sugars, reducing sugars and soluble nitrogen levels are, however, lower in resistant genotypes.

The analysis of total phenols and specific activities of polyphenol oxidase (PPO), peroxidase (PO), and catalase in *A. brassicae*- and *A. brassicicola*-infected mustard leaves of tolerant (RC-781, RH-8113) and susceptible (RH 30, Prakash) cultivars of mustard was done at 20 days interval after 40 days of sowing and showed that the total phenols increased initially and decreased with the age of the plant; these were markedly higher in tolerant as compared to the susceptible ones at all growth stages. After infection, their amount decreased in all genotypes. The specific activities of polyphenol oxidase remained higher, while that of peroxidase lower in tolerant than in the susceptible cultivars. In response to infection, although the activities of both the enzymes increased sharply in the cultivars, this increase was considerably higher in susceptible than in tolerant cultivars. Catalase activity was appreciably higher at initial stage and dropped markedly at later stages (Gupta et al. 1990).

Adjuction of *A. brassicae* culture filtrate into subcultures of hypocotyl explants callus on MS medium supplemented with 1 mg l^{-1} NAA and 1 mg l^{-1} BAP (MS N_1B_1), causing greater accumulation of total soluble and reducing sugars in susceptible *B. juncea* cv. Kranti than in resistant *B. napus* cv. GSH-1 and *B. juncea* cv. RH-781. Phenolic content and flavonols were higher in GSH-1 and RH-781 calli than that of cultivar Kranti. Protein content, though increased in all three cultivars, was more in resistant than susceptible cultivars (Kiran et al. 2003). Destruxin B detoxification of white mustard (*Sinapis alba*) through biotransformation can help discover the specific disease resistance enzyme(s)/gene(s). If the black spot disease resistance of white mustard is related with the presence of specific detoxifying enzyme(s), this trait can be transferred to susceptible brassicas and may improve their resistance to this destructive fungal disease (Fig. 9.1; Pedras et al. 2003).

9.6 Proteome-Level Resistance

Two *Brassica* lines derived from an interspecific cross between *B. napus* and *B. carinata* were evaluated for tolerance to *A. brassicae*. Pathogen-induced chlorosis and necrosis spread significantly in one line, whereas it remains localized in the other. Proteome-level changes in response to the fungal pathogen invasion were investigated using two-dimensional electrophoresis. Levels of 48 proteins were significantly affected at various time points in the tolerant line (41 upregulated and 7 downregulated). In contrast, in the susceptible line, the levels of 23 proteins were significantly affected with four increasing and 19 decreasing. The identities of 38 proteins and those identified from the tolerant line included enzymes involved in the generation of reactive

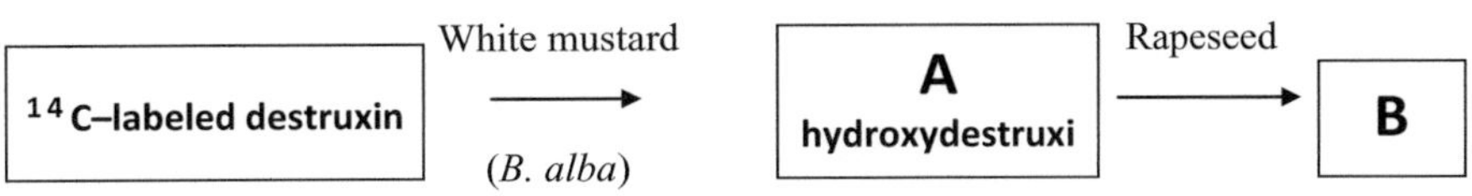

Fig. 9.1 Metabolism of destruxin B by white mustard and rapeseed

oxygen species (ROS), ROS-mediated signalling, auxin signal transduction and metabolic pathways. The changes in proteome levels suggest a role for ROS-mediated auxin signalling in this pathosystem, which was further investigated and confirmed using quantitative real-time PCR (Sharma et al. 2007). Chitinase-modifying proteins (*Cmps*) are secreted by fungal pathogens of crucifers, which interfere with fungalysin *Cmp* activity to improve plant resistance to multiple fungal diseases (Naumann and Wicklow 2013).

The *Arabidopsis thaliana* secretome analysed by the proteomic approach led to the identification of secreted proteins implicated in many aspects of cell biology. Oh et al. (2005) investigated the change in the *Arabidopsis* secretome in response to salicylic acid and identified several proteins involved in pathogen response. One of these, a secreted lipase with a GDSL-like motif designated as GDSL LIPASE 1(*GLIP1*), was further characterized for its function in disease resistance; *glip1* plants are markedly more susceptible to infection by *A. brassicicola* compared with the parental wild plants. The recombinants *GLIP1* protein possessed lipase and antimicrobial activities that directly disrupt fungal spore integrity. *GLIP1* appears to trigger systemic resistance signalling in plants when challenged with *A. brassicicola*, because pretreatment of *glip1* mutant with recombinant *GLIP1* protein inhibited *A. brassicicola*-induced cell death in both peripheral and distal leaves. Moreover, *glips1* shows altered expression of defence and ethylene-related genes. *GLIP1* transcription was increased by ethephon, the ethylene releaser, but not by salicylic acid or jasmonic acid. *GLIP1* in association with ethylene signalling may be a critical component in plant resistance to *A. brassicicola*.

Chitinase, capable of degrading the cell walls of invading phytopathogenic fungi, plays an important role in plant defence response, particularly when this enzyme is overexpressed through genetic engineering. *Brassica* plant (*B. juncea* L.) was transformed with chitinase gene tagged with an overexpressing promoter 35 S *CaMV*. In in vitro fungal growth inhibition assays, chitinase inhibits the fungal colony size by 12–56 % over the non-transgenic control. The bioassay under artificial epiphytotic conditions reveals the delay in the onset of disease as well as reduces the lesion number and size in 35 S chitinase *Brassica* as compared to the untransformed control plants (Mondal et al. 2003).

Genotypes EC-399299, EC-399296, EC-399313 and PHR-2 are comparatively more resistant to infection by *A. brassicae*. Increased level of PAL, PPO and peroxidase may play an important role in undertaking the defence mechanisms of *B. juncea* genotypes against *Alternaria* blight pathogenesis (Parihar et al. 2012). With 1 ppm concentration of chitosan sprayed on the leaves of *B. juncea*, superoxide dismutase and 5 ppm chitosan peroxide and phenylalanine ammonia lyase significantly increased resistance to *Alternaria* blight with respect to control. Phenolic content in the leaves of treated plants also increased with both the treatments of chitosan. Defence-related enzymes and phenolic content showed positive correlations, i.e. increase with the elicitor application. Since these enzymes and phenolics increased resistance to pathogens, they may provide protection against *Alternaria* blight (Neerja and Sohal 2012). After inoculation of *B. juncea* leaves with *A. brassicae*, activities of the cell wall-degrading enzymes, polygalacturonase (EC-3.2.1.15) and cellulase (EC-3.2.1.4), decreased in leaf blight-resistant cultivar RC-781 and increased in the susceptible cultivar Varuna up to 3 days. In both cultivars, 11 polypeptides are observed in leaves in the absence of *A. brassicae* inoculation. After inoculation, although there was no change in the polypeptide pattern in the resistant cultivar RC-781, but in the susceptible cultivar Varuna, four polypeptides (43.7–58.8 kDa) disappeared on the 3rd day after inoculation (Garg et al. 1999). Occurrence of a lipase in *A. brassicicola* spores plays a crucial role in the infection of cauliflower leaves. Addition of anti-lipase antibodies into *A. brassicicola* conidial suspension prior to inoculation resulted in reducing black spot lesions by 90 % on intact cauliflower leaves, but not on leaves from which surface wax had been removed. Spore surface-bound lipase is thought to interact closely with epicuticular leaf waxes for adhesion

Table 9.9 Specific activity of peroxidase enzyme in hypocotylar calli of *Brassica* species raised on MS medium supplemented with or without fungal culture filtrate (FCF) of *Alternaria brassicae* (Dhingra et al. 2004)

	Concentration of FCF (%)		
Species	Control	5	10
Brassica juncea cv. Kranti	11.21	13.77	16.88
Brassica juncea cv. RH-781	3.57	13.41	21.73
Brassica juncea cv. GSH-1	0.18	0.37	1.89

Enzymes unit = change in absorbance/3 min/mg/protein

Table.9.10 Specific activity of enzyme catalase in hypocotylar calli of *Brassica* species raised on MS medium supplemented with or without fungal culture filtrate (FCF) of *Alternaria brassicae* (Dhingra et al. 2004)

	Concentration of FCF (%)		
Species	Control	5	10
Brassica juncea cv. Kranti	0.79	0.83	0.87
Brassica juncea cv. RH-781	4.21	3.57	1.72
Brassica juncea cv. GSH-1	1.43	1.21	0.45

Enzymes unit = Change in absorbance/5 min/mg/protein

and/or penetration of fungal propagules during the early stages of host–parasite interactions (Berto et al. 1999).

Dhingra et al. (2004) reported higher peroxidase activities in two *B. juncea* cultivars than in *B. napus* cv. GSH-1. Between two *B. juncea* cultivars, enzyme activity in Kranti was more than three times than in RH-781. Addition of fungal culture filtrate (FCF) to the culture medium, although increased peroxidase activity in all cases, further increases the FCF concentration; the peroxidase activity increase was much lesser in Kranti than in RH-781 and GSH-1 (Table 9.9). The catalase activity is low in the calli of *B. juncea* cv. Kranti than in *B. juncea* cv. RH-781 and *B. napus* cv. GSH-1. Between the latter two cultivars, the former showed 3 times more activity than the latter. Catalase activity was not affected by FCF in the calli of Kranti, while activity decreased in other two cultivars; the decrease was more in GSH-1 than in RH-781 (Table 9.10). Polyphenol oxidase activity was low in the calli of *B. juncea* cv. Kranti than in *B. juncea* cv. RH-781 and *B. napus* cv. GSH-1 (Table 9.11). Addition of FCF to the culture medium, in general, increased PPO activity; the increase was much lesser in cv. Kranti than in the other two cultivars. Calli of GSH-1 showed more than twofold increase in PPO activity over that of RH-781 (Table 9.11). Inoculation of leaves with *A. brassicae* increased PPO activity in both resistant and susceptible cultivars of *B. juncea*, but the increase was much higher in RC-781 than in Varuna (Gupta 1995).

Table. 9.11 Specific activity of enzyme polyphenol oxidase in hypocotylar calli of *Brassica* species raised on MS medium supplemented with or without fungal culture filtrate (FCF) of *Alternaria brassicae* (Dhingra et al. 2004)

	Concentration of FCF (%)		
Species	Control	5	10
Brassica juncea cv. Kranti	4.12	5.37	6.54
Brassica juncea cv. RH-781	5.70	8.49	10.56
Brassica juncea cv. GSH-1	6.19	10.21	14.63

Enzymes unit = change in absorbance/5 min/mg/protein

9.7 Induced Resistance

Plant possesses a very vast array of defence strategies to combat pathogen's growth and colonization. One of the most efficient and immediate defence reactions of plants against the pathogen is the hypersensitive response, which is defined as 'the rapid death of plant cells in association with restriction of pathogen growth' (Goodman and Novacky 1994). It is characterized by the presence of brown, dead cells at the site of infection which limit the growth of biotrophic pathogens. Hypersensitive reaction is a highly concerted complex defence response which is followed by the production of reactive oxygen species (Lamb and Dixon 1997), modification of ion fluxes (Levine et al. 1996) and activation of defence through the synthesis of signalling molecules such as jasmonic acid, salicylic acid and protein kinase (Dangl et al. 1996; Dixon et al. 1994). Salicylic acid accumulation leads to the onset of systemic acquired resistance in the distal plant tissues (Ryals et al. 1996). These events are accompanied by the activation of several plant defence genes, local accumulation of pathogenesis-related (PR) proteins, activation of transcription factors, degradation of proteins by the polyubiquitin system and programmed cell death (Mishra et al. 2010).

A pre-inoculative foliar application of 5 mM β-aminobutyric acid (BABA) significantly inhibits the colonization of *A. brassicae* on leaves of *B. carinata* susceptible cultivar Car 6. BABA treatment leads to transient but significant increase in H_2O_2 level during early stages of pathogen colonization. A significant increase in superoxide dismutase (SOD), ascorbate peroxidase (APX) and guaiacol-dependent peroxidase (GDP) is known to inhibit the oxidative stress in BABA-treated plants in response to pathogen infection. BABA treatment also leads to proper balance of oxidant and antioxidants suitable for the expression of resistance resulting in curtail of pathogen ingress during early stages of colonization (Chavan et al. 2013).

Recent studies have revealed an important role of hormones in plant immunity. Cytokinins are phytohormones that are involved in various regulatory processes including plant defence. Zeatin, a cytokinin, antagonizes the effect of *A. brassicae* pathotoxin in cell culture of *B. juncea*. Phytohormones are also the inducers of MAP kinase signalling pathways, which are the important signalling modules in eukaryotic cells. Marmath et al. (2013) observed the role of exogenous application of zeatin on disease score, infection behaviour of *A. brassicae* and expression pattern of MARK-4 in the non-host *Sinapis alba*, susceptible *B. juncea* cv. Varuna, moderately tolerant *B. juncea* cv. Divya and transgenic *Brassica* to confer its role in plant immunity. High concentration of zeatin leads to increased defence responses by delaying the infection process as well as significantly reducing the disease score. Semi-quantitative RT-PCR reveals that zeatin also increases the expression of MAPK 4 at early hours of infection. Zeatin upregulates plant immunity via an elevation of MAPK 4 and clearly reflects that it antagonizes the effect of *A. brassicae*. The cross talk between zeatin and MAPK 4 signalling pathways may help plants fine-tune defence responses against *A. brassicae* in *Alternaria* blight.

9.8 Identification, Cloning and Sequencing of Resistant Genes

The antifungal effect of hevein, the chitin-binding lectin from rubber plant (*Hevea brasiliensis*), has been analysed in transgenic plants for potential control of *Alternaria* blight in Indian mustard. A cDNA encoding hevein was transferred into Indian mustard (*B. juncea* cv. RLM-198). Southern analysis of the putative transgenics shows integration of the transgene. Northern and Western analyses proved that the integrated transgene is expressed in the transgenics. In whole plant bioassay under glasshouse conditions, transgenics possess parameters that are associated with resistance such as longer incubation and latent period, smaller necrotic lesion size, lower disease intensity and delayed senescence (Kanrar et al. 2002).

β-Aminobutyric acid (BABA) pretreatment of *Brassica* plants provides protection against the necrotrophic pathogen *A. brassicae*. The achieved resistance level is much higher than that seen after salicylic acid (SA) and jasmonic acid (JA) pretreatments. BABA pretreatment to leaves, 1 day before inoculation, although it leads to an inhibition of the oxidative burst and a decrease in SA levels, neither influences lipoxygenase activity nor causes callose deposition at the site of inoculation. Expression of two marker genes of the SA and JA pathways, namely, *PR1* and *PDF1.2*, enhances in response to BABA pretreatment. BABA-induced resistance is mediated through an enhanced expression of pathogenesis-related protein genes, independent of SA and JA accumulation (Kamble and Bhargava 2007). The identification and cloning of *hsr203j* homologues from tolerant and susceptible genotypes of *B. juncea* through RT-PCR analysis have been accomplished. In silico analysis of the sequences isolated from susceptible and tolerant genotypes of *B. juncea* shows the presence of conserved abhydrolase domain having a role in cell death. Motif analysis indicates that motif 19 that functions in prenylation is found exclusively in tolerant genotypes and motif 12 having myristoylation site is found in susceptible genotypes. Various defence-related important *cis* and *trans* acting factors are also found in these homologues. This suggests that these *hsr203j* like homologues of *Brassica* play an important role in differential defence against *Alternaria* blight – a recalcitrant disease caused by *A. brassicae* (Mishra et al. 2010).

Cysteine-rich antimicrobial peptides isolated from plants have emerged as potential resource for protection of plants against phytopathogens. The significance of an antimicrobial peptide, *Pm AMPI*, isolated from Western white pine (*Pinus monticola*), in providing canola resistance against several phytopathogenic fungi, has been reported by Verma et al. (2012). The cDNA encoding *Pm AMPI* was successfully incorporated into the genome of *B. napus* and its in planta expression conferred greater protection against *A. brassicae*, *Leptosphaeria maculans* and *Sclerotinia sclerotiorum*. In vitro experiments with proteins extracted from transgenic canola expressing *Pm AMPI* demonstrated its inhibitory activity by reducing growth of fungal hyphae. In addition, the in vitro synthesized peptide also inhibited the growth of fungi. Generating transgenic crops expressing *Pm AMPI* may be an effective and versatile method to protect susceptible crops against multiple phytopathogens.

To develop resistance against *A. brassicae*, the barley antifungal genes, class II chitinase (*AAA56786*) and type I ribosome-inactivating protein (RIP; *AAA32951*), were co-expressed in Indian mustard via *Agrobacterium*-mediated transformation. The stable integration and expression of transgenes in T_0 plants were confirmed by Southern blot and Western analysis. The transgenic lines showing inheritance in Mendelian fashion (3:1) were further evaluated by in vitro studies and under greenhouse conditions for resistance against *A. brassicae*. The transgenic plants showed up to 44 % reduction in *A. brassicae* hyphal growth in in vitro antifungal assays. In greenhouse screening, the transgenic plants sprayed with *A. brassicae* spores showed resistance through delayed onset of the disease and restricted number, size and expansion of lesions as compared to wild-type plants. The expression of chitinase and RIP from a heterologous source in *B. juncea* provides subsequent protection against *Alternaria* leaf spot and can be helpful in increasing the production of Indian mustard (Chhikara et al. 2012).

Mutation in the gene *PAD3*, encoding a cytochrome *P450*, abolishes the biosynthesis of camalexin (Glazebrook and Ausubel 1994) and results in enhanced susceptibility to necrotrophic fungi (Thomma et al. 1999). The phytoanticipins glucosinolates (GS) are sulphonated thioglycosides comprising a common glycone moiety with a variable aglycone side chain and are considered the major secondary metabolites of Brassicaceae (Fahey et al. 2001). Upon tissue damage, GS come into contact with myrosinases, a specific class of β-thioglucosidases, which are stored separately in the cell. Hydrolysis of GS by myrosinases yields isothiocyanates (ITC), nitrile and epithionitriles. The most common breakdown products, ITC, exhibit toxicity towards several plant pathogens including bacteria, fungi, insects and nematodes (Fahey et al. 2001). It has been demonstrated that ITC produced by *Arabidopsis thaliana* significantly inhibit growth of some fungal pathogens in plants (Tierens et al. 2001), although recent results suggest that apart from their direct toxic effects, GS breakdown products may also act by modulation of plant defence signalling (Brader et al. 2006). During host infection, *A. brassicicola* is exposed to high levels of antimicrobial defence compounds such as indolic phytoalexins and glucosinolate breakdown products. To investigate the transcriptomic response of *A. brassicicola* when challenged with brassicaceous defence metabolites, suppression subtractive hybridization (SSH) was performed to generate two cDNA libraries from germinated conidia treated either with allyl isothiocyanate (AI-ITC) or with camalexin. Following exposure to AI-ITC, *A. brassicicola* displays a response similar to that experienced during oxidative stress. A substantial subset of differentially expressed genes is related to cell protection against oxidative damage. Treatment of *A. brassicicola* conidia with the phytoalexin camalexin appears to activate a compensatory mechanism to preserve cell membrane integrity, and among the camalexin-elicited genes, several are involved in sterol and sphingolipid biosynthesis. The transcriptomic analysis suggests that protection against the two tested compounds also involves mechanisms aimed at limiting their intracellular accumulation, such as melanin biosynthesis (in case of camalexin exposure only) and drug efflux. From the AI-ITC and the camalexin differentially expressed genes identified, 25 are selected to perform time course studies during interactions with brassicaceous hosts. In planta, upregulation of all the selected genes are observed during infection of *Raphanus sativus*, whereas only a subset are overexpressed during the incompatible interaction with *A. thaliana* ecotype Columbia (Sellam et al. 2007a; Fig. 9.2).

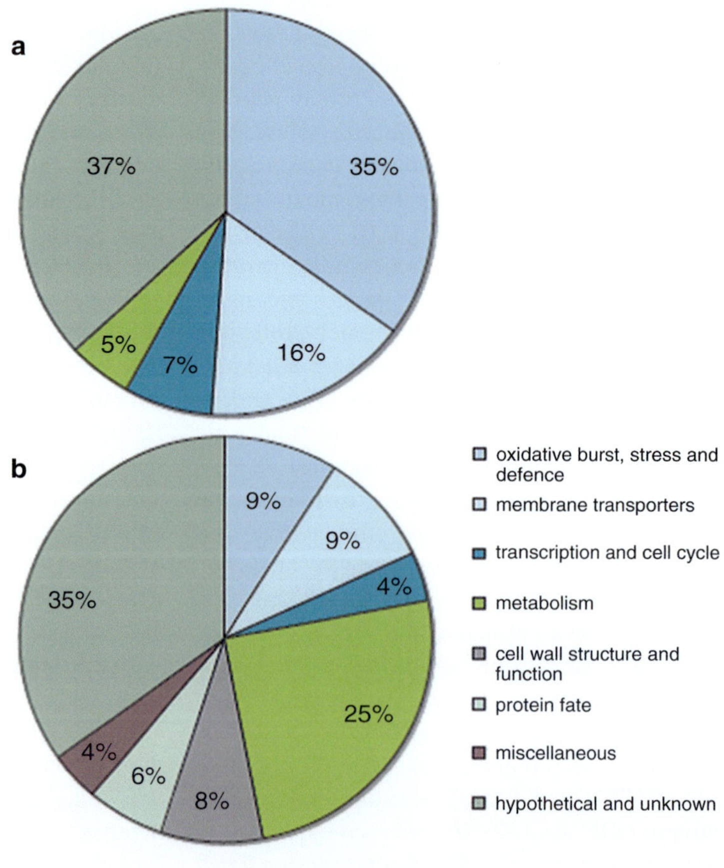

Fig. 9.2 Functional classification of upregulated genes in *Alternaria brassicicola*, AI-ITC (**a**) and camalexin; (**b**) treated conidia according to their putative biological function (Sellam et al. 2007a)

9.9 Elicitation of Phytoalexins

The accumulation of phytoalexins in brassicas after exposure to microorganisms and their role in disease resistance have been demonstrated by many workers. These are low molecular weight, antimicrobial compounds that are both synthesized by and accumulated in plants after exposure to microorganisms (Paxton 1981). So far, approximately a dozen different phytoalexins have been reported from crucifers (Bains and Tewari 1985; Browne et al. 1991; Devys et al. 1988; Takasugi et al. 1988). All phytoalexins identified from brassicas contain sulphur in their molecules. Conn et al. (1987, 1991) and Tewari et al. (1987, 1988) described elicitation of phytoalexins in *B. rapa* var. *oleifera* (canola), *B. rapa* ssp. *rapifera* (an accession of turnip), *B. napus*, *Camelina sativa*, *Eruca sativa* and *Capsella bursa-pastoris* similar to that of cyclobrassinin. However, the quantity produced in turnip is much greater. Jejelowo et al. (1991) studied the kinetics of phytoalexin elicitation in *C. sativa.* An accession of *E. sativa* was reported to show hypersensitive response to *A. brassicae* (Conn and Tewari 1986). Browne et al. (1991) described camalexin and methoxycamalexin phytoalexins from *C. sativa*, which are thiazolyl-substituted indole phytoalexins, and showed strong structural resemblance to the fungicide thiabendazole.

The antifungal effects of camalexin, brassinin and allyl (AI ITC)- and benzyl (Bz ITC)-isothiocyanates, at different development stages of *A. brassicae* and *A. brassicicola*, have been observed by Sellam et al. (2007b). Irrespective of the isolate, the phytoalexin camalexin exhibits the greater inhibitory effect with mean EC_{50} values ranging from 34 μm (germ-tube elongation) to 183 μm (mycelial growth). Germ-tube elongation is more sensitive compared to conidial germination and mycelial growth, with mean EC_{50} values of the former of 81 μm, 520 μm and 870 μm for brassinin, Bz ITC and AI ITC, respectively. However, the role of camalexin in disease resistance varies among different *Arabidopsis* population in nature and provides some clue to other possible determinants of resistance to *A. brassicicola* (Kagan and Hammerschmidt 2002).

The relationship between conidial concentration, germling growth and phytoalexin production by *C. sativa* leaves inoculated with *A. brassicae* has been studied by Jejelowo et al. (1991). The rapid rate of phytoalexin accumulation shortly after inoculation results into inhibition of fungal growth on the leaf surface. The phytoalexin slows germination and inhibits germ-tube growth of *A. brassicae* conidia in vitro.

Three phytoalexins, camalexin ($C_{11}H_8N_2S$), 6-methoxycamalexin ($C_{12}H_{10}N_2SO$) and N-methylcamalexin ($C_{12}H_{10}N_2S$), have been isolated from shepherd's purse (*Capsella bursa-pastoris*) challenged by *A. brassicae* (Fig. 9.3). Phytoalexin elicitation in shepherd's purse is associated with its resistance to *A. brassicae* (Jimenez et al. 1997).

Capsella bursa-pastoris shares one phytoalexin with *C. sativa* and the other one with rapeseed and is also highly resistant to *A. brassicae* (Conn et al. 1988). The highly resistant *Camelina sativa* elicits large quantities of two novel phytoalexins; it appears that in crucifers, both the quantity and quality of phytoalexins are important in regulating resistance to *A. brassicae.* These phytoalexins are highly antimicrobial, as shown by the *Cladosporium* sp. thin-layer chromatography bioassay (Tewari 1991a, b). Antifungal compounds have been found to accumulate when root slices of radish are inoculated

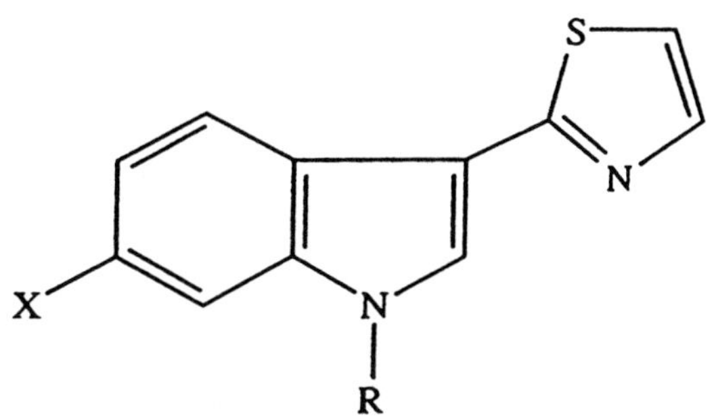

(1) R = X = H

(2) R = H,X = OCH_3

(3) R = CH_3,X = H

Fig.9.3 Structures of camalexin (1), 6-methoxycamalexin (2) and N-methylcamaxin (3)

with the spore suspension of *A. alternata*. Antifungal activity of the diffusate shows inhibitory action to spore germination, germ tube elongation and radial mycelial growth. The phytoalexins produced are fungistatic in nature (Kulshreshtha and Chauhan 1985). Chinese cabbage lines/hybrids on *A. brassicae* inoculation produce phytoalexins (Saharan et al. 2003).

9.10 Calcium Sequestration

The plant cell walls are rich in calcium, which is tightly bound to pectins. Calcium is known to be a factor in disease resistance. The insoluble calcium polypectates are resistant to hydrolysis by pectolytic enzymes produced by the pathogens (Vidhyasekaran 1988). Oxalic acid and possibly other organic acids produced by various pathogens sequester calcium, which may overcome this form of resistance (Punja and Jankins 1984; Rao and Tewari 1987). Examination of black spot lesions on rapeseed leaves by scanning electron microscopy in conjunction with energy-dispersive X-ray microanalysis has revealed sequestration of calcium by *A. brassicae*. Therefore, there are possibilities of enhancing resistance to *A. brassicae* in rapeseed by soil or foliar application of calcium compounds (Tewari 1991a, b). Foliar sprays of calcium compounds sequester the organic acids at the site of infection, and soil application has the potential of boosting calcium content of the plant.

9.11 Sources of Resistance

Digenomic *Brassica* species such as *B. napus* and *B. juncea* have better sources of resistance than the monogenomic (Tables 9.12 and 9.13) species like *B. rapa*. Some lines of *B. alba*, *B. carinata*, *B. spinescenes*, *B. maurorum*, *Eruca sativa*, *Camelina sativa*, *Capsella bursa-pastoris*, *Diplotaxis* sp. and *N. paniculata* have also shown resistance to *A. brassicae* in different areas of the world (Brun et al. 1987a, b; Conn et al. 1991; Grontoft 1986; Jejelowo et al. 1991; Saharan 1992, 1997; Tewari and Conn 1993; Hansen and Earle 1997; Chrungu et al. 1999; Sharma et al. 2002). Sources of resistance to *A. brassicae* and *A. brassicicola* in different host species from different areas of the world are given in Table 9.13. In India, *Brassica* genotypes, viz. CSR 43, CSR I42, CSR- I42-2, CSR 343, CSR 448, CSR 622,

Table 9.12 *Brassica* germplasm holdings at different organizations and research centres in the world

Sr. no.	Name of organization/research centre	Germplasm accessions (Total number)	References
1.	The National Bureau of Plant Genetic Resources (NBPGR), New Delhi, India	10301	Radhamani et al. (2013)
2.	ICAR-All India Coordinated Research Project on Rapeseed-Mustard, Centres	12778	AICRPRM (2011)
3.	ICAR-Directorate of Rapeseed-Mustard Research, Bharatpur (Raj) India	2452	Nanjundan et al. (2014)
4.	National Plant Germplasm System (USDA) Germplasm Resources Information Network (GRIN)	2958	http://www.ars-grin.gov/cgi-bin/npgs/html/taxon.pl?319659
5.	Biotechnology Department, Polytechnic University, Valencia, Caminode, Spain	150	Nuej et al. (1987)
6.	The Crucifer Genetics Cooperation (CrGC), Department of Plant Pathology, University of Wisconsin, Madison, USA	70 × 17 stocks in each *Brassica* crop	Williams (1988b)
7.	Wisconsin Fast Plants (WFP), Department of Plant Pathology, University of Wisconsin, Madison, USA	Rapid cycling of *Brassica* seed stocks	Williams (1988a)
8.	The Asian Vegetable Research and Development Centre (A VRDC), Shanhua, Taiwan	822	Anonymous (1984); Opena and Lo (1981)

Table 9.13 Sources of resistance to *Alternaria brassicae* and *Alternaria brassicicola* (Verma and Saharan 1994; to date)

Host species/genotype	Common name	References
A. brassicae		
Brassica rapa ssp. *oleifera*	Turnip rape	AICORPO (1989)
EC-242660-61, EC-242646, EC-253287, EC-253291	Toria	
B. rapa var. Yellow Sarson, PYS6, BINA 1,2		Kolte (1987); Rahman et al. (1987)
B. rapa rapifera, Edmonton ACC	Turnip	Conn et al. (1988)
B. carinata	Ethiopian mustard	AICORPO (1989);
PPSC1, HC-1, EC-25381, EC-253282,		Bhowmik and Munde (1987);
EC-253284, EC-253826, S67, PC3, PC5, CE9		Kidane and Bekele (1986)
B. napus spp. *oleifera*	Oilseed rape	AICORPO (1989); Conn et al. (1987);
Altex, Gulivar 1, Karat, Narde, Midas, Primer		Conn and Tewari (1989a); Kumar
Tower, WRGI5, Westar, Wei Bull- 541, 521		and Kumar, (1989); Rozej (1974)
Vuindsok, Regent 1, GS-1-1, GS7027, Vestal		Romero-Munoz and Jimenez
GSL-1501, 1506, 1508, 1513, HNS1, HNS3, Vestal		(1979); Saharan (1992, 1997); Stankova
EC-338986-2, EC-338996-1, EC-338997, EC-339000		(1972); Kolte et al. (2008); AICRPRM (2011)
GS-05-1		Kumar et al. (2014)
B. juncea	Indian mustard	AICORPO (1989); Kolte (1985a);
BECI07-109, 112, EC-I29126-1, PR8805, PHR 1		Saharan (1992, 1997); Kolte et al. (2000)
KRV Tall, RC-781, Divya, Kranti, PR- 8999, PR-9024; EC-399296, EC-399299, EC-399301, EC-399313, DRMR-2805		2008); Kumar (2008); Pratap et al. 2014
RH-8544, Pusa Swarnim, HC-9605		Kumar et al. (2014)
B. oleracea var. *alboglabra*	Chinese kale	Munde and Bhowmik (1985); Zaman and Biswas (1987)
B. oleracea var. *botrytis*	Cauliflower	Singh et al. (1987)
Lines 1 6 1 2, 1 6 1 4, Pusa Shubhra		
B. oleracea ssp. *gemmifera*	Brussels sprouts	Berry and Lennard (1988); Williams and Pink (1987)
Cambridge no.5		
B. alba	White mustard	Kolte (1987); Saharan (1992)
B. hirta		Brun et al. (1987a, b)
Camelina sativa	False flax	Tewari and Conn (1993);
Capsella bursa-pastoris	Shepherd's purse	Conn et al. (1991);
Neslia paniculata	Ball mustard	Jejelowo et al. (1991); Grontoft (1986)
Sinapis alba	White mustard	Kolte (1985b);
S10001, S1004-10		Rai et al. (1977)
A. brassicicola		
B. oleracea var. *Botrytis*	Cauliflower	Braverman (1971); Singh et al. (1987)
PI'S 231208-209, 217934, 231209, 267725,		
Pusa Shubhra		
B. oleracea var. *gemmifera*	Brussels sprouts	Braverman (1976, 1977)
B. rapa, Saori, Edononatsu		Doullah et al. (2006)

CSR 741, Gulivar, KRV-Tall, Midas, PHR 1, RC-781, TMV2, Tower and YRT3, were consistently found to have field resistance proven after testing for several years at different locations under uniform disease nursery trials (Kolte 1985b, 1987; Saharan 1984, 1992, 1997; Saharan and Chand 1988; Saharan et al. 2003, 2005; Verma and Saharan 1994). *Brassica coenospecies* are rich reservoir for genetic resistance to *A. brassicae* (Sharma et al. 2002; Table 9.14). *Alternaria brassicae* pathotype-specific resistance has been recorded in the genotype GS-05-1 of *B. napus* to pathotypes Abr 1 and Abr 5, whereas genotypes,RH-8544, Pusa Swarnim and HC-9605 of *B. juncea* showed moderately resistant reaction to three pathotypes (Kumar et al. 2014).

9.11.1 Sources of Resistance from Cruciferous Relatives

Sources with high level of resistance against *A. brassicicola* and *A. brassicae* have not been identified among the cultivated *Brassica* species although individual cabbage varieties differ considerably in their level of susceptibility to black spot (Otani et al. 2001). The highest level of *Alternaria* resistance among the oilseed *Brassica* crops is displayed by the Ethiopian mustard (*B. carinata*). Among the wild cruciferous plants closely related to the *Brassica* genus, the highest resistance against *A. brassicae* is found in white mustard (*Sinapis alba*) (Kolte 1985a; Brun et al. 1987a, b; Ripley et al. 1992; Sharma and Singh 1992; Verma and Saharan 1994; Hansen and Earle 1995, 1997). The highest overall *Alternaria* spp. resistance, however, have been identified in the crucifers more distant from the *Brassica*, such as *Camelina* (*Camelina sativa*; false flax), shepherd's purse (*Capsella bursa-pastoris*), taramira (*Eruca sativa*) and ball mustard (*Neslia paniculata*) (Conn and Tewari 1986; Conn et al. 1988; Tewari 1991b). Resistance against *A. brassicae* and *A. brassicicola* has been reported among other wild members of the Brassicaceae family (Table 9.14) (Sharma et al. 2002; Tewari and Conn 1993; Warwick 2011), viz. *Alliaria petiolata*, *Barbarea vulgaris*, *B. elongata*, *B. desnottessi*, *B. fruticulosa*, *B. maurorum*, *B. nigra*, *B. souliei*, *B. spinescens*, *C. sativa*, *C. bursa-pastoris*; *Coincya* spp., *Diplotaxis catholica*, *D. berthautii*, *D. creacea*, *D. erucoides*, *D. tenuifolia*, *Erucastrum gallicum*, *E. vesicaria* subsp. *sativa*, *Hemicrambe fruticulosa*, *H. matronalis*, *N. paniculata*, *R. sativus*, *S. alba* and *S. arvensis*. The completely immune plants remain symptom-free both under natural field infection and under controlled artificial inoculation (Sharma et al. 2002). Comparatively, broccoli and cauliflower varieties exhibit moderate *Alternaria* resistance, while the cabbage as susceptible.

Table 9.14 Classification of 38 *Brassica coenospecies* based on the reaction to *Alternaria brassicae* under in vitro and in vivo inoculation conditions (Sharma et al. 2002)

Resistant	Moderately resistant	Susceptible	Highly susceptible
Brassica desnottessi	*Brassica oleracea* (broccoli)	*B. oleracea* (cabbage)	*B. juncea*
Camelina sativa	*B. oleracea* (cauliflower)	*B. napus*	*B. rapa*
Coincya pseuderucastrum	*B. nigra*	*B. fruticulosa*	*B. carinata*
Diplotaxis berthautii	*B. barrelieri*	*D. gomez-campoi*	*B. cossoniana*
Diplotaxis catholica	*B. oxyrrhina*	*D. assurgens*	*B. spinescens*
Diplotaxis cretacea	*B. gravinae*	*D. tenuisiliqua*	*Coincya longirostra*
Diplotaxis erucoides	*Coincya rupestris*	*Erucastrum abyssinicum*	*D. viminea*
Eruca gallicum	*D. harra*	*E. canarianse*	*Enarthocarpus lyratus*
	D. muralis	*Moricandida arvensis*	*Raphanus sativus*
	D. siifolia	*Moricandia moricandioides*	
	Sinapis alba	*S. flexuosa*	
	S. pubescens		

9.12 Sources of Multiple Disease Resistance

Brassica alba and genotypes, viz. HC-1, PCC-2 (*B. carinata*), GSL-1501, GS-7027, Midas, Tower (*B. napus*), DIR-1507 and DIR-1522 (*B. juncea*), have been identified as sources of multiple disease resistance against white rust, *Alternaria* blight, downy mildew and powdery mildew diseases (Dang et al. 2000; Kumar and Saharan 2002). Earlier *B. napus* genotype GSL-1501 was reported resistant to white rust and powdery mildew (Gupta and Singh 1994). The variations in the incidence of powdery mildew scores among different *Brassica* genotypes differ considerably due to variations in date of sowing, stage of the crop at the time of observations and prevailing environmental conditions. Hence, early maturing genotypes or the genotypes planted early in the season would help in escaping the powdery mildew infection.

The genotypes EC-322090, EC-322092, EC-322093 and RC-781 of *B. juncea*; HC-1 and PCC-2 of *B. carinata*; and GS-7027, GSL-1, GSL-1501, Gulivar 1, HNS-4 and Tower sel.1 of *B. napus* are resistant to *A. candida* and *A. brassicae*, whereas genotypes EC-129126-1, EC-129126-2 and PR-8805 of *B. juncea*; Candle, Tobin and Torch of *B. rapa*; DHC-1, HDC-4, DHC-1960, HC-1 and PCC-2 of *B. carinata*; and GSL-1501, N 20-12-2 and Wester of *B. napus* are resistant to *A. candida* and *Erysiphe cruciferarum*. However, HC-1 and PCC-2 of *B. carinata* and GSL- 1501 (*B. napus*) are resistant to both *A. brassicae* and *E. cruciferarum*. GSL-1501 and Gulivar 1 of *B. napus* were found resistant to *A. candida* and *E. cruciferarum* by Gupta and Singh (1994). The genotypes Wester (*B. napus*) and Tobin and Candle (*B. rapa*) are reported completely free from *A. candida* and *E. cruciferarum* (Shivpuri et al. 1997). The variations in the observations may be attributed to the occurrence of new races of the pathogen and congenial weather factors (Table 9.15; Kumar and Saharan 2002).

9.13 Relationship between Major Foliar Diseases

The increase in level of resistance to different pathogens is positively correlated; the improvement in the accumulation of resistance to multiple diseases may be facilitated easily in agricultural or natural populations; negative association among the level of resistance will hinder plant defence against pathogens' attack and leads to difficulties in the development of multiple disease resistance. The non-availability of single resistance source line with different mode of inheritance further complicates the work of genetic amelioration for multiple disease resistance. The significant and positive association between disease scores of white rust, *Alternaria* blight and powdery mildew has been recorded except white rust stag head and powdery mildew infection in genotypes of different *Brassica* species. A high degree of correlation is observed between leaf phases of white rust and *Alternaria* blight (Table 9.16). Similarly, the

Table 9.15 Sources of multiple disease resistance in oilseed *Brassica* (Kumar and Saharan 2002)

Resistant to WR and AB	Resistant to WR and PM	Resistant to AB and PM	Resistant to WR, AB, PM and DM
B. juncea			
EC-322090, EC-322092, EC-322093, RC-781	EC-129126-1,EC-29126-2, PR-8805	None	DIR 1507, DIR 1522, *Brassica alba*
B. rapa			
None	Candle, Tobin, Torch	None	None
B. carinata			
HC-1, PCC-2	DHC-1, DHC-4, DHC-9601, HC-1, PCC-2	HC-1, PCC-2	HC-1, PCC-2
B. napus			
GS-7027, GSL-1, GSL-1501, Gulivar-1, HNS-4, Towel Sel-1	GSL-1501, N-20-12-2, Wester	GSL-1501	GSL-1501, GS-7027, Midas, Tower

WR white rust, *AB Alternaria* blight, *PM* powdery mildew, *DM* downy mildew

Table 9.16 Relationship among major foliar diseases (Kumar and Saharan 2002)

Disease	WR (L)	WR (S)	AB (L)	AB (S)	PM	ADL	ADI
B. juncea							
WR (L)	1.00						
WR (S)	0.32**	1.00					
AB (L)	0.39**	0.12	1.00				
AB (S)	0.14	0.09	0.75**	1.00			
PM	0.10	0.09	0.35**	0.30**	1.00		
ADL	0.81**	0.28*	0.69**	0.44**	0.50**	1.00	
ADI	0.13	0.11	0.61**	0.71**	0.82**	0.55**	1.00
B. rapa, B. carinata* and *B. juncea							
WR (L)	1.00						
WR (S)	0.19	1.00					
AB (L)	0.01	0.30	1.00				
AB (S)	0.21	0.30	0.96**	1.00			
PM	0.46**	0.05	0.07	0.15	1.00		
ADL	0.69**	0.25	0.53**	0.64**	0.79**	1.00	
ADI	0.43*	0.31	0.73**	0.79**	0.71**	0.94**	1.00
***Brassica* species**							
WR (L)	1.00						
WR (S)	0.32**	1.00					
AB (L)	0.39**	0.26**	1.00				
AB (S)	0.34**	0.27**	0.93**	1.00			
PM	0.33**	0.14	0.36**	0.33**	1.00		
ADL	0.82**	0.31**	0.68**	0.62**	0.68**	1.00	
ADI	0.42**	0.29**	0.75**	0.77**	0.85**	0.81**	1.00

WR (L) white rust at leaf, *WR (S)* white rust at stag head, *AB (L) Alternaria* blight at leaf, *AB (S) Alternaria* blight at siliquae, *PM* powdery mildew, *ADL* average disease on leaf, *ADI* average disease on inflorescence, *significant at $P=0.05$; **significant at $P=0.01$

positive genetic correlation is observed between *Hyaloperonospora parasitica* and *Leptosphaeria maculans* and *Hyaloperonospora parasitica* and *Albugo candida* in *B. rapa* (Mitchell-Olds et al. 1995). Sharma et al. (1991) reported significant positive correlation between *Alternaria* blight and black rot in cauliflower. The strong evidence for a positive genetic correlation in the level of resistance against *A. candida*, *A. brassicae* and *Erysiphe cruciferarum* may indicate the presence of plant resistance genes that provide defence against some fundamental characteristics common to different taxonomic orders of fungal pathogens as observed by Bruns et al. (1991). Genetic correlations may arise from several factors like (I) linkage disequilibrium, a nonrandom association among genotypes at several loci, which is rarely found in cross-pollinated species (Crow and Kimura 1970) (however, the linkage disequilibrium may arise from previous generations of selection or small population size), (II) pleiotropy and (III) tight linkage between resistance genes (Martin et al. 1993).

9.14 Development of Resistant Cultivars

Various methods including conventional as well as biotechnological approaches have been utilized either alone or in combination to incorporate the desired traits and to steer the genetic variability into new improved cultivars of crucifers. The conventional approach involves the identification and selection of disease-resistant genotypes and transfer of the desired traits into agronomically superior genotypes by hybridization. Hybridization combined with backcrossing, either singly or in combination with pedigree selection, is the most widely used and accepted conventional method to transfer disease resistance in crucifers.

Attempts to obtain resistance in agronomically superior genotypes of *B. juncea* through mutagenic agents such as gamma rays have met little success. Verma and Rai (1980) exposed the seeds of *B. juncea* cultivars Varuna, PR-5 and RS-3 to gamma rays and obtained a mutant with 60 KR doses which is promising in yield and resistant to *A. brassicae*. Rahman et al. (1987) selected mutant BINA 1 resistant to *A. brassicae* from *B. rapa* cultivar YS52 treated with EMS in Bangladesh. Das and Rahman (1989) obtained the highest frequency of variation for *Alternaria* blight in M2 by irradiating seeds of Yellow Sarson cultivar YS52 with 70 KR gamma rays. Sharma (1990) identified mutants in M2 population of cultivars Varuna and Kranti as resistant to *Alternaria* blight on the basis of small lesion size under natural epiphytotic conditions. MacDonald and Ingram (1985) selected and regenerated plants from calli resistant to partially purified culture filtrates of *A. brassicicola*. According to Katiyar and Chopra (1990), a somaclonal variant obtained from an exotic yellow-seeded accession of *B. juncea* retained resistance to disease similar to the parental material. The transfer of resistance from different sources in different *Brassica* crops is possible and is being done through the following conventional and modern techniques (Verma and Saharan 1994; Saharan 1992, 1997; Saharan et al. 2003, 2005):

1. Germplasm evaluation for sources of resistance at National and International levels
2. Selection for disease resistance through (a) pure line selection, (b) mass selection, (c) modified recurrent mass selection and (d) recurrent selection
3. Breeding for disease resistance by increasing the level of resistance through (a) multiple crosses, (b) recurrent selection, (c) diallel crossing and (d) selective mating system
4. Transfer of resistance by (a) intraspecific pedigree, backcross and modified recurrent mass selection methods and (b) interspecific genome substitution, chromosome substitution and gene introgression
5. Transfer of resistance through mutation breeding
6. Use of biotechnological and genetic engineering techniques such as (a) genome manipulation, (b) manipulation of cytoplasmic genomes, (c) use of transformation and foreign gene expression techniques and (d) embryo rescue techniques for wide hybridization

9.15 Strategies and Methods of Screening for Resistance

The prerequisites to obtain resistant (R) cultivars (cvs) are:

(a) Knowledge of the pathogenic variability
(b) Development of a screening method able to mimic the conditions required by the plants when exposed to natural sources of inoculum in the diverse field environment
(c) Availability of usable sources of resistance

Screening methods for disease resistance (R) should be developed within the framework of a general strategy for resistance. The changes in the frequency of virulence genes among the populations of pathogens inciting disease of the above-ground parts of plants are very frequent. Populations of such pathogens vary in time and space because of the airborne and seed-borne nature of inoculum which facilitates long-distance dispersal of their variants. As a result of these situations, breeding for resistance to foliar pathogens is, in general, more difficult than in the case of less mobile pathogens, e.g. soil-borne fungi which are, therefore, more stable. Screening as part of a strategy for developing R cultivars requires good planning and understanding of the processes involved in resistance. A screening programme should be initiated with a clear statement of the type of resistance, which is sought, i.e. complete resistance (no sporulation of the pathogen) or partial resistance (reduced sporulation of the pathogen) or both, and with at least some knowledge of pathogenicity and virulence patterns in pathogen. The durability of resistance can be practically tested only when the R cultivars are widely used in space and time. Multilocation cultivar testing or the challenge of cultivars to a large amount of pathogen population can help verify resistance and give timely warning of the possibility of resistance breakdown, but cannot actually be considered as a test for durability of resistance. Care must be taken in interpreting results of glasshouse or laboratory tests, as the expression of resistance in the field may be considerably different because of interaction between microorganism, pathogens and environmental conditions.

For foliar pathogens, the plant material must be adequately challenged with a single race or a pathotype at a realistic inoculum dose to allow disease development, but at the same time not obscure minor differences in host response required to identify partial resistance. The use of inoculum composed of a mixture of known races or unknown races/pathotypes in the naturally infested crop debris will not be adequate to achieve this objective. Illustrations of three cultivars each having a single gene for complete resistance to a given race would be identified only when inoculated singly with each isolate but not if the isolates were used in a mixture. The use of a single race provides the best condition for the selection of partial resistance in the presence of complete resistance, and the selected races should have the broadest possible virulence spectrum to suppress the expression of as many complete R genes as possible. Genotypes with resistance to one virulent race should then be systematically tested with a collection of other isolates. The identification of cultivars with complete resistance is a first step in the development of effective and durable resistance, and genetic analysis of resistance reaction is essential to reveal similarities and differences in the gene(s) that confer resistance in each genotype. Several genes known to be effective against a given race or genes effective against all prevalent races in a productive region are then recombined in a single cultivar. Large amount of genetic variability sources are available in cruciferous crops in the world, which can be exploited by researchers to develop a resistant cultivar against *Alternaria* blight in each *Brassica* crop (Table 9.17). Figure 9.4 illustrates a scheme to develop resistant cultivars of *Brassica* (Verma and Saharan 1994; Saharan et al. 2003, 2005). Methods of testing the resistance and screening techniques in the field and glasshouse are given in Chap. 12.

Table 9.17 Sources of cruciferous genetic variability in the world

Species	Exotic lines	Indigenous collections	Adv. breeding lines	Registered lines	Released varieties	Others	Total
Brassica juncea	189	636	656	19	73	263	1836
B. rapa var. Toria	12	67	59	–	14	74	226
B. rapa var. Yellow Sarson	07	78	50	01	12	24	172
B. rapa var. Brown Sarson	01	01		–	02	08	12
B. napus	61		30	09	06	14	120
Eruca sativa	27		19	–	03		49
B. carinata	02		10	03	05	01	21
Others				01	–	15	16
Total	299	782	824	33	115	399	2452

A. rapeseed–mustard germplasm at ICAR-DRMR, Bharatpur

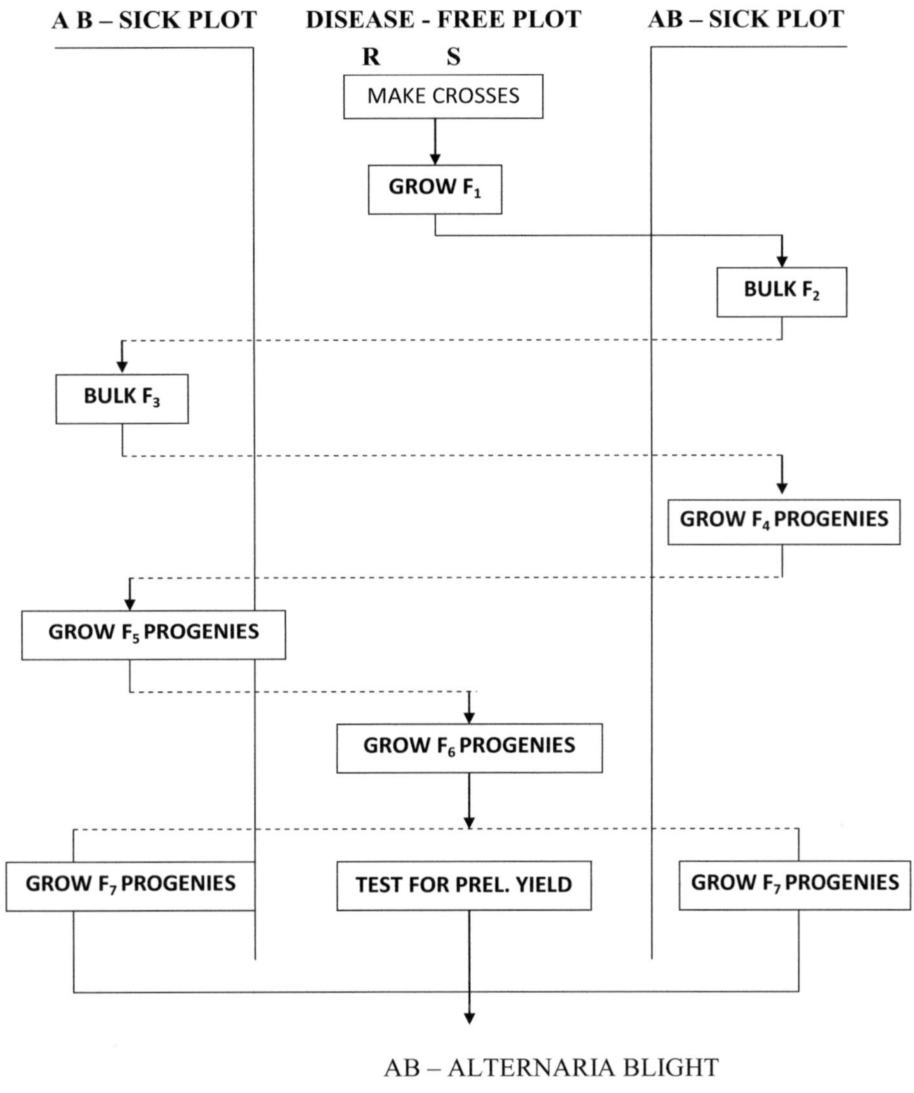

Fig. 9.4 Scheme to develop high-yielding disease-resistant cultivars (Saharan et al. 2003)

9.16 Bottlenecks in Resistance Breeding

Since resistance against *Alternaria* black spot is generally governed by polygenes, breeding for resistance should involve pyramiding minor genes to provide additive/polygene resistance. Rapid advances in techniques of tissue culture, protoplast fusion, embryo rescue and genetic engineering have made possible the transfer of disease resistance traits across the otherwise impassable self-incompatibility barriers. Disease-resistant transgenic plants which overexpress different antifungal compounds like pathogenesis-related (PR) proteins (chitinase, glucanase, osmotin, etc.) and ribosome-inhibiting proteins (RIPs) (thionins, defensins, and phytoalexins) to inhibit growth of the pathogen seem less efficacious (Zhou et al. 1999).

For introduction of *Camelina*-derived *A. brassicicola* resistance into commercial varieties, although somatic hybrids between *C. sativa* and *B. carinata* have been procured, the researchers failed to multiply the resulting hybrids

(Narasimhulu et al. 1994). Similarly, strategy of protoplast fusion between *C. sativa* and *B. oleracea* with subsequent hybrid regeneration has also proved unsuccessful (Hansen 1998). Brassica research groups, although attempted, did not succeed in introducing the *E. sativa*-derived black spot resistance into various species of cultivated crucifers (Fahleson et al. 1988; Sikdar et al. 1990; Sigareva and Earle 1997). First somatic hybrids to be obtained as a result of protoplast fusion were those of *B. napus* (rapeseed) and *S. alba* (Primard et al. 1988). None of the hybrids procured that way showed *A. brassicae* resistance comparable to that exhibited by *S. alba*. Chevre et al. (1991) used these species for interspecies crosses through somatic hybridization and biodirectional crosses using the embryo rescue technique; these researchers have succeeded in regeneration of *B. napus* plants carrying 38 chromosomes and displayed *A. brassicae* resistance at levels close to these of *S. alba*, *B. oleracea* var. *botrytis* or *B. carinata* (Ryschka et al. 1996). Seeds of developed intertribal somatic hybrids between *B. napus* and *C. sativa* (by means of protoplast electrofusion) exhibited phenotype intermediate to that of the parental species. Although the hybrid plants exhibited higher levels of linolenic and eicosanoic acids, their resistance against *Alternaria* still awaits confirmation (Jiang et al. 2009).

Researchers, in general, now realize that introduction of *Alternaria* resistance genes into commercial crucifer cultivars is dependent on cumulation of horizontal resistance genes (Sharma et al. 2002). It is, therefore, imperative to identify sources of horizontal resistance among the various *Brassica* species and subsequently combine them to increase durable *Alternaria* resistance. Strong cross-incompatibility, polygenic background of the resistance (additive and dominant gene interactions) and the differences in ploidy (differing number of chromosomes) between different *Brassicaceae* species render the *Alternaria* resistance transfer from the wild to the cultivated species difficult. Additionally, researchers often need to employ advanced in vitro hybridization techniques, including somatic hybridization, embryo and ovary rescue or protoplast fusion (Nowicki et al. 2012).

9.17 Biotechnological Approaches

Besides utilizing conventional methods of selection and sexual reproduction, biotechnological techniques involving tissue culture and genetic transformation-based methods have revolutionized the search and development of economically promising disease-resistant crops. These tools have enabled the researchers to combine the traits from distantly related *Brassica* species for the development of superior *Alternaria* blight-tolerant varieties of *Brassica* (Agnihotri et al. 2009; Aneja et al. 2012; Aneja and Agnihotri 2013; Saharan et al. 2003)

9.17.1 In Vitro Embryo Rescue

Embryo rescue, documented for the first time in the eighteenth century, is one of the most successful techniques for the production of interspecific and intergeneric hybrids from naturally incompatible crosses that do not result in the production of viable offspring. The technique involves excision of the ovary/ovule/embryo resulting from an interspecific cross and its culturing and maturing on suitable media for the production of a hybrid plant with desired traits. The technique has proven very efficient in transferring *Alternaria* blight tolerance in oilseed *Brassicas*. In vitro ovary culture was used to transfer the resistance against *Alternaria* blight from *B. tournefortii* to *B. juncea* cv. RH 30 (Yadav et al. 1991). Agnihotri et al. (1991) developed intergeneric hybrids of *B. campestris* and *B. spinescens* through sequential ovary, ovule and embryo culture. The hybrid plants were then multiplied using somatic embryogenesis. The resistance trait from *Sinapis alba* cv. Carine was transferred to *B. napus* cv. Brutor using in vitro fertilized ovary culture (Chevre et al. 1994). Intergeneric hybrids with *Alternaria* blight resistance were developed between *Erucastrum cardaminoides* and *B. oleracea* var. *alboglabra* using sequential ovary and ovule culture (Mohanty et al. 2009). Gupta et al. (2010) also used interspecific hybridization in combination with in vitro ovule culture to incorporate high tolerance to *Alternaria* blight and white rust from *B. carinata* cv. Kiran to low erucic acid TERl (OE) M21 lines; hybrids with high tolerance to *Alternaria* blight and white rust contained negligible amount of erucic acid (1.3 %) and high oleic acid (42.5 %).

9.17.2 Somatic Hybridization

Somatic hybridization refers to the enzymatic removal of the cell wall and isolation of protoplasts which are then fused together to produce hybrids. Due to the absence of the cell wall, there occurs a non-specific fusion between the two protoplasts without any barrier to intergeneric/interspecific crosses. It also bypasses both the pre- and postfertilization barriers. This technique has been utilized effectively in *Brassica*, but the studies are limited to *B. napus* and *B. oleracea.* Efforts have been made to transfer *Alternaria* resistant trait from *Moricandida arvensis* to *B. oleracea* (Toriyama et al. 1987) and from *S. alba* to *B. napus* (Primard et al. 1988). Sharma and Singh (1992) attempted the transfer of *Alternaria* resistance trait from *B. carinata* to *B. juncea.* Leaf mesophyll protoplasts from *M. arvensis* and *B. napus* were hybridized to produce disease-resistant hybrid plants (O'Neill et al. 1996). Jouradan and Salazar (1993) resynthesized *B. carinata* by protoplast fusion between *B. nigra* and *B. oleracea*, and the hybrids thus obtained were fertile and grew into robust plants. Hybridization between *S. alba* and *B. oleracea* and between *Camelina sativa* and *B. oleracea* has also shown to produce resistant hybrids (Hansen and Earle 1997; Hansen 1998). Sigareva and Earle (1999) developed somatic hybrids between *S. alba* and *B. oleracea* by protoplast fusion followed by embryo rescue and recovered four highly resistant hybrid progenies after repeated back crosses.

9.17.3 Somaclonal Variations

These are regarded as the impulsive, inheritable changes in gene expression which often appear under in vitro conditions especially in tissue culture experiments. Sharma and Singh (1995) used somaclonal variations for incorporation of disease resistance/tolerance against *Alternaria* blight. Resistance in *Alternaria* blight-susceptible *B. juncea* varieties was achieved by inducing variations through mutagenesis (exposing seeds to gamma radiations) (Verma and Rai 1980) and by treating the microspore-derived embryos with chemical mutagens (Agnihotri et al. 2009) including ethyl methane-sulphonate (EMS) and ethyl nitrosourea (ENU).

9.17.4 Genetic Transformation

Production of transgenic plants through genetic transformation forms another biotechnological approach to integrate the disease-resistant genes from resistant/tolerant genotypes to the economically important susceptible ones. The genetic transformations in *B. juncea* have been reported to achieve delayed onset of the *Alternaria* blight. Kanrar et al. (2002) isolated cDNA sequence for 'hevein-rubber tree lectin' from *Hevea brasiliensis* and transferred it into *B. juncea* cv. RLM-198 through binary vector *pBinAR*. The transgenic plants thus produced supported delayed onset of *Alternaria* blight with considerably low disease intensity. Mondal et al. (2003) identified the role of 'chitinase' in plant defence responses and transformed *B. juncea* (cv. RLMI98) with chitinase gene tagged with overexpressing promoter 35S *CaMV*. The transformed plants resulted in delayed disease onset and about 12–56 % reduction in fungal colony size. Osmotin protein, known for its role in signal transduction, when transferred to *B. juncea*, also showed tolerance to *A. brassicae* in transgenic plants (Taj et al. 2004). In another study, transgenic Indian mustard expressing '*class I glucanase*' gene under the control of *CaMV* 35S promoter was developed (Mondal et al. 2007). For undertaking the genetic transformation, cDNA encoding '*class I glucanase*' gene was isolated from tomato and cloned into *Bam*HI-*Sal I* restriction site of binary vector, *pBinAR*, flanking *CaMV* 35S constitutive promoter and an *ocs* terminator region. The resulting recombinant $pBinG_B$ clone was transferred into *Agrobacterium tumefaciens* strain GV 2260, which was further used to transform *B. juncea* (cv. RLM 198). Recent study by Verma et al. 2012 reported the introgression of cysteine-rich antimicrobial peptide, *PmAMP1*, from *Pinus monticola* into *B. napus*, thus conferring enhanced protection against *A. brassicae*, *Leptosphaeria maculans* and *Sclerotinia sclerotiorum.* Combined expression of type I ribosome-inactivating protein (RIP; *AAA32951*) and barley '*class II chitinase*' gene in *B. juncea* not only reduced the growth of fungal hyphae by 44 % but also affected the onset of disease (Chhikara et al. 2012). Efforts have also been made to study the signal transduction pathways in *Brassica* species. Taj et al. (2011) reported interaction of MAPK 3 and Lox genes during pathogenesis of *A. brassicae*, which is suggested to play a role in biosynthesis of jasmonic acid and jasmonic acid-induced expression of defence genes in *B. juncea.*

9.17.5 Molecular Markers

As the efficient management of plant diseases relies on the accurate identification of the plant pathogen, knowledge of the variability in the pathogen population, both at the pathological and genetic levels, is very important. Characterization of variability in the pathogens including *Alternaria* spp. has been done earlier using traditional methods of morphological markers (Goyal et al. 2011), biochemical tests (Vishwanath and Kolte 1997) and cultural and pathogenic assays (Goyal et al. 2011; Singh et al. 2011). These markers differentiate the pathogens on the basis of their growth behaviour, physiological characters and microscopic appearance. However, these criteria are greatly influenced by different factors including age and quality of the inoculum, age and type of host, culture media used and incubation conditions. Also, these techniques are more laborious and time-consuming and often give non-reproducible results. To overcome these limitations of traditional markers, detection of genetic variability through DNA fingerprinting technology by the use of molecular markers has come to the aid of plant breeders and pathologists. The main advantage of using these fingerprinting techniques is that they do not require any prior knowledge or availability of variable sequences in the database, as they deal with the genome-wide biodiversity.

Various molecular markers being used for assessing variability include internal transcribed spacer (ITS) regions, restriction fragment length polymorphism (RFLP), randomly amplified polymorphic DNA (RAPD), amplified fragment length polymorphism (AFLP), inter-simple sequence repeats (ISSRs), microsatellite, sequence-tagged sites (STS) and single-nucleotide polymorphisms (SNPs). Due to the presumed dependency on traditional characterization, only a few studies are reported on molecular characterization of *Alternaria* species pathogenic to oilseed brassicas. The ITS regions (also called the conserved regions) in the fungal genome are considered the most popular locus for DNA-based mycological studies at species level. Chou and Wu (2002) and Berbee et al. (2003) studied the ITS region of rDNA of *A. brassicae* and *A. brassicicola* to evaluate the phylogeny of the pathogen. Molecular techniques, like RAPD, that do not require any prior knowledge of the DNA sequences have been efficiently used to analyse the genetic variations in *Alternaria* species (Cooke et al. 1998; Sharma and Tewari 1995, 1998; Gherbawy 2005). However, due to the limitation of reproducibility in RAPD-generated data, the assaying of genetic variability in *Alternaria* species has shifted to more sensitive techniques like AFLP (Bock et al. 2002) and microsatellite markers. Twelve polymorphic microsatellite markers were isolated from *A. brassicicola*. The number of alleles detected in 12 loci ranges from 2 to 10 (mean 3.5). Cross-species amplification shows that the designated primers are specific to the pathogen (Avenot et al. 2005). Though these molecular tools have revealed huge amount of genetic variations in *Alternaria* at the subgeneric as well as subspecies level, this field still remains largely unexplored and demands an extensive study into the variability at different geographic levels among different hosts.

9.17.6 Induction of Systemic Resistance

Induced host resistance refers to the activation of resistance in otherwise susceptible plants, through biotic as well as abiotic agents (Stitcher et al. 1997; Kessman et al. 1994), without changing their basic genetic make-up. Foliar application of β-aminobutyric acid induces resistance against *A. brassicae* in *B. carinata* (Chavan et al. 2013). *GLIP1* protein in association with ethylene signalling is a critical component in plant resistance to *A. brassicicola* (Oh et al. 2005). The resistance/defence-related genes in the vulnerable plants can be activated by inoculating the plant either by an avirulent form of the pathogen or by limited inoculation with the pathogen (Deverall 1995). The infecting avirulent pathogen triggers natural defence responses in the plant through the release of the elicitors, which then result in the expression of novel anti-pathogenic proteins. In order to harp the benefits of induced host resistance and build up a stable, long-term resistance mechanism in the host plant against the pathogen, there is a need to identify the pathogen and understand its behaviour under diversified conditions. In case of *Alternaria* blight, considerable morphological and pathological variations have been observed in *A. brassicae* under different environmental conditions (Goyal et al. 2011; Kaur et al. 2007; Singh et al. 2007; Patni et al. 2005; Meena et al. 2005, 2012; Ansari et al. 1989; Awasthi and Kolte 1989; Saharan et al. 2015). Vishwanath et al. (1999) reported the induction of resistance in susceptible *B. juncea* cv. PR-15 against extremely virulent *A. brassicae* isolate A (AbA) and moderately virulent isolate C (AbC) from *B. carinata* cv. PPCS-1 by inoculating with an avirulent isolate D (AbD) isolated from the same *B. carinata* cultivar. About 60 % reduction in the disease severity was observed against AbA and AbC by the induction of resistance. However, due to the limited studies in molecular characterization of the pathogen, the clear outcome is still awaited.

Variations in *A. brassicae* isolates have been recorded in traits including colony morphology, colony diameter ranging from 32 to 68 mm, surface texture from velvety to woolly, colour from cream to dark olive green and sporulation from very less to intense. The host–pathogen interaction studies using four varieties of *B. juncea* and two of *B. rapa* have shown pathological variation among the isolates that differ in their aggressiveness from avirulent (few) to highly virulent (Aneja et al. 2012; Aneja and Agnihotri 2013). A thorough knowledge of diversity in the pathogen will not only help in characterizing and analysing it phylogenetically but also in inducing durable resistance against *Alternaria* blight, which can act as an environment-friendly substitute to the traditional use of fungicides.

9.17.7 Genetic Engineering

Development of *Alternaria*-resistant genetically modified plants (GM) is an excellent approach to manage *Alternaria* diseases of crucifers. Resistant genes from other plants or bacteria encoding enzymes like chitinase or glucanase can be introduced in crucifers. Chitinase and glucanase, respectively, breakdown chitin and glucan, which are essential components of fungal cell wall. Introduction of plant genes to enhance innate plant defence mechanisms like activating phytoalexins, inhibiting proteinase and detoxifying phytotoxins is another useful way of developing resistant varieties. Invoking hypersensitive reaction when a fungal pathogen is attacking can also be an innovative approach.

9.18 Factors Affecting Plant Disease Resistance

Different biology of the several types of plant pathogens also presents a substantial problem in developing GM-resistant plants. Firstly, the kind of organisms causing disease is taxonomically highly diverse; the major groups include cellular pathogens (bacteria, fungi and the algal oomycetes) and molecular pathogens (viruses). These are physiologically very different from each other, and therefore, no single gene product can be expected to have a direct toxic effect on all types of pathogens. Secondly, pathogens use two major life strategies, namely, biotrophic and necrotrophic. Biotrophic pathogens essentially act as a sink for the host's anabolic assimilates and therefore keep it alive. Meanwhile, necrotrophic pathogens consume the host tissues on invasion. Hemi-biotrophs combine both strategies in their life cycle. Consequently, plants have developed quite different ways for dealing with these two strategies.

9.19 Accessing and Exploiting Genetic Diversity

The advent of molecular genetics provided new opportunities for mapping and tracking genes of agronomic interest, leading to more efficient marker-assisted selection. Whole genome sequencing, starting with *Arabidopsis* and rice as models for dicotyledons and monocotyledons, respectively, and followed by a rapidly increasing number of crop plant genomes, has led to a quantum leap not only in understanding plant genetic diversity but also for accessing this enormous resource. For many crop plant species, for example, tomato, barley, maize, wheat and various *Brassica* species, either the entire genome or the generic parts of the genome are now emerging. As bioinformatic tools for analysing the exponential increase in genome data improve, the practical utility of such data will also be enhanced. This will extend the options for breeding disease resistance.

The presence of conserved motifs in plant resistance (R genes), such as the nucleotide-binding site–leucine-rich repeat (NBS-LRR) domains, has facilitated the identification of gene families and resistance gene analogues in other plants. The *Arabidopsis* genome has around 150 NBS-LRR encoding genes and rice c. 400 (McHale et al. 2006). Studies of genome structure have shown that many putative R genes are clustered and have undergone duplication and evolution due to diversifying selection. Functional analysis of these entire candidate genes is a demanding task, but improvement in plant transformation protocols and high-throughput gene attenuation methods, such as RNA interference (RNAi) and virus-induced gene silencing (VIGS), should accelerate the identification of novel genes of practical utility.

References

Agnihotri A, Lakshmikumaran MS, Jagannathan V, Shivanna KR (1991) Wide hybridization for improvement in cultivated Brassicas. Acta Horticult 289:213–214

Agnihotri A, Gupta K, Deepak P, Sarkar G, Mehra VS, Zargar SM (2009) Genetic enhancement in rapeseed-mustard for quality and disease resistance through *in vitro* techniques. In: Proceedings of 16th Australian Research Assembly on Brassicas on changing foods, changing climate, changing canola held at Ballarat (Australia), 14–16 September 2009

AICORPO (1989) All India Co-ordinated Research Project on Rapeseed-Mustard, Hisar

AICRPRM (2011) Annual progress report. All India Co-ordinated Research Project on Rapeseed-Mustard. Directorate of Rapeseed-Mustard Research, Bharatpur

Aneja JK, Agnihotri A (2013) Alternaria blight of oilseed brassicas: epidemiology and disease control strategies with special reference to use of biotechnological approaches for attaining host resistance. J Oilseed Brassica 4:1–10

Aneja JK, Goel P, Agnihotri A (2012) Diversity assessment in *Alternaria brassicae*. In: 1st National Brassica conference on production barriers and technological options in Oilseed Brassica, CCS Haryana Agriculture University, Hisar, 2–3 March 2012, p 149. (Abstr.)

Anonymous (1984) Annual report. Asian Vegetable Research and Development Centre (AVRDC), Shanhua, Taihnan, Taiwan

Ansari NA, Khan MW, Muheet A (1989) Effect of some factors on growth and sporulation of *Alternaria brassicae* causing Alternaria blight of rapeseed and mustard. Acta Bot Ind 17:49–53

Avenot HF, Dongo A, Bataillé N, Iacomi-Vasilescu B, Hamon B, Peltiers D, Simoneau P (2005) Isolation of 12 polymorphic micro-satellite loci in the phytopathogenic fungus *Alternaria brassicicola*. Mol Ecol Notes 5:948–950

Awasthi RP, Kolte SJ (1989) Variability in *Alternaria brassicae* affecting rapeseed and mustard. Indian Phytopathol 42:275. (Abstr.)

Bains PS, Tewari JP (1985) Purification and properties of the phytoalexin (s) produced by *Alternaria brassicae*. Phytopathology 75:1298. (Abstr.)

Berbee ML, Payne BP, Zhang GJ, Roberts RG, Turgeon BG (2003) Shared ITS DNA substitutions in isolates of opposite mating type reveal a recombining history for three presumed asexual species in the filamentous ascomycete genus *Alternaria*. Mycol Res 107:169–182

Berry LA, Lennard JH (1988) Foliar infection of *Brassica* plants by *Alternaria brassicicola* in relation to leaf surface characteristics. Cruciferae Newsl 13:104–106

Berto P, Comménil P, Belingheri L, Dehorter B (1999) Occurrence of a lipase in spores of *Alternaria brassicicola* with a crucial role in the infection of cauliflower leaves. FEMS Microbiol Lett 180:183–189

Bhowmik TP, Munde PN (1987) Identification of resistance in rapeseed and mustard against *Alternaria brassicae* (Berk.) Sacc and some resistant sources. Beitr Trop Landwirtsch Veterinarmed 25:49–53

Bock CH, Thrall PH, Brubaker CL, Burdon JJ (2002) Detection of genetic variation in *Alternaria brassicicola* using AFLP fingerprinting. Mycol Res 106:428–434

Brader G, Mikkelsen MD, Halkier BA, Palva ET (2006) Altering glucosinolate profiles modulates disease resistance in plants. Plant J 46:758–767

Braverman SW (1971) Reaction of broccoli and cauliflower introductions to *Alternaria brassicicola*. Plant Dis Rep 55:454–457

Braverman SW (1976) Resistance to *Alternaria brassicicola* in brussels sprouts introductions. Proc Am Phytopathol Soc 3:302

Braverman SW (1977) Reaction of Brussels sprouts introductions to artificial inoculation with *Alternaria brassicicola*. Plant Dis Rep 61:360–361

Browne LM, Conn KL, Ayer WA, Tewari JP (1991) The camalexins: new phytoalexins produced in the leaves of *Camelina sativa* (Cruciferae). Tetrahedron 47:3909–3914

Brun H, Plessis J, Renard M (1987a) Resistance of some crucifers to *Alternaria brassicae* (Berk.) Sacc. In: Proceedings of 7th international rapeseed congress, Poznan, pp 1222–1227

Brun H, Renard M, Plessis J (1987b) Resistance of some crucifers to *Alternaria brassicae* (Berk.) Sacc. In: Proceedings of 7th international rapeseed congress, Poznan, p 247

Brun H, Renard M, Tribodet M, Plessis I, Tanguy X (1989) Contribution to the genetic control of rapeseed diseases. Phytoma 404:36–37; 40–41

Bruns TD, White TJ, Taylor JW (1991) Fungal molecular systematic. Annu Rev Ecol Syst 22:525–564

Chavan V, Bhargava S, Kamble A (2013) Temporal modulation of oxidant and antioxidative responses in *Brassica carinata* during β-aminobutyric acid-induced resistance against *Alternaria brassicae*. Physiol Mol Plant Pathol 83:35–39

Chevre AM, Eber F, Brun H, Plessis J, Primard C, Renard M (1991) Cyto-genetic studies of *Brassica napus* – *Sinapsis alba* hybrids from ovary culture and protoplast fusion. Attempts to introduce *Alternaria* resistance into rapeseed. Proc Int Rapeseed Conf 8:346–351

Chevre AM, Eber F, Margale E, Kerlan MC (1994) Comparison of somatic and sexual *Brassica napus*-*Sinapis alba* hybrids and their progeny by cytogenetic studies and molecular characterization. Genome 37:367–371

Chhikara S, Chaudhury D, Dhankher OP, Jaiwal PK (2012) Combined expression of a barley class II chitinase and type I ribosome inactivating protein in transgenic *Brassica juncea* provides protection against *Alternaria brassicae*. Plant Cell Tiss Org Cult 108:83–89

Chou HH, Wu WS (2002) Phylogenetic analysis of internal transcribed spacer regions of the genus *Alternaria* and the significance of filament-beaked conidia. Mycol Res 106:164–169

Chrungu B, Verma N, Mohanty A, Pradhan A, Shivanna KR (1999) Production and characterization of interspecific hybrids between *Brassica maurorum* and crop brassicas. Theor Appl Genet 98:608–613

Comstock RE, Robinson HF (1952) Estimation of average dominance of genes. In: Heterosis. Iowa State College Press, Ames, pp 494–516

Conn KL (1986) Leaf epicuticular wax of canola: ultrastructure, chemistry and interaction with *Alternaria brassicae*. M. Sc. Thesis, University of Alberta, Edmonton, 159 p

Conn KL, Tewari JP (1985) The ultra-structure of leaf epicuticular wax in canola. Phytopathology 75:1377. (Abstr.)

Conn KL, Tewari JP (1986) Hypersensitive reaction induced by *Alternaria brassicae* in *Eruca sativa*, an oil-yielding crucifer. Can J Plant Pathol 8:348. (Abstr.)

Conn KL, Tewari JP (1989a) Interactions of *Alternaria brassicae* conidia with leaf epicuticular wax of canola. Mycol Res 93:240–242

Conn KL, Tewari JP (1989b) Ultra-structure of epicuticular wax in canola. Zeitschrift fur Naturforschung (Section C Biosci) 44:705–711

Conn KL, Tewari JP, Hadziyev D (1984) The role of epicuticular wax in canola in resistance to *Alternaria brassicae*. Phytopathology 74:851. (Abstr.)

Conn KL, Tewari JP, Dahiya JS (1987) Resistance to *Alternaria brassicae* and phytoalexin-elicitation in rapeseed and some other crucifers. Phytopathology 77:1703. (Abstr.)

Conn KL, Tewari JP, Dahiya JS (1988) Resistance to *Alternaria brassicae* and phytoalexin-elicitation in rapeseed and other crucifers. Plant Sci 55:21–25

Conn KL, Jejelowa OA, Tewari JP, Bains PS (1991) *Camelina sativa* phytoalexins (Camalexin and methoxy camalexin) provide resistance against *Alternaria brassicae*. Can J Plant Pathol 13:274 (Abstr.)

Cooke DEL, Forster JW, Jenkins PD, Jones DG, Lewis DM (1998) Analysis of intraspecific and interspecific variation in the genus *Alternaria* by the use of RAPD-PCR. Ann Appl Biol 132:197–209

Crow JF, Kimura M (1970) An introduction to population genetics theory. Burgess Publishing Company, Minneapolis, 591 p

Dang JK, Sangwan MS, Mehta N, Kaushik CD (2000) Multiple disease resistance against four fungal foliar

diseases of rapeseed-mustard. Indian Phytopathol 53:455–458
Dangl JL, Dietrich RA, Richberg MH (1996) Death don't have no mercy: cell death programs in plant-microbe interactions. Plant Cell 8:1793–1807
Das ML, Rahman A (1989) Alternaria resistance of *Brassica campestris* L. improved by induced mutations. Mutat Breed Newsl 33:9–10
Deverall BJ (1995) Plant protection using natural defense systems of plants. Adv Plant Pathol 11:211–228
Devys M, Barbier M, Loiselet I, Rouxel T, Sarniguet A, Kollmann A, Bousquet JF (1988) Brassilexin a novel sulfur-containing phytoalexin from *Brassica juncea* L. Cruciferae. Tetrahedron Lett 29(49):6447–6448
Dhingra A, Portis AR Jr, Daniell H (2004) Enhanced translation of a chloroplast-expressed *RbcS* gene restores small subunit levels and photosynthesis in nuclear *RbcS* antisense plants. Proc Natl Acad Sci U S A 101:6315–6320
Dixon RA, Harrison MJ, Lamb CJ (1994) Early events in the activation of plant defense responses. Annu Rev Phytopathol 32:479–501
Doullah MAU, Meah MB, Okazaki K (2006) Development of an effective screening method for partial resistance to *Alternaria brassicicola* (dark leaf spot) in *Brassica rapa*. Eur J Plant Pathol 116:33–43
Fahey JW, Zalcmann AT, Talalay P (2001) The chemical diversity and distribution of glucosinolates and isothiocyanates among plants. Phytochemistry 56:5–51
Fahleson J, Rahlen L, Glimelius K (1988) Analysis of plants regenerated from protoplast fusions between *Brassica napus* and *Eruca sativa*. Theor Appl Genet 76:507–512
Garg S, Dhawan K, Chawla HKL, Nainawati HS (1999) *Alternaria brassicae* induced changes in the activity of cell wall degrading enzymes in leaves of *Brassica juncea*. Biol Plant 42:475–478
Gates FE, Comstock RE, Robinson HF (1957) Generalized genetic variance and covariance formulae for self fertilized crop assuming linkages. Genetics 42:749–763
Gherbawy YAMH (2005) Genetic variation among isolates of *Alternaria* spp. from selected Egyptian crops. Arch Phytopathol Plant Protect 38:77–89
Glazebrook J, Ausubel FM (1994) Isolation of phytoalexin deficient mutants of *Arabidopsis thaliana* and characterization of their interactions with bacterial pathogens. Proc Natl Acad Sci U S A 91:8955–8959
Gomez-Campo C, Prakash S (1999) Origin and domestication. In: Gomez-Campo C (ed) Biology of *Brassica coenospecies*. Elsevier, Amsterdam, pp 33–58
Goodman RN, Novacky AJ (1994) The hypersensitive response reaction in plants to pathogens, a resistance phenomenon. American Phytopathological Society, Press, St Paul
Goyal P, Chahar M, Mathur AP, Kumar A, Chattopadhyay C (2011) Morphological and cultural variation in different oilseed Brassica isolates of *Alternaria brassicae* from different geographical regions of India. Indian J Agric Sci 81:1052–1058
Grontoft M (1986) Resistance to *Alternaria* spp. in oilseed crops. Sveriges Utsadesforen-ings Tidskrift 96:263–269
Gupta V (1995) Changes in activities of oxidative enzymes and polypeptide patterns of *Brassica juncea* in relation to Alternaria blight. M.Sc. Thesis, HAU, Hisar
Gupta RBL, Singh M (1994) Sources of resistance to white rust and powdery mildew of mustard. Int J Trop Plant Dis 12:225–227
Gupta SK, Kumar P, Yadava TP, Saharan GS (1984) Changes in phenolic compounds, sugars and total nitrogen in relation to Alternaria leaf blight in Indian mustard. Haryana Agric Univ J Res 14:535–537
Gupta SK, Gupta PP, Yadava TP (1987a) Leaf surface constituents of *Brassica* species in relation to Alternaria leaf blight (*Alternaria brassicae* (Berk.) Sacc. and *A. brassicicola* Schw.) Wilts. In: Proceedings of 7th international rapeseed congress, Poznan, pp 1241–1247
Gupta SK, Gupta PP, Yadava TP (1987b) Leaf surface constituents of *Brassica* species in relation to Alternaria leaf blight (*Alternaria brassicae* (Berk.) Sacc.) and *A. brassicicola* (Schw.) Wilts. In: Proceedings of 7th international rapeseed congress, Poznan, p 252
Gupta SK, Gupta PP, Yadava TP (1987c) Leaf surface constituents of mustard genotypes in relation to Alternaria leaf blight disease. Indian J Mycol Plant Pathol 17:332–334
Gupta SK, Gupta PP, Yadav TP, Kaushik CD (1990) Metabolic changes in mustard due to Alternaria leaf blight. Indian Phytopathol 43:64–69
Gupta K, Saharan GS, Singh D (2002) Effective selection criteria to assess disease tolerance in Indian mustard (*Brassica juncea* L. Czern. & Coss.). J Mycol Plant Pathol 32:72–80
Gupta K, Prem D, Agnihotri A (2010) Pyramiding white rust resistance and Alternaria blight tolerance in low erucic acid *Brassica juncea* using *Brassica carinata*. J Oilseed Brassica 1:55–65
Hansen LH (1998) Intertribal somatic hybridization between rapid cycling *Brassica oleracea* (L.) and *Camelina sativa* (L) Cranz. Euphytica 104:173–179
Hansen LH, Earle ED (1995) Transfer of resistance to *Xanthomonas campestris* pv. *campestris* (L.) by protoplast fusion. Theor Appl Genet 91:1293–1300
Hansen LH, Earle ED (1997) Somatic hybrids between *Brassica oleracea* (L.) and *Sinapis alba* (L.) with resistance to *Alternaria brassicae* (Berk.) Sacc. Theor Appl Genet 94:1078–1085
Holloway FJ, Brown GA, Baker EA, Macey MJK (1977) Chemical composition and ultrastructure of the epicuticular wax in three lines of *Brassica napus* (L.). Chem Phys Lipids 19:114–127. http://www.ars-grin.gov/cgi-bin/npgs/html/taxon.pl?319659
Jejelowo OA, Conn KL, Tewari JP (1991) Relationship between conidial concentration, germling growth, and phytoalexin production by *Camelina sativa* leaves inoculated with *Alternaria brassicae*. Mycol Res 95:928–934
Jiang JJ, Zhao XX, Tian W, Li TB, Wang YP (2009) Intertribal somatic hybrids between *Brassica napus* and *Camelina sativa* with high linolenic acid contents. Plant Cell Tiss Org Cult 99:91–95

Jimenez LD, Ayer WA, Tewari JP (1997) Phytoalexins produced in the leaves of *Capsella bursa-pastoris* (shepherd's purse). Phytoprotection 78:99–103

Jouradan PS, Salazar E (1993) *Brassica carinata* resynthesized by protoplast fusion. Theor Appl Genet 86:567–572

Kagan IA, Hammerschmidt R (2002) *Arabidopsis* ecotype variability in camalexin production and reaction to infection by *Alternaria brassicicola*. J Chem Ecol 28:2121–2140

Kamble A, Bhargava S (2007) β-Amino butyric acid-induced resistance in *Brassica juncea* against the necrotrophic pathogen *Alternaria brassicae*. J Phytopathol 155:152–158

Kanrar S, Venkateswari JC, Kirti PB, Chopra VL (2002) Transgenic expression of hevein, the rubber tree lectin, in Indian mustard confers protection against *Alternaria brassicae*. Plant Sci 162:441–448

Katiyar RK, Chopra VL (1990) Somaclonally induced earliness in *a Brassica juncea* germplasm accession with field resistance to important diseases. Plant Breed 104:262–264

Kaur S, Singh G, Banga SS (2007) Documenting variation in *Alternaria brassicae* isolates based on conidial morphology, fungicidal sensitivity and molecular profile. In: GCIRC proceedings of 12th international rapeseed congress, vol 4, Wuhan, 26–30 March 2007, pp 87–89

Kessman H, Staub T, Hofmann C, Maetzke T, Herzog G, Ward E, Uknes S, Ryals J (1994) Induction of systemic acquired disease resistance in plants by chemicals. Annu Rev Phytopathol 32:439–460

Kidane A, Bekele E (1986) Major diseases of brassica and preliminary observations on yield loss due to Alternaria leaf spot. In: Proceedings of IDRC (Canada) 3rd Oil Crops Network Workshop held in Addis Ababa, 6–10 October 1986, pp 53–57

Kiran, Dhingra HR, Mehta N, Sangwan MS (2003) Effect of culture filtrate of *Alternaria brassicae* on biochemical constituents of calli of Brassicas. J Mycol Plant Pathol 43:659–669

Kolte SJ (1985a) Disease management strategies for rapeseed-mustard crops in India. Agric Rev 6:81–88

Kolte SJ (1985b) Diseases of annual edible oilseed crops, vol 2. CRC Press, Boca Raton, 135p

Kolte SJ (1987) Important diseases of rapeseed and mustard in India. Present research progress and future research needs. In: Proceedings of IDRC (Canada) 3rd Oil Crops Network Workshop, Addis Ababa, 6–10 October 1986, pp 91–106

Kolte SJ, Awasthi RP, Vishwanath (2000) *Divya* mustard: a useful source to create Alternaria black spot tolerant dwarf varieties of oilseed brassicas. Plant Varieties Seeds 13:107–111

Kolte SJ, Nashaat NI, Kumar A, Awasthi RP, Chauhan JS (2008) Towards improving the genetic base of rapeseed-mustard through an Indo-UK research collaboration. Indian J Plant Genet Resour 21:132–137

Krishnia SK, Saharan GS, Singh D (2000) Genetics of Alternaria blight resistance in inter and intra-specific crosses of *Brassica juncea* and *Brassica carinata*. Ann Biol 16:212–216

Kulshreshtha BM, Chauhan RKS (1985) Production of phytoalexins in the roots of radish (*Raphanus sativus*) after inoculation with *Alternaria alternata*. Indian Phytopathol 38:286–291

Kumar B (2008) Assessment of slow blighting (*Alternaria brassicae*) resistance and yield in Indian mustard [*Brassica juncea* (L.) Czern. & Coss.]. Indian Phytopathol 61:171–183

Kumar B, Kolte SJ (2001) Progression of Alternaria blight of mustard in relation to components of resistance. Indian Phytopathol 54:329–331

Kumar B, Kolte SJ (2006) Development of Alternaria blight in genotypes of Indian mustard (*Brassica juncea* (L.) Czern & Coss.) under field conditions. Indian Phytopathol 59:314–317

Kumar PR, Kumar P (1989) Genetic improvement of rapeseed-mustard, achievements, critical gaps and future priorities. J Oilseeds Res 6:211–219

Kumar S, Saharan GS (2002) Sources of multiple disease resistance in *Brassica* spp. J Mycol Plant Pathol 32:184–188

Kumar B, Singh R (2006) Development of Alternaria blight in Indian mustard genotypes under field conditions. Ann Plant Prot Sci 14:415–418

Kumar A, Katoch A, Sharma PN, Kumar V, Kumar A (2014) Pathogenic and genetic variability in *Alternaria brassicae* infecting rapeseed-mustard and evaluation of resistance sources. Indian Phytopathol 67:257–262

Lamb C, Dixon RA (1997) The oxidative burst in plant disease resistance. Annu Rev Plant Physiol Plant Mol Biol 48:251–275

Levine A, Pennell RI, Alvarez ME, Palmer R, Lamb C (1996) Calcium-mediated apoptosis in a plant hypersensitive disease resistance response. Curr Biol 6:427–437

MacDonald MV, Ingram DS (1985) *In vitro* selection for resistance to *Alternaria brassicicola* in *Brassica napus* spp. oleifera (winter oilseed rape) using partially purified culture filtrates. Cruciferae Newsl 10:97–100

Marmath KK, Giri P, Taj G, Pandey D, Kumar A (2013) Effect of zeatin on the infection process and expression of MAPK-4 during pathogenesis of *Alternaria brassicae* in non-host and host *Brassica* plants. Afr J Biotechnol 12:2164–2174

Martin GB, de Vicente MC, Tanksley SD (1993) Higher solution linkage analysis and physical characterization of the Pto bacterial resistance locus in tomato. Mol Plant-Microbe Interact 6:26–34

Matzinger DF, Cockerham CC (1963) Simultaneous selfing and partial diallel set crossing I: estimation of genetic and environmental parameters. Crop Sci 3:309–314

McHale L, Tan X, Koehl P, Michelmore RW (2006) Plant NBS-LRR proteins: adaptable guards. Genome Biol 7:212

Meena PD, Chattopadhyay C, Kumar VR, Meena RL, Rana US (2005) Spore behaviour in atmosphere and trends in variability of *Alternaria brassicae* population in India. In: Proceedings of global conference II on plant health-global wealth, Journal of Mycology &

Plant Pathology, MPUAT, Udaipur, 25–29 November 2005, (Abstr.)

Meena PD, Rani A, Meena R, Sharma P, Gupta R, Chowdappa P (2012) Aggressiveness, diversity and distribution of *Alternaria brassicae* isolates infecting oilseed Brassica in India. Afr J Microbiol Res 6:5249–5258

Mishra A, Pandey D, Goel A, Kumar A (2010) Molecular cloning and *in silico* analysis of functional homologues of hypersensitive response gene(s) induced during pathogenesis of *Alternaria* blight in two genotypes of *Brassica*. J Proteomic Bioinformat 3:244–248

Mitchell-Olds T, James RV, Palmer MJ, Williams PH (1995) Genetics of *Brassica rapa* (syn. *campestris*). 2. Multiple disease resistance to three fungal pathogens: *Peronospora parasitica, Albugo candida* and *Leptosphaeria maculans*. Heredity 75:362–369

Mohanty A, Chrungu B, Verma N, Shivanna KR (2009) Broadening the genetic base of crop brassicas by production of new intergeneric hybrid. Czech J Genet Plant Breed 45:117–122

Mondal KK, Chatterjee SC, Viswakarma N, Bhattacharya RC, Grover A (2003) Chitinase-mediated inhibitory activity of *Brassica* transgenic on growth of *Alternaria brassicae*. Curr Microbiol 47:171–173

Mondal KK, Bhattacharya RC, Koundal KR, Chatterjee SC (2007) Transgenic Indian mustard (*Brassica juncea*) expressing tomato glucanase leads to arrested growth of *Alternaria brassicae*. Plant Cell Rep 26:247–252

Munde PN, Bhowmik TP (1985) A source of morphological resistance to leaf blight diseases of rapeseed and mustard caused by *Alternaria brassicae*. Curr Sci 54:514–515

Nanjundan J, Singh K, Singh KH, Singh D (2014) Catalogue on rapeseed-mustard germplasm. Directorate of Rapeseed-Mustard Research, Bharatpur, 180 p

Narasimhulu SB, Kirti PB, Bhat SR, Prakash S, Chopra VL (1994) Inter-generic protoplast fusion between *Brassica carinata* and *Camelina sativa*. Plant Cell Rep 13:657–660

Naumann TA, Wicklow DT (2013) Chitinase modifying proteins from phylogenetically distinct lineages of *Brassica* pathogens. Physiol Mol Plant Pathol 82:1–9

Neerja S, Sohal BS (2012) Induced resistance in *Brassica juncea* L. against *Alternaria blight* by foliar application of arachidonic acid. Plant Dis Res 27:153–157

Nowicki M, Nowakowska M, Niezgoda A, Kozik EU (2012) Alternaria black spot of crucifers: symptoms, importance of disease, and perspectives of resistance breeding. Veg Crops Res Bull 76:5–19

Nuej F, Diez MJ, Ferrando C, Cuartero J, Costa J (1987) Germplasm resources of *Brassica* and *Raphanus* from Spain. Cruciferae Newsl 12:4

O'Neill CM, Murata T, Morgan CL, Mathia RJ (1996) Expression of the C3-C4 intermediate character in somatic hybrids between *Brassica napus* and the C3-C4 species *Moricandida arvensis*. Theor Appl Genet 93:1234–1241

Oh IS, Park AR, Bae MS, Kwon SJ, Kim YS, Lee JE, Kang NY, Lee S, Cheong H, Park OK (2005) Secretome analysis reveals an *Arabidopsis* lipase involved in defense against *Alternaria brassicicola*. Plant Cell 10:2832–2847

Opena RT, Lo SE (1981) Breeding for heat tolerance in heading Chinese cabbage. In: eds. Talekar NS, Griggs TD (ed) Chinese cabbage proceedings of the first International symposium Asian Vegetable Research and Development Centre, Shanhua, Taihnan, Taiwan, pp 431–441

Otani H, Kohnobe A, Narita M, Shiomi H, Kodama M, Kohmoto K (2001) A new type of host-selective toxin, a protein from *Alternaria brassicicola*. In: Keen NT, Mayama S, Leach JE, Tsuyumu S (eds) Delivery and perception of pathogen signals in plants. American Phytopathological Society, Press, St. Paul, pp 68–76

Parihar PS, Prakash O, Punetha H (2012) Changes in metabolites of *Brassica juncea* (Indian mustard) during progressive infection of *Alternaria brassicae*. Nat Sci 10:39–42

Patni CS, Bhatiya BS, Kolte SJ, Awasthi RP (2005) Pathogenic variability of *Alternaria brassicae* causing Alternaria blight of mustard. Indian Phytopathol 58:545

Paxton JD (1981) Phytoalexins- a working redefinition. Phytopathol Zeitschrift 101:106–109

Pedras MSC, Jha M, Ahiahonu PWK (2003) The synthesis and biosynthesis of phytoalexins produced by cruciferous plants. Curr Org Chem 7:1635–1647

Pratap P, Meena PD, Singh BK, Meena HS, Meena SS, Sharma P, Majumdar R, Singh D (2014) Development and evaluation of Alternaria blight tolerant lines in Indian mustard (*Brassica juncea*). J Oilseed Brassica 5:141–148

Primard C, Vedel F, Mathieu C, Pelletier G, Chevre AM (1988) Inter-specific somatic hybridization between *Brassica napus* and *Brassica hirta* (*Sinapis alba L.*). Theor Appl Genet 75:546–552

Punja ZK, Jankins SF (1984) Light and scanning electron microscopic observations of calcium oxalate crystals produced during growth of *Sclerotium rolfsii* in culture and in infected tissue. Can J Bot 62:2028–2032

Radhamani J, Singh R, Srinivasan K, Tyagi RK (2013) Conservation of trait-specific germplasm of Brassicas in National gene bank. National Bureau of Plant Genetic Resources, New Delhi, 114p

Rahman A, Das ML, Howlider MAR, Mansur MA (1987) Promising mutants in *Brassica campestris*. Mutat Breed Newsl 29:14–15

Rai B, Kolte SJ, Tewari AN (1977) Evaluation of oleiferous Brassica germplasm for resistance to Alternaria leaf blight. Indian J Mycol Plant Pathol 6:76–77

Rao DV, Tewari JP (1987) Production of oxalic acid by *Mycena citricolor,* causal agent of the American leaf spot of coffee. Phytopathology 77:780–785

Ripley V, Thorpe M, Iler S, Mizier K, Beversdorf WD (1992) Isozyme analysis as a tool for introgression of *Sinapis alba* germplasm into *Brassica napus*. Theor Appl Genet 84:403–410

Romero MF, Jimenez DRM (1979) Black spot: a disease of turnip rape recently recorded in Spain. Annales del Institute Nacional de investigaciones Agrarias, Proccion Vegetal 9:11–31

Rozej A (1974) Susceptibility of different varieties of winter rape (*Brassica napus* L. var. Oleifera) to infection with *Alternaria brassicae* (Berk.) Sacc. Roczniki Nauk Rolniczych, A 100:67–71

Ryals JA, Neuenschwander UH, Willits MG, Molina A, Steiner HY, Hunt MD (1996) Systemic acquired resistance. Plant Cell 8:1809–1819

Ryschka U, Schumann G, Klocke E, Scholze P, Neumann M (1996) Somatic hybridization in *Brassicaceae*. Acta Horticult 407:201–208

Saharan GS (1984) A review of research on rapeseed-mustard pathology in India. Paper presented in the Annual Rabi Oilseed Workshop held at Jaipur, 6–10 August 1984

Saharan GS (1991) Assessment of losses, epidemiology and management of black spot disease of rapeseed-mustard. In: Proceedings of GCIRC 8th international rapeseed congress, vol 2, 9–11 July Saskatoon, (Abstr.) p 84, 465–470

Saharan GS (1992) Disease resistance. In: Labana KS, Banga SS, Banga SK (eds) Breeding oilseed brassicas. Narosa Publishing House, New Delhi, pp 181–200

Saharan GS (1997) Disease resistance. In: Kalia HR, Gupta SK (eds) Recent advances in oilseed brassicas. Kalyani Publishers, Ludhiana, pp 233–259

Saharan GS, Chand JN (1988) Chapter 3: Diseases of rapeseed and mustard. In: Diseases of oilseed crops (in Hindi). Directorate of Publication, Haryana Agric. Univ. Press, Hisar, pp 84–91

Saharan GS, Kadian AK (1983) Analysis of components of horizontal resistance in rapeseed and mustard cultivars against *Alternaria brassicae*. Indian Phytopathol 36:503–507

Saharan GS, Krishania SK (2001) Multiple disease resistance in rapeseed and mustard. In: Nagarajan SS, Singh DP (eds) Role of resistance in intensive agriculture. Kalyani Publications, New Delhi, pp 98–108

Saharan GS, Mehta N, Sangwan MS (2003) Nature and mechanism of disease resistance to Alternaria blight in rapeseed-mustard system. Annu Rev Plant Pathol 2:85–125

Saharan GS, Mehta N, Sangwan MS (2005) Fungal diseases of rapeseed-mustard. In: Diseases of oilseed crops. Indus Publishing Company, New Delhi, pp 15–86

Saharan GS, Mehta N, Verma PR, Meena PD (2015) Analysis of variability determinants in *Alternaria-Brassica-* interactions. J Oilseed Brassica 6:126–141

Sellam A, Dongo A, Guillemette T, Hudhomme P, Simoneau P (2007a) Transcriptional responses to exposure to the brassicaceous defence metabolites camalexin and allyl-isothiocyanate in the necrotrophic fungus *Alternaria brassicicola*. Mol Plant Pathol 8:195–208

Sellam A, Iacomi-Vasilescu B, Hudhomme P, Simoneau P (2007b) *In vitro* antifungal activity of brassinin, camalexin and two isothiocyanates against the crucifer pathogens *Alternaria brassicicola* and *Alternaria brassicae*. Plant Pathol 56:296–301

Sharma TR (1990) Induction and transfer of resistance to Alternaria blight of mustard caused by *Alternaria brassicae* and *A. brassicicola*. Ph.D. Thesis, Himachal Pradesh Krishi Vishva Vidyalaya, Palampur

Sharma TR, Singh BM (1992) Transfer of resistance to *Alternaria brassicae* in *Brassica juncea* through interspecific hybridization among brassicas. J Genet Breed 46:373–378

Sharma TR, Singh BM (1995) Generations and evaluations of somaclones of *Brassica juncea* for resistance to *Alboga candida* and *Alternaria brassicae*. Proc Indian Natl Sci Acad B Biol Sci 61:155–161

Sharma TR, Tewari JP (1995) Detection of genetic variation in *Alternaria brassicae* by RAPD fingerprints. J Plant Biochem Biotechnol 4:105–107

Sharma TR, Tewari JP (1998) RAPD analysis of three *Alternaria* species pathogenic to crucifers. Mycol Res 102:807–814

Sharma BR, Dhiman JS, Thakur JC, Singh A, Bajaj K (1991) Multiple disease resistance in cauliflower. Adv Horticult Sc 5:30–34

Sharma G, Kumar VD, Haque A, Bhat SR, Prakash S, Chopra VA (2002) *Brassica coenospecies*: a rich reservoir for genetic resistance to leaf spot caused by *Alternaria brassicae*. Euphytica 125:411–417

Sharma N, Rahman MH, Strelkov S, Thiagarajah M, Bansal VK, Kav NNV (2007) Proteome-level changes in two *Brassica napus* lines exhibiting differential responses to the fungal pathogen *Alternaria brassicae*. Plant Sci 172:95–110

Shivpuri A, Chipa HP, Gupta RBL, Sharma KN (1997) Field evaluation of mustard genotypes against white rust, powdery mildew and stem rot. Ann Arid Zone Res 36:387–390

Sigareva MA, Earle ED (1997) Direct transfer of a cold-tolerant Ogura male sterile cytoplasm into cabbage (*Brassica oleracea* ssp. *Capitata*) via protoplast fusion. Theor Appl Genet 94:213–220

Sigareva MA, Earle ED (1999) Camalexin induction in intertribal somatic hybrids between *Camelina sativa* and rapid-cycling *Brassica oleracea*. Theor Appl Genet 98:164–170

Sikdar SR, Chatterjee G, Das S, Sen SK (1990) "Erussica" the intergeneric fertile somatic hybrid developed through protoplast fusion between *Eruca sativa* Lam. and *Brassica juncea* (L.) Czern & Coss. Theor Appl Genet 79:561–567

Singh H, Singh D (1989) Studies on genetic control of resistance to *Alternaria brassicae* in Indian mustard. In: XII Eucarpia Congress, pp 11–13

Singh D, Lodhi GP, Singh H (1986) Studies on genetic variability for various attributes in different mating system F2 and F3 populations of rapeseed. Oil Crops Newsl 58:60

Singh R, Trivedi BM, Gill HS, Sen B (1987) Breeding for resistance to black rot, downy mildew and curd blight in Indian cauliflower. Cruciferae Newsl 12:96–97

Singh DN, Singh NK, Srivastava S (1999) Biochemical and morphological characters in relation to Alternaria blight resistance in rapeseed-mustard. Ann Agric Res 20:472–477

Singh D, Singh R, Singh H, Yadav RC, Yadav N, Barbetti MJ, Salisbury P, Nimbal S, Chattopadhyay C, Kumar

A (2007) Cultural and morphological variability in *Alternaria brassicae* isolates of Indian mustard [*Brassica juncea* (L.) Czern & Coss.]. In: Proceedings of the 12th international rapeseed congress, vol 4, Wuhan, 26–30 March, pp 158–160

Singh R, Singh D, Singh H, Caxia L, Sivasithampram K, Salisbury P, Barbetti MJ (2011) Pathogenic variability in *Alternaria brassicae* incitant of Alternaria blight of *Brassica* species. In: Proceedings of the 13th international rapeseed congress, Prague, 5–9 June, pp 1265–1270

Skoropad WP, Tewari JP (1977) Field evaluation of the role of epicuticular wax in rapeseed and mustard in resistance to Alternaria black spot. Can J Plant Sci 57:1001–1003

Stankova J (1972) Varietal variability of winter rape in its susceptibility to dark leaf spot and the factors influencing the development of the disease. Rostlinna Vyroba 18:625–630

Stitcher L, Mauch-Mani B, Metraux JP (1997) Systemic acquired resistance. Annu Rev Phytopathol 35:235–270

Taj G, Kumar A, Bansal KC, Garg GK (2004) Introgression of osmotin gene for creation of resistance against Alternaria blight by perturbation of cell cycle machinery. Indian J Biotechnol 3:291–298

Taj G, Agarwal P, Grant M, Kumar A (2011) Co-expression and *in-silico* interaction studies for inter-linking the activation of mapk3 and lox genes during pathogenesis of Alternaria blight (*Alternaria brassicae*) in *Brassica juncea*. J Oilseed Brassica 2:13–20

Takasugi M, Monde K, Katsui N, Shirata A (1988) Novel sulfur-containing phytoalexins from the Chinese cabbage *Brassica campestris* L. ssp. *Pekinensis* (Cruciferae). Bull Chem Soc Jpn 61:285–289

Tewari JP (1991a) Current understanding of resistance to *Alternaria brassicae* in crucifers. In: Proceedings of GCIRC 8th international rapeseed congress, vol 2, Saskatoon, 9–11 July, p 84 (Abstr.), 471–476

Tewari JP (1991b) Structural and biochemical bases of the black spot disease of crucifers. Adv Struct Biol 1:325–349

Tewari JP, Conn KL (1993) Reactions of some wild crucifers to *Alternaria brassicae*. IOBC/WPRS Bull 16:53–58

Tewari JP, Skoropad WP (1976) Relationship between epicuticular wax and black spot caused by *Alternaria brassicae* in three lines of rapeseed. Can J Plant Sci 56:781–785

Tewari JP, Conn KL, Dahiya JS (1987) Resistance to *Alternaria brassicae* in crucifers. In: Proceedings of 7th international rapeseed congress, Poznan, p 48 (Abstr.), 1085–1089

Tewari JP, Conn KL, Dahiya JS (1988) Dynamic response of rapeseed and other crucifers to *Alternaria brassicae* and *Rhizoctonia solani*. In: Proceedings of the 5th international congress of plant pathology, Kyoto, p 242 (Abstr.)

Thomma BPHJ, Nelissen I, Eggermont K, Broekaert WF (1999) Deficiency in phytoalexin production causes enhanced susceptibility of *Arabidopsis thaliana* to the fungus *Alternaria brassicicola*. Plant J 19:163–171

Tierens KF, Thomma BPHJ, Brouwer M, Schmidt J, Kistner K, Porzel A, Mauch-Mani B, Cammue BP, Broekaert WF (2001) Study of the role of antimicrobial glucosinolate-derived isothiocyanates in resistance of *Arabidopsis* to microbial pathogens. Plant Physiol 125:1688–1699

Toriyama K, Hinata K, Kameya T (1987) Production of somatic hybrid plants "Brassicomoricandia", through protoplast fusion between *Moricandida arvensis* and *Brassica oleracea*. Plant Sci 48:123–128

Tripathi NN, Kaushik CD, Yadava TP, Yadav AK (1980) Alternaria leaf spot resistance in raya. Haryana Agric Univ J Res 10:166–168

Verma VD, Rai B (1980) Note on induced mutagenesis for spotting out the source of resistance to Alternaria leaf-spot in Indian mustard. Indian J Agric Sci 50:278–280

Verma PR, Saharan GS (1994) Monograph on alternaria diseases of crucifers, Saskatoon Research Centre technical bulletin, 1994–6E. Agriculture and Agri-Food Canada, Saskatoon, 162p

Verma SS, Yajima WR, Rahman MH, Shah S, Liu JJ, Ekramoddoullah AK, Kav NN (2012) A cysteine-rich antimicrobial peptide from *Pinus monticola* (*PmAMP1*) confers resistance to multiple fungal pathogens in canola (*Brassica napus*). Plant Mol Biol 79:61–74

Vidhyasekaran P (1988) Physiology of disease resistance in plants, vol II. CRC Press Inc, Boca Raton, 128p

Vishwanath K, Kolte SJ (1997) Variability in *Alternaria brassicae:* response to host genotypes, toxin production and fungicides. Indian Phytopathol 50:373–381

Vishwanath, Kolte SJ, Singh MP, Awasthi RP (1999) Induction of resistance in mustard *(Brassica juncea)* against Alternaria black spot with an avirulent *Alternaria brassicae* isolate-D. Eur J Plant Pathol 105:217–220

Warwick SI (2011) Brassicaceae in agriculture. In: Schmidt R, Bancroft I (eds) Genetics and genomics of the brassicaceae, Plant genetics and genomics: crop models 9. LLC, New York, pp 33–65

Williams PH (1988a) Wisconsin fast plants. Cruciferae Newsl 13:7

Williams PH (1988b) Activities of the crucifer genetics cooperative in 1988. Cruciferae Newsl 13:13

Williams J, Pink DAC (1987) Development of an *in vitro* screen in brussels sprout for resistance to *Alternaria brassicae*. Cruciferae Newsl 12:91

Yadav RC, Sareen PK, Chowdhury JB (1991) Inter-specific hybridization in *Brassica juncea* × *Brassica tournefortii* using ovary culture. Cruciferae Newsl 14/15:84

Zaman MW, Biswas KP (1987) *Brassica napus*: a potential new oilseed crop for Bangladesh. Cruciferae Newsl 12:36–37

Zhou N, Tootle TL, Glazebrook J (1999) *Arabidopsis PAD3*, a gene required for camalexin biosynthesis, encodes a putative cytochrome P450 monooxygenase. Plant Cell 11:2419–2428

Phytotoxins

10

10.1 Introduction

The long search for a 'toxin' produced by plant pathogens met with disappointment until attention was centred on a host-specific toxin (HST). The disappointment that spanned almost half a century resulted mostly from the study of toxins that function in various types of symptom development, but are not the initial inciting agents of disease. 'Microorganism X (but not others) produces substance Y which damages plant or plant group A (but not others), and only A is parasitized'. This statement by Wood (1973) showed serious need for toxin study. The finding of a group of highly host selective HSTs, valuable both as tools for academic research on plant host–parasite interactions and as markers for epidemiological survey of pathogens in fields, confirms that the search for toxins is an important part of modern plant pathology.

The present knowledge of HSTs has come almost entirely from the so-called saprophytic pathogens. This group includes familiar fungi, such as *Alternaria* and *Helminthosporium*, and others less well known. It is just within the last 30 years (1985 to date) that we have really begun to work with HSTs in *Alternaria* species infecting crucifers. Until recently, our attention for the toxins was biased towards the physiological or biochemical mechanisms of their selectivity and elucidation of their chemical structures. However, an understanding of the pathological roles of toxins in relation to epidemiology of diseases as well as successful pathogenesis appears to be the critical need and should be the ultimate objective of toxin study for plant pathologists (Nishimura and Kohmoto 1983).

Many pathogenic *Alternaria* species produce toxins, which facilitate their necrotrophic lifestyle. Prior to colonization, necrotrophs must kill their host cells at a distance by producing both toxins and lytic enzymes often by triggering genetically programmed apoptotic pathways or by directly causing cell damage resulting in necrosis. Many species of *Alternaria* produce toxins with broad ranges, but some agronomically important species produce very host-specific toxins with a narrow range often to the cultivar level. *Alternaria* is an important pathogen of many crucifers, and because several of its species produce carcinogenic, teratogenic and mutagenic mycotoxins, researchers around the globe are using this necrotrophic pathogen to study molecular mechanisms of plant defence (Oliver and Ipcho 2004; Thaler et al. 2004; Rawe and Kliebenstein 2010). Using *A. brassicicola* as model representative for their basic research on virulence, Lawrence et al. (2008) generalized the role of toxins in the pathogenesis of genus *Alternaria*, and Chung (2012) studied the stress response and pathogenicity of the necrotrophic fungal pathogen *Alternaria alternata*. This chapter deals with nature and structure of *Alternaria*

G.S. Saharan et al., *Alternaria Diseases of Crucifers: Biology, Ecology and Disease Management*,
DOI 10.1007/978-981-10-0021-8_10

toxins, other metabolites and their effect on pathogenesis of *Alternaria* spp. at physiological, biochemical and molecular level along with the role in host resistance.

10.2 Historical Developments

The present knowledge of phytotoxins from *Alternaria* species infecting crucifers received impetus from the age-old discovery of non-specific toxins (tab toxins, fusaric acid, pyricularin) and host-specific toxins (victorin, *Periconia circinata* toxin, *Alternaria kikuchiana* toxin). Out of the four species of *Alternaria* infecting crucifers, much of the information on phytotoxins has been generated on *A. alternata* pathogenesis, that too with other than crucifers host pathosystem. However, research on phytotoxins in the *Alternaria*–crucifers host pathosystem was initiated during 1980–1990. Credit goes to Tewari and Pedras from Canada who along with their associates did excellent work on toxins in *Alternaria*–crucifers host pathosystem.

Alternaria brassicae has a multitoxin system and produces at least three phytotoxins. One phytotoxin is the cyclodepsipeptide, destruxin B, having molecular formula $C_{30}H_{51}N_5O_7$ and a molecular weight of 593 (Ayer et al. 1987; Ayer and Pena-Rodriguez 1987a, b; Bains 1989; Bains and Tewari 1986, 1987, 1989). The effects of destruxin B are similar to *A. brassicae* on various brassicas (Plates 10.1 and 10.2; Tewari 1991). Destruxin B is also produced by *Metarhizium anisopliae* and some other fungi and has insecticidal properties as well (Gupta et al. 1989). Destruxin B is a tripeptide based on isoleucine, proline and β-alanine. The second phytotoxin, homodestruxin B with molecular formula $C_{31}H_{53}N_5O_7$ (Ayer and Pena-Rodriguez 1987a, b), affects both hosts and non-hosts of *A. brassicae* (Bains 1989).

Reversed-phase chromatography on Sep-Pak C_{18} cartridges followed by high-performance liquid chromatography (HPLC) has successfully been used for the separation and detection of destruxins from *M. anisopliae* (Samuels et al. 1988). Using a modification of this method, three destruxins, viz. destruxin B, destruxin B2 and homodestruxin B, have been isolated from *A. brassicae* (Buchwaldt and Jensen 1991; Buchwaldt et al. 1991). According to Buchwaldt and Green (1992), destruxin B, the major phytotoxin produced by *A. brassicae*, is not host-specific as reported earlier (Bains and Tewari 1987). This toxin causes necrosis and chlorosis on 30 species of host and non-host plants. Two minor destruxins, homodestruxin B and destruxin B2, are phytotoxic to leaves of *B. napus*. However, there are significant differences between taxonomic plant groups in their sensitivity to destruxin B. *Brassica* species are most

Plate 10.1 Symptoms caused by inoculation of *Brassica napus* cv. Altex leaf with *Alternaria brassicae* (left leaf). The right leaf is the control (Tewari 1991)

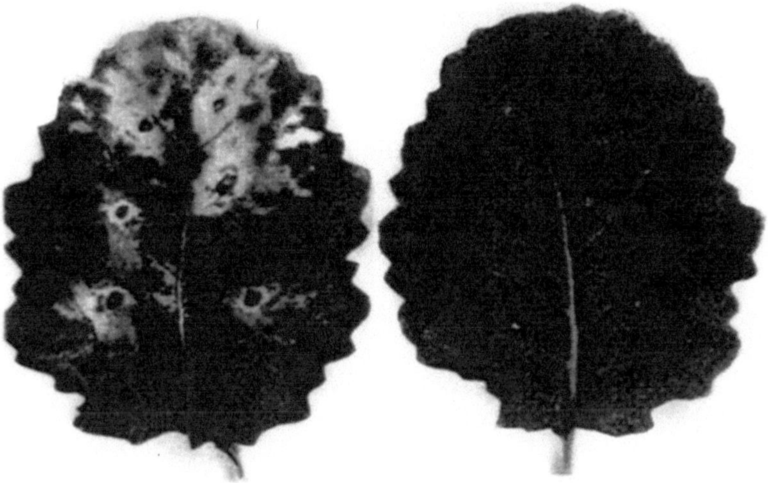

Plate 10.2 Symptoms caused by the application of destruxin B on *Brassica napus* cv. Altex leaf (left leaf). Compare with Plate 10.1. The right leaf is the control (Tewari 1991)

sensitive to the toxin, and sensitivity decreases as relatedness of plant groups become more distant. The dilution endpoint of destruxin B is 0.2–3.8 μg/ml for the most sensitive host species and 15–120 μg/ml for the least sensitive. The sensitivity of non-host species is between 15 and 750 μg/ml; destruxin B shows a high degree of biological activity (Bains 1989; Bains and Tewari 1989). In detached rapeseed leaf bioassay, the limit of detection of destruxin B lies between 15 and 30 μg/ml, and it is possible to distinguish pathogen susceptibility differences among host species (Bains and Tewari 1987). The symptoms caused by destruxin B appear to be light dependent. Destruxin B appears to be a virulence factor, contributing to the aggressiveness of *A. brassicae* by conditioning the host tissue and, thereby, determining the susceptibility of the host. Homodestruxin B is able to affect both the host and non-hosts of *A. brassicae* and, therefore, functions as a non-host-specific toxin. Homodestruxin B differs from destruxin B by the replacement of a methyl with an ethyl group (Bains et al. 1993). Somaclonal variation produced in cultured cells and tissues provides a good system for selection of new disease-resistant genotypes of brassicas using destruxin B. As a host-specific toxin, destruxin B is the primary determinant of the black spot disease of the Brassicaceae. It should be possible to use destruxin B as a selection agent for disease resistance in tissue culture systems. Tissue culture host systems with cell walls would be suitable for this purpose because the protoplasts are insensitive to the toxin (Bains and Tewari 1989).

In the presence of destruxin B, leaf discs of the susceptible host *B. rapa* var. Yellow Sarson lose significantly more electrolytes than the moderately resistant *B. rapa* var. *rapifera* (turnip). The effect takes up to 48 h to become apparent, and the lowest concentration of toxin detectable by this bioassay method is 10 μg/ml. The delayed electrolyte leakage indicates that perhaps the target site of the toxin in the host cell is not situated directly on the cell membrane, because of the delay for the permeability change to occur and the electrolyte leakage to start. The mesophyll protoplasts of both susceptible (*B. rapa* var. Yellow Sarson) and resistant (*Camelina sativa*) hosts are not affected by destruxin B, even at a concentration of 100 μg/ml. This indicates that the target site of the toxin must be in the cell wall or the periplasmic space. Germination and tube growth of pollen grains of the susceptible host (*B. rapa* var. Yellow Sarson) have been found to be very sensitive to destruxin B concentration as low as 2.5 μg/ml for 30 min. Complete inhibition of these processes occurs at a concentration of 10.0 μg/ml (Bains and Tewari 1989). These processes are not completely inhibited in the resistant host (*Camelina sativa*), even at a toxin concentration of 100 μg/ml. Past 1990, phytotoxin research mushroomed to reach to the present status.

10.3 Metabolites from *Alternaria*

Some metabolites from *Alternaria* fungi are toxic to plants and animals and are designated as phytotoxins and mycotoxins, respectively. These mainly include nitrogen-containing metabolites, steroids, terpenoids, pyranones, quinones and phenolics. *Alternaria* metabolites exhibit a variety of biological activities such as phytotoxic, cytotoxic and antimicrobial properties, which have drawn the attention of many chemists, pharmacologists and plant pathologists in research programmes as well as in application studies. Porritoxin 24 from endophytic *Alternaria* species has been reported as the candidate of cancer chemotherapeutic agent. Depudecin, an inhibitor of histone deacetylase (HDAC) from *A. brassicicola*, also shows its antitumor potency. Some *Alternaria* metabolites such as tenuazonic acid, maculosin and tentoxin act as the herbicide candidate. In the early 1990s, about 70 metabolites from *Alternaria* fungi were reviewed. Several reviews on *Alternaria* phytotoxins have been published over the last few decades. In recent years, more than 268 metabolites with bioactivities from *Alternaria* fungi have been isolated and structurally characterized (Lou et al. 2013).

10.3.1 Classification and Occurrence

The metabolites from *Alternaria* fungi can be grouped into several categories which include nitrogen-containing compounds, steroids, terpenoids, pyranones (pyrones), quinones and phenolics (Table 10.1). Several metabolites are unique to one *Alternaria* species, but most metabolites are produced by more than one species. The most widespread metabolite is alternariol which has been isolated from a few *Alternaria* species. Some metabolites have also been isolated from other fungal genera and even from higher plants. Typical examples include AAL toxins from *Fusarium* species, helvolic acid from *Aspergillus* species and *Pichia* species, paclitaxel from yew trees (*Taxus* spp.) and resveratrol from a variety of plant species including *Vitis vinifera*, *Polygonum cuspidatum* and *Glycine max* (Lou et al. 2013).

The *Brassica pathogen*, *Alternaria brassicae* when grown in liquid-still culture produces host-specific phytotoxins; cyclodepsipeptides; drimane sesquiterpenes deoxyuvidin B (4), alberassitriol (8) and isoalbrassitriol (9); brassicadiol (10); and a C_{15} prenylated pentaketide. These compounds are not phytotoxic to canola (Ayer and Pena-Rodriguez 1987a, b). *Alternaria brassicae* produces host-specific, ABR toxins (protein) from germinating spores on host leaves (Otani et al. 1998; Parada et al. 2008). AB toxin from *A. brassicicola* germinating spores is induced by a host-derived oligosaccharide (Oka et al. 2005).

Metabolites from cultures of *A. brassicicola* are phomapyrone F (4.1 mg), phomapyrone A (3.30 mg), phomapyrone G (5, 1 mg), brassicicolin A (1, 30 mg), brassicicene H (8, <1 mg), brassicicene G (7, 4 mg), brassicicene I (9, 1 mg) and infectopyrone G (6, 2 mg). The three new fusicoccane diterpenes, brassicicenes G (7), H (8) and I (9), were named by analogy to previous designations (MacKinnon et al. 1999).

From the least polar and non-phytotoxic fractions, phomapyrones A (3), F (4), G (5) and infectopyrone (6) were isolated and structures were determined by comparison of their chromatographic and spectroscopic data with those of authentic samples isolated from *Leptosphaeria maculans* and from *L. biglobosa* (Pedras et al. 1994; Pedras and Biesenthal 2001; Pedras and Chumala 2005).

From the most phytotoxic fractions, F4 and F5, brassicicolin A (1) was obtained as a colourless and chromatographically homogeneous oil (HPLC R_t, 27.3 min); HR ESI-MS analysis indicated a molecular formula of $C_{32}H_{48}N_2O_{14}$, and the FT IR spectrum contained strong bands at 3400, 2145 and 1760 cm^{-1}, suggesting the presence of hydroxyl, isonitrile and carbonyl groups, respectively.

The ^1H NMR spectrum was complex suggesting that the sample was actually a mixture of closely related compounds. Likewise, the ^{13}C NMR spectral data showed that each cluster of signals contained four resonances indicating the presence of four isomers. Comparison of these

Table 10.1 Isolated metabolites from *Alternaria* pathogenic to crucifers (Lou et al. 2013)

Metabolic class	Metabolite name	*Alternaria* species
Terpenoids	Brassicicene A (106)	*A. brassicicola*
	Brassicicene B (107)	*A. brassicicola*
	Brassicicene C (108)	*A. brassicicola*
	Brassicicene D (109)	*A. brassicicola*
	Brassicicene E (110)	*A. brassicicola*
	Brassicicene F (111)	*A. brassicicola*
	Brassicicene G (112)	*A. brassicicola*
	Brassicicene H (113)	*A. brassicicola*
	Brassicicene I (114)	*A. brassicicola*
	Abscisic acid=ABA (115)	*A. brassicae*
	Tricycloalternarenes (TCAs)	*A. alternata*
Pyranones	Phomapyrone A	*A. brassicicola*
	Phomapyrone G	*A. brassicae*
Nitrogen-containing metabolites	Destruxin B (50)	*A. brassicae*
	Homodestruxin B (51)	*A. brassicae*
	Demethyldestruxin B (52)	*A. brassicae*
	Brassicicolin A (58)	*A. brassicae*
Steroids	3β-Hydroxy-ergosta 5,8 (9), 22-trien-7-one (66)	*A. brassicicola*ML-P08
	3β,5α-Dihydroxy-ergosta 7, 22-dien-6-one (67)	*A. brassicicola* ML-P08
	Cerevisterol (68)	*A. brassicicola* ML-P08
Pyranones	Herbarin A (132)	*A. brassicicola* ML P08
	Altechromone A (151), 2,5-dimethyl-7	*A. brassicicola* ML P08
	Hydroxychromone (152)	*Alternaria* sp.
	Phomapyrone F (153)	*A. brassicicola*
Miscellaneous (polyketide)	Depudecin (257)	*A. brassicicola*

spectroscopic data and specific optical activity ([α] $_D$=+16 (c 1.0, CH_2Cl_2); lit.[α] $_D$ =16.1(c, 1.86 CH_2 Cl_2)) with those previously reported (Gloer et al. 1988), along with chemical hydrolysis results, indicated that this metabolite was brassicicolin A (1), a derivative of mannitol containing two acetyl, two 2-isocyano-3-methylbutanoyl and two 2 hydroxy-3-methylbutanoyl substituents. Fraction F5 was further separated to yield a chromatographically homogenous material ($C_{26}H_{41}NO_{13}$) shown to contain a mixture of nonsymmetrical mannitol derivates in which one of the 2-isocyano-3-methylbutanoyl groups was missing. The ^{13}C NMR spectrum displayed clusters of four to six peaks, suggesting that the fraction contained a mixture of at least four compounds. Given the complexity of this fraction and the presence of inseparable epimers, no structures were assigned to each specific component.

Three related metabolites were isolated from fractions displaying low phytotoxicity. The ^1H NMR spectrum of each of these metabolites showed the presence of an isopropyl, a methoxymethylene and two methyl substituents suggesting that these metabolites had fusicoccane structures, similar to those previously reported for brassicicenes A-F (MacKinnon et al. 1999). The most oxidized metabolite was obtained as a white solid whose molecular formula was determined to be $C_{21}H_{32}O_5$ (six-double bond equivalent) based on NMR and HRE1-MS data. The ^1H NMR spectrum displayed four methyl doublets, a methoxyl group and one olefinic proton.

Metabolites from leaves of *B. juncea* inoculated with *A. brassicicola* were detected by Pedras et al. (2009). No phytoalexins were detected 24 h post inoculation; four phytoalexins were detected 48 h post inoculation and were quantified using calibration curves constructed with synthetic samples (Pedras et al. 2007). The phytoalexin spirobrassinin was produced in larger amounts than brassicic, cyclobrassinin or rutalexin. As well, indolyl-3-acetonitrile was detected in higher concentration in leaves (0.8–4.0 nmol/g fresh weight) of infected plants than in control leaves (<0.1 nmol/g fresh weight). Furthermore, HPLC-MS analysis of leaf extracts did not detect any of the metabolites produced by *A. brassicicola* in liquid cultures.

Nine novel compounds closely related to ACTG toxin, termed as tricycloalternarenes, were isolated from a strain of *A. alternata*, from *B. sinensis*, which was earlier used for the production of non-specific phytotoxin, tentoxin (Nussbaum et al. 1999). Liebermann et al. (2000) also reported isolation of 11 new bicycloalternarenes as well as ACTG toxins A and B. *Alternaria brassicae* (Berk.) Sacc. produces four cyclic depsipeptide phytotoxins including destruxin B, homodestruxin, destruxin B2 and desmethyldestruxin B host-specific proteinaceous ABR toxin (Ayer and Pena-Rodriguez 1987a, b; Buchwaldt and Green 1992; Montemurro and Visconti 1992; Agarwal et al. 1994; Parada et al. 2008). In addition to proteinaceous AB toxin (Otani et al. 1998) and brassicillin A, a very selective phytotoxin metabolite in liquid cultures (Pedras et al. 2009), *A. brassicicola* also produces depsipeptides and fusicoccin-like toxin compounds (Cooke et al. 1997; MacKinnon et al. 1999).

10.4 Effect on Plants at Physiological, Biochemical and Molecular Level

The metabolites from plant pathogenic fungi are usually toxic to plants and are called phytotoxins. They are further divided into host-specific and non-host-specific toxins. The host-specific toxins (HSTs) are toxic only to host plants of the fungus that produces the toxin. Another definition seems to be more acceptable that the host-specific toxins are toxic to plants that host the pathogen, but have lower phytotoxicity on non-host plants. Most HSTs are considered to be the pathogenicity factors, which the pathogenic fungi producing them require to invade tissue and induce disease. All isolates of the pathogen that produce an HST are pathogenic to the specific host. All isolates that fail to produce HSTs lose their pathogenicity to the host plants. Plants that are susceptible to the pathogen are sensitive to the toxin. Such correlations between HST production and pathogenicity in the pathogens, and between toxin sensitivity and disease-producing ability in plants, provide persuasive evidence that HSTs can be responsible for host-specific infection and disease development. Johnson et al. (2000, 2001) revealed that the genus involved in HST synthesis such as the cyclopeptide synthetase gene, whose product catalysed AM toxin production in *A. alternata* apple pathotype, might reside on a conditionally dispensable (CD) chromosome. The loss of the CD chromosome led to loss of both toxin production and pathogenicity without affecting fungal growth. On the other hand, the exact roles of non-specific toxins in pathogenesis are largely unknown, but some are thought to contribute to the features of virulence, such as the symptom development, and in plant pathogen propagation. The virulence and host specificity of these pathogens are based on the production of the distinctive HSTs. Most of *Alternaria* HSTs are nitrogen-containing metabolites causing black spot disease in brassicas, destruxin A, destruxin B, homodestruxin B and desmethyl-destruxin B. Destruxins are group of HSTs produced both in vitro and in planta by *A. brassicae*, the causal agent of *Alternaria* black spot disease of rapeseed and canola. These cyclodepsipeptides exhibit a wide variety of biological activities such as antitumor, antiviral, insecticidal, cytotoxic, immunosuppressant and anti-proliferative effect except their phytotoxicity. Interaction between *Alternaria* species and cruciferous plants was studied in detail by Pedras et al. (2009). Necrotrophic phytopathogens such as *A. alternata* and *A. brassicae* are known to synthesize phytotoxins that damage plant tissues and facilitate colonization, while in response to pathogens' attack, crucifers biosynthesize phytoanticipins and phytoalexins. Phytoalexins are secondary metabolites produced de novo by plants in response to diverse forms of stress including microbial infection, UV irradiation and heavy metal salts, whereas phytoanticipins are constitutive defences whose concentrations can increase upon stress. To the detriment of cruciferous plants, the phytopathogens can overcome phytoanticipins and phytoalexins by producing detoxifying enzymes; phytoalexin brassinin is detoxified into 3-indolyl-methanamine and N″-acetyl-3-indolylmethanamine by the pathogen *A. brassicae* (Fig. 10.1). Very interestingly, cruciferous plants (*B. napus* and *Sinapis alba*) can convert host-specific toxins, destruxin B and homodestruxin B, into less phytotoxins, hydrodestruxin B and hydroxyl-homodestruxin B, respectively (Fig. 10.2; Lou et al. 2013).

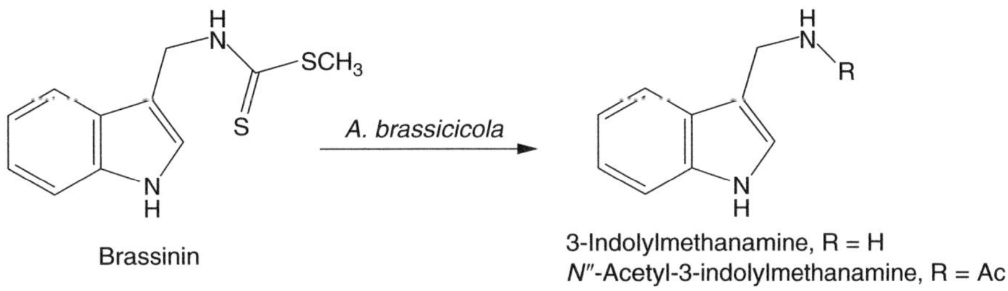

Fig. 10.1 Detoxification pathway of the phytoalexin brassinin by the pathogen *A. brassicicola* (Lou et al. 2013)

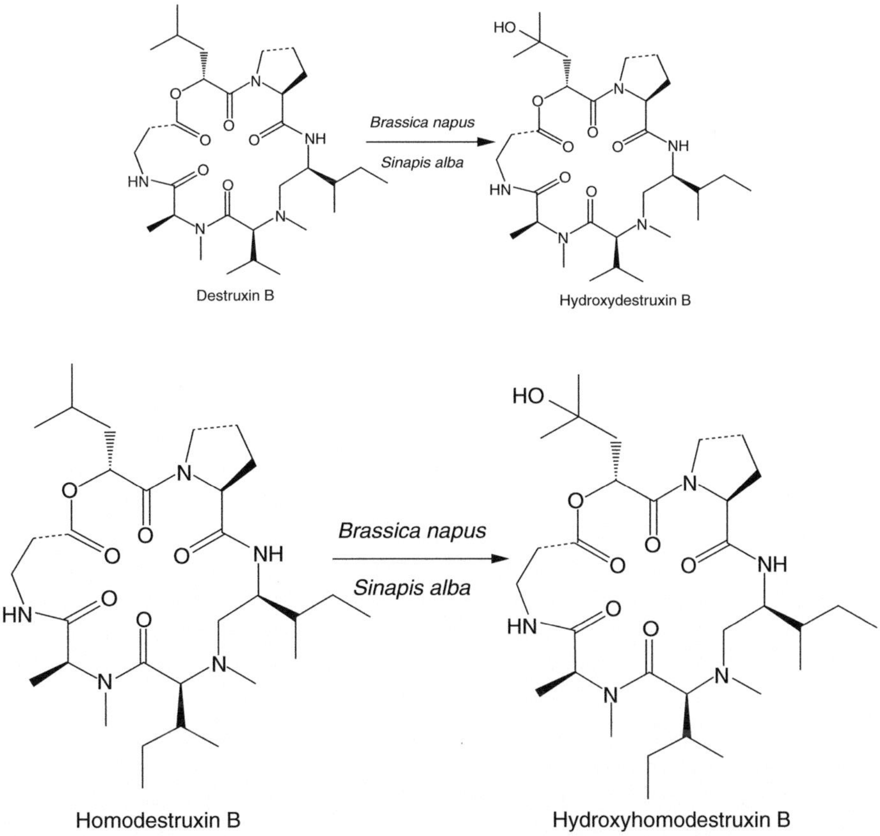

Fig. 10.2 Detoxification pathway of the phytotoxin B and homodestruxin B by the hosts *Brassica napus* and *Sinapis alba* (Lou et al. 2013)

Non-host-specific *Alternaria* phytotoxins can affect many plants regardless of whether they are a host or non-host of pathogen. Non-host-specific nitrogen-containing phytotoxins include tenuazonic acid, porritoxin and tentoxin. Tentoxin, a cyclic tetrapeptide from *A. alternata,* inhibits chloroplast development, which phenotypically manifests as chlorotic tissue. Tentoxin was suggested to exert its effect on chlorophyll accumulation through over-energization of thylakoids (Lou et al. 2013).

Destruxin B purified from culture filtrates (CFs) of *A. brassicae* induces chlorosis on host leaves at 50–100 μg ml^{-1} and chlorosis or necrosis on non-host leaves at 250–500 μg ml^{-1} (Bains and Tewari 1987). Destruxin B was detected in spore germination fluids (SGFs) on host and non-host leaves, but not in sufficient amount to exert toxicity to host plants. When spores of non-pathogenic *A. alternata* are combined with destruxin B at 100 μg ml^{-1} and inoculated on leaves, destruxin does not affect the infection behaviour of the spores. Interestingly, SGF on host leaves allows non-pathogenic spores to colonize host leaves (Bains 1989). Moreover, a high molecular weight fraction (>5 kDa) without destruxin B obtained by ultrafiltration of SGF has host-specific toxic activity and infection-inducing activity. According to Parada et al. (2007), destruxin B is not an HST and does not induce the accessibility of the host plant, which is essential for colonization of the pathogen. In addition, the results with SGF imply that a high molecular weight HST(s) is involved in the host–pathogen interaction.

According to Khandelwal et al. (2002), P^{53}-like protein is overexpressed in toxin-treated and nutritionally depleted calli. Similar changes are seen in senescent damaged *Brassica* species indicating involvement of P^{53}-dependent pathways. The effects of destruxin B, a host-specific toxin of *A. brassicae* that causes black spot disease in oilseed brassicas, were studied on in vitro pollen germination and pollen tube growth of *Brassica campestris* var. Brown Sarson, *B. juncea* cvs. Westar and Cresor, *B. nigra* and *Sinapis alba.* Pollen grains of *B. nigra*, *B. juncea* and *B. campestris* were the most sensitive and those of *S. alba* the least sensitive to the toxin. Effects of toxin on leaves of these species and the degree of sensitivity of leaves of different species are comparable to that of their pollen grains. The results on the responses of pollen grains as well as leaves to the toxin are in agreement with the degree of susceptibility/resistance of these species to *A. brassicae*, indicating that the genes imparting susceptibility/resistance are expressed in the pollen, a prerequisite for pollen selection. The toxin fed to the cut end of isolated inflorescence axis is readily taken up by the developing pollen and results in the inhibition of germination of susceptible pollen (Hodgkin and Macdonald 1986; Shivanna and Sawhney 1993).

10.4.1 Physiological Level

The effect of phytotoxins on plants at the physiological level is characterized by the malfunctioning of many physiological processes including respiration, transpiration, photosynthesis, translocation and growth and development. It is also characterized by the appearance of specific symptoms including wilting, growth suppression, chlorosis, necrosis and spotting of aerial portions.

Alternaria HSTs are classified into three groups in terms of primary site of action. AM toxin targets the chloroplast and plasma membrane, ACR (L) toxin targets mitochondria, and ACT, ACTG, AK and AF target the plasma membrane. Toxins produced by *A. brassicae* cause a brown necrotic spots on leaves and brown streaks on stem leading to yield losses. In general, respiration increases once parasitic relationship is established. *Alternaria* pathogens infect green aerial tissues and reduce photosynthetic activity; produce cytokinins, which lead to the formation of Green Island below the germinating conidia on senescing tissues; and cause deformation in chloroplast and mitochondria (Chaturvedi 1972; Zheng et al. 2006). In *B. juncea*, *A. brassicae*, HST destruxin B, inhibits almost all macromolecular biosynthesis, promotes ion leaching and causes aberrations in mitochondria and chloroplast.

10.4.2 Biochemical Level

Infection of plants by pathogenic fungi may trigger several biochemical defence responses including enzyme synthesis, cell wall deposition of lignin and suberin and accumulation of specific metabolites (Daayf and Platt 2000; Abdel-Farid et al. 2009). The mechanisms of host-selective pathogenesis are not well understood at the biochemical level, even in cases where the toxin site of action is known. The first physical barrier between plants and pathogen invasion is the cuticle (Schweizer et al. 1996; Fan and Köller 1998) and cell wall, which inhibit both initiation and spread of infection. One characteristic feature of many phytopathogenic organisms is their ability to produce an array of extracellular and highly stable enzymes capable of degrading the complex polysaccharides of the plant cell wall and membrane constituents. These enzymes usually are produced inductively in infected tissues; generally, they are extracellular and highly stable. Inoculation of *A. brassicae* on the leaves of *B. juncea* blight resistant cultivar RC-781 decreases the activities of cell wall-degrading polygalacturonase and cellulose enzymes, but increases their activities in a susceptible cultivar Varuna up to 3 days of infection (Garg et al. 1999). Infection of *A. brassicicola* shows a differential and sequential induction of two classes of cutinolytic esterases; one class is expressed only in short duration contact (24 h) with intact cutin, but not induced by cutin monomer. The second class, however, is induced by cutin monomers only in prolonged exposure with intact cutin. This differential behaviour indicates a sequential recognition of cutin as a barrier to be penetrated and utilized as a carbon source in saprophytic stages (Yao and Köller 1995; Fan and Koller 1998; Bakar et al. 2005). The small (7–10 KD) lipid transfer proteins (LTPs) are expected to be involved in wax transport because of their increased expression during drought condition (Beisson et al. 2003; Cameron et al. 2006). During *A. brassicicola* infection, the *ltpgl* (GPI-anchored LTP) knockout mutant showed increased susceptibility (Lee et al. 2009), but the genetically unidentified cutinase-deficient mutants were non-pathogenic (Tanabe et al. 1988).

Results of a transcript profiling study of *hsr203J*, a known marker gene for HR, showed that the expression of *hsr*203J gene is more in *A. brassicae*-tolerant than in *A. brassicae-susceptible* genotype of *B. juncea* (Mishra et al. 2011). In Chinese cabbage, *A. brassicae* infection significantly increases glucosinolates (aliphatic and indole) and anthocyanins, but decreases sucrose levels (Ludwig-Müller et al. 1999; Rostás et al. 2002; Abdel-Farid et al. 2009).

Chitin, an important component of fungal cell wall, is absent in plants, but chitin-degrading chitinase enzymes are present in abundant quality which emphasizes that chitinase can directly affect the viability of fungal pathogens (Boller 1995; Stacey and Shibuya 1997; Shibuya and Minami 2001). This is proved by overexpression of chitinase in transgenic plants, which are generally more resistant to fungal pathogens. Further, it was observed that more chitinase is produced and accumulates at the site of fungal infection (Majeau et al. 1990; Roby et al. 1990; Wubben et al. 1992). It is reported that co-expression of chitinase and ribosome-inactivating protein in *B. juncea* provides more protection against *A. brassicae* (Chhikara et al. 2012).

AAL toxin, structurally related to sphinganine, a member of sphinganine-analogue mycotoxin (SAMS), is an inhibitor of sphinganine-N-acyltransferase (Gilchrist et al. 1992; Abbas et al. 1994) enzyme important in sphingolipid biosynthesis, leading to the accumulation of free sphingoid bases (Brandwagt et al. 2000; Spassieva et al. 2002; Gechev and Hille 2005). These long-chain sphingoid bases (LCBs) are determinant in the occurrence of programmed cell death (PCD) in susceptible plants (Shi et al. 2007; de Zelicourt et al. 2009). Sphinganine-analogue mycotoxin (SAMS) inhibits ceramide biosynthesis. Biochemical and molecular data demonstrated that PCD triggered by AAL toxin is also associated with H_2O_2 (Gechev et al. 2004) as *atr* mutant of *Arabidopsis* which shows enhanced tolerance to H_2O_2 and ROS-induced cell death (Gechev and Hille 2005; Gechev et al. 2008). Interaction with ROS regulates PCD (Gechev and Hille 2005). Brassinin hydrolase, a detoxifying enzyme (BHAb) of a crucifer phytoalexin brassinin, also plays an important role in the development of *Alternaria* blight caused by *A. brassicicola* in *Brassica* (Pedras et al. 2009). Imazaki et al. (2010) reported that *A. alternata* pathotypes contain abundant peroxisomes, which are very important in both tissue colonization and pathogenesis. Other functions of peroxisomes include fatty acid metabolism, acetyl-CoA generation, secondary metabolism, cell wall biogenesis and peroxisome homeostasis.

10.4.3 Molecular Level

The interaction between plants and pathogens at molecular level is specific, complex and dynamic. Many responses are evoked in plants upon encountering pathogens, but relatively very few have been studied in detail. Key events in plant–pathogen interaction include perception of pathogen on the plant or cell surface by receptors/sensors and transduction of these perceptions through various transcription factors and target genes which are involved in coordination of the appropriate responses (Hammond-Kosack and Jones 1996). Different types of both plant and pathogen genes have been shown to be involved in plant and necrotrophic fungal interaction. These responses range from genes that encode proteins like receptor/receptor kinase, cell wall-degrading enzymes, toxins and transporter proteins to those involved in signal transduction cascades such as mitogen-activated protein (MAP) kinases and various transcription factors like WRKY and NAC (Lawrence et al. 2008; Wang et al. 2009; Amselem et al. 2011). Ghose et al. (2008) also studied the differential profiling of selected defence-related genes induced on challenge by *A. brassicicola* in resistant white mustard and their comparative expression pattern in susceptible Indian mustard. These genes have similarity with receptor-like protein kinase genes, genes involved with calcium-mediated signalling, salicylic acid-dependent genes and other functional genes in *Arabidopsis.* After the attack, the fungal cell wall is hydrolyzed by an enzyme chitinase, and fragment of cell wall chitin is perceived by a receptor known as CERK 1 or Lys M RLK 1. A mutation in Lys M receptor-like protein (*Lys M RLK1*) gene blocks the induction of almost all chito-oligosaccharide-responsive genes and leads to increased susceptibility to fungal pathogens indicating that *LysM RLK1* is essential for chitin signalling in plants (Wan et al. 2008). Another receptor gene, PSK receptor, PSK2 and PSKR I (phytosulphokine receptor), also induced in *Arabidopsis* leaves following *A. brassicicola* inoculation, also shows the role of PSK receptor in pathogenesis (Loivamaki et al. 2010).

Molecular mechanism of *Alternaria* blight in *Brassica* shows involvement of chlorotic and necrotic toxins and phytohormone. *Alternaria brassicae* produces a chlorotic toxin destruxin B that plays an important role in signal transduction leading to programmed cell death and suppressing the defence system (Taj et al. 2004). Differential expression of cell cycle proteins in toxin-treated leaves and calli and overexpression of p53 suggest that the toxin-mediated perturbations in cell cycle eventually cause p53-induced programmed cell death (PCD) (Khandelwal et al. 2002). Interestingly, *A. brassicae* pathotoxin-induced cell death pathway was antagonized by a phytohormone zeatin, in cell culture of *B. juncea* (Pandey et al. 2001). The antagonistic effect of these two structurally different entities strongly suggests the role of interactive signalling pathways in pathogenesis of *Alternaria* blight in *Brassica* species. Introgression of osmotin, a known PR-5 protein, causes perturbation in resistance to biotic and abiotic stresses (Taj et al. 2004). The effect of toxin and zeatin treatments on the *B. juncea* leaf proteome was investigated by using two-dimensional electrophoresis and LCMS techniques; results showed that 20 proteins were uniquely expressed in toxin-treated leaf, while 27 proteins were expressed in together with toxin and zeatin. LCMS technique has also been used to identify a total of 15 proteins with differential expression in toxin-treated leaves. The proteins identified in response to the toxin are glycosyl hydrolases, subtilisin-like proteases, P-nitrophenyl phosphatase, malate dehydrogenase, heat shock, ribulose 1, 5-bisphosphate carboxylase, cucumisin-like serine proteases, globulin-like protein and ATP synthase. Sharma et al. (2007) also studied the proteome-level changes in *A. brassicae–B. napus* and suggested role of ROS-mediated auxin signalling in the pathosystem.

Mitogen-activated protein kinase cascades are also standard players in the signal transaction literature for diverse organisms (Madhani and Fink 1998; Cobb 1999). Activation of MAPK confers resistance to both bacterial and fungal pathogens (Sheen et al. 2002a, b). To decipher the MAP kinase machinery components in *B. juncea*, RT-PCR amplification of all 20 known MAPKs has been done. Among the MAP kinases 8, 12 and 18 showed no expression, and expression of MAP kinases 3, 10 and 14 was validated with the Northern Blot. MAPK 3 gene is important and directly correlated with the transcription factors, expressed in compatible interaction of *A. brassicae* and *B. juncea* (Taj et al. 2011). Expression of MAPKK 4, MAPKK 5, MAPKK 9, MAPKKK 1, Wrky 22 and Wrky 29 has also been validated by real-time PCR. Expression analysis of MAP 2 K9 and MAPK 6 is also governed during the pathogenesis of *Alternaria* blight in *Arabidopsis thaliana* where simultaneous increased levels up to middle stage of disease progression were observed (Kannan et al. 2011).

Inculcating the resistance mechanism in *Sinapis alba*, against *A. brassicicola*, the downstream signalling of MAPK 6 can be shown, which was found to be activated after ethylene treatment (Taj et al. 2010) and might be a significant upregulation step of NDR1/HIN-l-like, NHL25 and PR genes reported during *S. alba* infection (Varet et al. 2002); these genes were downstream targets of MAPK 6 in *Arabidopsis* which are controlled by salicylic acid (SA) (Desikan et al. 1999; Ghose et al. 2008). Activation of more than one member of MAPK by *Alternaria* suggests that MAPK cascades act as points of convergence and divergence in signalling and might play a pivotal role in deciding whether disease should progress or defence pathways defeat the pathogen.

The involvement of phytohormone-dependent pathway is well documented in plant–pathogen interaction. Also well documented is the role of jasmonic acid (JA)-dependent signalling pathway of necrotrophic pathogen and SA-dependent signalling pathway of biotrophic pathogen. Mutants of JA and/or ethylene (ET) signalling pathways and JA-insensitive *coi-l* of *Arabidopsis thaliana* show increased susceptibility (Thomma et al. 1999b) and resistance against *A. brassicicola* (Thomma et al. 1999a; Kachroo et al. 2001; Nandi et al. 2005; Mang et al. 2009); systemic expression of the JA-inducible *PDF1.2* gene is also reduced after infection with *A. brassicicola.* This reflects the necessity of JA-mediated responses for expression of this trait (Glazebrook 1999). In contrast, the SA-insensitive mutant *npr-l* and SA-depleted *nah-G* line have no effect on the resistance phenotype (Thomma et al. 1998), indicating no direct involvement of SA as a signalling molecule. In another study, induction of SA signalling marker, PR 1, and enhanced biosynthesis of the antifungal compound camalexin upon infection by *A. brassicicola* in *Arabidopsis* raises the possibility of crosstalk between these different signalling networks (Doares et al. 1995; Kunkel and Brooks 2002; Kachroo et al. 2003a, b). Jasmonic acid also helps in the nodulation of MAP kinase 4 and MAP kinase 6 during phytopathogenesis of *Alternaria* blight in *A. thaliana.*

Furthermore, biosynthesis of antifungal compound camalexin which plays a role against *Alternaria* pathogen has been found to be positively controlled by SA, ROI and ET. JA-regulated *LOX* gene is also known to play a role during *A. brassicae* pathogenesis (Taj et al. 2011). During incompatible interaction of *Arabidopsis*, transgenic plants harbouring CaLOX 1-OX *loxl* mutant, and *A. brassicicola*, it was discovered that *loxl* mutant plants are more susceptible than wild types, CaLOX1-OX transgenics and CaLOX1 plants (Hwang and Hwang 2010). Transcription of the plant defence genes *PDF1.2* and *Thi2.1* is enhanced in response to *Botrytis cinerea* and *A. brassicicola* infection and is dependent on ET, JA and OA signals (Epple et al. 1995; Penninckx et al. 1996; Kachroo et al. 2003a, b). Resurrection I (*rst1*) mutant plants show more resistance to the necrotrophic fungi *B. cinerea* and *A. brassicicola* by suppressing pathogen growth, sporulation and disease symptoms, which might be due to altered cuticular waxes (Chen et al. 2005), because the amount of cutin monomers, phytoalexin accumulation and basal expression of the *PDF1.2* gene were significantly enhanced in infected leaves (Mang et al. 2009). Unlike *rst I* mutation, other mutations including mpk 4, bik 1 and wrky 33 cause susceptibility to the necrotrophic pathogens, *A. brassicicola* and *B. cinerea* (Petersen et al. 2000; Wiermer et al. 2005; Veronese et al. 2006; Zheng et al. 2006; Mang et al. 2009). DEAD-box RNA helicase also plays a role during *Arabidopsis–A. brassicicola* interaction. Transgenic *Arabidopsis* plant that overexpresses the OsBIRH1 gene (DEAD-box RNA helicase protein) also shows enhanced expression of PR-1, PR-2, PR-5 and PDF 1.2 and disease resistance against *A. brassicicola* (Li et al. 2008).

During the course of infection, various genes of pathogen are also expressed and play a role in fungal pathogenesis; *AaFus3* gene, which encodes for *FUS3MAPK* in *A. alternata*, is required for conidial development and penetration of the fungus in plant (Lin et al. 2010). Another study shows that deletion of AbPro1 (transcription factor) and *AbNIK* 1 (two-component histidine kinase) gene of *A. brassicicola* results in 70 % and complete loss of virulence, respectively (Cho et al. 2009). Cyanide hydratase, arsenic ATPase and formate dehydrogenase are some other genes of *A. brassicicola*, which are being expressed and have a role in fungal pathogenesis (Cramer and Lawrence 2004). Unfolded protein response (UPR) pathway also regulates the fungal pathogenesis, and *AbHac A* gene encodes the major UPR transcriptional regulator; loss of UPR in mutants of *A. brassicicola* resulted in complete loss of virulence (Joubert et al. 2011).

10.5 Role of Toxins in the Infection Process

In general, *Alternaria* species are foliar pathogens that cause relatively slow destruction of host tissues through the reduction of photosynthetic potential. An infection leads to the formation of necrotic lesions, which sometimes have a target-like appearance due to growth interruptions caused by unfavourable conditions. The fungus resides in the centre of the lesion, which is surrounded by an un-invaded chlorotic halo, a symptom that is commonly observed for the infection process of necrotrophic pathogens. This zone is created by the diffusion of fungal metabolites like toxins (Agarwal et al. 1997; Tewari 1983). Members of the genus *Alternaria* frequently cause quiescent infections in which the fungus enters the tissue where it remains dormant until conditions favour infections. *Alternaria* species generally do not affect water or nutrient transport throughout the plant, because they do not specifically target roots or vessels (Rotem 1994).

Alternaria has no known sexual stage or overwintering spores, but the fungus can survive as mycelium or spores on decaying plant debris for a considerable time, or as a latent infection in seeds (Rotem 1994). The fungus can attack the seedling once the seed has germinated if seed-borne infection exists. In other cases, once the spores are produced, they are mainly spread by wind onto plant surfaces where infection can occur. Typically, weakened tissues either due to stresses, senescence or wounding are more susceptible to *Alternaria* infection than healthy tissues. The observation that saprobic *Alternaria* species can become parasitic when they meet a weakened host illustrates that the distinction between saprophytic and parasitic behaviour is not always evident.

Despite the taxonomic and pathogenic differences between *Alternaria* species, these cause similar infection patterns. Dormant spores have heavily melanized walls that, under favourable conditions, produce one or more germ tubes. Subsequently, the germ tubes penetrate stomata, cuticle or wounds with or without the formation of small appressoria. In less virulent species, wounds and stomata are targeted, while more virulent species can also penetrate directly (Rotem 1994): enzymatic processes in *Alternaria* infections are essentially similar to those in other diseases. The cuticle, which consists of a combination of cutin (a hydroxyl fatty acid polyester) and waxes, comprises the first line of defence to be overcome by directly penetrating fungal pathogens. In the case of *A. brassicicola*, the differential expression of cutinase genes was monitored between saprophytic and pathogenic stages of the fungus (Yao and Köller 1995), Furthermore, it was found that different cutinolytic enzymes are sequentially induced upon landing on and penetration of the cabbage leaf (Fan and Köller 1998). Constitutively produced cutinases are expressed during the initial contact of *A. brassicicola* with the cuticle. After reaching subcuticular layer, different cutinases that are active during saprophytic growth are induced (Trail and Köller 1993; Yao and Köller 1994). These cutinases are inducible by cutin monomers. This implies a switch between the parasitic and the saprophytic stage of the fungal pathogen. So far, however, it has not been demonstrated that any extracellular hydrolase is crucially involved in fungal pathogenesis.

In addition to cutinases, lipases might also contribute to the establishment of infection. *A. brassicicola* was found to produce a lipase that acts as a virulence factor (Berto et al. 1997). Anti-lipase antibodies were found to have an inhibitory effect on the in vivo infection of cauliflower leaves, as symptom development was inhibited in a dose-dependent manner despite normal germination of spores. However, the antibodies did not have any effect on fungal infection of dewaxed leaves, which suggests that this lipase has an early pathogenic activity during penetration (Berto et al. 1999).

About one-third of the total cell wall components in dicotyledonous plants are pectic polysaccharides. These components can be hydrolyzed by fungal galacturonidases. On the other hand, the tangerine pathotype of *A. alternata* did not show this dependency, possibly because this particular pathogen largely depends on toxin production for the colonization of its host (Isshiki et al. 2001).

For a specific *A. alternata* endoglucanase, it was demonstrated that its production is triggered by a pathogen-induced pH increase on the host (Eshel et al. 2002b). Correlation studies between enzyme production and symptom development suggest that endoglucanases and also exoglucanases are involved in *A. alternata* pathogenicity (Eshel et al. 2000, 2002a, b).

10.6 Toxin Biosynthesis

Most phytotoxins employed by a fungal invader are chemically diverse secondary metabolites: low molecular weight components that are not required for normal growth or reproduction. Often, these toxins are produced as families of related compounds. Based on selectivity, phytotoxins can be divided into two categories: non-host-specific toxins and host-specific toxins. In general, non-host-specific toxins have relatively mild phytotoxic effects, affect a broad spectrum plant species and are thought to be an additional factor of disease for penetration mechanisms and enzymatic processes. Although they generally act as virulence factors and intensify disease symptom severity, they are not absolutely required for establishing disease, since they are also toxic to plant species outside the host range of the pathogen. They merely precondition the host for disease (Ballio 1991).

In *Alternaria*, many non-host-specific toxins have been identified, although the precise action of only a few has been studied in detail. Brefeldin A (dehydrogenase), curvularin, tenuazonic acid, tentoxin and zinniol are examples of toxins that are produced by several *Alternaria* species. They exert their phytotoxic activity through different modes. Brefeldin A causes disassembly of the Golgi complex and acts as an inhibitor of secretion, while curvularin is an inhibitor of cell division through its disturbance of the microtubule assembly. Tenuazonic acid inhibits protein synthesis and zinniol affects membrane permeabilization (Fujiwara et al. 1988; Meronuck et al. 1972; Robeson and Strobel 1981; Thuleau et al. 1988). Tentoxin is produced by *A. alternata* and acts as a photophosphorylation inhibitor through specific binding to chloroplast ATP synthetase, causing the inhibition of ATP hydrolysis and ATP synthesis (Steele *et al.* 1978), because these toxins often target basic cellular processes and are often powerful mycotoxins.

Host-specific toxins are involved in the development of a few, destructive diseases. They generally display severe effects on a rather narrow species range that serves as host to the fungus, and they are indispensable for disease. Along with species of the genus *Cochliobolus* (*Helminthosporium*), *Alternaria* species are known to produce host-specific toxins (Markham and Hille 2001; Wolpert et al. 2002). There are at least 12 host-specific toxin-producing plant pathogenic *Alternaria* species, most of which appear to be variants of *A. alternata*. It was proposed that these variants should be considered pathotypes of *A. alternata* (Nishimura and Kohmoto 1983), a hypothesis that is supported by molecular analysis (Kusaba and Tsuge 1994, 1995, 1997). The toxins that are produced by these pathotypes are chemically diverse, ranging from low molecular weight secondary metabolites to peptides. In a study on pathogenic (host-specific toxin-producing) and non-pathogenic (non-host-specific toxin-producing) *A. alternata* species, it was established that all pathotypes tested carried (small) extra chromosomes, whereas the non-pathogenic ones did not carry these extra chromosomes (Akamatsu et al. 1999). In fungal species, one can regularly find extra chromosomes in certain subsets of individuals, so-called supernumerary chromosomes. These chromosomes are not essential because they are not required for normal growth, but they can carry extra traits. The supernumerary chromosomes confer an adaptive advantage to the individual in some habitats; they are referred to as conditionally dispensable chromosomes. The production of the cyclic AM toxin peptide by the apple pathogen *A. alternata* f. sp. *mali* is determined by such *a* chromosome (Johnson et al. 2000, 2001). The AK and AF toxin biosynthesis genes and their homologues, identified in the Japanese pear and strawberry pathotypes of *A. alternata*, respectively, generally clustered together on a small chromosome (Hatta et al. 2002; Tanaka and Tsuge 2000; Tanaka et al. 1999). Physical clustering is a phenomenon that is commonly found for genes involved in the production of secondary metabolites in fungi (Keller and Hohn 1997).

AK and AF toxins are related to ACT toxin, produced by the tangerine pathotype of *A. alternata*, and carry an epoxydecatrienoic ester backbone. Interestingly, the AF- and ACT-producing strains were also found to be pathogenic on pear lines that are sensitive to AK toxin, but not vice versa (Kohmoto et al. 1993; Maekawa et al. 1984). Four genes involved in the biosynthesis of AK toxin have been cloned from the Japanese pear pathotype genome of *A. alternata*, and homologues of three of these genes have been identified in the strawberry and tangerine pathotypes. This suggests that these homologues are involved in the production of common moieties between these toxins (Hatta et al. 2002; Masunaka et al. 2000; Tanaka and Tsuge 2000; Tanaka et al. 1999). Together with the existence of such homologous toxin biosynthesis genes between pathotypes, the physical clustering of toxin genes on a single chromosome suggests that these genes are acquired through horizontal gene transfer (Tanaka et al. 1999; Walton 2000). This hypothesis is supported by the observation that homologues of toxin biosynthesis genes have not been found in strains from other pathotypes or non-pathogenic *A. alternata* isolates, making it unlikely that mutations or internal genetic rearrangements account for the toxin-producing ability (Hatta et al. 2002; Masunaka et al. 2000; Tanaka et al. 1999). There is no advantage for the clustering of genes during gene transmission from one generation to the next (vertical gene transfer), because the entire genome is transferred as a unit. However, during horizontal gene transfer, a relatively small contiguous DNA sequence is generally transferred. Since the total set of biosynthesis genes should be transmitted from one species to the other in order to provide the recipient with a complete biosynthesis pathway, this chance is higher if genes are clustered. An originally saprophytic fungus like *Alternaria* could have acquired pathogenic capacity in this way (Walton 2000), a hypothesis which is supported by the finding that single pathotype populations do not form monophyletic groups (Kusaba *and* Tsuge 1994, 1995, 1997). The occurrence of horizontal gene transfer is well accepted for prokaryotes, and even for fungal mitochondrial genes, but experimental evidence for the horizontal gene transfer of fungal nuclear genes is scarce. However, the difference between patterns of repeated DNA sequences on certain supernumerary chromosomes and the rest of the chromosomes in the same genome suggests that both types have a different evolutionary history (Covert 1998). Furthermore, it was demonstrated for the fungus *Colletotrichum gloeosporioides* that a small chromosome could be transferred between two genetically isolated strains, thus demonstrating that horizontal gene transfer can occur (He et al. 1998).

10.7 Mode of Action of Host-Specific Toxins

Although the site of action of different *Alternaria* toxins varies, in the end, they all trigger host cell death. AF, ACT and ACTG toxins act at the plasma membrane and cause permeabilization (Otani et al. 1995). AM toxin not only affects the plasma membrane but also acts on chloroplasts, while ACT and AT toxins were found to affect mitochondria (Otani et al. 1998). ACR toxin induces swelling, and other morphological modifications of mitochondria, and increases NADH oxidation, which is followed by plasma membrane disorders leading to electrolyte leakage and necrosis (Akimitsu et al. 1989). For most of these toxins, however, their mechanism of action is barely understood.

AAL toxin is an aminopentol ester, an analogue of the sphingosine precursor sphinganine, which is produced by the tomato pathogen *A. alternata* f. sp. *lycopersici*. This toxin closely resembles the mycotoxin fumonisin B1 that was identified as a *Fusarium moniliforme* toxin and is also produced by AAL toxin-producing *A. alternata* species (Chen et al. 1992). Interestingly, fumonisin B1 is also selectively toxic to AAL toxin-sensitive tomato genotypes (Gilchrist et al. 1992). *Alternaria alternata* mutants affected in AAL toxin production also lose their pathogenicity, demonstrating the requirement of this toxin for pathogenicity (Akamatsu et al. 1997). Like fumonisin B1, AAL toxin is an inhibitor of sphingolipid (ceramide) biosynthesis (Abbas et al. 1994; Spassieva et al. 2002). The application of AAL toxin leads to an accumulation of sphingoid base precursors, a depletion of complex sphingolipids, and subsequently to the cell death of sensitive tomato species (Abbas et al. 1994; Wang et al. 1996b). Similar effects of the toxin are observed on mammalian cells (Wang et al. 1996a). In eukaryotic cells, ceramides modulate a number of biochemical and cellular stress responses such as cell cycle arrest and apoptosis (Hannun 1996). The AAL toxin-induced cell death observed in plant cells shares characteristics with genetically controlled cellular suicide, or apoptosis, as observed in animals, thus suggesting an active participation of the host in AAL toxin-mediated cell death (Wang et al. 1996b). Indeed, transgenic tomato plants expressing a viral anti-apoptotic gene were protected against AAL toxin-induced cell death and pathogen infection (Lincoln et al. 2002). The use of the host's suicide programme as a compatibility factor was not only shown for AAL toxin-producing *A. alternata* strains, but seems to be a general strategy of necrotrophic pathogens (Dickman et al. 2001; Lincoln et al. 2002).

In plants, insensitivity to AAL toxin and resistance to AAL-producing *A. alternata* f. sp. *lycopersici* is determined by the codominant *Alternaria* stem cancer (*Asc-1*) gene, a homologue of a yeast longevity assurance gene of which homologues are also found in nematodes and humans (Brandwagt et al. 2000, 2002; Spassieva et al. 2002). Although the precise mode of action of *Asc-1* has not yet been determined, based on studies with human and nematode homologues in yeast, it is believed that *Asc-1* relieves a toxin-induced block on sphingolipid synthesis through salvage of the ER to Golgi transport of GPI-anchored proteins (Brandwagt et al. 2000).

In addition to a poor knowledge of the mechanism of action of different host-specific toxins, the basis of host specificity is only understood for a few of them. For destruxin B, a host-selective phytotoxin produced by *A. brassicae* (Pedras et al. 2002), the mechanism for host selectivity was recently revealed. The selective phytotoxicity appears to be due to a fast and efficient detoxification, by sequential hydroxylation and glycosylation reactions, in tissues of resistant species. These reactions were also found to occur in susceptible species, but at a slower rate, providing an explanation for selective toxicity (Pedras et al. 2001). Similarly, toxin detoxification has been demonstrated to determine resistance against the host-specific toxin-producing fungus *Cochliobolus carbonum* (Multani et al. 1998).

For ACR toxin, specificity seems to be determined by differential post-transcriptional processing of a mitochondrial gene. This gene is present in the mitochondrial DNA of toxin-sensitive as well as toxin-resistant species, but the transcript of the gene is shorter in resistant than in sensitive mitochondria. In the end, an oligomeric protein is produced in toxin-sensitive mitochondria, whereas the transcript is not translated in resistant mitochondria. Although the gene is likely to encode a mitochondrial membrane protein, no function has been assigned yet (Ohtani et al. 2002).

10.8 Role of Toxin in Host Defence against *Alternaria* species

Destruxin B is a host-specific *Alternaria* phytotoxin as mentioned before, whose selective phytotoxicity is due to the detoxification in tissues of resistant species (Pedras et al. 2001). In addition, the hydroxylation product that is generated during detoxification was found to induce phytoalexin biosynthesis in resistant, but not in susceptible, plant species. This suggests that resistance depends on phytotoxin detoxification and simultaneous phytoalexin elicitation (Pedras et al. 2001). Phytoalexins have also been implicated in pathogen containment for other *Alternaria*/host interactions. In *Camelina sativa* (false flax), resistance against *A. brassicae* has been correlated with production of the phytoalexin camalexin (Jejelowo et al. 1991). Moreover, *Arabidopsis* synthesizes camalexin upon pathogen challenge. Since *Arabidopsis* is amendable to phytopathological studies as well as molecular genetics, it has been established as a model for dissecting plant defence mechanisms. The *Arabidopsis pad3* mutant (phytoalexin deficient) was found to be deficient in camalexin biosynthesis and to display severely enhanced susceptibility to *A. brassicicola*, demonstrating that camalexin contributes to *Alternaria* resistance (Thomma et al. 1999b). Although camalexin was found to exhibit direct antimicrobial activity to this fungus in vitro (Thomma et al. 1999b), it cannot be excluded that camalexin also contributes to *Alternaria* resistance in an indirect way, as camalexin was found to inhibit production of the host-specific toxin destruxin B in *A. brassicae* (Pedras et al. 1998).

In a study of 24 different *Arabidopsis* ecotypes, it was not possible to discover a direct relationship between camalexin production upon *A. brassicicola* inoculation on one hand and resistance against this pathogen on the other (Kagan and Hammerschmidt 2002) This is perhaps not surprising, since defence against *A. brassicicola* in *Arabidopsis* appears to depend on multiple defence components, the relative importance of which can vary between ecotypes. Previously, jasmonate has also been implicated in defence against *Alternaria*, as the jasmonate-insensitive *Arabidopsis* mutant *coi1* was found to display enhanced susceptibility to this pathogen (Thomma et al. 1998, 1999a, 2000). Jasmonate is a plant hormone that, along with salicylic acid and ethylene, plays a role in the activation of defence responses in plants (Thomma et al. 2001). However, mutants affected in either salicylic acid or ethylene signalling is as resistant as wild-type plants, leading to the conclusion that neither salicylic acid nor ethylene is directly involved in resistance against *Alternaria* in *Arabidopsis* (Thomma et al. 1998, 1999a). Jasmonate was found not to directly induce camalexin biosynthesis, and in addition, jasmonate insensitivity does not lead to camalexin deficiency. It was therefore concluded that both camalexin and jasmonate mediate host resistance against *Alternaria* through separate but parallel pathways in *Arabidopsis* (Thomma et al. 1998, 1999b; 2001). Further, support for this observation came from the finding that the *Arabidopsis* mutation *esa1* affects resistance against *A. brassicicola* through a severe reduction in both camalexin production and jasmonate-dependent gene induction (Tierens et al. 2002). It is currently not clear, however, which jasmonate-inducible effect or molecules are responsible for containment of *Alternaria* in *Arabidopsis.*

In a cDNA microarray analysis on a set of 2375 *Arabidopsis* genes, 168 genes were found to be upregulated and 39 were found to be repressed 72 h after inoculation with *A. brassicicola.* About 10 % of these 207 genes can be implicated in plant defence and about one-third in cell maintenance and development (Schenk et al. 2000). However, a precise role for any of the responsive genes in the defence of *Arabidopsis* against *A. brassicicola* has not yet been established. Studies, using ethylene- and jasmonate-insensitive

Arabidopsis mutants, concluded that a set of PR genes encoding plant defensin, and hevein-like protein, and basic chitinase are not markedly involved in *Alternaria* resistance (Thomma et al. 1998, 1999a). In tomato, PR proteins have been implicated in *Alternaria* resistance through the observation that resistance in tomato against *A. solani* correlates with high and rapid accumulation of PR proteins. However, it is speculated that those PR proteins, many of which possess hydrolytic activity, are not necessarily directly involved in arresting the pathogen, but rather release elicitors from the pathogen cell wall, thus triggering a HR (Lawrence et al. 2000).

The production of reactive oxygen species (ROS) is one of the earliest defence reactions of plants against pathogen attack. In the presence of tobacco plant extracts, *A. alternata* was found to synthesize mannitol, which can act as a potent quencher of reactive oxygen species. This suggests that *Alternaria* uses the active oxygen quencher mannitol to overcome host defence. On the other hand, pathogen attack was found to induce mannitol dehydrogenase expression in tobacco, suggesting that plants counter this fungal defence system by catabolizing fungal mannitol (Jennings et al. 2002). Support for this hypothesis also came from the fact that the constitutive expression of a mannitol dehydrogenase gene in tobacco confers resistance against *A. alternata*, but not against non-mannitol-secreting fungi (Jennings et al. 2002).

Using a transgenic approach, enhanced levels of resistance against *Alternaria* species were also obtained in a number of cases. It was previously discussed that the expression of anti-apoptotic genes provides disease resistance, against *Alternaria* species, among others, through blocking a programmed cell death response (Dickman et al. 2001; Lincoln et al. 2002). Constitutive expression of a radish defensin gene, a rubber tree chitin-binding lectin and a human lysozyme, respectively, conferred enhanced resistance against *Alternaria* spp. in tobacco, Indian mustard and carrot (Kanrar et al. 2002; Takaichi and Oeda 2000; Terras et al. 1995). In addition, constitutive expression of an endochitinase gene isolated from the biocontrol fungus *Trichoderma harzianum* has resulted in enhanced *Alternaria* resistance in transgenic tobacco, potato and broccoli (Lorito et al. 1998; Mora and Earle 2001). In addition, the treatment of plants with biocontrol organisms has in some cases resulted in enhanced resistance against *Alternaria* pathogens (Morita et al. 2003; Ton et al. 2002).

Many aspects of the interaction between *Alternaria* species and their respective hosts are currently under investigation. The function and biosynthesis of host-specific toxins, melanins in particular, have gained much attention, and considerable advances have been made. Finally, although the general strategies employed by the pathogenic strains to attack their hosts are similar, the responses of the various hosts against these attacks are very diverse (Thomma 2003).

White mustard cells can transform a very toxic compound called destruxin B (^{14}C labelled), produced by the microorganism that causes the plant disease, into a non-toxic compound called hydroxydestruxin B. Hydroxydestruxin B (^{14} C labelled) was further biotransformed to the β-d-glucosyl derivative at slower rates. The structures of hydroxydestruxin B and β-D-glucosyl hydroxydestruxin B were deduced from their spectroscopic data (NMR, HRMS, FTIR) and confirmed by total chemical synthesis (Fig. 10.3). Although these hydroxylation and glucosylation reactions occurred in both resistant (*S. alba*) and susceptible (*B. napus*, *B. juncea* and *B. rapa*) species, hydroxylation was the rate-limiting step in the susceptible, whereas glucosylation was the rate-limiting step in the resistant species. Remarkably, it was observed that the hydroxyl destruxin B induced the biosynthesis of phytoalexins in black spot resistant, but not in susceptible species. This appears to be a unique example of phytotoxin detoxification and simultaneous phytoalexin elicitation by the detoxification product. *Sinapis alba* can overcome the fungal invader through detoxification of destruxin B coupled with production of phytoalexins (Figs. 10.4 and 10.5). Since the hydroxylation reaction happens at a very fast rate and it leads to a less toxic product, it may prevent the plant cells from being killed by the pathogen. Canola plants can become resistant to the black spot disease

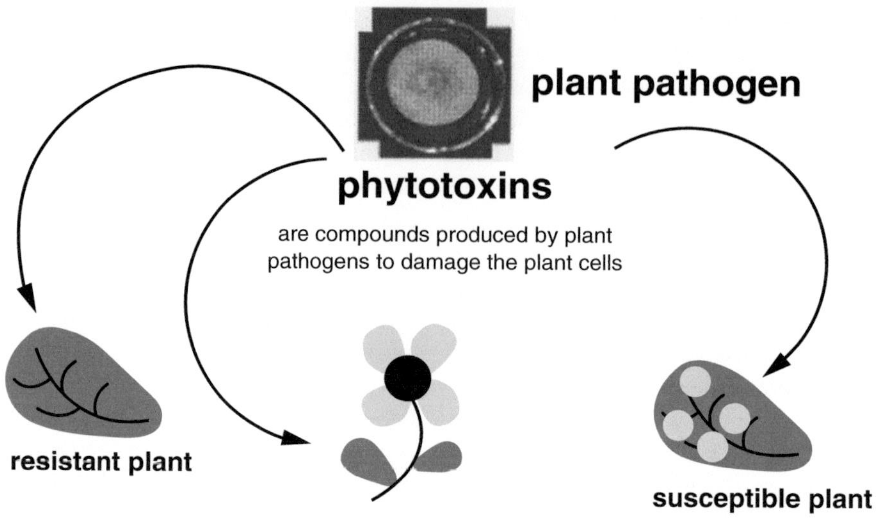

Fig. 10.3 Phytotoxin are produced by microbial plant pathogens

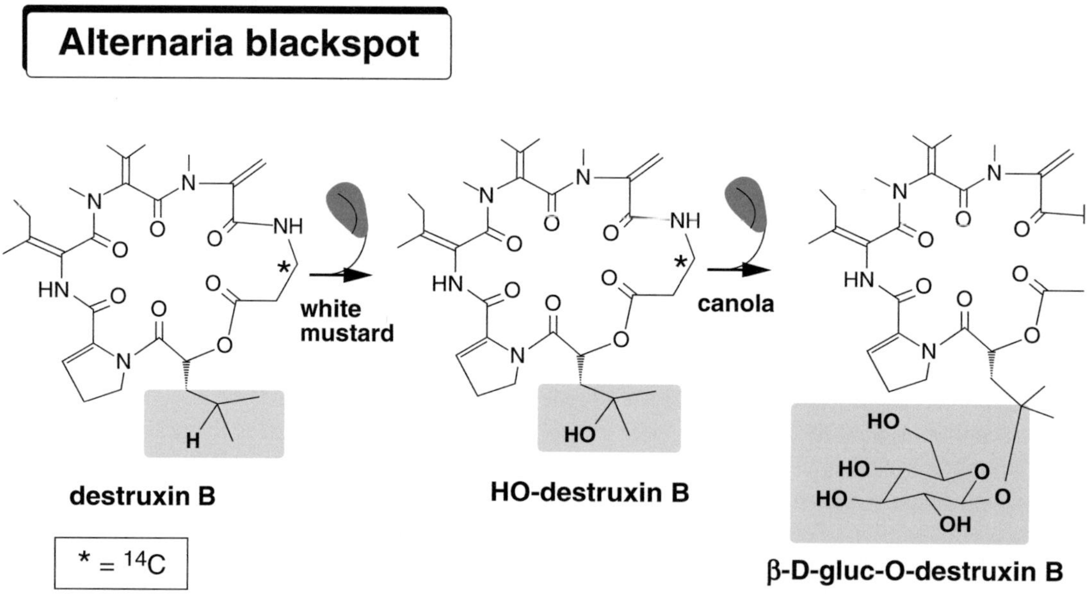

Fig. 10.4 Enzymatic reactions of plants to the pathogen causing *Alternaria* black spot

upon incorporation of this detoxification trait (Pedras et al. 1999, 2001).

In unique interactions, three different rapeseed pathogens convert the phytoalexin cyclobrassinin into three different phytoalexins, brassilexin, dioxibrassinin and brassicanal A, and ultimately each phytoalexin into non-toxic products. Such pathogen may have acquired a more effective mechanism for overcoming this chemical by adopting phytoalexin biosynthetic pathways operating in planta. This strategy appears quite plausible, especially considering that most fungal pathogens have been coevolving with plants for numerous generations (Pedras and Okanga 1999; Fig. 10.6). Each detoxification pathway is (i) specific to a group of pathogens and (ii) the enzymes

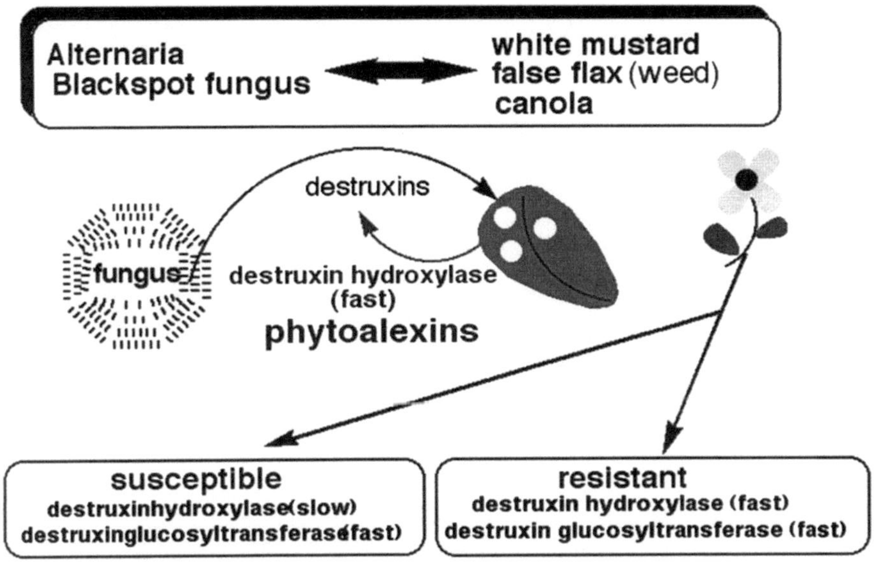

Fig. 10.5 Defence responses of plants resistant and susceptible to *Alternaria* black spot

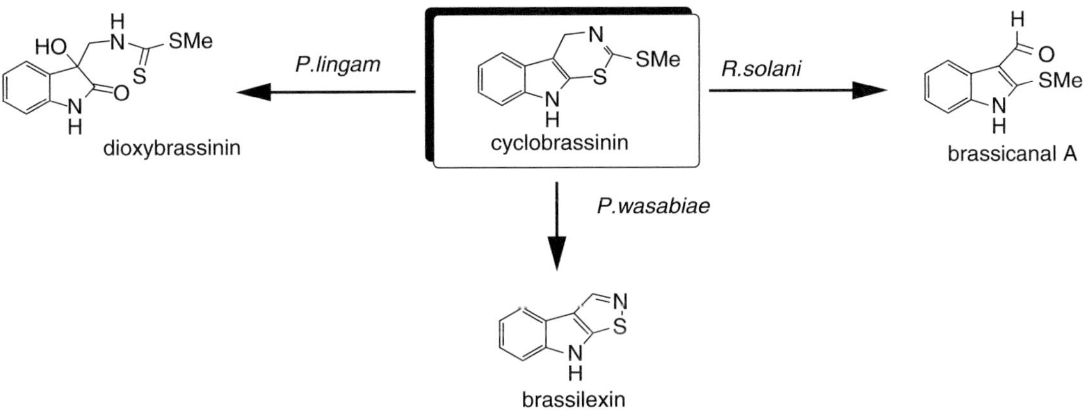

Fig. 10.6 Transformation of the phytoalexin cyclobrassinin by different pathogens of *Brassica*

involved in these transformations are quite selective (Pedras 1998; Pedras et al. 2000).

Pedras et al. (2009) have established that A. *brassicicola* produces brassicicolin A (1) (a mixture of epimers, Gloer et al. 1988), as the major host-selective phytotoxin. Brassicicolin A (1) appeared to be more phytotoxic to the susceptible cruciferous species *B. juncea* than to the tolerant *B. napus*. Previous claims that *A. brassicicola* produced destruxins (Cooke et al. 1997; Evans et al. 1996) could not be confirmed; in spite of the availability of several destruxins in metabolite library (Pedras et al. 2000), none of these phytotoxins could be detected in either extracts of liquid cultures or in infected leaf tissues. In addition, the four polyketides, phomapyrones A (3), F (4) and G (5) and infectopyrone (6), previously isolated from *Leptosphaeria maculans* and *L. biglobosa* (Pedras et al. 1994; Pedras and Biesenthal 2001; Pedras and Chumala 2005), were also isolated from *A. brassicicola*. Clones of *A. brassicicola* (altr203 x n03 and altrO14 x eO1) were recently reported (Cramer et al. 2006) to encode enzymes involved in the biosynthesis of undetermined polyketides, which could be the phomapyrones 3 and 5. In this context, it is pertinent to point out that genome sequences currently available in public databases have indicated the existence of numerous 'orphan pathways', i.e., gene clusters for which the encoded natural product is unknown (Gross 2007). Therefore, it is expected that further new metabolites and/or phytotoxins could potentially be isolated from fungal species previously investigated, such as *A. brassicicola* (Pedras et al. 2009).

Phytoalexin production in leaves of *B. juncea* infected with *A. brassicicola* led to the detection and isolation of two known phytoalexins and a phytoanticipin: cyclobrassinin (11) and spirobrassinin (12) and indolyl-3-acetonitrile (15), respectively. Since indolyl-3-acetonitrile (15) was occasionally detected in control leaves, it is most likely a phytoanticipin. In addition, other minor phytoalexins (13 and 14) and phytoanticipins (16 and 17) were detected in fractions of leaf extracts. Since the product of detoxification of brassinin (10), and *N*′-acetyl-3-indolylmethanamine (19), was also isolated from infected leaves, but never detected in control leaves, it is suspected that brassinin (10) does not accumulate in sufficient quality in infected tissues because it is quickly detoxified to 19, similar to the detoxification that occurs in *A. brassicicola* cultures. To support this hypothesis, Pedras et al. (2009) determined that *A. brassicicola* detoxifies brassinin (10) to indolyl-3-methanamine (18) and N′-acetyl-3-indolylmethanamine (19) in fungal cultures. Although brassinin (10) is not a strong antifungal against *A. brassicicola*, this phytoalexin is a precursor of brassilexin (14) and other phytoalexins of *B. juncea* (Pedras et al. 2007). It is very probable that its detoxification makes the plant more vulnerable to fungal invasion, and the brassinin-detoxifying enzymes produced by *A. brassicicola* might contribute to its virulence. The comparison of the in vitro antifungal activity of the phytoalexins present in infected tissues of *B. juncea* and that of camalexin (20, not produced in *Brassica* species) suggests that camalexin might be a more effective plant defence against *A. brassicicola*.

10.9 In Silico Protein–Protein Interaction

Since cell perceives the signal through the cell surface receptors including receptor-like kinases (RLK), receptor-like proteins (RLP), and receptor-like extracellular binding proteins, they are known to interact with MAMPs (microbe-associated molecular pattern) or PAMPS (pathogen-associated molecular pattern) and effectors. Receptor-like kinases interact with pathogen-associated molecular pattern, which are small peptides like chitin oligosaccharide (Hamel and Beaudoin 2010) and flagellin (Pitzschke et al. 2009). Destruxin B, which is a small peptide, produced by *A. brassicae*, is found to interact with Lys M receptor-like kinase and WAK receptor through in silico interaction. The results of docking interaction of destruxin B with different receptor-like kinases from different species indicate that Lys M receptor-like kinase in *Arabidopsis* showed better interaction with it. A mutation in Lys M receptor-like protein (Lys M RLK I) gene blocked the induction of almost all chito-oligosaccharide-responsive genes and led to increased susceptibility to fungal pathogens indicating that Lys M RLK1 is essential for chitin signalling in plants (Wan et al. 2008).

Arabidopsis genome has more than 80 MAPKKKs, but very few members of MAPKKKs have been studied. To continue the downstream cascade and to take part in the defence, docking was performed with MAPKKKs (MAPKKK I, MAPKKK 2, MAPKKK I2 and MAPKKK 19) and receptor kinases which showed better interaction with destruxin B. It was found that MAPKKK 19 showed better interaction with Lys M receptor-like kinase (*Ricinus communis*); WAK receptor-like kinase (*A. thaliana*) and MAPKKK 2 were also found to interact better with CHRK 1 receptor-like kinase (*Senecio squalidus*).

The activity of MAPK 4 has been identified in plant defence against biotic stress (Desikan et al. 2001). Out of all 10 MAPKKs known, the activity of some MAPKKs in plant defence has already been identified against biotic stress, viz. MAPKK 3 and MAPKK 5 (Asai et al. 2002). Docking results of all 10 MAPKKs with MAPK 4 showed that MAPKK 5, MAPKK 8, MAPKK 3 and MAPKK 9 have better interaction with MAPK 4. There are several evidences in the literature that showed MAPKK 3, MAPKK 5, MAPKK 1 and MAPKK 2 are the upstream kinase of MAPK 4 (Popescu et al. 2009; Andreasson and Ellis 2010) and MAPKK 9 is the upstream kinase of MAPK 3/MAPK 6 (Tena et al. 2011). According to the superimposition results, MAPKK 9 and MAPKK 5 also belong to the same group C. So it could be hypothesized that MAPKK 9 might be the upstream kinase not only for MAPK 3/MAPK 6 but also for MAPK 4. Recently one group also tried to determine the interaction of all MAPKs with all upstream MAPKKs through functional protein microarrays and found that MAPK 4 interacts with MAPKK 1, MAPKK 2, MAPKK 3, MAPKK 5 and MAPKK 7 (Popescu et al. 2009). Tena et al. (2011) showed that activation of MAPK 4 cascade is required for sustainable activation of defence pathway. Since our docking results do not show the better interaction of MAPK 4 with MAPKK1/MAPKK 2, it implies that this interaction occurs through scaffold protein or after post-translational modification of the protein. At the early stage of infection of *Alternaria* blight, all three MAPKs (MAPK 3, MAPK 6 and MAPK 4) were expressed. It should be further investigated by the wet lab experiment if all these three kinases (MAPK 3, MAPK 4 and MAPK 6) are activated by specific MAPKKs or by a common MAPKK. To determine the upstream members of MAPKKs in the signalling module, the docking was again performed with the same members of MAPKKKs family which have already been docked with receptor-like kinases, and results suggested that MAPKK9 is interacting with MAPKKK 2/MAPKKK I9. To evaluate the downstream targets of MAPK 4, the docking performed with some WRKY transcription factors in plant defence, viz. WRKY 21, 25, 29, 33, 40, 53 and 69 (Popsecu et al. 2008; Tena et al. 2011; Asai et al. 2002; Fill et al. 2009; Cristina et al. 2010; Anderson et al. 2005), indicated that WRKY 25 and 40 could be the better interacting partner of MAPK 4. Several molecular and biochemical studies have shown that MAPK 4 interacts with all above-listed WRKY proteins, except WRKY 29, a downstream partner of MAPK 3/MAPK6; it can be assumed that this might be due to the specificity provided by the scaffold proteins. The best-studied example of scaffold protein is interaction of MAPK 4 with WRKY 33 via MKS I protein (Fill et al. 2009). By protein microarray and Y2H studies, WRKY 40 and WRKY 25 were reported to be the downstream interacting protein of MAPK 4 (Popsecue et al. 2008; Anderson et al. 2005). Superimposition results also group WRKY 40 in a same group which is also supported by the docking results. Marmath et al. (2013) described the role of MAPK 4 in plant defence against *Alternaria* blight and predicted some potential candidate for MAPK 4 cascade. Observations lead to propose the model of MAPK cascade involving MAPK 4 in the pathogenesis of *Alternaria* blight disease.

References

Abbas HK, Tanaka T, Duke SO, Porter JK, Wray EM, Hodges L, Sessions AE, Wang E, Merrill AH Jr, Riley RT (1994) Fumonisin-and AAL-toxin-induced disruption of sphingolipid metabolism with accumulation of free sphingoid bases. Plant Physiol 106:1085–1093

Abdel-Farid IB, Jahangir M, van den Hondel CAMJJ, Kim HK, Choi YH, Verpoorte R (2009) Fungal infection-induced metabolites in *Brassica rapa*. Plant Sci 176:608–615

Agarwal A, Garg GK, Singh US, Mishra DP (1994) Detection and role of chlorotic toxin and phytohormone in the pathogenesis of Alternaria blight in *Brassica napus*. Curr Sci 66:442–443

Agarwal A, Garg GK, Devi S, Mishra DP, Singh US (1997) Ultra-structural changes in *Brassica* leaves caused by *Alternaria brassicae* and destruxin B. J Plant Biochem Biotechnol 6:25–28

Akamatsu H, Itoh Y, Kodama M, Otani H, Kohmoto K (1997) AAL-toxin-deficient mutants of *Alternaria alternata* tomato pathotype by restriction enzyme-mediated integration. Phytopathology 87:967–972

Akamatsu H, Taga M, Kodama M, Johnson R, Otani H, Kohmoto K (1999) Molecular karyotypes for *Alternaria* plant pathogens known to produce host-specific toxins. Curr Genet 35:647–656

Akimitsu K, Kohmoto K, Otani H, Nishimura S (1989) Host-specific effects of toxin from the rough lemon pathotype of *Alternaria alternata* on mitochondria. Plant Physiol 89:925–931

Amselem J, Cuomo CA, van Kan JAL, Viaud M, Benito EP, Couloux A, Coutinho PM, de Vries RP, Dyer PS,

Fillinger S, Fournier E, Gout L, Hahn M, Kohn L, Lapalu N, Plummer MK, Pradier JM, Quevillon E, Sharon A, Simon A, Have A, Tudzynski B, Tudzynski P, Wincker P, Andrew M, Anthouard V, Beever RE, Beffa R, Benoit I, Bouzid O, Brault B, Chen Z, Choquer M, Collemare J, Cotton P, Danchin EG, Da Silva C, Gautier A, Giraud C, Giraud T, Gonzalez C, Grossetete S, Guldener A, Henrissat B, Howlett BJ, Kodira C, Kretschmer M, Lappartient A, Leroch M, Levis C, Mauceli E, Neuveglise C, Oeser B, Pearson M, Poulain J, Poussereau N, Quesneville H, Rascle C, Schumacher J, Segurens B, Sexton A, Silva E, Sirven C, Soanes DM, Talbot NJ, Templeton M, Yandava C, Yarden O, Zeng Q, Rollins JA, Lebrun MH, Dickman M (2011) Genomic analysis of the necrotrophic fungal pathogens *Sclerotinia sclerotiorum* and *Botrytis cinerea*. PLoS Genet 7, e1002230. doi:10.1371/journal.pgen.1002230

Anderson DR, Hegde S, Reinhard E, Gomez L, Vernier WF, Lee L, Liu S, Sambandam A, Snider PA, Masih L (2005) Aminocyanopyridine inhibitors of mitogen activated protein kinase-activated protein kinase 2 (MK-2). Bioorg Med Chem Lett 15:1587–1590

Andreasson E, Ellis B (2010) Convergence and specificity in the *Arabidopsis* MAPK nexus. Trends Plant Sci 15:106–113

Asai T, Tena G, Plotnikova J, Willmann MR, Chiu WL, Gomez-Gomes L, Boller T, Ausubel FM, Sheen J (2002) MAP kinase signaling cascade in *Arabidopsis* innate immunity. Nature 415:977–983

Ayer WA, Pena-Rodriguez LM (1987a) Metabolites produced by *Alternaria brassicae*, the black spot pathogen of Canola, Part 2, Sesquiterpenoid metabolites. J Nat Prod 50:408–417

Ayer WA, Pena-Rodriguez LM (1987b) Metabolites produced by *Alternaria brassicae*, the black spot pathogen of canola. J Nat Prod 50:400–407

Ayer WA, Bains PS, Pena-Rodriguez LM, Tewari JP (1987) Production of a host specific phytotoxin by *Alternaria brassicae*. In: Proceedings of the 7th international rapeseed congress, vol 257. Poznan, Poland, (Abstr.), pp 1256–1261

Bains PS (1989) Purification, chemical characterization, host-specificity, bioassay, mode of action, and herbicidal use of the toxin produced by *Alternaria brassicae*. Ph.D. thesis, University of Alberta, Edmonton, Alberta, Canada, 137 p

Bains PS, Tewari JP (1986) Isolation and characterization of the toxin produced by *Alternaria brassicae*. In: Proceedings of crucifer genetic workshop 3, University of Guelph, Canada. Crucifer Genetic Cooperative, 66 p

Bains PS, Tewari JP (1987) Purification, chemical characterization and host specificity of the toxin produced by *Alternaria brassicae*. Physiol Mol Plant Pathol 30:259–271

Bains PS, Tewari JP (1989) Bioassay and mode of action of an *Alternaria brassicae* toxin. Can J Plant Pathol 11:184–189

Bains PS, Tewari JP, Ayer WA (1993) A note on phytotoxicity of homodestruxin B- a compound produced by *Alternaria brassicae*. Phytopathology 74:157–160

Bakar FDA, Murad AMA, Hamid AA, Zamrod Z, Mahadi NM, Sullivan P (2005) Induction and expression of cutinase activity during saprophytic growth of the fungal plant pathogen, *Glomerella cingulata*. Asia Pacific J Mol Biol Biotechnol 13:63–69

Ballio A (1991) Non-host-selective fungal pathogens: biochemical aspects and their mode of action. Experientia 47:783–790

Berto P, Belingheri L, Dehorter B (1997) Production and purification of a novel extracellular lipase from *Alternaria brassicicola*. Biotechnol Lett 19:533–536

Berto P, Comménil P, Belingheri L, Dehorter B (1999) Occurrence of a lipase in spores of *Alternaria brassicicola* with a crucial role in the infection of cauliflower leaves. FEMS Microbiol Lett 180:183–189

Beisson F, Koo AJK, Ruuska S, Schwender J, Pollard M, Thelen JJ, Paddock T, Salas JJ, Savage L, Milcamps A, Mhaske VB, Cho Y, Ohlrogge JB (2003) *Arabidopsis* genes involved in acyl lipid metabolism. A 2003 census of the candidates, a study of the distribution of expressed sequence tags in organs, and a web-based database. Plant Physiol 132:681–697

Boller T (1995) Chemoperception of microbial signals in plant cells. Annu Rev Plant Physiol Plant Mol Biol 46:189–214

Brandwagt BF, Mesbah LA, Takken FLW, Laurent PL, Kneppers TJA, Hille J, Nijkamp HJJ (2000) A longevity assurance gene homolog of tomato mediates resistance to *Alternaria alternata* f. sp. *lycopersici* toxins and fumonisin B-1. Proc Natl Acad Sci U S A 97:4961–4966

Brandwagt BF, Kneppers TJA, Nijkamp HJJ, Hille J (2002) Over expression of the tomato Asc-1 gene mediates high insensitivity to AAL toxins and fumonisin B-1 in tomato hairy roots and confers resistance to *Alternaria alternata* f. sp. *lycopersici*. *Nicotiana umbratica* plants. Mol Plant-Microbe Interact 15:35–42

Buchwaldt L, Green H (1992) Phytotoxicity of destruxin B and its possible role in the pathogenesis of *Alternaria brassicae*. Plant Pathol 41:55–63

Buchwaldt L, Jensen JS (1991) HPLC purification of destruxins produced by *Alternaria brassicae* in culture and in leaves of *Brassica napus*. Assignment of the IH-13-NMR spectra by ID-and-2D techniques. Photochemistry 30:2311–2316

Buchwaldt L, Jensen JS, Green H (1991) HPLC purification of destruxins from *Alternaria brassicae* and examination of their role in pathogenesis. Can J Plant Pathol 13:272 (Abstr.)

Cameron KD, Teece MA, Smart LB (2006) Increased accumulation of cuticular wax and expression of lipid transfer protein in response to periodic drying events in leaves of tree tobacco. Plant Physiol 140:176–183

Chaturvedi C (1972) A new leaf spot of Poplar incited by *Alternaria alternata* (Fr.) Keissler. Indian Phytopathol 25:316–318

Chen J, Mirocha CJ, Xie W, Hogge L, Olson D (1992) Produc-tion of the mycotoxin fumonisin B1 by *Alternaria alternata* f.sp. lycopersici. Appl Environ Microbiol 58:3928–3931

Chen LY, Price TV, Silvapulle MJ (2005) Dark leaf spot (*Alternaria brassicicola*) on Chinese cabbage: spatial patterns. Aust J Agric Res 56:699–714

Chhikara S, Chaudhury D, Dhankher OP, Jaiwal PK (2012) Combined expression of a barley class II chitinase and type I ribosome inactivating protein in transgenic *Brassica juncea* provides protection against *Alternaria brassicae*. Plant Cell Tissue Org Cult 108:83–89

Cho Y, Kim KH, La Rota M, Scott D, Santopietro G, Callihan M, Mitchell TK (2009) Identification of virulence factors by high throughput targeted gene deletion of regulatory genes in *Alternaria brassicicola*. Mol Microbiol 72:1316–1333

Chung KR (2012) Stress response and pathogenicity of the necrotrophic fungal pathogen *Alternaria alternata*. Scientifica. http://dx.doi.org/10.6064/2012/635431

Cobb MH (1999) MAP kinase pathways. Prog Biophys Mol Biol 71:479–500

Cooke DEL, Jenkins PD, Lewis DM (1997) Production of phytotoxic spore germination liquids by *Alternaria brassicae* and *A. brassicicola* and their effect on species of the family Brassicaceae. Ann Appl Biol 131:413–426

Covert SF (1998) Supernumerary chromosomes in filamentous fungi. Curr Genet 33:311–319

Cramer R, Lawrence C (2004) Identification of *Alternaria brassicicola* genes expressed *in planta* during pathogenesis of *Arabidopsis thaliana*. Fungal Genet Biol 41:115–128

Cramer RA, La Rota CM, Cho Y, Thon M, Craven KD, Knudson DL, Mitchell TK, Lawrence CB (2006) Bioinformatic analysis of expressed sequence tags derived from a compatible *Alternaria brassicicola–Brassica oleracea* interaction. Mol Plant Pathol 7:113–124

Cristina M, Rodriguez S, Petersen M, Mundy J (2010) Mitogen activated protein kinase signaling in plants. Annu Rev Plant Biol 61:621–649

Daayf F, Platt HW (2000) Changes in metalaxyl resistance among glucose phosphate isomerase genotypes of *Phytophthora infestans* in Canada during 1997–1998. Am J Potato Res 77:311–318

de Zelicourt A, Montiel G, Pouvreau JB, Thoiron S, Delgrange S, Simier P, Delavault P (2009) Susceptibility of *Phelipanche* and *Orobanche* species to AAL-toxin. Planta 230:1047–1055

Desikan R, Clarke A, Hancock JT, Neill SJ (1999) H_2O_2 activates a MAP kinase-like enzyme in *Arabidopsis thaliana* suspension cultures. J Exp Bot 50:1863–1866

Desikan RS, Mackerness AH, Hancock JT, Neill SJ (2001) Regulation of the *Arabidopsis* transcriptome by oxidative stress. Plant Physiol 127:159–172

Dickman MB, Park YK, Oltersdorf T, Li W, Clemente T, French R (2001) Abrogation of disease development in plants expressing animal antiapoptotic genes. Proc Natl Acad Sci U S A 98:6957–6962

Doares SH, Syrovets T, Weiler EW, Ryan CA (1995) Oligo galacturonides and chitosan activate plant defensive genes through the octadecanoid pathway. Proc Natl Acad Sci U S A 92:4095–4098

Epple P, Apel K, Bohlmann H (1995) An *Arabidopsis thaliana* thionin gene is inducible via a signal transduction pathway different from that for pathogenesis-related proteins. Plant Physiol 109:813–820

Eshel D, Ben-Arie R, Dinoor A, Prusky D (2000) Resistance of gibberellin-treated persimmon fruit to *Alternaria alternata* arises from the reduced ability of the fungus to produce endo-1,4-β-glucanase. Phytopathology 90:1256–1262

Eshel D, Lichter A, Dinoor A, Prusky D (2002a) Characterization of *Alternaria alternata* glucanase genes expressed during infection of resistant and susceptible persimmon fruits. Mol Plant Pathol 3:347–358

Eshel D, Miyara I, Ailing T, Dinoor A, Prusky D (2002b) pH regulates endoglucanase expression and virulence of *Alternaria alternata* in persimmon fruit. Mol Plant-Microbe Interact 15:774–779

Evans N, Mc Roberts N, Hill RA, Marshall G (1996) Phytotoxin production by *Alternaria linicola* and phytoalexin production by the linseed host. Ann Appl Biol 129:415–431

Fan CY, Köller W (1998) Diversity of cutinases from plant pathogenic fungi: differential and sequential expression of cutinolytic esterases by *Alternaria brassicicola*. FEMS Microbiol Lett 158:33–38

Fill BK, Petersen K, Petersen M, Mundy J (2009) Gene regulation by MAP kinase cascades. Curr Opin Plant Biol 12:615–621

Fujiwara T, Oda K, Yokota S, Takatsuki A, Ikehara Y (1988) Brefeldin A causes disassembly of the golgi complex and accumulation of secretary proteins in the endoplasmic reticulum. J Biol Chem 263:18545–18552

Garg S, Dhawan K, Chawla HKL, Nainawatee HS (1999) *Alternaria brassicae* induced changes in tile activity of cell wall degrading enzymes in leaves of *Brassica juncea*. Biol Plants 30:387–392

Gechev TS, Hille J (2005) Hydrogen peroxide as a signal controlling plant programmed cell death. J Cell Biol 168:17–20

Gechev TS, Gadjev IZ, Hille J (2004) An extensive microarray analysis of AAL-toxin-induced cell death in *Arabidopsis thaliana* brings new insights into the complexity of programmed cell death in plants. Cell Mol Life Sci 61:1185–1197

Gechev T, Ferwerda M, Mehterov N, Laloi C, Qureshi MK, Hille J (2008) *Arabidopsis* AAL-toxin-resistant mutant *atr1* shows enhanced tolerance to programmed cell death induced by reactive oxygen species. Biochem Biophys Res Commun 375:639–644

Ghose K, Dey S, Barton H, Loake GJ, Basu D (2008) Differential profiling of selected defense-related genes induced on challenge with *Alternaria brassicicola* in resistant white mustard and their comparative expression pattern in susceptible Indian mustard. Mol Plant Pathol 9:763–775

Gilchrist DG, Ward B, Moussato V, Mirocha CJ (1992) Genetic and physiological response to fumonisins and AAL-toxin by intact tissue of a higher plant. Mycopathologia 117:57–64

Glazebrook J (1999) Genes controlling expression of defense response in *Arabidopsis*. Curr Opin Plant Biol 2:280–286

Gloer JB, Poch GK, Short DM, McCloskey DV (1988) Structure of brassicicolin A: a isocyanide antibiotic from the phylloplane fungus *Alternaria brassicicola*. J Org Chem 53:3758–3761

Gross H (2007) Strategies to unravel the function of orphan biosynthesis pathways: recent examples and future prospects. Appl Microbiol Biotechnol 75:267–277

Gupta S, Roberts DW, Renwick JAA (1989) Preparative isolation of destruxins from *Metarhizium anisopliae* by high performance liquid chromatography. J Liq Chromatogr 12:383–395

Hamel LP, Beaudoin N (2010) Chit oligosaccharide sensing and downstream signaling: contrasted outcomes in pathogenic and beneficial plant–microbe interactions. Planta 232:787–806

Hammond-Kosack KE, Jones JDG (1996) Resistance gene-dependent plant defense responses. Plant Cell 8:1773–1791

Hannun YA (1996) Functions of ceramide in coordinating cellular responses to stress. Science 274:1855–1859

Hatta R, Ito K, Hosaki Y, Tanaka T, Tanaka A, Yamamoto M, Akimitsu K, Tsuge T (2002) A conditionally dispensable chromosome controls host-specific pathogenicity in the fungal plant pathogen *Alternaria alternata*. Genetics 161:59–70

He C, Rusu AG, Poplawski AM, Irwin JA, Manners JM (1998) Transfer of a supernumerary chromosome between vegetatively incompatible biotypes of the fungus *Colletotrichum gloeosporioides*. Genetics 150:1459–1466

Hodgkin T, MacDonald MV (1986) The effect of a phytotoxin from *Alternaria brassicicola* on *Brassica* pollen. New Phytol 104:631–636

Hwang IS, Hwang BK (2010) The pepper 9-lipoxygenase gene CaLOX1 functions in defense and cell death responses to microbial pathogens. Plant Physiol 152:948–967

Imazaki A, Tanaka A, Harimoto Y, Yamamoto M, Akimitsu K, Park P, Tsuge T (2010) Contribution of peroxisomes to secondary metabolism and pathogenicity in the fungal plant pathogen *Alternaria alternata*. Eukaryot Cell 9:682–694

Isshiki T, Pearson B, Holbrook S, Doe CQ (2001) *Drosophila neuroblasts* sequentially express transcription factors which specify the temporal identity of their neuronal progeny. Cell 106:511–521

Jejelowo OA, Conn KL, Tewari JP (1991) Relationship between conidial concentration, germling growth and phytoalexin production by *Camelina sativa* leaves inoculated with *Alternaria brassicae*. Mycol Res 95:928–934

Jennings DB, Daub ME, Pharr DM, Williamson JD (2002) Constitutive expression of a celery mannitol dehydrogenase in tobacco enhances resistance to the mannitol-secreting fungal pathogen *Alternaria alternata*. Plant J 32:41–49

Johnson RD, Johnson L, Itoh Y, Kodama M, Otani H, Kohmoto K (2000) Cloning and characterization of a cyclic peptide synthetase gene from *Alternaria alternata* apple pathotype whose product is involved in AM-toxin synthesis and pathogenicity. Mol Plant-Microbe Interact 13:742–753

Johnson LJ, Johnson RD, Akamatsu H, Salamiah A, Otani H, Kohmoto K, Kodama M (2001) Spontaneous loss of a conditionally dispensable chromosome from the *Alternaria alternata* apple pathotype leads to loss of toxin production and pathogenicity. Curr Genet 40:65–72

Joubert A, Simoneau P, Campion C, Bataille-Simoneau N, IacomiVasilescu B, Poupard P, François JM, Georgeault S, Sellier E, Guillemette T (2011) Impact of the unfolded protein response on the pathogenicity of the necrotrophic fungus *Alternaria brassicicola*. Mol Microbiol 79:1305–1324

Kachroo P, Shanklin J, Shah J, Whittle EJ, Klessig DF (2001) A fatty acid desaturase modulates the activation of defense signaling pathways in plants. Proc Natl Acad Sci U S A 98:9448–9453

Kachroo A, Lapchyk L, Fukushige H, Hildebrand D, Klessig D, Kachroo P (2003a) Plastidial fatty acid signaling modulates salicylic acid- and jasmonic acid-mediated defense pathways in the *Arabidopsis ssi2* mutant. Plant Cell 15:2952–2965

Kachroo P, Kachroo A, Lapchyk L, Hildebrand D, Klessig DF (2003b) Restoration of defective cross talk in *ssi2* mutants: role of salicylic acid, jasmonic acid, and fatty acids in *SSI2*-mediated signaling. Mol Plant-Microbe Interact 16:1022–1029

Kagan IA, Hammerschmidt R (2002) *Arabidopsis* ecotype variability in camalexin production and reaction to infection by *Alternaria brassicicola*. J Chem Ecol 28:2121–2140

Kanrar S, Venkateswari JC, Kirti PB, Chopra VL (2002) Transgenic expression of hevein, the rubber tree lectin, in Indian mustard confers protection against *Alternaria brassicae*. Plant Sci 162:441–448

Kannan P, Pandey D, Gupta AK, Punetha H, Taj G, Kumar A (2011) Expression analysis of MAP 2K 9 and MAP K 6 during pathogenesis of Alternaria blight in *Arabidopsis thaliana* ecotype Columbia. Mol Biol Rep 39:4439–4444

Keller NP, Hohn TM (1997) Metabolic pathway gene clusters in filamentous fungi. Fungal Genet Biol 21:17–29

Khandelwal A, Kumar A, Banerjee M, Garg GK (2002) Effect of *Alternaria* pathotoxin(s) on expression of p53- like apoptotic protein in calli and leaves of *Brassica campestris*. Indian J Exp Biol 40:89–94

Kohmoto K, Itoh Y, Shimomura N, Kondoh Y, Otani H, Kodama M, Nishimura S, Nakatsuka S (1993) Isolation and biological activities of two host-specific toxins from the tangerine pathotype of *Alternaria alternata*. Phytopathology 83:495–502

Kunkel BN, Brooks DM (2002) Cross talk between signaling pathways in pathogen defense. Curr Opin Plant Biol 5:325–331

Kusaba M, Tsuge T (1994) Nuclear ribosomal DNA variation and pathogenic specialization in *Alternaria* fungi known to produce host-specific toxins. Appl Environ Microbiol 60:3055–3062

Kusaba M, Tsuge T (1995) Phylogeny of *Alternaria* fungi known to produce host-specific toxins on the basis of variation in internal transcribed spacers of ribosomal DNA. Curr Genet 28:491–498

Kusaba M, Tsuge T (1997) Mitochondrial DNA variation in host-specific toxin-producing pathogens in the genus *Alternaria*. Annu Phytopathol Soc Japan 63:463–469

Lawrence CB, Singh NP, Qiu J, Gardner RG, Tuzun S (2000) Constitutive hydrolytic enzymes are associated with polygenic resistance of tomato to *Alternaria solani* and may function as an elicitor release mechanism. Physiol Mol Plant Pathol 57:211–220

Lawrence CB, Mitchel TK, Craven KD, Cho Y, Cramer RA Jr, Kim KH (2008) At Death's Door: *Alternaria* pathogenicity mechanisms. Plant Pathol J 24:101–111

Lee SB, Go YS, Bae HJ, Park JH, Cho SH, Cho HJ, Lee DS, Park OK, Hwang I, Suh MC (2009) Disruption of glycosylphosphatidylinositol-anchored lipid transfer protein gene altered cuticular lipid composition, increased plastoglobules, and enhanced susceptibility to infection by the fungal pathogen *Alternaria brassicicola*. Plant Physiol 150:42–54

Li HY, Xiao S, Chye ML (2008) Ethylene- and pathogen-inducible *Arabidopsis* acyl-CoA binding protein 4 interacts with an ethylene-responsive element binding protein. J Exp Bot 59:3997–4006

Liebermann B, Nussbaum RP, Günther W (2000) Bicycloalternarenes produced by the phytopathogenic fungus *Alternaria alternata*. Phytochemistry 55:987–992

Lin CH, Yang SL, Wang NY, Chung KR (2010) The *FUS3 MAPK* signaling pathway of the citrus pathogen *Alternaria alternata* functions independently or cooperatively with the fungal redox-responsive AP1 regulator for diverse developmental, physiological and pathogenic processes. Fungal Genet Biol 47:381–391

Lincoln JE, Richael C, Overduin B, Smith K, Bostock R, Gilchrist DG (2002) Expression of the antiapoptotic baculovirus p35 gene in tomato blocks programmed cell death and provides broad-spectrum resistance to disease. Proc Natl Acad Sci U S A 99:15217–15221

Loivamaki M, Stuhrwohldt N, Deeken R, Steffens B, Roitsch T, Hedrich R, Sauter M (2010) A role for PSK signaling in wounding and microbial interactions in *Arabidopsis*. Physiol Plant 139:348–357

Lorito M, Woo SL, Garcia I, Colucci G, Harman GE, Pintor-Toro JA, Filippone E, Muccifora S, Lawrence CB, Zoina A, Tuzun S, Scala F, Fernandez IG (1998) Genes from mycoparasitic fungi as a source for improving plant resistance to fungal pathogens. Proc Natl Acad Sci U S A 95:7860–7865

Lou J, Fu L, Peng Y, Zhou L (2013) Metabolites from *Alternaria* fungi and their bioactivities. Molecules 18:5891–5935

Ludwig-Müller J, Pieper K, Ruppel M, Cohen JD, Epstein E, Kiddle G, Bennett R (1999) Indole glucosinolate and auxin biosynthesis in *Arabidopsis thaliana* L. glucosinolate mutants and the development of the clubroot disease. Planta 208:409–419

MacKinnon SL, Keifer P, William AA (1999) Components from the phytotoxic extract of *Alternaria brassicicola*, a black spot pathogen of canola. Photochemistry 51:215–221

Madhani HD, Fink GR (1998) The riddle of MAP kinase signaling specificity. Trends Genet 14:151–155

Maekawa N, Yamamoto M, Nishimura S, Kohmoto K, Kuwata M, Watanabe Y (1984) Studies on host-specific AF-toxins produced by *Alternaria alternata* strawberry pathotype causing Alternaria black spot of strawberry. I. Production of host-specific toxins and their biological activity. Ann Phytopathol Soc Japan 50:600–609

Majeau N, Trudel J, Asselin A (1990) Diversity of cucumber chitinase isoforms and characterization of one seed basic chitinase with lysozyme activity. Plant Sci 68:9–16

Mang HG, Laluk KA, Parsons EP, Kosma DK, Cooper BR, Park HC, AbuQamar S, Boccongelli C, Miyazaki S, Consiglio F, Chilosi G, Bohnert HJ, Bressan RA, Mengiste T, Jenks MA (2009) The *Arabidopsis* RESURRECTION1 gene regulates a novel antagonistic interaction in plant defense to biotrophs and necrotrophs. Plant Physiol 151:290–305

Markham JE, Hille J (2001) Host-selective toxins as agents of cell death in plant–fungus interactions. Mol Plant Pathol 2:229–239

Marmath KK, Giri P, Sharma S, Taj G, Kumar A (2013) *In-silico* interaction studies of *Alternaria brassicae* toxin destruxin b and potential partners of MAPK4 cascade. Int J Agric Environ Biotechnol 6:203–210

Masunaka A, Tanaka A, Tsuge T, Peever TL, Timmer LW, Yamamoto M, Yamamoto H, Akimitsu K (2000) Distribution and characterization of AKT homologs in the tangerine pathotype of *Alternaria alternata*. Phytopathology 90:762–768

Meronuck RA, Steele JA, Mirocha CJ, Christensen CM (1972) Tenuazonic acid, a toxin produced by *Alternaria alternata*. Appl Microbiol 23:613–617

Mishra MK, Suresh N, Bhat AM, Suryaprakash N, Kumar SS, Kumar A, Jayarama (2011) Genetic molecular analysis of *Coffea arabica* (Rubiaceae) hybrids using SRAP markers. Rev Biol Trop 59:607–617

Montemurro N, Visconti A (1992) *Alternaria* metabolites – chemical and biological data. In: Chelkovski J, Visconti A (eds) *Alternaria* biology, plant diseases and metabolites. Elsevier, Amsterdam, pp 449–557

Mora MM, Earle ED (2001) Resistance to *Alternaria brassicicola* in transgenic broccoli expressing a *Trichoderma harzianum* endochitinase gene. Mol Breed 8:1–9

Morita S, Azuma M, Aoba T, Satou H, Narisawa K, Hashiba T (2003) Induced systemic resistance of Chinese cabbage to bacterial leaf spot and Alternaria leaf spot by the root endophytic fungus, *Heteroconium chaetospira*. J Gen Plant Pathol 69:71–75

Multani DS, Meeley RB, Paterson AH, Gray J, Briggs SP, Johal GS (1998) Plant-pathogen microevolution: molecular basis for the origin of a fungal disease in maize. Proc Natl Acad Sci U S A 95:1686–1691

Nandi A, Moeder W, Kachroo P, Klessig DF, Shah J (2005) *Arabidopsis* ssi2-conferred susceptibility to *Botrytis cinerea* is dependent on EDS5 and PAD4. Mol Plant-Microbe Interact 18:363–370

Nishimura S, Kohmoto K (1983) Host-specific toxins and chemical structures from *Alternaria* species. Annu Rev Phytopathol 21:87–116

Nussbaum RI, Guenther W, Heinze S, Liebermann B (1999) New tricycloalternarenes produced by the phytopathogenic fungus *Alternaria alternata*. Phytochemistry 52:593–599

Ohtani K, Yamamoto H, Akimitsu K (2002) Sensitivity to *Alternaria alternata* toxin in citrus because of altered mitochondrial RNA processing. Proc Natl Acad Sci U S A 99:2439–2444

Oka K, Akamatsub H, Kodamab M, Nakajimab H, Kawadab T, Otani H (2005) Host-specific AB-toxin production by germinating spores of *Alternaria brassicicola* is induced by a host-derived oligosaccharide. Physiol Mol Plant Pathol 66:12–19

Oliver RP, Ipcho SVS (2004) *Arabidopsis* pathology breathes new life into the necrotrophs-vs-biotrophs classification of fungal pathogens. Mol Plant Pathol 5:347–352

Otani H, Kohmoto K, Kodama M (1995) Alternaria toxins and their effects on host plants. Can J Bot 73:8453–8458

Otani H, Kohnobe A, Kodama M, Kohmoto K (1998) Production of a host-specific toxin by germinating spores of *Alternaria brassicicola*. Physiol Mol Plant Pathol 52:285–295

Pandey D, Kumar A, Garg GK (2001) Antagonism between *Alternaria brassicae* toxin and zeatin in cell culture of *Brassica juncea* (L.) Czern and Coss. cv. Divya. Physiol Mol Biol Plants 7:181–184

Parada RY, Oka K, Yamagishi D, Kodama M, Otani H (2007) Destruxin B produced by *Alternaria brassicae* does not induce accessibility of host plants to fungal invasion. Physiol Mol Plant Pathol 71:48–54

Parada RY, Sakuno E, Mori N, Oka K, Egusa M, Kodama M, Otani H (2008) *Alternaria brassicae* produces a host-specific protein toxin from germinating spores on host leaves. Phytopathology 98:458–463

Pedras MSC (1998) Phytoalexins from crucifers: mimicking or overcoming plant chemical defenses. Recent Res Dev Phytochem 2:259–267

Pedras MSC, Biesenthal CJ (2001) Isolation, structure determination, and phytotoxicity of unusual dioxopiperazines from the phytopathogenic fungus *Phoma lingam*. Phytochemistry 58:905–909

Pedras MSC, Chumala PB (2005) Phomapyrones from blackleg causing phytopathogenic fungi: isolation, structure determination, biosyntheses and biological activity. Phytochemistry 66:81–87

Pedras MSC, Okanga FI (1999) Strategies of cruciferous pathogenic fungi: detoxification of the phytoalexin cyclobrassinin by mimicry. J Agric Food Chem 47:1196–1202

Pedras MSC, Morales VM, Taylor JL (1994) Phomapyrones: three metabolites from the blackleg fungus. Phytochemistry 36:1315–1318

Pedras MSC, Kahn AQ, Taylor JL (1998) The phytoalexin camalexin is not metabolized by *Phoma lingam*, *Alternaria brassicae*, or phytopathogenic bacteria. Plant Sci 139:1–8

Pedras MSC, Zaharia JL, Smith KC, Gai Y, Ward DE (1999) Alternaria blackspot phytotoxins: new strategies for determining specific disease resistance traits. In: Proceedings of the 10th international rapeseed congress, Canberra, Australia, 26–29 September

Pedras MSC, Biesenthal CJ, Zaharia IL (2000) Comparison of the phytotoxicity of the phytotoxin destruxin B and four natural analogs. Plant Sci 156:185–192

Pedras MSC, Zaharia JL, Gai Y, Zhou Y, Ward DE (2001) *In planta* sequential hydroxylation and glycosylation of a fungal phytotoxin: avoiding cell death and overcoming the fungal invader. Proc Natl Acad Sci U S A 98:747–752

Pedras MSC, Zaharia JL, Ward DE (2002) The destruxins: synthesis, biosynthesis, biotransformation, and biological activity. Phytochemistry 59:579–596

Pedras MSC, Gadagi RS, Jha M, Sarma-Mamillapalle VK (2007) Detoxification of the phytoalexin brassinin by isolates of *Leptosphaeria maculans* pathogenic on brown mustard involves an inducible hydrolase. Phytochemistry 68:1572–1578

Pedras MSC, Chumala PB, Jin W, Islam MS, Hauck DW (2009) The phytopathogenic fungus *Alternaria brassicicola*: phytotoxin production and phytoalexin elicitation. Phytochemistry 70:394–402

Penninckx IAMA, Eggermont K, Terras FRG, Thomma BPHJ, De Samblanx GW, Buchala A, Metraux JP, Manners JM, Broekaert WF (1996) Pathogen-induced systemic activation of a plant defensin gene in *Arabidopsis* follows a salicylic acid-independent pathway. Plant Cell 8:2309–2323

Petersen M, Brodersen P, Naested H, Andreasson E, Lindhart U, Johansen B, Nielsen HB, Lacy M, Austin

MJ, Parker JE (2000) *Arabidopsis* MAP Kinase 4 negatively regulates systemic acquired resistance. Cell 103:1111–1120

Pitzschke A, Schikora A, Hirt H (2009) MAPK cascade signalling networks in plant defence. Curr Opin Plant Biol 12:421–426

Popescu SC, Popescu GV, Synder M, Dinesh-Kumar SP (2008) MAPK target networks in *Arabidopsis thaliana* revealed using functional protein microarrays. Genes Dev 23:1–13

Popescu SC, Popescu GV, Bachan S, Zhang Z, Gerstein M, Snyder M, Dinesh-Kumar SP (2009) MAPK target networks in *Arabidopsis thaliana* revealed using functional protein microarrays. Genes Dev 23:80–92

Rawe HC, Kliebenstein DJ (2010) *All Mold Is Not Alike*: the importance of intra-specific diversity in necrotrophic plant pathogens. PLoS Pathog 6, e1000759

Robeson DJ, Strobel GA (1981) Alpha-beta-dehydrocurvularin and curvularin from *Alternaria cinerariae*. Z Naturforsch 36:1081–1083

Roby D, Broglie K, Cressman R, Biddle P, Chet I, Broglie R (1990) Activation of a bean chitinase promoter in transgenic tobacco plants by phytopathogenic fungi. Plant Cell 2:999–1008

Rostás M, Bennett R, Hilker M (2002) Comparative physiological responses in Chinese cabbage induced by herbivory and fungal infection. J Chem Ecol 28:2449–2463

Rotem J (1994) The genus *Alternaria*: biology, epidemiology and pathogenicity. APS Press, St Paul, 326 p

Samuels RI, Charnley AK, Renolds SE (1988) Application of reversed-phase HPLC in separation and detection of the cyclodepsipeptide toxins produced by the entomopathogenic fungus *Metarhizium anisopliae*. J Chromatogr Sci 26:15–19

Schenk PM, Kazan K, Wilson I, Anderson JP, Richmond T, Somerville SC, Manners JM (2000) Coordinated plant defense responses in *Arabidopsis* revealed by microarray analysis. Proc Natl Acad Sci U S A 97:11655–11660

Schweizer P, Jeanguenat A, Whitacre D, Métraux JP, Mosinger E (1996) Induction of resistance in barley against *Erysiphe graminis* f. sp. *hordei* by free cutin monomers. Physiol Mol Plant Pathol 49:103–120

Sharma N, Rahman MH, Strelkov SE, Thiagarajah M, Bansal VK, Kav NNV (2007) Proteome-level changes in two *Brassica napus* lines exhibiting differential responses to the fungal pathogen *Alternaria brassicae*. Plant Sci 172:95–110

Sheen VL, Feng Y, Graham D, Takafuta T, Shapiro SS, Walsh CA (2002a) Filamin A and filamin B are co-expressed within neurons during periods of neuronal migration and can physically interact. Hum Mol Genet 11:2845–2854

Sheen J, He P, Shan L, Xiong Y, Tena G, Yoo S, Cho Y, Boudsocq M, Lee H (2002b) Signaling specificity and complexity of MAPK cascades in plant innate immunity. Biol Plant-Microbe Interact 6:1–10

Shi LH, Bielawski J, Mu JY, Dong HL, Teng C, Zhang J, Yang XH, Tomishige N, Hanada K, Hannun YA, Zuo JR (2007) Involvement of sphingoid bases in mediating reactive oxygen intermediate production and programmed cell death in *Arabidopsis*. Cell Res 17:1030–1040

Shibuya N, Minami E (2001) Oligosaccharide signaling for defence responses in plant. Physiol Mol Plant Pathol 59:223–233

Shivanna KR, Sawhney VK (1993) Pollen selection for Alternaria resistance in oilseed brassicas: responses of pollen grains and leaves to a toxin of *A. brassicae*. Theor Appl Genet 86:339–344

Spassieva SD, Markham JE, Hille J (2002) The plant disease resistance gene *Asc-1* prevents disruption of sphingolipid metabolism during AAL-toxin-induced programmed cell death. Plant J 32:561–572

Stacey G, Shibuya N (1997) Chitin recognition in rice and legumes. Plant Soil 194:161–169

Steele JA, Durbin RD, Uchytil TF, Rich DH (1978) Tentoxin uncompetitive inhibitor of lettuce chloroplast coupling factor-1. Biochim Biophys Acta 501:72–82

Taj G, Kumar A, Bansal KC, Garg GK (2004) Introgression of osmotin gene for creation of resistance against Alternaria blight by perturbation of cell cycle machinery. Indian J Biotechnol 3:291–298

Taj G, Agarwal P, Grant M, Kumar A (2010) MAPK machinery in plants. Plant Signal Behav 5:1370–1378

Taj G, Agarwal P, Grant M, Kumar A (2011) Co-expression and *in-silico* interaction studies for inter-linking the activation of *mapk3* and lox genes during pathogenesis of Alternaria blight (*Alternaria brassicae*) in *Brassica juncea*. J Oilseed Brassica 2:13–20

Takaichi M, Oeda K (2000) Transgenic carrots with enhanced resistance against two major pathogens, *Erysiphe heraclei* and *Alternaria dauci*. Plant Sci 153:135–144

Tanabe K, Nishimura S, Kohmoto K (1988) Pathogenicity of cutinase- and pectic enzymes deficient mutants of *Alternaria alternata* Japanese pear pathotype. Ann Phytopathol Soc Japan 54:552–555

Tanaka A, Tsuge T (2000) Structural and functional complexity of the genomic region controlling AK-toxin biosynthesis and pathogenicity in the Japanese pear pathotype of *Alternaria alternata*. Mol Plant-Microbe Interact 13:975–986

Tanaka A, Shiotani H, Yamamoto M, Tsuge T (1999) Insertional mutagenesis and cloning of the genes required for biosynthesis of the host-specific AK-toxin in the Japanese pear pathotype of *Alternaria alternata*. Mol Plant-Microbe Interact 12:691–702

Tena G, Boudsocq M, Sheen J (2011) Protein kinase signaling networks in plant innate immunity. Curr Opin Plant Biol 14:519–529

Terras FRG, Eggermont K, Kovaleva V, Raikhel NV, Torrekens S, Van Leuven F, Osborn RW, Kester A, Rees SB, Vanderleyden J, Cammue BPA, Broekaert WF (1995) Small cysteine-rich antifungal proteins from radish. Their role in host defense. Plant Cell 7:573–588

Tewari JP (1983) Cellular alterations in the blackspot of rapeseed caused by *Alternaria brassicae*. Phytopathology 73:831 (Abstr.)

Tewari JP (1991) Structural and biochemical bases of the black spot disease of crucifers. Adv Struc Biol 1:325–349

Thaler JS, Owen B, Higgins VJ (2004) The role of the jasmonate response in plant susceptibility to diverse pathogens with a range of lifestyles. Plant Physiol 135:530–538

Thomma BPHJ (2003) Pathogen profile- *Alternaria* spp.: from general saprophyte to specific parasite. Mol Plant Pathol 4:225–236

Thomma BP, Eggermont K, Penninckx IA, Mauch-Mani B, Vogelsang R, Cammue BPA, Broekaert WF (1998) Separate jasmonate-dependent and salicylate-dependent defense-response pathways in *Arabidopsis* are essential for resistance to distinct microbial pathogens. Proc Natl Acad Sci U S A 95:15107–15111

Thomma BPHJ, Eggermont K, Tierens FMJ, Broekaert WF (1999a) Requirement of functional EIN2 (ethylene insensitive 2) gene for efficient resistance of *Arabidopsis thaliana* to infection by *Botrytis cinerea*. Plant Physiol 121:1093–1101

Thomma BPHJ, Nelissen I, Eggermont K, Broekaert WF (1999b) Deficiency in phytoalexin production causes enhanced susceptibility of *Arabidopsis thaliana* to the fungus *Alternaria brassicicola*. Plant J 19:163–171

Thomma BPHJ, Eggermont K, Broekaert WF, Cammue BPA (2000) Disease development of several fungi on *Arabidopsis* can be reduced by treatment with methyl jasmonate. Plant Physiol Biochem 38:421–427

Thomma BPHJ, Penninckx IAMA, Broekaert WF, Cammue BPA (2001) The complexity of disease signaling in *Arabidopsis*. Curr Opin Immunol 13:63–68

Thuleau P, Graziana A, Rossignol M, Kauss H, Auriol P, Ranjeva R (1988) Binding of the phytotoxin zinniol stimulates the entry of calcium into plant-protoplasts. Proc Natl Acad Sci U S A 85:5932–5935

Tierens FMJ, Thomma BPHJ, Bari RP, Garnier M, Eggermont K, Brouwer M, Penninckx IAMA, Broekaert WF, Cammue BPA (2002) Esa1, an *Arabidopsis* mutant with enhanced susceptibility to a range of necrotrophic fungal pathogens, shows a distorted induction of defense responses by reactive oxygen generated compounds. Plant J 29:131–140

Ton J, Van Pelt JA, Van Loon LC, Pieterse CM (2002) Differential effectiveness of salicylate-dependent and jasmonate/ethylene dependent induced resistance in *Arabidopsis*. Mol Plant-Microbe Interact 15:27–34

Trail F, Köller W (1993) Diversity of cutinases from plant pathogenic fungi: purification and characterization of two cutinases from *Alternaria brassicicola*. Physiol Mol Plant Pathol 42:205–220

Varet A, Parker J, Tornero P, Nass N, Nürnberger T, Dangl JL, Scheel D, Lee J (2002) NHL25 and NHL3, two NDR1/HIN1-like genes in *Arabidopsis thaliana* with potential role(s) in plant defense. Mol Plant-Microbe Interact 15:608–616

Veronese P, Nakagami H, Bluhm B, Abuqamar S, Chen X, Salmeron J, Dietrich RA, Hirt H, Mengiste T (2006) The membrane-anchored *Botrytis*-induced kinase1 plays distinct roles in *Arabidopsis* resistance to necrotrophic and biotrophic pathogens. Plant Cell 18:257–273

Walton JD (2000) Horizontal gene transfer and the evolution of secondary metabolite gene clusters in fungi: an hypothesis. Fungal Genet Biol 30:167–171

Wan JR, Zhang XC, Neece D, Ramonell KM, Clough S, Kim SY, Stacey MG, Stacey G (2008) A LysM receptor-like kinase plays a critical role in chitin signaling and fungal resistance in *Arabidopsis*. Plant Cell 20:471–481

Wang H, Li J, Bostock RM, Gilchrist DG (1996a) Apoptosis: a functional paradigm for programmed plant cell death induced by a host-selective phytotoxin and invoked during development. Plant Cell 8:375–391

Wang W, Jones C, Ciacci-Zanella J, Holt T, Gilchrist DG, Dickman MB (1996b) Fumonisins and *Alternaria alternata lycopersici* toxins: sphinganine analog mycotoxins induce apoptosis in monkey kidney cells. Proc Natl Acad Sci U S A 93:3461–3465

Wang L, Tsuda K, Sato M, Cohen JD, Katagiri F, Glazebrook J (2009) *Arabidopsis* CaM binding protein CBP60g contributes to MAMP-induced SA accumulation and is involved in disease resistance against *Pseudomonas syringae*. PLoS Pathol 5, e1000301

Wiermer M, Feys BJ, Parker JE (2005) Plant immunity: the EDS1 regulatory node. Curr Opin Plant Biol 8:383–389

Wolpert TJ, Dunkle LD, Ciuffetti LM (2002) Host-selective toxins and avirulence determinants: what's in a name? Annu Rev Phytopathol 40:251–285

Wood RKS (1973) Specificity in plant diseases. In: Byrde RJW, Cutting CV (eds) Fungal pathogenicity and the plant response. Academic, New York, pp 1–16

Wubben JP, Joosten MHAJ, Van Kan JAL, De Wit PJGM (1992) Subcellular localization of plant chitinases and 1,3-beta-glucanases in *Cladosporium fulvum* (syn. *Fulvia fulva*) infected tomato leaves. Physiol Mol Plant Pathol 41:23–32

Yao C, Köller W (1994) Diversity of cutinases from plant pathogenic fungi: cloning and sequence analysis of a cutinase gene from *Alternaria brassicicola*. Physiol Mol Plant Pathol 44:81–92

Yao C, Köller W (1995) Diversity of cutinases from plant pathogenic fungi: different cutinases are expressed during saprophytic and pathogenic stages of *Alternaria brassicicola*. Mol Plant-Microbe Interact 8:122–130

Zheng ZY, Abu Qamar S, Chen ZX, Mengiste T (2006) *Arabidopsis* WRKY33 transcription factor is required for resistance to necrotrophic fungal pathogens. Plant J 48:592–605

11 Disease Management

11.1 Introduction

No single method or approach in current use is feasible, viable, stable, effective and economical in dealing with any host–pathogen system. Therefore, it is necessary to integrate all the methods of plant disease control, which are available, including cultural, chemical, biological and host resistance to effectively manage *Alternaria* diseases of brassicaceous plants (Eastburn 1989; Saharan 1992; Verma and Saharan 1994).

11.2 Cultural Measures

Among the common cultural practices, use of clean, large, healthy seed, seed treatment, long crop rotation (4 years under Canadian conditions), sanitation, weed control, planting shallow (2 cm) at recommended times; use of balanced nutrients, proper plant density (45 cm row spacing), proper drainage in the fields, plant debris management; use of tolerant/resistant cultivars and application of chemicals at proper time with adequate foliage coverage have been advocated to control *Alternaria* diseases of brassicaceous plants (Channon and Maude 1971; Chupp 1925; Chupp and Sherf 1960; Daebeler et al. 1981, 1987; Dixon 1981; Kolte 1985; Lapis and Ricaforte 1974; Saharan and Chand 1988; Singh 1987; Thomas 1984; Verma and Saharan 1994; Kharbanda and Tewari 1996; Kumar and Kumar 2006; Peruch and Michereff 2007). Seed infection can largely be controlled by hot water treatment at 50 °C for 18 min (Schimmer 1953) or by soaking *Brassica* seeds in a 0.2 % aqueous suspension of Thiram for 24 h at 30 °C (Maude 1978; Maude and Shuring 1968; Maude et al. 1969). In tropical countries including India, aging of *Brassica* seeds for 5–6 months at temperatures above 35 °C eliminates seed infection and seed-borne inoculum (Chahal 1981; Kolte 1985). *Alternaria* pod blight of radish is significantly less in the early transplanting crop (15 November) under Indian conditions (Sandhu et al. 1985). In Sweden, *A. brassicae* occurs more frequently in areas with both summer and winter oilseed crops than in areas with mainly summer crops (Morner 1980). In India, early planting of toria (*B. rapa*) in September usually escapes *Alternaria* infection. In the Lujaska area of Leningrad (USSR), there is lower incidence of *A. brassicae* on cabbage, if transplanting is done between 20 and 30 June. Application of fertilizers containing humus (45 % peat and 5 % mullcin, plus 3.9 g ammonium nitrate, 3.4 g potassium chloride, and 8.1 g superphosphate/100), added to 110 g organic matter, reduces the disease better than the organic manure alone (Kupryanova 1957). *Alternaria* black spot of broccoli can be managed by incorporating infected leaf debris in the soil, at least 10 cm depth for no less than a 60 days interval in subsequent *Brassica* plantings (Peruch and Michereff 2007).

G.S. Saharan et al., *Alternaria Diseases of Crucifers: Biology, Ecology and Disease Management*,
DOI 10.1007/978-981-10-0021-8_11

To avoid storage, and transit losses in cabbage and cauliflower, heads should be handled carefully to avoid bruising and surface moisture allowed to evaporate before storage. The storage house should be kept at 33–34 °F and ample ventilation provided to reduce humidity (Walker 1927).

11.3 Seed Treatment

11.3.1 Hot Water Treatment

Excellent control of *A. brassicae* in white cabbage and cauliflower seeds has been obtained in Danish experiments from 1934 to 1936 by 30-min immersion in water heated to 45 °C or for 20 min at 50 °C, or even 30 min at 40 °C. This treatment improves germination by up to 13 % along with elimination of other fungi such as *Penicillium* and *Mucor* spp. (Nielsen 1936). Hot water seed treatment at 50 °C for 30 min controls the disease in cabbage (Walker 1952). According to Ellis (1968a, b), hot water treatment for 25 min at 50 °C eliminates *Alternaria* infection from brassicaceous seeds. Seed treatment at 50 °C for 20 min is highly effective in controlling the seed-borne fungi including *A. brassicae* of raya (*B. juncea*) without any pronounced effect on the germination of the seeds (Randhawa and Aulakh 1984).

11.3.2 Chemical Treatment

For controlling seed-borne *A. brassicae, A. brassicicola, A. raphani* and *A. alternata* infections in brassicaceous plants, seed dressings with Benlate, Dithane M-45, Dithane Z-78, Difolatan, Dichlofluanid, Delan C, Bristan, Bavistin, Brassicol, Captan, Fenpropimorph, Granosan M, Quinolate, Morpan, Rovral, Thiram, Vitavax, Thiride and Zineb have been found to be very effective (Atkinson 1950; Chahal et al. 1977; Champion et al. 1979; Chirco and Harman 1979; Ellis 1968a, b; Gupta and Saxena 1984; Holtzhausen 1978; Humpherson-Jones et al. 1980, 1981, 1983, 1984; Kanwar and Khanna 1979; Kumar and Singh 1986; Maude and Humpherson-Jones 1980a, b, 1984; Maude et al. 1972, 1984; Mridha and Safa 1985; Randhawa and Aulakh 1982; Verma and Saharan 1994; Wu and Lu 1984; Latif et al. 2006).

Soaking *Brassica* seeds in an aqueous suspension of Thiram (0.2 %) for 24 h at 30 °C eradicates *Alternaria* infection (Maude 1978; Maude and Shuring 1968; Maude et al. 1969). In New South Wales, two samples of infected seed were treated with iprodione at 1.25 g a.i. Kg^{-1} seed, and the level of seed-borne *A. brassicae* was reduced from 21 to 2 % and from 14 to 4.5 %, respectively. The same treatment eliminates *A. brassicae* from pieces of trash mixed with seed (Stovold et al. 1987). Seed treatment with iprodione, which virtually eradicates seed-borne infection of *Alternaria* spp. in the Brassicaceae, is now used to treat basic seed stocks in England (Maude and Humpherson-Jones 1980a, b). This practice is most effective in crops grown in isolation from maturing brassicaceous seed crops. However, where seed crops are grown successively and in close proximity, as it often occurs in the main seed-producing areas, the infection may arise from spores released from diseased seeding crops during cutting and threshing and disseminate by the wind to young crops destined for seed production in the following year (Humpherson-Jones and Maude 1982). In such situations, additional control measures are necessary.

According to Maude and Suett (1986), use of a prototype fluidized-bed seed treater to apply iprodione in a polymer film coat to cabbage seeds infected with *A. brassicae* and *A. brassicicola* has been found effective. Analysis of seed-to-seed variability in dosing demonstrated the greater accuracy in fungicide application and control of *A. brassicae* by film coating than by slurry or dust methods. Application of polymer film coat slightly reduces superficial inoculum of *A. brassicicola.*

Seed treatment of mustard with Bavistin, Difolatan (1.5 g/Kg) and Dithane M-45 (2 g/ Kg^{-1}) eradicates *A. brassicae* infection. A minimum period of 24 h is necessary for obtaining effective control of the pathogen in seeds (Kumar and Singh 1986). Seed treatment with iprodione (1.35–2.5 g a.i./Kg^{-1}) and Fenpropimorph (0.625–2.5 g a.i./Kg^{-1}) virtually eliminated *A. brassicicola* in naturally infected *Brassica* seeds without affecting seed germination in laboratory and seedling tests (Maude 1983, 1986). In Bangladesh, seed treatment of mustard with Rovral (0.25 %) enhances seed germination remarkably (Latif et al. 2006).

In Finland, *A. brassicicola* on Chinese cabbage (*B. pekinensis*) is effectively controlled by dressing seed with Thiram or powdered *Streptomyces griseoviridis* (Valkonen and Koponen 1990). According to Crisan and Pall (1986), seed treatment with Vitavax, Quinolate, Dichlofluanid, Zineb and Morpan completely inhibits *A. alternata* infection in cauliflower seed. In India, seed-borne infections of *A. alternata* in *Eruca sativa* have been controlled by treating the seed with Agrosan GN and Dithane M-45 at 0.3 % (Gupta and Saxena 1984).

11.3.3 Bioagent Treatment

Seed treatment with Mycostop, a powdery formulation of spores, and mycelium of *Streptomyces griseoviridis* controls *A. brassicicola* damping-off in cauliflower and cabbage (Fig. 11.1; White et al. 1990). Biocontrol of seed-borne *A. raphani* and *A. brassicicola* in radish has been obtained through other antagonists including *Chaetomium globosum*, *Trichoderma harzianum*, *T. koningii* and *Fusarium* spp. (Plate 11.1; Vannacci and Harman 1987). Garlic extract is most effective in controlling the seed-borne fungi of mustard (Latif et al. 2006).

11.4 Chemical Control

Inhibition of *A. alternata*, *A. brassicae* and *A. brassicicola* growth in culture and spore germination by Bavistin, Blitane, Blitox-50, Benlate, Brestan-60, Captan, Captafol, Ceresen dry, Copper oxychloride, Carbendazim, Cupramar, Cuman L, Dithane M-45, Dithane Z-78, Difolatan, Emisan, Fytolan, Fentinhydroxide, Fostyl-al dinobuton, Griseofulvin, Rovral, Methofuroxan, Mycostatin, Polyoxin B, Polyoxin D, Thiram, Tridimefon, Wettable sulphur, Zincop, Quintozene, and Ziram have been observed by several workers (Ansari et al. 1990; Chand and Jatian 1969; Maude 1976; Prasada et al. 1970; Singh and Bhowmik 1985; Singh and Rai 1982; Tewari and Skoropad 1979).

In vitro and in vivo evaluation of large number of chemicals (Table 11.1) against *Alternaria* species of Brassicaceae have been carried out (Verma and Saharan 1994). Some of the fungicides, which have been found to be effective in controlling the *Alternaria* diseases and increasing the yield under field conditions, are Apron 35 SD (6 g/Kg^{-1}), Antracol (0.02 %), Baycor (0.2 %), Benlate (0.1 %), Blitane (0.2 %), Blitox-50 (0.3 %), Bordeaux mixture (0.3 %), Boric acid powder (0.53 %), Brestan (0.1 %), Captan (0.2 %), Captafol (0.3 %), Cuman L (0.1 %), Difolatan (0.2 %), Dithane M-45 (0.2 %), Dithane

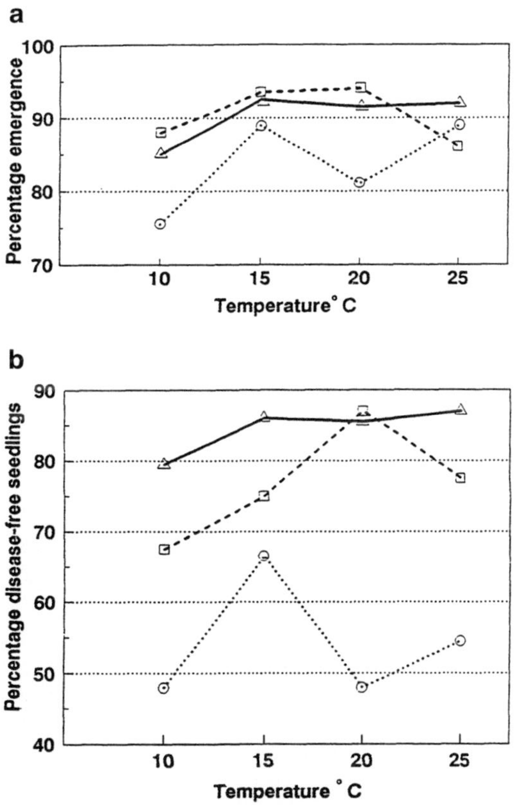

Fig. 11.1 The effect of *Streptomyces griseoviridis* preparation on (**a**) percentage emergence and (**b**) percentage disease-free seedlings of cabbage cultivar Celtic. O….O untreated seed; □….□ seed treated with *S. griseoviridis* preparation; and Δ…….Δ seed treated with iprodione (White et al. 1990)

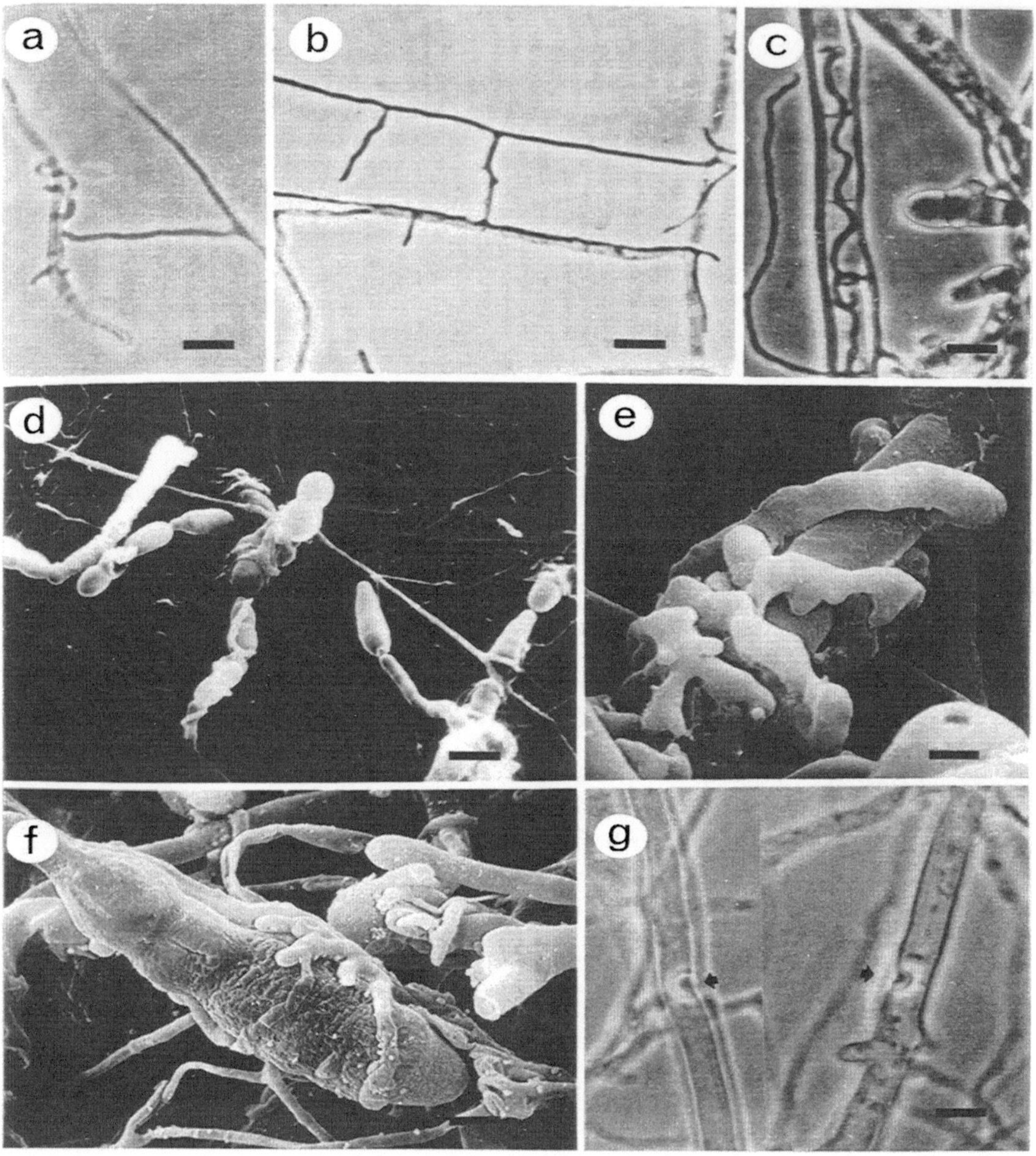

Plate 11.1 Biocontrol of seed-borne *Alternaria raphani* and *A. brassicicola:* (**a**) coiling of *Trichoderma harzianum* ATCC 56678 hyphae around *A. brassicicola* mycelium. Bar = 70 μm; (**b**) hyphae of *T. harzianum* 420 growing towards hyphae of *A. raphani.* Bar = 80 μm; (**c**) parallel growth of hyphae of *Fusarium* sp. along *A. brassicicola* hyphae. Bar = 25 μm; (**d**) scanning electron micrographs of *Chaetomium globosum* hyphae coiling around conidia of *A. brassicicola.* Bar = 35 μm; (**e**) scanning electron micrographs of *C. globosum* hyphae coiling around hyphae of *A. brassicicola.* Bar = 7 μm; (**f**) scanning electron micrographs of *C. globosum* hyphae coiling around *A. raphani* conidia. Bar = 25 μm; and (**g**) reaction zones (*arrows*) of *A. brassicicola* under the stimulus of *C. globosum.* Bar = 25 μm (Vannacci and Harman 1987)

Table 11.1 Chemicals tested against species of *Alternaria* attacking Brassicaceae (Verma and Saharan 1994; to date)

Fungicides		
Acetone	Fentin acetate	Syllit
Actidione	Ferbam	Tebuconazole (Nativo)
Agrosan GN	Fermate	Tetrahydropyrimidine
Alar	Folicur (tebuconazole)	Thiabendazole
Antracol	Folpet	Thiophanate methyl
Arasan	Flutriafol	Thiovit
Azoxystrobin (Amistar)	Formaldehyde	Thiram
Azadirachtin (Neemarin)	Granosan	Tillex
Bafin	Guazatin	Topsin M
BAS 480 F	Halogenated derivatives	Triademefon
Baycor	Imazalil	Triapenthenol
Bayleton	Indofil M-45	Triarimol
Bavistin	Indofil Z-78	Tribasic copper sulphate
Benlate	Iprodione (Rovral)	Tribasic copper zinc
Benz (1,2) isoxazoles	Karbam white	Tridemorph
Bioquin	Kavach	Trifloxystrobin
Blitox	Lunasan	Trimiltox forte
Bordeaux mixture	Malic acid	Vinclozolin (Ronilan)
Boric acid	Mancozeb	Wettable sulphur
Boscalid	Maneb	Zato 50 WG
Brassicol	Manzate	Zincop
Brestan	Merpan	Zineb
Bromosan	Metalaxyl	Zinc sulphate
Calixin	Metiram	Ziram
Captaf	Miltox	**Antibiotics**
Captafol	4- Nitrosopyrazole	Griseofulvin
Captan	Nuarimole	Mycostatin
Carbendazim	Ozone	Mycothricin
Carboxiin (Vitavax)	Panogen	Polyoxin B and D
Ceresen	Parzate	**Insecticides**
Chlorothalonil (Daconil, Bravo)	Penconazole (Topaz)	Cypermethrin
Copper oxychloride (Blitox)	Pentachlorophenol	Deltamethrin
Copper sulphate	Phygon	Dimecron
Cuman L	P-Methoxytetrachlorophenol	Fenvalerate
Cupravit	Prochloraz	Flucythrinate
Cupric acetate	Propiconazole	Metasystox
Cuprox	Propineb	Methyl demeton
Cycloheximide	Pyraclostrobin	Permethrin
Delan C	Pyrena compounds	Phosphamedon
Dichlofluanid	zQuinolate	Rogar
Difolatan	Quinone	**Nutrients**
Difenoconazole (Score)	Ridomil MZ	N, P, K, Ca, $Cacl_2$
Dithane D-14	Ronilan	$CuSO_4$, $CO(NO_3)_2$, Fe EDTA,
Dithane M-45 (Mancozeb)	Semesan	Mn SO_4, Na_2 BO_7, Zn SO_4
Dithane Z-78	Signum 334WG	Thiourea
Duter	Sisthane	Borax
Edifenphos	Sodium fluoride	S, SO_2
Euparen	Spergon	Zn
Fenarimol	S-triazines	
Fenpropimorph	Sumilex (Procymidone)	

Z-78 (0.2 %), Daconil (0.2 %), Indofil M-45 (0.02 %), Duter (0.2 %), Folicur (500 g a.i./ha), Kavach (0.02 %), Ridomil MZ (0.2 %), Rovral (0.2 %), Procymidone (0.750 Kg a.i./ha), Prochloraz (500 g a.i./ha), TPTH (fentin hydroxide and triphenyltin hydroxide) (0.2 %) (Ansari et al. 1990; Chahal et al. 1977; Chand and Jatian 1969; Domsch 1957; Gaur and Ahmed 1980; Gupta et al. 1977, 1985; Howlidar et al. 1985; Kaspers and Siebert 1989; Keyworth 1969; Kolte 1985; Kolte and Tewari 1978; Lapis and Ricaforte 1974; Maude and Dudley 1972; Noon et al. 1988; Saharan 1991, 1992; Shivpuri et al. 1988; Singh 1986; Singh and Bhowmik 1985; Singh and Sobti 1980; Singh and Singh 2005; Verma and Saharan 1994; Yadav 2003) (Fig. 11.1).

Fungicides Mancozeb and Carbendazim caused 100 % reduction in mycelial growth of *Alternaria brassicae* in vitro, while 1 % (w/v) aqueous bulb extract of *Allium sativum* and leaf extract of *Acacia nilotica* caused significant reductions. In dual culture, GR isolate of *Trichoderma viride* performed the best among the test isolates of *Trichoderma*, causing 82 % reduction in mycelial growth of *A. brassicae*. Performance of isolates SI-2, P and SI-1 of *T. viride* was at par ($P<0.01$) with that of GR isolate (Meena et al. 2004). Conidial suspension and culture filtrate (3:1) of *T. viride* cause inhibition of radial growth of *A. brassicae* (74 %) and *A. brassicicola* (77 %) in dual cultures (Reshu and Khan 2012). Three systemic fungicides, Topsin-M (thiophanate methyl, 70%WP), Ridomil MZ (mancozeb, 64 %+metalaxyl, 8 % WP) and Bavistin (carbendazim, 50%WP), alone and in combination with four non-systemic fungicides, Captaf (captan, 50%WP), Indofil M-45 (mancozeb, 75 % WP), Indofil Z-78 (zineb-78%WP) and Thiram (thiram 75%WP) were evaluated both in vitro and in vivo for their effectiveness to manage *Alternaria* blight of rapeseed–mustard caused by *Alternaria brassicae*. All the fungicides were evaluated for their efficacy at various concentrations, viz. 50, 100, 150, 200 and 500 ppm, and were sprayed in the field at 0.2 % a.i. l^{-1}. All the fungicides significantly reduced the severity of the disease, but Ridomil MZ was most effective. Topsin-M at conc. of 500 ppm was most effective in reducing radial growth of the pathogenic fungi (74.2 %). Ridomil MZ reduced disease severity by 32 % and was followed in effectiveness by combination of Bavistin+Captaf (26.5 %). Maximum yield was obtained in plots sprayed with Bavistin+Captaf (1198 kg ha^{-1}) followed by Bavistin+Indofil Z-78 (1172 kg ha^{-1}). It was worth noting that highest net profit as well as the highest cost–benefit ratio was obtained with Bavistin+Indofil Z-78 (1:3.2) followed by Bavistin+Captaf (1:1.3) (Khan et al. 2007). Lower concentration of SO_2 (143 μg SO_2 m^{-3}) stimulates *Alternaria* blight of mustard, but higher concentration (214.5 μg SO_2 m^{-3}) suppresses the disease (Khan and Khan 2010). Six new fungicides, viz. azoxystrobin (Amistar), trifloxystrobin+tebuconazole ready-mix compound (Nativo), difenoconazole (Score), penconazole (Topas), mancozeb (Dithane M-45) and azadirachtin (Neemarin) were tested against *Alternaria* blight of rapeseed–mustard. Both in vitro and in vivo studies showed almost the same order of efficacy of the six different fungicides used. All fungicides were significantly effective in reducing disease severity and increased seed yield. However, the lowest disease severity (%), highest seed yield and significant increase in test weight were recorded from two sprays of Nativo (1 g l^{-1}) (Mondal et al. 2007b).

Three to four sprays of dithiocarbamate fungicides, copper oxychloride and Bordeaux mixture have been found effective for the control of *Alternaria* disease of rapeseed–mustard (Kolte 1985). The time of first spray, the interval and number of sprays would depend upon the type of crop. In *B. rapa* and *B. juncea* crops in India, a significant control of the disease was achieved if the first spray was given at 60–75 days after sowing; 2–4 more sprays at 10–15 days intervals, depending on the maturity of the crop, increased the control (Kaushik et al. 1983, 1984; Kolte 1985; Tripathi et al. 1987). The economics, in relation to number of sprays, have been calculated at Hisar, India, for the most effective five fungicides in rapeseed–mustard; it revealed that four sprays each of Dithane M-45, Dithane Z-78, Difolatan, Duter and Blitox-50 were required to minimize the disease and increased the yield.

However, the maximum net profit was obtained with four sprays of Dithane M-45 starting after 30–45 days of sowing and repeated at an interval of 15 days. Optimum growth stage of *B. juncea* in reducing black spot infection with higher yields through fungicidal sprays is 30–45 days after sowing. In Bangladesh, *Alternaria* blight of mustard was controlled by spraying Rovral (0.25) at 30, 40, 50 and 60 days after sowing (Sultana et al. 2009) (Tables 11.2, 11.3, and 11.4) (Kaushik et al. 1983, 1984; Saharan and Chand 1988; Tripathi et al. 1987; Saharan 1991).

Integrated control of *Alternaria* and aphid population was obtained with a mixture of three fungicides, Dithane M-45, Difolatan, Blitox and three insecticides, Rogar, Metasystox and Dimecron (Table 11.5). The maximum net profit was obtained with sprays of Difolatan + Metasystox followed by Difolatan + Dimecron, Blitox + Metasystox and Dithane M-45 + Metasystox (Saharan 1992; Tripathi et al. 1985). According to Thind and Jhooty (1988), Captafol (0.2 %) provides maximum disease control and persists longer on all brassicas. Mancozeb, Zineb and Copper oxychloride at 0.3 % exhibit highly differential efficacy against this disease on test brassicas. In mustard, Shivpuri et al. (1988) obtained best control of *Alternaria* blight and a significant increase in yield with Rovral followed by Captafol and Dithane M-45 (0.2 %). Sinha and Prasad (1989) found Captafol was best, followed by Dithane M-45, for the control of *Alternaria* blight of cauliflower to provide maximum yield. Similarly, Captafol applied at 0.2 % (w/v) four times at an interval of 15 days also reduced disease intensity from 69.8 to 15.6 % in *B. juncea,* 79.6 to 20.5 % in *B. rapa* variety Yellow Sarson and 74.6 to 19.6 % in *B. rapa* variety Brown Sarson; a significant increase in yield was also obtained (Saharan 1991). The optimum time for application of fungicides to control *Alternaria* disease of *B. juncea* has been found to be 30–45 days after sowing. Fungicidal sprays at 60 days after sowing are unable to arrest the disease.

In Bangladesh, among the three fungicides tested, Dithane M-45 proved to be the best for controlling *Alternaria* blight of *B. rapa* variety Sonari Sarisha. Maximum reduction in disease intensity and maximum increase in yield was obtained with seven sprays starting at 30 days after seeding. Disease control by Trimiltox forte and Cuprovit was also satisfactory (Meah et al. 1988).

In England, iprodione (0.5 Kg a.i./ha) or procymidone (0.75 Kg a.i./ha) has been shown to protect winter oilseed rape and turnip rape against *A. brassicae* for a period of up to 7 weeks following application at 95 % petal fall (GS 4.9) (Evans and Gladders 1981; Evans et al. 1983, 1984, 1988; Gladders and Rhodes 1985). Cox et al. (1981, 1983) in the UK and France also reported good control of this disease with a single spray of iprodione at the early pod stage. Control was obtained with high-volume sprays and with volumes as low as 22 L/ha applied by aircraft.

Flowable formulations are superior to wettable powder (Humpherson-Jones et al. 1981). Seed yield increase of up to 20–30 % together

Table 11.2 Optimum time for spraying Difolatan (2 g product/l water) for the control of *Alternaria* leaf blight of raya at different locations in 1979–1980 (Saharan 1984)

	Hisar		Ludhiana		Kanpur	
Sprays after sowing (days)	Disease intensity (%)	Yield (Qtls)	Disease intensity (%)	Yield (Qtls)	Disease intensity (%)	Yield (Qtls)
30–45	33.7	21.6	46.0	19.0	20.4	8.4
60	42.9	19.7	46.4	17.8	24.2	7.7
75	50.5	19.5	53.4	19.9	30.7	7.3
82	57.8	19.2	–	–	36.7	6.6
90	54.7	19.8	62.3	17.8	43.3	6.4
105	–	–	73.8	17.0	–	–
120	–	–	78.9	14.7	–	–
Control	63.4	18.3	84.4	9.1	46.9	6.1

Table 11.3 The efficacy and economics of fungicidal spray on *Alternaria* leaf spot of raya at Hisar in 1977–1978 (Saharan 1984)

Fungicides	No. of spray	Disease intensity (%)	Yield/ha (Qtls)	Net profit / ha (Rupees)
Dithane M-45	1	47	12	53
	2	45	13	72
	3	40	14	372
	4	29	15	803
Dithane Z-78	1	51	13	212
	2	42	13	70
	3	40	14	312
	4	33	15	751
Difolatan	1	42	12	12
	2	37	13	152
	3	32	15	639
	4	22	16	792
Duter	1	50	13	248
	2	38	14	444
	3	33	14	451
	4	28	16	748
Blitox-50	1	53	13	211
	2	47	14	357
	3	43	13	187
	4	32	15	748
Control	–	62	12	–

Table 11.4 Optimum growth stage of *B. juncea* for reducing black spot through fungicidal sprays (Saharan 1991)

Fungicidal[a] sprays after days of sowing	Disease intensity (%)	Yield (Q/ha)
30	15.5	21.5
45	19.0	21.3
60	40.6	18.2
75	49.5	17.0
90	65.5	15.0
Control	75.5	12.0
C.D. at 5 %	3.33	–

[a]Captafol (0.2 %) was sprayed

with low residual levels of inoculum on the seed has been reported following application of iprodione (Ogilvy 1984). Two sprays of Prochloraz at 500 g a.i./ha, the first at stem extension and second at mid-late flowering, have also proved effective in controlling this disease in winter oilseed rape together with a significant increase in yield (Marshall and Harris 1984). Procymidone, Iprodione and Prochloraz fungicides have also been reported effective against *Alternaria* spp. on winter oilseed rape in Poland (Bonin and Fratczak 1987) and Chinese cabbage in Denmark (Anonymous 1988). Tebuconazole suppresses the *Alternaria* blight of rape in Lithuania more effectively than prochloraz. The highest efficacy is recorded when the fungicides are applied after the first symptoms of the disease are spotted on siliques (BraZauskiene and Petraitiene 2004).

In India, iprodione spray (500 g a.i./ha) during the early pod stage is most effective in reducing *Alternaria* blight of mustard with increase in yield by 24–59 % (Mukherjee et al. 2003). In Czechoslovakia, *A. brassicae* and *A. brassicicola* on cauliflower were successfully controlled, and yield increased by 2–22 times after spraying with a mixture of Thiram + Benlate alternated with Mancozeb + insecticide (Kudela et al. 1978).

For the control of *Alternaria* blight on radish seed crop, Dithane-M-45 (0.25 %) proved most effective, followed by Bordeaux mixture (0.4 %) in increasing seed weight, improving seed germination, and in reducing seed infection (Hussaini and Singh 1989). However, Mondal et al. (1989) found Rovral (0.2 %) best followed by Dithane M-45, Dithane Z-78 and Captan for the control of *Alternaria* blight of radish.

In the UK, Channon (1970) found Propineb and Mancozeb (0.2 %) sprays consistently effective against *A. brassicicola* on cabbage in glasshouse. Similarly, three sprays of iprodione at 0.5–1 Kg a.i./ha rate, at 3 weeks interval starting at the young green pod stage till harvesting, controlled pod infection of *A. brassicicola* and increased yields (Humpherson-Jones and Maude 1982). In India, Singh et al. (1989) found three sprays any of Mancozeb, Cuman L and Zineb found best for the control of *A. brassicicola* leaf spot of cabbage.

Storage rot of white cabbage due to *A. brassicicola* can also be effectively controlled with application of 5 % iprodione mixed in talc powder (Kear et al. 1977). Similarly, a post-harvest iprodione drench on stored winter cabbage controls *Alternaria* spp. (Geeson and Browne 1979). On the other hand, Tasca and Trandaf (1984)

Table 11.5 The effect of fungicide and insecticide mixture on *Alternaria* leaf spot and aphid population on Brown Sarson (BSH-1) at Hisar in 1977–1978 (Saharan 1984)

Mixture of fungicide and insecticide	Aphid population/10 twigs	Disease intensity (%)	Yield/ha (Qtals)	Net profit/ha (Rupees)
Dithane M-45 + Rogar	72	36	9.0	875
Dithane M-45 + Metasystox	57	25	9.7	1088
Dithane M-45 + Dimecron	75	42	8.8	766
Difolatan + Rogar	75	31	8.4	459
Difolatan + Metasystox	62	17	10.7	1285
Difolatan + Dimecron	72	33	10.4	1279
Blitox + Metasystox	70	48	9.9	1151
Blitox + Rogar	107	49	8.8	748
Blitox + Dimecron	67	56	8.9	836
Control	820	72	5.6	–
C.D. at 5 %	6.8	6.6	–	–

reported pre-harvest treatment of cabbage with Mancozeb (0.2 %) or Folpet (0.2 %) for protection from *A. brassicae* storage rots. Ronilan-smoke, a vinclozolin formulation used as smoke pellets, in *Brassica* storage rooms also controls *A. brassicae* (Jennrich 1985).

Foliar diseases of Brussels sprouts in the UK are effectively controlled with the application of mixtures of iprodione, triadimefon/Carbendazim and metalaxyl/Mancozeb. The timing of the sprays is more important than the number of applications. Two sprays, one in August and another one in September, are enough to reduce fungicide input (Brokenshire 1987; Davies et al. 1986).

11.4.1 Effect of Fungicides on Host Growth

Fungicides are frequently applied for crop protection in Brassicas from flowering to maturity. These fungicides may interfere with microsporogenesis and on pollen production, which are sometimes less vigorous/non-viable or may affect pollen tube growth through the style and thus interfere with fruit set and economic yield adversely (Williams et al. 1987). However, these aspects have not received due attention. Available information indicates that Benomyl (methyl 1-(butyl carbamoyl)-2-benzimidazole-carbamate) inhibits pollen germination and/or tube elongation in apple and cucumber. He et al. (1996) reported that lower concentration of Benomyl has limited effects on pollen germination, and tube elongation, while higher concentration inhibits germination appreciably and causes greater number of ruptured and swollen tubes in *Tradescantia virginiana.*

Application of Dithane M-45 also inhibits both pollen germination and tube growth of *B. juncea*. The germination inhibition is more pronounced in RH-8812 than RH-30. However, it inhibits germination completely in RH-30 at 0.2 % concentration, while a substantial number (18.85 %) of pollen germinate in RH-8812 at same concentration. Lower concentration (0.025 %) of fungicide decreases tube length significantly (38.84 %) in RH-30, which remains unaffected in RH-8812 by the same concentration of fungicide (Table 11.6). Pollen germination also decreases with increasing concentration of Ridomil MZ-72, and it is completely inhibited at a concentration of 0.15 % and above. The overall reduction in germination is more severe in RH-8812 and RH-30. However, converse is more true with regard to tube length (Table 11.6) (Jain et al. 2000). Spraying Ridomil MZ-72 at 0.0125 % may not have deleterious effect on pollen biology, and consequently, crop yield as has been suggested by Kolte et al. (1981).

Table 11.6 The effect of Dithane M-45 and Ridomil MZ-72 on in vitro pollen germination (%), and tube length (μm) in two cultivars of *B. juncea* (Jain et al. 2000)

Fungicide conc. (%)	Pollen germination (transformed values)						Pollen tube length					
	Dithane M-45			Ridomil MZ-72			Dithane M-45			Ridomil MZ-72		
	RH-30	RH-8812	Mean	RH-30	RH-8812	Mean	RH-30	RH-8812	Mean	RH-30	RH-8812	Mean
Control	58.4 (72.5)	55.0 (67.0)	56.7 (69.7)	58.4 (72.5)	55.0 (67.0)	56.7 (69.7)	193.6	105.3	149.5	193.6	105.3	149.5
0.025	56.2 (69.1)	38.9 (39.4)	47.6 (54.2)	41.2 (43.4)	19.8 (11.7)	30.5 (27.5)	118.4	105.3	111.9	73.2	85.1	79.2
0.05	37.2 (36.7)	39.0 (39.1)	38.1 (37.9)	40.3 (41.9)	18.1 (9.7)	29.2 (25.8)	114.9	96.9	105.9	62.2	75.5	68.9
0.10	32.6 (29.4)	28.2 (22.3)	30.4 (25.8)	28.2 (22.3)	16.8 (8.7)	22.5 (15.5)	114.1	86.0	100.1	48.2	74.6	61.4
0.15	32.1 (28.3)	24.3 (16.9)	28.2 (22.6)	0.0 (0.0)	0.0 (0.0)	0.0 (0.0)	95.6	72.8	84.2	0.0	0.0	0.0
0.20	0.0 (0.0)	18.8 (10.5)	9.4 (5.2)	0.0 (0.0)	0.0 (0.0)	0.0 (0.0)	0.0	68.4	34.2	0.0	0.0	0.0
Mean	36.1 (39.3)	34.0 (32.7)	–	28.0 (30.0)	18.3 (16.2)	–	106.1	89.1	–	62.9	56.7	–
C.D. ($P<0.05$)	Cultivar (cv.)			2.29						5.7		
	Fungicide (F)			1.87						3.3		
	Cv×F			4.57						8.1		

Figures in parentheses are mean of actual values, while others are angular transformed ones of the same

11.5 Biological Control

11.5.1 Plant Extracts as Fungitoxicants

Essential oil from the roots of radish inhibits (1:2500) *A. brassicae* (Nehrash 1961). The deproteinized leaf extracts of *Acacia nilotica, Enicostema hyssopifolium, Mimosa hamata* and *Vitis vinifera* have shown fungistatic activity against *A. brassicicola* (Umalkar et al. 1977). The extracts prepared from the leaves of *Lawsonia alba,* roots of *Datura stramonium* and inflorescence of *Mentha piperita* have fungitoxic activity against *A. brassicae* isolated from cauliflower leaves. Extracts of the ferns *Adiantum caudatum, Diplazium esculentum* and *Pteris vittata* reduce growth and germination of *A. brassicicola* (Yasmeen and Saxena 1990; Ahmed and Agnihotri 1977).

Plant extracts of roots, leaves, stems, inflorescence and fruits of several species (Table 11.7) have shown fungicidal activities in in vitro and in vivo conditions against two major *Alternaria* spp., viz. *A. brassicae* and *A. brassicicola* causing blight or black spot diseases of crucifers. In in vitro conditions, almost all of the plant extracts restrict mycelial growth, sporulation and conidial germination. Some of them have shown efficacy under field conditions by restricting the infection points, spot size and disease intensity (Khurana et al. 2005; Patni et al. 2005; Muto et al. 2006; Ho et al. 2006, 2007; Patni and Kolte 2006; Guleria and Kumar 2010; Guleria et al. 2010; Baka 2010;

Table 11.7 Plant extracts tested against *Alternaria* spp. attacking crucifers

Plant species	Concentration (%)	References
Acacia nilotica	–	Umalkar et al. (1977)
Encostema hyssopifolium	–	Umalkar et al. (1977)
Mimosa hamata	–	Umalkar et al. (1977)
Vitis vinifera	–	Umalkar et al. (1977)
Lawsonia alba	–	Umalkar et al. (1977); Yasmeen and Saxena (1990)
Datura stramonium	–	Umalkar et al. (1977)
Mentha piperita	–	Yasmeen and Saxena (1990)
Adiantum caudatum	–	Ahmed and Agnihotri (1977)
Diplazium esculentum	–	Ahmed and Agnihotri (1977)
Pteris vittata	–	Ahmed and Agnihotri (1977)
Allium sativum	5	Khurana et al. (2005)
Azadirachta indica	10	Sasode et al. (2012)
Lawsonia inermis	5	Meena and Sharma (2012)
Erythrina chiaposana	5	Meena and Sharma (2012)
Ricinus communis	5	Meena and Sharma (2012)
Zingiber officinale	5	Meena and Sharma (2012)
Euphobia pulchrima	5	Patni et al. (2005)
Rumax dentatus	5	Patni et al. (2005)
Urtica dioica	5	Patni et al. (2005)
Eucalyptus globules	5	Patni et al. (2005)
Solanum nigrum	5	Muto et al. (2006)
Rhus chinensis	–	Ho et al. (2006)
Ocimum sanctum	5	Patni and Kolte (2006)
Anagallis arvensis	5	Patni and Kolte (2006)
Polygonum perfoliatum	1	Ho et al. (2007)
Agave americana	1	Guleria and Kumar (2010)
Solanum xanthocarpum	1	Guleria et al. (2010)
Lavandula pubescens	10	Baka (2010)
Calotropis procera	15	Singh et al. (2013)

Meena and Sharma 2012; Sasode et al. 2012; Singh et al. 2013).

Eucalyptus spray gives significantly lesser number of spots per leaf (2.05), minimum size of spots (1.28 mm), minimum sporulation intensity (1.22×10^5) and minimum disease index followed by *Calotropis*, *Ocimum* and *Polyanthai* extracts at 5 % (Patni and Kolte 2006). *Alternaria* blight of mustard can be managed by spraying aqueous extract of *Azadirachta indica*, *Allium sativum* and *Zingiber officinale* at 5, 10, 15 and 20 % along with higher yield. However, spraying of 15 % *A. indica* extract gives highest yield with best cost–benefit ratio (Mahapatra and Das 2013). While, according to Meena et al. (2004, 2008), application of bulb extract of *Allium sativum* at 45 and 75 d.a.s. gave highest seed yield at Sewar, India. The *Allium sativum* extract is most effective in controlling seed-borne fungi of mustard also (Latif et al. 2006).

11.5.2 Antagonists for Biocontrol

Saprophytic phylloplane fungi such as *Aureobasidium pullulans* and *Epicoccum nigrum* are pathogenic to *A. brassicicola* (Pace and Campbell 1974). The *Verticillium* state of *Nectria inventa* Pethybridge, a destructive mycoparasite, is one of the dominant phylloplane fungi of rapeseed (Tsuneda and Skoropad 1978b). Among the leaf surface mycoflora, the most antagonistic fungi are *E. purpurascens, A. pullulans* and *Cladosporium cladosporioides* in the case of *A. brassicae.* The metabolites of *Acremonium roseogriseum, Aspergillus terreus* and *C. cladosporioides* inhibit *A. brassicae.* The most significant effects are observed when the spores of the leaf surface fungi, and their metabolites, are sprayed prior to inoculation of pathogen on leaves (Rai and Singh 1980).

Pre-treatment application with the spore suspension of *Streptomyces rochei* or its diffusate results in a marked reduction of leaf blight intensity caused by *A. brassicae* and *A. brassicicola* on Brown Sarson, *B. rapa* variety *Dichotoma* (Sharma and Gupta 1978a, b, 1979). With the decline in the population of *S. rochei,* there is a rise in the population of *Alternaria* from December onwards indicating the possibility of antagonism. Several factors, such as climatic conditions, age of the host plant, maturity of the leaves and variety of the host, have profound influence on the deposition of conidia of both the pathogens and *S. rochei* (Sharma and Gupta 1980). Jayant and Sinha (1981) reported that *S. hygroscopicus* is strongly antagonistic to *A. brassicae* and *A. brassicicola.* When the culture filtrate is sprayed over the Sarson plants, a week before or a week after spraying the spore suspension of *A. brassicae* and *A. brassicicola,* germination and disease development are reduced. Whenever the spore population of *Streptomyces arabicus* is higher on leaves of Yellow Sarson and Taramira, the conidial population of *A. brassicae* and *A. brassicicola* declines. The population of antagonistic *S. arabicus* was higher in the young leaves as compared to old ones (Sharma et al. 1984). The antifungal substance in the diffusate of the antagonist is thermolabile (Sharma et al. 1985). A pigmented and xylose-utilizing strain of *S. bobili* is found to be active against *A. brassicae, A. brassicicola* and *A. raphani* (Sharma and Sinha 1989).

An isolate of *Streptomyces* spp. obtained from light-coloured, Finnish Sphagnum horticultural peat has proved effective biological control agent against plant pathogens (Tahvonen 1982a, b). Treating cauliflower seeds with *Trichoderma viride* and *Streptomyces* spp. isolates inhibits or reduces damping-off caused by *A. brassicicola* (Tahvonen 1982b, 1988). Seed dressing with Mycostop, a powdery formulation prepared from spores and mycelium of *Streptomyces,* was successful in controlling damping-off (80–90 %) from seeds artificially infected with *A. brassicicola.* The dressing remains effective on seeds stored under dry conditions for 5–6 weeks, but subsequently decreases slowly. *Streptomyces* dressing controls in a manner comparable to chemical dressing with Thiram, preventing damping-off caused by *Alternaria* fungi in seedlings, which are grown from commercial seed lots of different origin (Fig. 11.2; Tahvonen 1985; Tahvonen and Avikainen 1987). White et al. (1990) reported that Mycostop, a biofungicide based on a selected *S. griseoviridis* isolate from Finnish Sphagnum peat, introduced either by seed treatment or soil treatment, controls *A. brassicicola* of cauliflower and cabbage (Fig. 11.1).

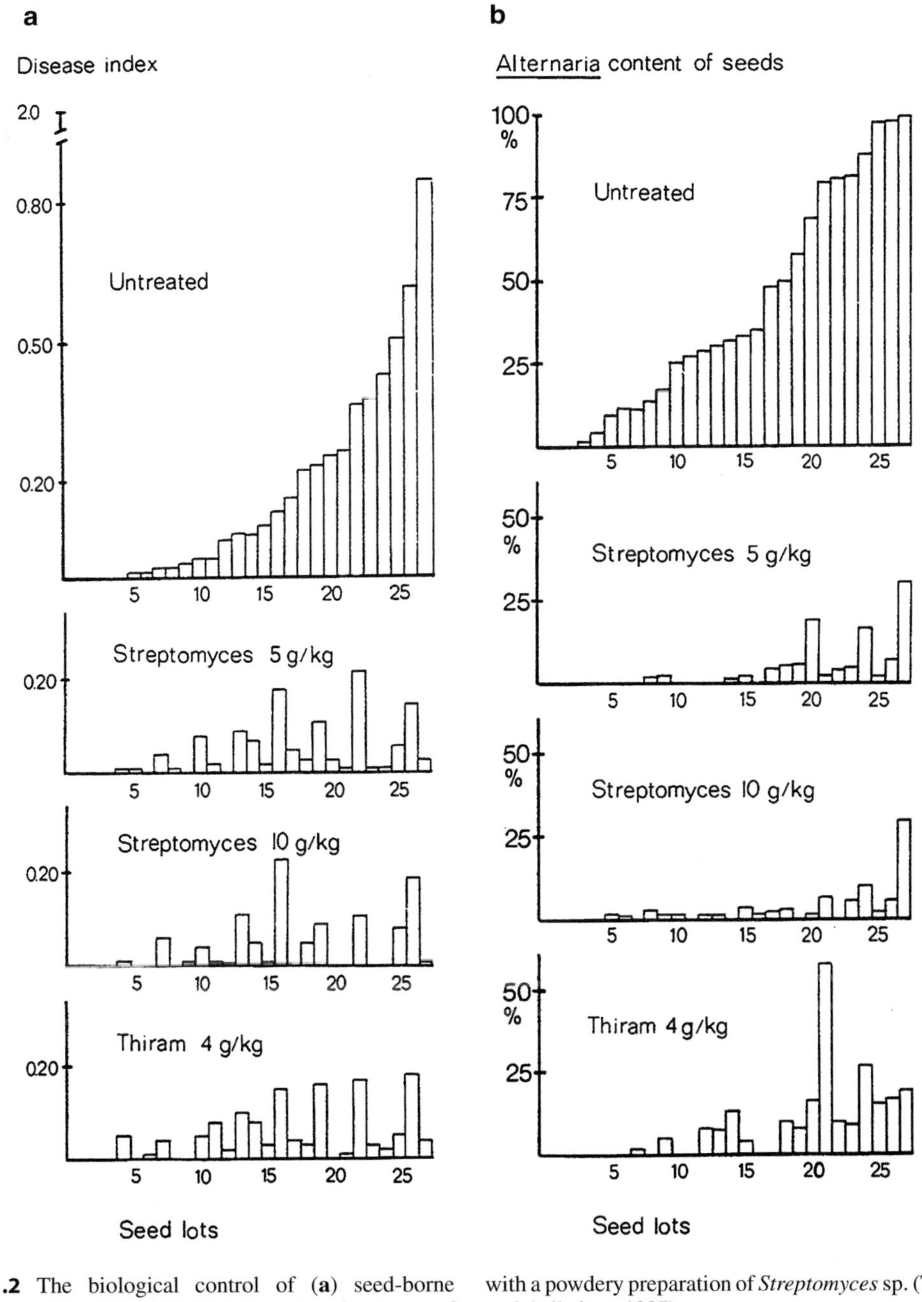

Fig. 11.2 The biological control of (**a**) seed-borne *Alternaria brassicicola* and (**b**) *A. brassicicola* content of seedlings grown from naturally infected *Brassica* seed with a powdery preparation of *Streptomyces* sp. (Tahvonen and Avikainen 1987)

Brassicaceous seeds treated with either *Gliocladium virens*-19, *Trichoderma harzianum*-22, *T. harzianum*-50, *Penicillium corylophilum*-36 and *P. oxalicum*-76 gave significantly higher emergence of healthy seedlings. The hyphae of antagonistic fungi were able to adhere to conidia or coil around or penetrated germ tubes or hyphae of *A. brassicicola.* Conidia of *A. brassicicola* shrivel and plasmolyze in the presence of antagonistic fungi (Plate 11.2; Wu and Lu 1984). For biocontrol of seed-borne *A. raphani* and *A. brassicicola* of radish, antagonists like *Chaetomium globosum, T. harzianum, T. koningii* and *Fusarium* spp. treated plants showed consistently increased numbers of healthy seedlings (Vannacci and Harman 1987). Wu and Lu (1984) found *Trichoderma, Gliocladium* and *Penicillium* spp. parasitizing *A. brassicicola.* The application of *A. alternata,* prior to inoculation with *A. brassicae,* reduced the level of *A. brassicae* on oilseed rape by about 60 % and subsequent to inoculation by about 26 % (Mee 1991). Spraying of *T. viride* conidial suspension causes reduction in *Alternaria* blight of mustard on leaves (76 %) and pods (68 %), respectively. The bioagents survive on phylloplane up to 30 days at 80–90 % RH with 20–35 °C temperature (Reshu and Khan 2012). In Poland, Madej (1986) reported *Gonatobotrys simplex* as a hyperparasite of *A. brassicae.*

11.5.3 Mechanisms of Biocontrol

Antagonistic mechanism of biocontrol has been studied in a few selected host–parasite systems. The parasitism of *A. brassicae* by the *Verticillium state* of *Nectria inventa* occurs either by penetration or contact without penetration. Parasitic hyphae

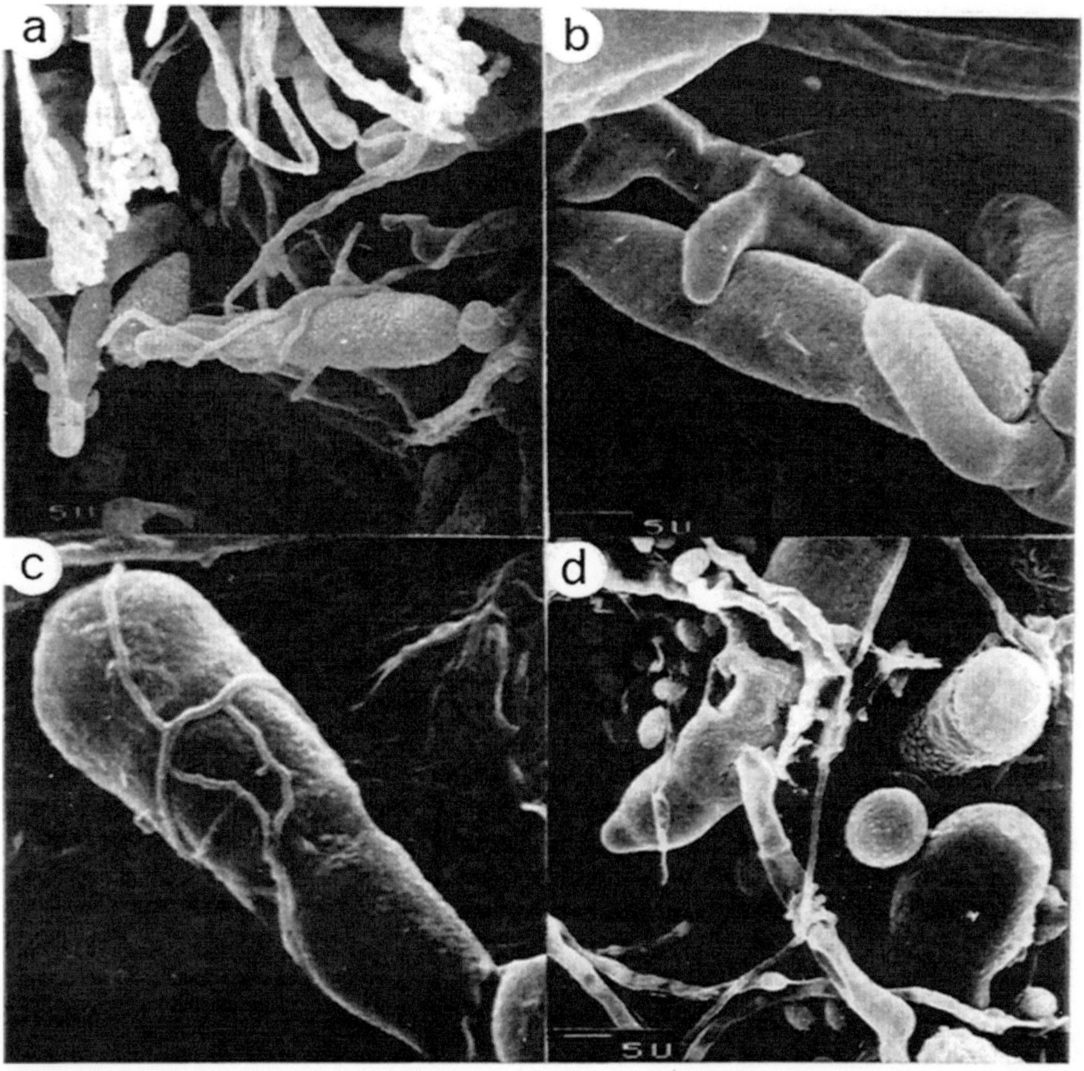

Plate 11.2 The effectiveness of antagonistic fungi against *Alternaria brassicicola*; (**a**) *Penicillium corylophilum*-36 and (**b**) *Trichoderma harzianum*-22 coiled around the conidia on agar medium; (**c**) *T. harzianum*-22 coiled around a conidium on a cabbage seed; and (**d**) *P. citrinum* coiled around the condial germ tube on a cabbage root (Wu and Lu 1984)

induce abnormal responses in host cells upon contact. A reaction consisting largely of an electron-dense transparent matrix and dispersed tubule-like electron-dense material develops between the cell wall and the invaginated plasma membrane. The tubule-like elements subsequently aggregate to form electron-dense deposits below the cell wall. The affected cell forms a septal plug, accumulates membranes and finally degenerates. Hyphae of *N. inventa* penetrate the conidial cells of *A. brassicae* mainly by a process, which appears primarily enzymatic in nature. The cytoplasm of the penetrated cell becomes progressively less dense, and the cell eventually appears empty (Tsuneda 1977; Tsuneda and Skoropad 1980; Tsuneda et al. 1976). Penetration of *A. brassicae* hyphae causes separation of the cells; penetration of conidia occurs most frequently at the septum or at the basal pore in juvenile conidia (Tsuneda and Skoropad 1977a, b). Conidia of *N. inventa* require at least 24 h to initiate germination and 4 days to parasitize *A. brassicae* on intact leaves (Plates 11.3, 11.4, 11.5, 11.6, and 11.7; Tsuneda and Skoropad 1978a, b). Conidia of *A. brassicae* release various amino acids and sugars when they are exposed to an alternate dry–wet conditions; the longer the drying period, the larger the amount of leakage is. Among the amino acids, glutamine is exuded in the largest amount followed by aspartic acid and glutamic acid. The sugar fraction consists mainly of glucose and fructose. These leaked nutrients stimulate germination and growth of *N. inventa*. As a result, the germination of *Alternaria* conidia is drastically reduced to less than 5 %, and they are eventually destroyed by *N. inventa* (Tsuneda and Skoropad 1978c). The treatment of oilseed rape seeds with *Bacillus* strains UCMB-5113 and UCMB-5036 results in significant protection against all fungal pathogens tested. The protection seems to be a combination of lower infection frequency and less severe disease symptoms on treated plants. Phenological analysis of plants for fitness effects due to *Bacillus* inoculation showed a significantly higher survival rate for the seeds treated with UCMB-5033, UCMB-5036 and UCMB-5113 and providing a growth-promoting effect. No visible signs of infection such as chlorosis or yellowing are observed on plants that germinated after bacterial treatment. The germination efficiency was also same as the seed harvested from different plants including control (Danielsson et al. 2006). *Bacillus subtilis* strain UK-9, an isolate from reclaimed soils, was studied for its biological control activity against *Alternaria* leaf spot of mustard. In dual culture, production of antifungal metabolites by the bacteria caused morphological alterations of vegetative cells, and spores, disruption and lysis of their cell wall. The antagonists reduced spore germination on leaves, and disease incidence of the pathogen in plant as well demonstrated plant growth-promoting ability (Sharma and Sharma 2008). Many antagonistic species produce antibiotic substances, which have the ability to produce

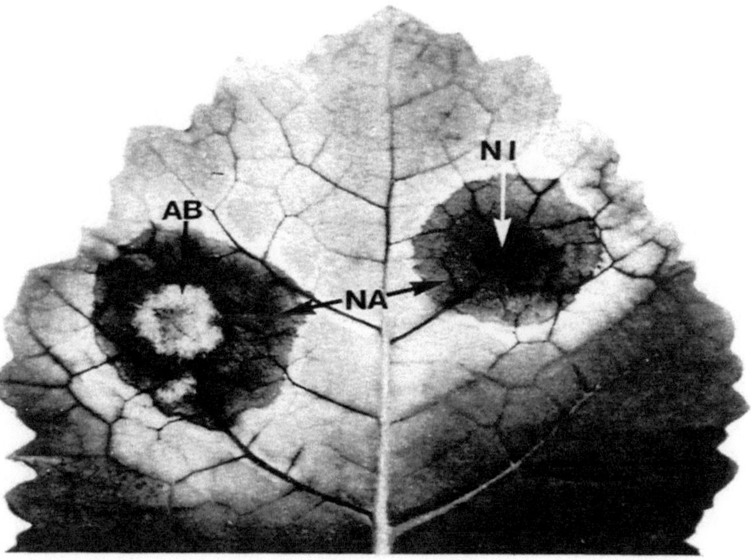

Plate 11.3 Excised non-dried rapeseed leaf inoculated with *Alternaria brassicae* alone (left half of leaf) and with the mixture of *A. brassicae* and *Nectria inventa* (right half of leaf) at 1 week after inoculation. In the necrotic areas (NA) caused by *A. brassicae,* note the presence of a white mycelia mass of *A. brassicae* with conidia (AB) on the left half and its absence on the right half of the leaf with abundant conidia of *N. inventa* (NI). × 1.5 (Tsuneda and Skoropad 1978b)

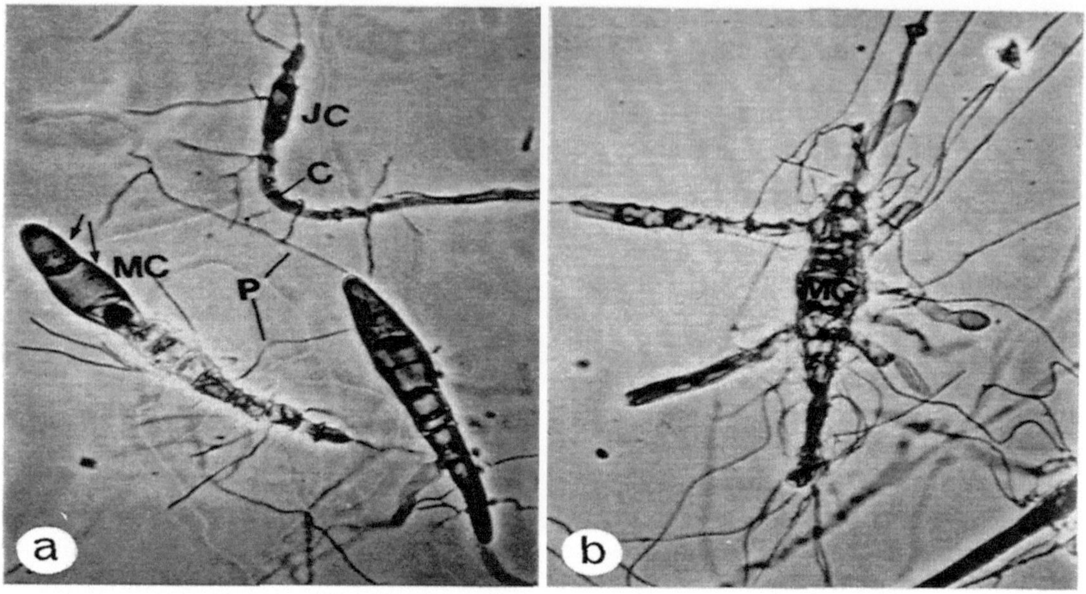

Plate 11.4 Phase contrast light micrographs of the conidia of *Alternaria brassicae* parasitized by *Nectria inventa*: (**a**) healthy-appearing conidium and infected mature and juvenile conidia. Note non-infected cells (*arrows*) in the heavily infected conidium. × 1,200; and (**b**) profuse growth of parasitic hyphae around a host conidium. × 1,200. Legend: *P* = parasite, *C* = host condiophore, *MC* = mature host conidium, *JC* = juvenile host conidium (Tsuneda et al. 1976)

enzymes causing the lysis of cell wall components of the pathogenic fungus. This helps the antagonist to penetrate the host hypha and grow as a hyperparasite (Tapio and Pohto-Lahdenpera 1989).

11.5.4 Biological Control vs. Biochemical Changes

Indian mustard seeds were treated separately with three bioagents, viz. *Trichoderma harzianum*, *Pseudomonas fluorescens* and *Bacillus subtilis* and grown in experimental fields, followed by spraying with these antibiotics at 30 and 60 days after sowing (DAS). The use of biocontrol agents showed enhanced growth in comparison to control and decreased disease index on leaves as well as on pods. These biocontrol agents also enhanced the content of dry matter, total phenol, ortho-dihydroxyphenol, starch, total soluble sugars, reducing sugars, total lipids and different membrane lipids in the leaves. However, the total protein content decreased. It suggests that treatment with biocontrol agents initiates in the plants a number of biochemical changes, which can be considered to be a part of plant defence responses (Sharma et al. 2010a). Biochemical analysis reveals that application of biocontrol agents results in increase in lipid and protein content in seeds collected from treated plants. The proportion of various lipidic fractions, i.e. phospholipids, glycolipids and sterol contents in seeds, also increased with a corresponding decrease in total glycerides. The proportion of 18:3, 20:1 and 22:1 fatty acids increased, while that of 18:1 and 18:2 fatty acids decreased in seeds with application of biocontrol agents. There are both qualitative and quantitative differences in the banding patterns of albumin and globulin proteins after application of biocontrol agents. The biochemical alterations in the host induced by treatment with biocontrol agents can be associated with defence mechanisms and enhanced growth of the plant (Sharma et al. 2010b).

11.6 Host Resistance

Four species of *Alternaria*, i.e. *A. brassicae, A. brassicicola, A. raphani* and *A. alternata*, are pathogenic on brassicas. Out of these, *A. brassicae* and *A. brassicicola* are widespread and economically most important since they cause losses of

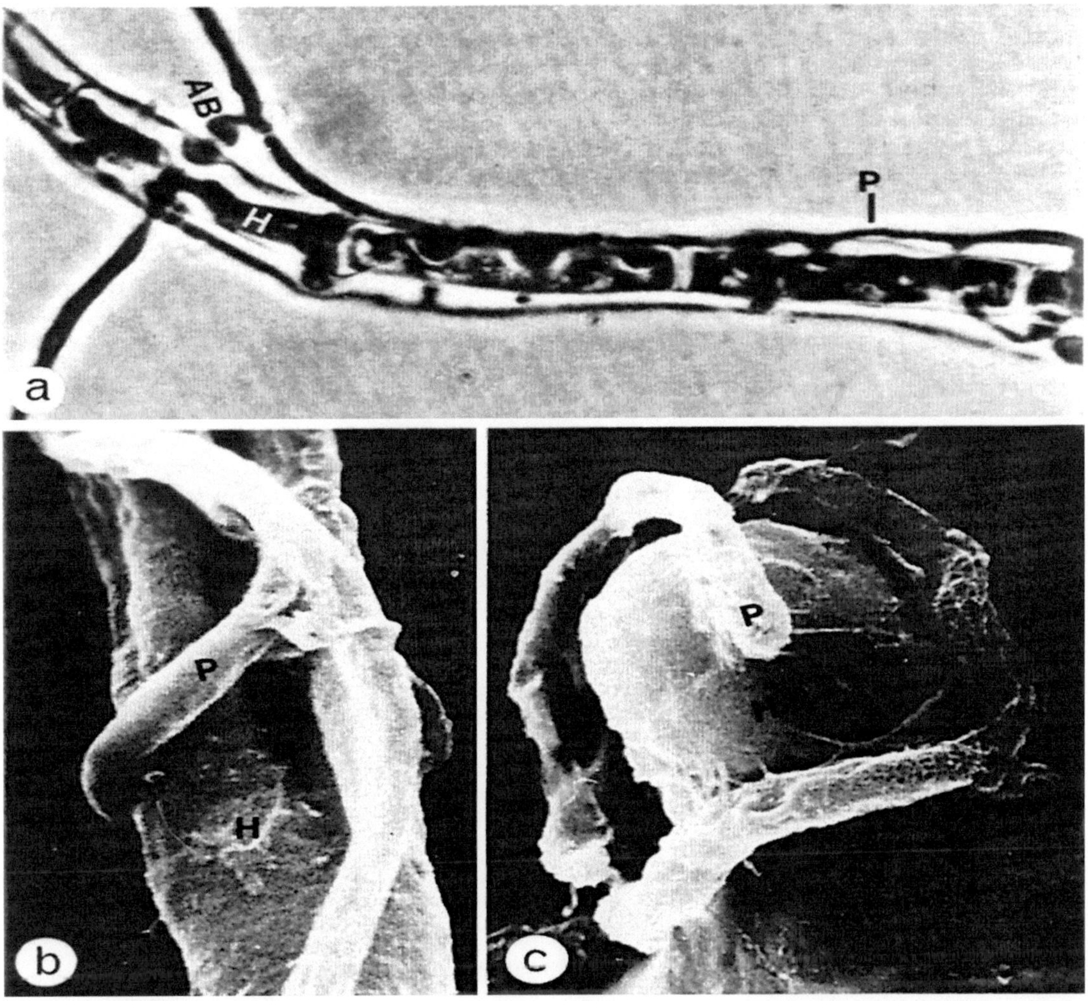

Plate 11.5 Hyphae of *Alternaria brassicae* parasitized by *Nectria inventa;* (**a**) light micrograph of parasite hyphae parallel to a host hypha. Note the swollen appressorium-like body of the parasite. × 4300; and (**b, c**) scanning electron micrographs of the parasite coiling around host hyphae. ×15,000. Legend: *P* = parasite, *H* = host, *AB* = appressorium-like body of parasite, *HB* = hyphal branch of the host (Tsuneda et al. 1976)

more than 70 %. All commercial brassicas are essentially susceptible to these pathogens. However, there are some differences in their degree of susceptibility. *Brassica napus* and *B. carinata* are less susceptible than *B. rapa* and *B. juncea*. In addition to these, some wild crucifers have high degree of resistance to *A. brassicae,* and attempts are being made to transfer this resistance to cultivated brassicas. *Alternaria* blight management strategies are to incorporate the host resistance as one of the components into the cultivated brassicas. One should have the knowledge of pathogenic variability development and standardization of effective germ-plasm screening techniques from diverse sources to broaden the genetic base: availability of usable source of resistance, nature and mechanism of resistance. The various techniques are available to screen germ plasm/varieties/lines, etc., against *A. brassicae* in the field, screen house/glass house and in the laboratory for resistance. Various studies have also been conducted which indicate the existence of *A. brassicae* pathotypes (Saharan 1992; Saharan and Kadian 1983b; Saharan et al. 2003).

Inheritance of resistance in inter- and intraspecific crosses of *B. juncea* x *B. carinata* genotypes reveals polygenic, additive gene effects, dominance

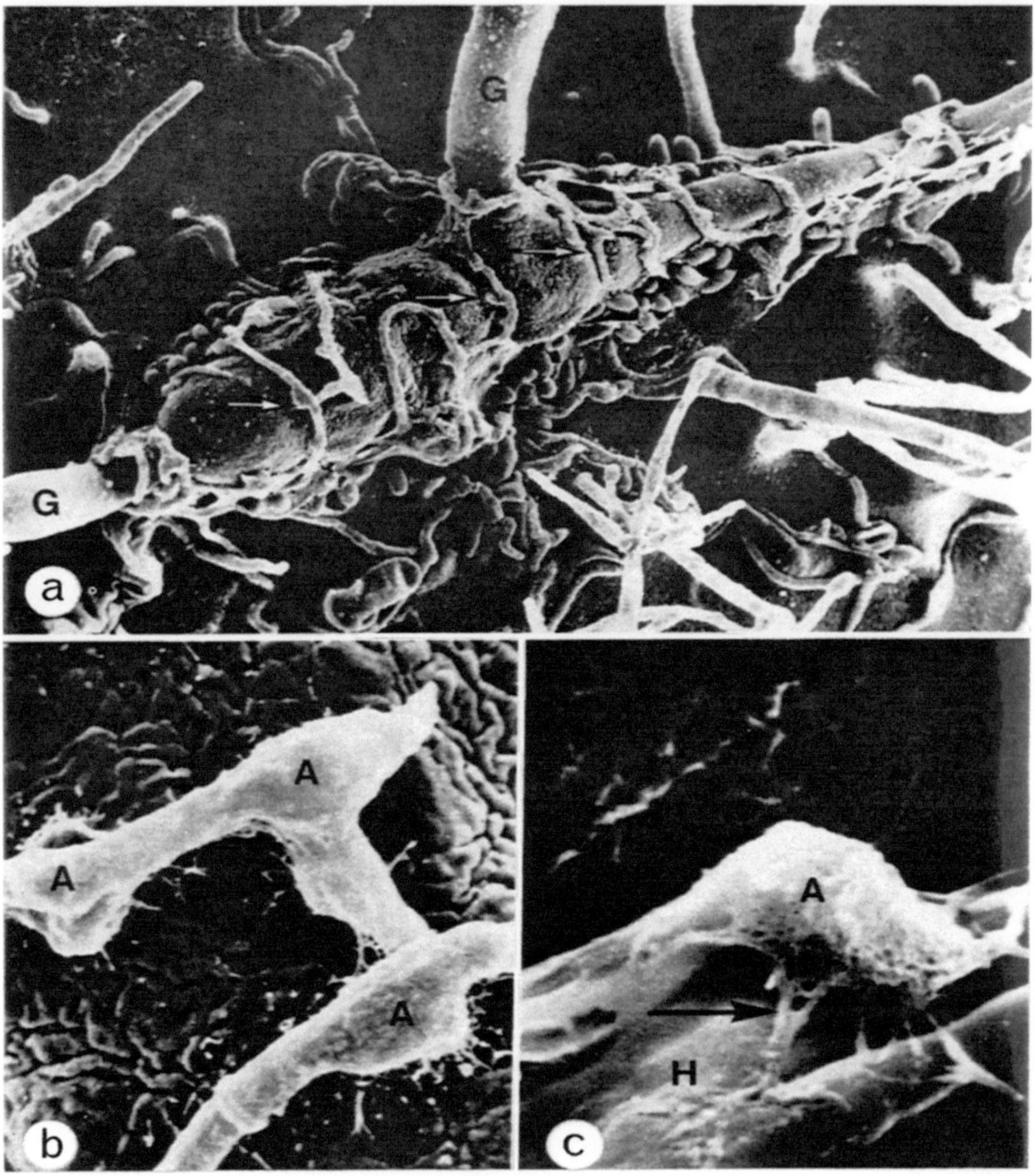

Plate 11.6 Scanning electron micrograph of mycoparasite *Nectria inventa* hyphae growing on *Alternaria brassicae;* (**a**) parasite hyphae occurring predominantly in the septal area (*arrows*) and the basal portions of the germ tubes (G) of a host conidium. × 1800; (**b**) appressorium-like bodies (A) formed on the host conidum. Note presence of adhesive material under these bodies. × 11,000; and (**c**) appressorium-like body (A) with fibrous adhesive material (*arrow*) formed on host hypha (H). × 25,000 (Tsuneda and Skoropad 1977a)

gene effects, additive x dominance epistasis and non-allelic interactions. Inter-mating between resistance plants helps in increasing the level of resistance through resistance gene pyramiding. Disease tolerance attributes in *B. juncea* genotypes have been identified. Identification of different components of horizontal resistance in *B. napus* cv. Tower and *B. juncea* cv. RC-781 can be exploited to breed for horizontal/slow blighting resistance. Rapid progress has been made in identification, cloning and sequencing of resistant genes in *B. juncea*, *B. napus*, *R. sativus* and *A. thaliana* (Krishnia et al.

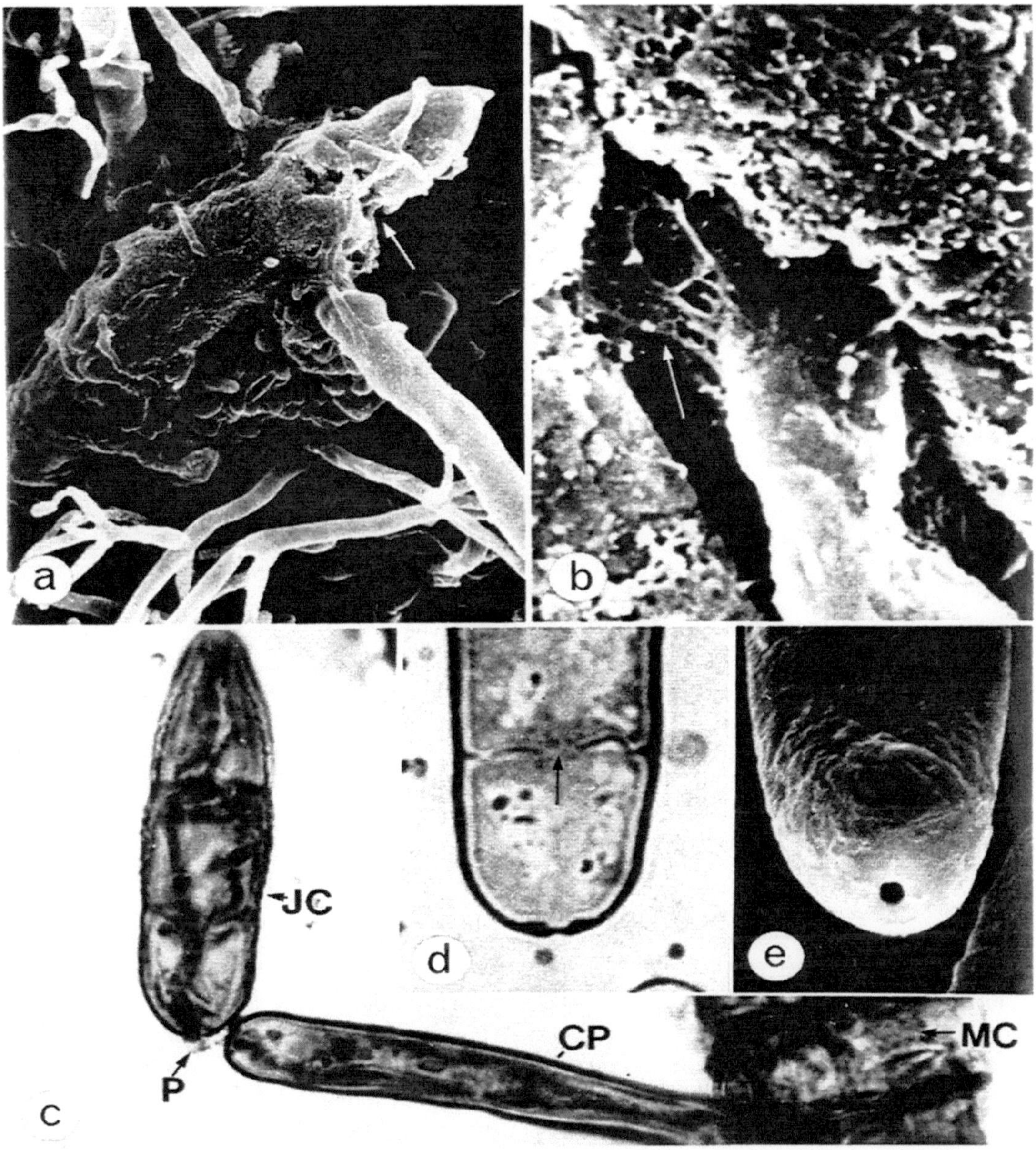

Plate 11.7 The *Alternaria brassicae–Nectria inventa* host–parasite interface; (**a**) Mature conidium of *A. brassicae* penetrated by hyphae of *N. inventa*. Note collapsed cell wall of the conidium (*arrow*)×1.800; (**b**) enlarged view of penetration site. A large hole develops in the wall of the host cell, and a meshwork of material appears at the penetration site (*arrow*)×27,000; (**c**) penetration of a juvenile conidium (JC) of *A. brassicae* by *N. inventa* (P). The host conidium is penetrated through the basal pore. MC, infected cell of a mature conidium. CP, conidiophore produced by the mature conidium×2500; (**d**) light micrograph of a thin section of a normal juvenile conidium of *A. brassicae* showing the basal pore and septal pore (*arrow*)×2800; and (**e**) Scanning electron micrograph of a basal pore in a juvenile conidium of *A. brassicae*×3500 (Tsuneda and Skoropad 1977a)

2000; Saharan and Krishnia 2001; Gupta et al. 2002; Saharan and Kadian 1983a; Kumar and Kolte 2001, 2006; Kanrar et al. 2002; Kamble and Bhargava 2007; Mondal et al. 2007a; Minami et al. 2008; Mishra et al. 2010; Verma et al. 2012; Chhikara et al. 2012; Sellam et al. 2007).

The effect of host-specific phytotoxins produced by *Alternaria* species on cruciferous plants at physiological, biochemical and molecular levels has been determined. The role of phytotoxins, and other metabolites produced during *Alternaria brassicae* interactions on pathogenesis, and host defence along with activity of MAPK 4 cascade gene in plant defence against biotic stress has been identified (Bains and Tewari 1989; Verma and Saharan 1994; Lou et al. 2013; Pedras et al. 2009; Thomma 2003; Marmath et al. 2013).

The nature and mechanisms of resistance in cruciferous crops are multilayered. It is in the form of morphological barriers/features like fewer numbers of stomata with narrow aperture in some genotypes of *B. napus* and *B. juncea.* The presence of epicuticular wax layers in *Brassica* genotypes provides a hydrophobic leaf surface, which hinders *Alternaria* spore germination and infection. Large number of biochemical constituents governs biochemical resistance. Proteome and enzyme activity in resistant genotypes activates resistant genes. Hypersensitive response genes for induced resistance have been identified. Expression of MAPK 4 (mitogen-activated protein kinase) cascade gene antagonizes the effects of pathogens in resistant genotypes. Elicitation of phytoalexins by *Brassica* species inhibits fungal growth on host leaf surface. Calcium sequestration character of brassicas gives ample possibilities of enhancing resistance to *Alternaria* spp. infection. Sources of resistance in cruciferous species (closely as well as distantly related) against *Alternaria* spp. have been identified, but very few have been explored for commercial use to develop resistant cultivars. The information generated so far on parameters of host resistance is being utilized through conventional and biotechnological approaches (Verma and Saharan 1994; Saharan et al. 2003: Saharan and Kadian 1983a; Kumar and Kolte 2006; Gupta et al. 1990; Dhingra et al. 2004; Oh et al. 2005; Mishra et al. 2010; Marmath et al. 2013; Conn et al. 1991; Sellam et al. 2007; Aneja and Agnihotri 2013).

11.7 Fungicidal Resistance

Mathematical and experimental studies have shown that parasitic and saprophytic fitness of a resistant subpopulation is crucial for its competition with other subpopulations. Only highly fit resistant individuals can be maintained in or even dominate the population, causing loss of control of the corresponding fungicides. It has been frequently observed that iprodione-resistant isolates of different plant pathogenic fungi have higher osmotic sensitivity than do sensitive isolates. This osmotic sensitivity is considered to be responsible for the lower pathogenicity and lower fitness of the resistant isolates. Huang and Levy (1995) investigated the parasitic and saprophytic fitness, the resistance to other fungicides and the osmotic sensitivity of 25 laboratory mutant isolates of *A. brassicicola* exhibiting iprodione resistance and the corresponding wild types from which they were isolated. The resistance of *Alternaria brassicicola* isolates to iprodione exhibits a polymodal distribution. Four distinct groups were detected demonstrating ED_{50s} of 0–7, 10–40, 70–80 and 300–516 μg a.i. per ml. Most resistant isolates produce smaller colonies on unamended potato dextrose agar. Fewer conidia are produced per colony for some resistant isolates than for their corresponding wild types, while other resistant isolates produced more conidia per unit of colony area. The resistant isolates either produce smaller lesions or sporulate less or both than the sensitive isolates on untreated broccoli leaf discs. Only resistant isolates are able to produce lesions on leaf discs sprayed with 50 μg a.i. of Iprodione per ml. Conidia of resistant isolates germinated in distilled water containing iprodione at a concentration as high as 250 μg a.i. per ml, while the germination of sensitive isolates was greatly inhibited in distilled water containing only 5 μg a.i. of Iprodione per ml. Iprodione-resistant isolates are cross-resistant to vinclozolin and dichloran. Resistant isolates exhibited a greater osmotic sensitivity than sensitive isolates; however, osmotic sensitivity is independent on the degree of resistance (Huang and Levy 1995).

Three sterol biosynthesis inhibiting fungicides, two of the triazole group (flutriafol and difenoconazole) and one of the imidazole group (prochloraz), performed equally well in reducing the mycelial growth of *A. brassicae*, *A. brassicicola* and *A. japonica*. Out of the two selected dicarboximides, Iprodione proved to be more effective than Procymidone. Most of the tested isolates are also very sensitive to fludioxonil (phenylpyrroles family). However, in vitro tests for mycelial growth, conidial germination and germ-tube elongation revealed the existence of *A. brassicicola* isolates highly resistant (EC_{50} > 100 mg/l) to both dicarboximides (e.g. iprodione and procymidone) and phenylpyrroles (e.g. fludioxonil). These resistant isolates did not exhibit lower sporulation capacities or reduced aggressiveness towards host plants compared to sensitive *A. brassicicola* isolates (Iacomi-Vasilescu et al. 2004).

11.8 Integrated Disease Management

It includes all means of disease and crop management, viz. cultural, nutritional, biological, biochemical, biotechnological, chemical, host resistance and genetic engineering in an integrated way including pest management (Mukerji et al. 1999; Verma and Saharan 1994; Saharan and Mehta 2002; Sharma and Kolte 1994; Kolte 2005; Godika et al. 2001; Pathak and Godika 2010; Meena et al. 2011, 2012, 2013; Mehta 2014). Following cultural practices have been advocated for better management of *Alternaria* disease of cruciferous crops. Cultural control of crucifer diseases is largely a matter of sanitation and manipulating the environment to the advantage of the host and to the detriment of the pathogen. In addition, the judicious use of fertilizer should be adopted. The various cultural practices help in reducing the diseases are:

1. Crop debris containing resting structures of the pathogens should be burnt or destroyed after the harvesting of the crop to restrict the sources of primary inoculum.
2. Deep summer ploughing in May–June should be done to destroy the resting structures of the pathogen.
3. Crop rotation with non-cruciferous crops for at least 3 years should be followed in the case of severe disease epidemic areas.
4. Timely sowing; application of recommended dosages of fertilizers; use of nutrients like boron, potassium, phosphorus, sulphur, zinc, biofertilizers, biocontrol agents; maintaining optimum plant population; timely thinning; timely irrigation, weeding and control of insect pests should be followed.
5. Sowing time should be adjusted depending upon weather conditions and disease-prevailing conditions in a particular area. Early sowing (up to 20 October) of the crop escapes the major diseases including *Alternaria* blight under North Indian conditions.

The multiple disease control strategy is mainly dependent on the balanced fertilizer application (N100;P40:K40), and available level of tolerance in host varieties, early sowing time (first week of October) and seed treatment with Apron 35 SD followed by use of fungicidal spray (Metalaxyl + mancozeb = Ridomil MZ 72 @0.25 %) for *Alternaria* blight, white rust + downy mildew complex control or carbendazim spray at 0.05 % for Sclerotinia rot and powdery mildew control (Kolte 2005; Tables 11.8, 11.9, and 11.10).

Godika et al. (2001) reported that the combination of boron (0.53 %) with boric acid or zinc (0.22 %) spray through zinc oxide showed

Table 11.8 Effect of fungicidal seed treatment on plant stand of mustard plants (Kolte 2005)

	Plant stand (%)[a]		
Fungicides	*S. rolfsii*	*R. solani*	*F. oxysporum*
Mancozeb	67.1 (84.2)	7.9 (4.0)	63.0 (79.2)
Thiophanate methyl	63.6 (79.5)	35.8 (35.2)	64.6 (81.5)
Thiram	67.3 (85.2)	21.5 (17.5)	62.2 (78.0)
Carbendazim	62.2 (78.0)	29.6 (33.5)	36.3 (69.0)
Metalaxyl (Apron 35SD)	66.4 (83.7)	6.5 (1.70)	50.3 (58.2)
Check (untreated)	23.3 (15.7)	20.6 (12.7)	22.9 (16.2)
CD at P=0.5	7.8	21.1	11.4

[a]Figures in parentheses are actual % plant stand and others are arcsine transformed values (Khan and Kolte 2002)

Table 11.9 Integrated disease management module (seed treatment, spray schedule and fertilizer doses for the control of DM, WR and AB) and its significance in achieving higher yield of mustard during 1999–2000 to 2001–2002[a] (Kolte 2005)

Treatments	DM cotyledons index (%)	DM leaf 60 DAS (%)	WR leaf index (%) 100DAS	AB leaf 100 DAS (%)	AB pod 2 WBM (%)	DM + WR staghead (incidence)	DM + WR staghead (severe)	1000 grain weight (g)	Yield potential (kg/ha)	Yield increase over check %
$D_1V_1P_1P_1$	17.71	33.58	29.34	45.45	46.92	1.92	4.64	3.74	1901.6	72.03
$D_1V_1P_0P_0$	62.28	36.84	40.42	59.41	52.74	2.13	4.71	3.36	1105.1	
$D_2V_1P_1P_1$	14.99	22.06	35.47	52.77	39.88	19.01	15.75	3.35	902.7	37.08
$D_2V_1P_0P_0$	44.90	24.06	43.31	58.97	47.17	16.11	21.50	3.04	658.0	
$D_1V_2P_1P_1$	17.71	34.13	28.44	46.14	46.65	1.76	3.71	3.76	1918.1	84.06
$D_1V_2P_0P_0$	63.06	37.57	40.64	59.65	51.75	2.64	5.06	3.45	1042.7	
$D_2V_2P_1P_1$	14.99	19.10	35.24	52.77	38.95	20.19	16.79	3.33	904.8	112.79
$D_2V_2P_0P_0$	42.03	21.50	–	59.70	49.16	16.85	20.24	2.99	425.2	
CD at 5 %	16.06	13.24	10.47	10.59	8.23	19.45	13.77	0.51	900.4	

[a]Data indicated in the table are averages of 3 years crop seasons 1999–2000 to 2001–2002;
DM downy mildew, *WR* white rust, *AB Alternaria* blight
D_1 = Sowing date 20 October; D_2 = Sowing date 20 November V_1 = Varuna, V_2 = Kranti
P_0P_0 = No plant protection chemical treatment
P_1 P_1 = Recommended plant protection practices i.e. NPK = 100:40; 40 kg ha^{-1}
Seed treatment with Apron 35 SD (6 g/kg)
Ridomil MZ at 0.25spray at 50 DAS followed by mancozeb spray at 0.2 % at 70 and 90 DAS

Table 11.10 Some micronutrients as possible inducer for multiple disease resistance in rapeseed–mustard (Kolte 2005)

Treatments (%)	Conc.	Phytoalexin inhibition zone (mm)	Disease index: Under artificial conditions DM	WR	AB	Underfield conditions WR	AB	Yield/ 20 plants (gm)
Fe EDTA	0.2	95.0**	66.9**	77.0*	39.8	36.7**	40.2**	85.6
$MnSO_4$	0.2	95.9**	43.9**	56.3**	64.5	28.6**	50.2	88.9
$CaCl_2$	1.0	90.9**	88.8	56.9**	24.4**	25.6**	46.2*	82.4
$ZnSO_4$	0.5	119.9**	4.1**	37.0**	39.2	48.3	49.7	109.0*
$CuSO_4$	0.1	119.0**	9.5**	4.0**	54.6	21.9**	47.7*	111.3*
$CO(NO_3)_2$	0.5	69.2**	59.3**	53.8**	28.3**	29.1**	46.6*	118.5*
Na_2BO_7	0.5	53.5**	31.8**	18.4**	23.0**	29.3**	50.2	11.4*
Distilled water	8.5	92.6	83.6	36.9	47.7	53.5	77.9	–

**Significant at $P<0.01$, *Significant at $P<0.05$
DM downy mildew, *WR* white rust, *AB Alternaria* blight

synergistic effect in the effectiveness of mancozeb and gave 16–20 % improvement in disease control in comparison with such treatments when used separately. It indicates the importance of integrating plant nutrients such as boron, potassium and zinc with foliar sprays of fungicides for management of the *Alternaria* disease in oilseed rape.

The use of biofertilizers (*Azatobacter* and PSB) and *Trichoderma* along with the recommended doses of fertilizers enhances the plant growth and improves yield along with less incidence of *Alternaria* blight and white rust. Organic manure results better growth, enhanced yield and reduced disease incidence in mustard crop. Plant growth and yield of mustard is also promoted by basal use of elemental sulphur as a nutritional supplement and spray of thiourea (0.1 %) at 50 % flowering stage of the crop, along with adoption of recommended doses of fertilizers (NPK, 80:40;40 kg/ha) (Pathak and Godika 2010).

Mancozeb recorded the lowest mean severity (leaf: 33.1 %; pod: 26.3 %) of *Alternaria* blight with efficacy of garlic bulb extract alone (leaf=34.4 %; pod=27.3 %) or in combination with cow urine (leaf=34.2 %; pod=28.6 %) being statistically at par with the recommended chemical fungicide. Chemicals also proved effective in reducing *Alternaria* blight severity on leaves and pods of Indian mustard (leaf=36.3–37.9 %; pod=27.5–30.1 %). The effective treatments besides providing significant reduction in disease severity also increased dry seed yield of the crop (mancozeb=2052 kg ha^{-1}; garlic=2006 kg ha^{-1}; control=1561 kg ha^{-1}) (Tables 11.11, 11.12, and 11.13; Meena et al. 2011).

Seed treatments with freshly prepared *Allium sativum* bulb aqueous extract (1 % w/v) resulted in significantly higher initial plant stands, across locations and years. Seed treatment with *A. sativum* bulb extract, followed by its use as a foliar spray, results in significantly reduced *Alternaria* leaf, and pod blight, white rust, fewer stagheads incidence per plot, reduced downy mildew, *Sclerotinia* rot incidence and reduced powdery mildew severity. It also provides significantly higher seed yields compared to the control and was at par with chemical fungicides. It was also effective as the combination of seed treatment with *Trichoderma harzianum* and foliar spraying with *Pseudomonas fluorescens* and *T. harzianum*. Economic returns were higher when using biorational treatments (*A. sativum* bulb extract, *T. harzianum*, *P. fluorescens*) compared with chemical fungicides. The combination of seed treatments with *T. harzianum* followed by its use as a foliar spray (17.22), and the similar combination of seed treatments, and foliar spraying with the *A. sativum* bulb extract (17.18) gave higher cost–benefit ratio (Tables 11.14, 11.15, and 11.16; Meena et al. 2013).

Table 11.11 Effect of different chemicals, plant extracts and bioagents on *Alternaria* leaf blight severity (Meena et al. 2011)

Treatment	% *Alternaria* leaf blight severity[a]												
	Bharatpur		Ludhiana		Hisar		Dholi		Pantnagar		Mean		Pooled mean
	2005	2006	2005	2006	2005	2006	2005	2006	2005	2006	2005	2006	
Mancozeb	14.5 (22.4)	11.1 (19.3)	52.6 (63.0)	32.4 (34.7)	22.2 (27.9)	20.1 (26.4)	33.8 (31.0)	20.3 (26.8)	44.9 (50.2)	57.6 (49.4)	33.6 (34.9)	28.3 (31.4)	30.9 (33.1)
Cow urine	16.6 (24.1)	13.8 (21.7)	56.8 (70.0)	45.8 (42.6)	43.3 (41.0)	36.7 (37.2)	35.5 (33.7)	18.7 (25.6)	50.0 (58.5)	65.7 (54.1)	40.4 (39.1)	36.1 (36.3)	38.3 (37.7)
Garlic + cow urine	16.2 (23.7)	15.6 (23.2)	55.9 (68.6)	45.0 (43.1)	22.3 (27.9)	14.6 (22.4)	29.9 (25.0)	16.0 (23.5)	48.8 (56.7)	64.7 (53.6)	34.6 (35.5)	31.2 (33.0)	32.9 (34.3)
Garlic	15.5 (23.2)	12.2 (20.4)	56.1 (69.0)	36.6 (37.2)	27.3 (31.3)	20.7 (27.0)	29.3 (24.0)	20.3 (26.8)	47.6 (54.7)	63.6 (52.9)	35.2 (35.9)	30.7 (32.9)	32.9 (34.4)
P. fluorescens	22.3 (28.2)	19.9 (26.5)	54.9 (67.0)	34.2 (35.8)	32.33 (34.6)	24.1 (29.4)	34.6 (32.3)	13.3 (21.3)	51.0 (60.2)	67.1 (54.9)	39.0 (38.4)	31.7 (33.6)	35.4 (36.0)
E. globosus	21.7 (27.8)	16.5 (23.9)	55.7 (68.2)	39.9 (39.1)	38.3 (38.2)	29.3 (32.8)	32.8 (29.3)	22.3 (28.2)	46.5 (52.9)	61.3 (51.5)	39.0 (38.4)	33.9 (35.1)	36.4 (36.8)
T. harzianum	17.6 (24.8)	11.8 (20.1)	53.5 (64.6)	33.5 (35.4)	35.0 (36.3)	27.3 (31.0)	28.9 (23.3)	15.0 (22.8)	50.0 (58.7)	68.3 (55.7)	37.0 (37.1)	31.2 (33.1)	34.1 (35.1)
Calcium sulphate	28.6 (32.3)	17.1 (24.4)	58.2 (72.2)	37.6 (37.8)	25.0 (30.0)	16.7 (24.0)	30.4 (25.7)	24.7 (29.8)	48.1 (55.6)	70.5 (57.1)	38.1 (37.9)	33.3 (34.6)	35.7 (36.3)
Borax	26.5 (31.0)	15.5 (23.2)	58.8 (73.3)	43.8 (41.4)	33.3 (35.3)	25.1 (30.1)	33.2 (30.0)	21.7 (27.0)	50.6 (59.5)	73.9 (59.3)	40.5 (39.4)	36.0 (36.3)	38.2 (37.9)
Potash + sulphur	23.5 (29.0)	17.3 (24.5)	54.9 (67.0)	38.6 (38.4)	28.7 (31.9)	20.7 (27.0)	34.6 (32.3)	25.3 (30.2)	49.8 (58.2)	73.1 (58.7)	38.3 (38.0)	35.0 (35.8)	36.6 (36.9)
Zinc sulphate	21.4 (27.5)	18.3 (25.3)	56.9 (70.3)	35.5 (36.6)	36.7 (37.3)	28.6 (32.3)	35.7 (34.0)	19.3 (26.1)	49.8 (58.3)	73.9 (59.2)	40.1 (39.1)	35.1 (35.9)	37.6 (37.5)
Control	40.6 (39.6)	20.9 (27.2)	68.3 (78.4)	47.6 (43.6)	44.3 (41.6)	36.8 (37.3)	38.4 (38.7)	27.7 (31.6)	56.1 (67.4)	81.3 (64.4)	49.5 (44.8)	42.9 (40.9)	46.2 (42.8)
LSD ($P<0.05$)	2.1	5.5	0.5	0.8	7.1	6.6	2.6	2.9	0.9	1.8	3.3	3.2	2.2

[a]Figures in parenthesis are angular transformed values and others are actual percent disease severity; mean of three replications 2005 = 2004–2005 season; 2006 = 2005–2006 season

Table 11.12 Effect of different chemicals, plant extracts and bioagents on *Alternaria* pod blight severity (Meena et al. 2011)

	% *Alternaria* pod blight severity[a]											
	Bharatpur	Ludhiana		Hisar		Dholi		Pantnagar		Mean		Pooled mean
Treatments	2005	2005	2006	2005	2006	2005	2006	2005	2006	2005	2006	
Mancozeb	7.2 (15.6)	50.3 (59.2)	30.1 (33.2)	5.3 (13.3)	3.7 (10.9)	18.1 (9.7)	20.0 (26.5)	44.9 (50.2)	19.1 (25.9)	25.2 (28.3)	18.2 (24.4)	21.7 (26.3)
Cow urine	7.6 (16.0)	52.6 (63.2)	44.2 (41.6)	22.2 (27.9)	18.1 (25.2)	23.2 (15.7)	10.0 (18.4)	50.0 (58.5)	23.5 (29.0)	31.1 (32.9)	23.9 (28.7)	27.5 (30.8)
Garlic + cow urine	6.1 (14.1)	54.7 (66.7)	44.0 (41.5)	16.7 (24.1)	4.3 (12.0)	14.5 (6.3)	16.7 (24.0)	48.8 (56.7)	22.2 (28.1)	28.2 (30.6)	21.8 (26.7)	24.9 (28.6)
Garlic	5.0 (12.9)	53.3 (64.4)	35.2 (36.4)	8.3 (16.8)	6.7 (14.9)	13.6 (5.7)	19.3 (26.1)	47.6 (54.7)	21.0 (27.3)	25.6 (28.4)	20.5 (26.3)	23.0 (27.3)
P. fluorescens	7.4 (15.6)	54.3 (66.0)	33.7 (35.4)	12.7 (20.8)	9.8 (18.2)	19.9 (11.7)	8.3 (16.7)	51.0 (60.2)	29.9 (33.1)	29.1 (31.2)	20.4 (26.1)	24.7 (28.7)
E. globosus	8.6 (17.0)	54.3 (66.0)	38.7 (38.5)	18.7 (25.6)	12.8 (21.0)	16.4 (8.0)	15.7 (23.2)	52.9 (46.5)	19.5 (26.2)	30.2 (32.1)	21.7 (27.3)	25.9 (29.7)
T. harzianum	5.9 (13.9)	52.8 (63.5)	32.1 (39.5)	16.7 (24.1)	13.7 (21.6)	13.3 (5.3)	10.0 (18.4)	50.0 (58.7)	29.7 (33.0)	27.7 (30.2)	21.4 (27.0)	24.6 (28.6)
Calcium sulphate	6.7 (15.0)	55.8 (68.4)	36.5 (37.2)	6.7 (14.9)	4.3 (12.0)	13. 6 (5.7)	16.0 (23.5)	48.1 (55.6)	27.0 (31.3)	25.1 (28.8)	20.9 (26.2)	23.6 (27.5)
Borax	7.5 (15.8)	50.9 (60.3)	42.0 (40.8)	13.3 (21.4)	10.5 (18.9)	18.8 (10.7)	12.3 (20.5)	50.6 (59.5)	28.8 (32.4)	28.2 (30.8)	23.4 (28.2)	25.8 (29.5)
Potash + sulphur	6.2 (14.3)	55.7 (68.3)	37.5 (37.7)	8.7 (17.1)	5.8 (13.9)	15.6 (7.3)	15.7 (23.3)	49.8 (58.2)	29.0 (32.6)	27.2 (29.6)	22.0 (27.1)	21.6 (28.4)
Zinc sulphate	5.1 (12.9)	53.3 (64.3)	34.5 (35.9)	16.7 (24.1)	13.3 (21.4)	24.8 (17.7)	15.0 (22.7)	49.8 (58.3)	30.4 (33.5)	29.9 (31.8)	23.3 (28.5)	26.6 (30.1)
Control	15.7 (23.3)	59.9 (75.0)	46.8 (43.2)	26.7 (31.2)	19.3 (26.1)	27.7 (21.7)	27.7 (31.7)	56.1 (67.4)	36.1 (37.0)	37.2 (37.1)	32.5 (34.5)	34.8 (35.8)
LSD ($P<0.05$)	3.7	0.7	0.3	4.9	4.1	2.9	2.7	0.9	1.5	4.4	4.4	2.6

[a]Figures in parenthesis are angular transformed values and others are actual percent disease severity; mean of three replications 2005 = 2004–2005 season; 2006 = 2005–2006 season

Table 11.13 Effect of different non-toxic chemicals, plant extracts and bioagents on *Alternaria* leaf blight severity (Meena et al. 2011)

Treatments	Yield (kg/ha)[a]												
	Bharatpur		Ludhiana		Hisar		Dholi		Pantnagar		Mean		Pooled mean
	2005	2006	2005	2006	2005	2006	2005	2006	2005	2006	2005	2006	
Mancozeb	1889	4183	2796	1287	1605	1705	1178	1200	2738	1936	2041	2062	2052
Cow urine	1195	3911	2657	730	1520	1635	1000	1333	2264	1480	1727	1817	1773
Garlic + cow urine	1534	3778	2729	737	1595	1715	1356	1488	2412	1675	1925	1879	1902
Garlic	2050	4233	2603	1130	1585	1698	1156	1200	2664	1744	2011	2001	2006
P. fluorescens	1156	3932	2603	1212	1560	1670	822	1266	2338	1478	1695	1911	1803
E. globosus	1578	3856	2544	1060	1530	1645	1178	911	2701	1916	1906	1878	1892
T. harzianum	1645	4278	2739	1257	1550	1660	1556	1377	2206	1503	1939	2015	1977
Calcium sulphate	1489	3711	2657	1110	1610	1720	1356	1000	2309	1501	1884	1808	1846
Borax	1539	3606	2603	775	1570	1685	844	1244	1894	1418	1690	1745	1718
Potash + sulphur	1250	3750	2618	1080	1580	1680	800	866	2005	1427	1650	1761	1706
Zinc sulphate	1445	3694	2704	1175	1545	1655	722	911	2168	1434	1716	1774	1745
Control	1117	3500	2563	725	1510	1650	700	822	1813	1215	1540	1582	1561
LSD ($P<0.05$)	274	458	NS	83	49	49	469	30	114	32	211	287	169

[a]Figures are average of three replications 2005 = 2004–2005 season; 2006 = 2005–2006 season

Table 11.14 The effect of different treatments on the initial plant stand of Indian mustard (Meena et al. 2013)

	Initial plant stand (25 days after sowing)		
Treatments	2006–2007[a]	2007–2008[b]	Pooled mean
Garlic bulb extract 1 % w/v (ST)	270.8[pq]	267.6[p]	269.2[p]
Apron 35 SD 6 g/kg (ST)	261.8[pq]	249.3[qrst]	255.5[pq]
Carbendazim 1 g a.i. (ST)	262.5[pq]	244.9[rst]	253.7[qr]
Apron 35 SD 6 g/kg + carbendazim 1 g a.i. (ST)	274.6 [p]	259.1[pqrs]	266.8[pq]
Trichoderma harzianum 10 g/kg (ST)	267.4[pq]	260.5[pqr]	264.0[pq]
T. harzianum (ST) + *P. fluorescens* 10 ml/l (FS)	271.2[pq]	250.3[pq]	260.7[pq]
T. harzianum (ST) + *T. harzianum* 10 ml/l (FS)	259.5[qr]	258.2[pqrs]	258.9[pq]
Garlic bulb extract (ST) + garlic bulb extract (FS)	270.3[pq]	263.9[pq]	267.1[pq]
Apron 35 SD (ST) + Ridomil MZ 72 WP 2 g/l (FS)	267.5[pq]	255.0[pqrs]	261.3[pq]
Carbendazim (ST) + Ridomil MZ 72 WP 2 g/l (FS)	267.0[pq]	254.2[pqrs]	260.6[pq]
Control	248.3[qr]	233.2[t]	240.7[r]
LSD ($P<0.05$)	13.4	16.4	15.2

Cultivar Varuna, *ST* seed treatment; *FS* foliar spray; [ab]Number of plants per plot of 5 m × 3 m
Same letter in superscript indicates no significant difference among data within the column
[a]Mean of three replications at each of the following locations: Faizabad, Morena, Navgaon and Kanpur
[b]Mean of three replications at each of the following locations: Faizabad, Morena, Navgaon and Jagdalpur

Table 11.15 The effect of different treatments on *Alternaria* blight severity (percent) in Indian mustard 90 days after sowing (Meena et al. 2013)

Treatment	2006–2007[a]	2007–2008[b]	2008–2009[c]	Pooled mean
Garlic bulb extract 1 % w/v (ST)	30.9[q] (27.3)	31.4[q] (25.6)	27.1 (23.0)	30.2[q] (25.3)
Apron 35 SD 6 g/kg (ST)	31.5[q] (28.3)	30.9[q] (26.0)	27.4 (23.4)	30.6[pq] (25.9)
Carbendazim 1 g a.i. (ST)	30.1[q] (31.2)	31.2[q] (26.3)	27.9 (23.8)	30.3[q] (25.4)
Apron 35 SD 6 g/kg + carbendazim 1 g a.i. (ST)	28.9[pq] (24.1)	31.1[q] (26.9)	27.3 (23.1)	29.8[q] (24.7)
Trichoderma harzianum 10 g/kg (ST)	30.2[q] (26.4)	31.9[q] (27.3)	27.5 (23.5)	30.5[q] (25.7)
T. harzianum (ST) + *P. fluorescens* 10 ml/ l (FS)	29.2[pq] (24.8)	30.6[q] (25.7)	25.8 (20.5)	29.1[pq] (23.7)
T. harzianum (ST) + *T. harzianum* 10 ml/l (FS)	29.2[pq] (24.8)	29.5[q] (22.7)	25.0 (19.5)	28.2[pq] (22.3)
Garlic bulb extract (ST) + garlic bulb extract (FS)	26.1[p] (20.1)	26.3[p] (19.8)	26.4 (21.7)	26.9[p] (20.5)
Apron 35 SD (ST) + Ridomil MZ 72 WP 2 g/l (FS)	26.2[p] (20.3)	26.6[p] (19.6)	25.5 (19.9)	26.1[p] (19.3)
Carbendazim (ST) + Ridomil MZ 72 WP 2 g/l (FS)	30.0[q] (26.5)	29.9[q] (24.2)	27.1 (22.5)	29.6[q] (24.4)
Control	36.9[r] (36.5)	37.5[r] (36.1)	28.0 (24.3)	34.6[r] (32.3)
LSD ($P<0.05$)	3.4	2.7	NS	3.1

The figures in parentheses are the actual means of disease severity, and the other figures are angular transformed values
ST seed treatment, *FS* foliar spray, *cv.* varuna, *NS* not significant: same letter in superscript indicates no significant difference among data within the column
[a]Mean of three replications at each of the following locations: Sri Ganganagar, Faizabad, Morena, Pantnagar, Ludhiana, Navgaon and Kanpur
[b]Mean of three replications at each of the following locations: Sri Ganganagar, Faizabad, Morena, Pantnagar, Jagdalpur, Kanpur, Hisar and Dholi
[c]Mean of three replications at each of the following locations: Sri Ganganagar, Faizabad, Morena, Pantnagar, Ludhiana, Navgaon and Jagdalpur

Table 11.16 The effect of different treatments on *Alternaria* blight severity (percent) in Indian mustard 120 days after sowing (Meena et al. 2013)

Treatment	2006–2007[a]	2007–2008[b]	2008–2009[c]	Pooled mean
Garlic bulb extract 1 % w/v (ST)	34.2^{r} (32.4)	30.9^{r} (27.2)	24.4 (19.3)	30.8^{q} (26.3)
Apron 35 SD 6 g/kg (ST)	33.6^{r} (31.5)	30.5^{r} (27.1)	23.8 (18.6)	30.5^{q} (25.7)
Carbendazim 1 g a.i. (ST)	33.4^{qr} (31.1)	31.1^{r} (27.7)	24.3 (19.4)	30.7^{q} (26.1)
Apron 35 SD 6 g/kg + carbendazim 1 g a.i. (ST)	34.4^{r} (32.7)	30.9^{r} (27.7)	24.8 (20.3)	31.1^{qr} (26.9)
Trichoderma harzianum 10 g/kg (ST)	33.4^{qr} (31.2)	30.9^{r} (27.8)	25.3 (20.9)	31.0^{q} (26.9)
T. harzianum (ST) + *P. fluorescens* 10 ml/ l (FS)	34.1^{r} (32.1)	31.6^{r} (28.6)	25.1 (20.3)	31.3^{qr} (27.0)
T. harzianum (ST) + *T. harzianum* 10 ml/l (FS)	33.7^{r} (31.4)	28.6^{q} (24.1)	24.0 (19.0)	29.9^{q} (24.8)
Garlic bulb extract (ST) + garlic bulb extract (FS)	30.5^{pq} (27.5)	25.2^{p} (20.1)	23.0 (17.8)	27.8^{p} (21.8)
Apron 35 SD (ST) + Ridomil MZ 72 WP 2 g/l (FS)	29.8^{p} (26.5)	25.6^{p} (20.2)	22.5 (17.3)	27.5^{p} (21.3)
Carbendazim (ST) + Ridomil MZ 72 WP 2 g/l (FS)	34.8^{r} (33.4)	31.3^{r} (28.1)	24.4 (19.4)	31.3^{qr} (27.0)
Control	36.4^{r} (35.8)	35.0^{s} (33.0)	25.4 (20.6)	33.1^{r} (29.8)
LSD ($P<0.05$)	3.0	1.4	NS	2.1

The figures in parentheses are the actual means of disease severity, and the other figures are angular transformed values

ST seed treatment, *FS* foliar spray, *cv.* varuna, *NS* not significant

Same letter in superscript indicates no significant difference among data within the column

[a]Mean of three replications at each of the following locations: Sri Ganganagar, Faizabad, Morena, Pantnagar, Ludhiana, Navgaon and Kanpur

[b]Mean of three replications at each of the following locations: Sri Ganganagar, Faizabad, Pantnagar, Jagdalpur, Kanpur, Hisar and Dholi

[c]Mean of three replications at each of the following locations: Sri Ganganagar, Faizabad, Morena, Pantnagar, Ludhiana and Jagdalpur

References

Ahmed SR, Agnihotri JP (1977) Antifungal activity of some plant extracts. Indian J Mycol Plant Pathol 7:180–181

Aneja JK, Agnihotri A (2013) Alternaria blight of oilseed brassicas: epidemiology and disease control strategies with special reference to use of biotechnological approaches for attaining host resistance. J Oilseed Brassica 4:1–10

Anonymous (1988) The control of Alternaria on Chinese cabbage. Gartner Tidende 104:392

Ansari NA, Khan MW, Muheet A (1990) Evaluation of some fungicides for seed treatment and foliar application in management of damping-off of seedlings and blight of rapeseed caused by *Alternaria brassicae*. Mycopathologia 110:163–167

Atkinson RG (1950) Studies on the parasitism and variation of *Alternaria raphani*. Can J Res Sect C 28:288–317

Bains PS, Tewari JP (1989) Bioassay and mode of action of an *Alternaria brassicae* toxin. Can J Plant Pathol 11:184–189

Baka ZAM (2010) Antifungal activity of six Saudi medicinal plant extracts against five phyopathogenic fungi. Arch Phytopathol Plant Protect 43:736–743

Bonin K, Fratczak E (1987) Control of fungal diseases in winter oil seed rape. In: Proceedings of the 7th international rapeseed congress, Poznan, pp 1281–1287

Brazauskiene I, Petraitiene E (2004) Effects of fungicide application timing on the incidence and severity of Alternaria blight (*Alternaria brassicae*) and on the productivity of spring oilseed rape (*Brassica napus* L. ssp. *oleifera annual* Metzg.). Agron Res 2:121–133

Brokenshire T (1987) The control of button diseases of brussels sprouts. In: Proceedings of crop protection, Northern Britain, pp 405–410

Chahal AS (1981) Seed-borne infection of *Alternaria brassicae* in Indian mustard and its elimination during storage. Curr Sci 50:621–623

Chahal AS, Kang MS, Chohan JS (1977) Alternaria blight – a serious disease of rapeseed and mustard. Progress Farm 12:21–22

Champion R, Brunet D, Anselme C, Bourdin J, Benhier G (1979) Fungicidal seed disinfection of market garden plants: carrot (*Daucus carota*), cabbage (*Brassica olerace*a Bullata), Valerianella olitoria and radish (*Raphanus sativus*). Rev Hortic 196:23–34

Chand JN, Jatian SS (1969) Efficacy of different fungicides against Alternaria blight of radish (*Raphanus sativus* L.). JNKVV Res J 3:132–133

Channon AG (1970) Tests of fungicides against *Alternaria brassicicola* and studies on some factors enhancing attack by the fungus. Ann Appl Biol 65:481–487

Channon AG, Maude RB (1971) Chapter 16: Vegetables. In: Western JH (ed) Diseases of crop plants. Macmillan, London, pp 323–363

Chhikara S, Chaudhury D, Dhankher OP, Jaiswal PK (2012) Combined expression of a barley class II chitinase and type I ribosome inactivating protein in transgenic *Brassica juncea* provides protection against *Alternaria brassicae*. Plant Cell Tissue Org Cult 108:83–89

Chirco EM, Harman GE (1979) The effects of *Alternaria brassicicola* infection on *Brassica* seed vigor and viability. J Seed Technol 3:12–22

Chupp C (1925) Alternaria leaf spot and blight of crucifers. In: Manual of vegetable garden diseases. The MacMillan Company, New York, pp 146–150

Chupp C, Sherf AF (1960) Chapter 8: Crucifer diseases. In: Vegetable diseases and their control. Ronald Press Company, New York, pp 237–288

Conn KL, Jejelowa OA, Tewari JP, Bains PS (1991) *Camelina sativa* phytoal- exins (Camalexin and methoxy camalexin) provide resistance against *Alternaria brassicae*. Can J Plant Pathol 13:274 (Abstr.)

Cox TW, Swash D, Paviot J (1981) The control of *Alternaria brassicae* and *Sclerotinia sclerotiorum* on oilseed rape with Iprodione. In: Proceedings of the British Crop Protection conference, pests and diseases, Brighton, vol 2, pp 513–520

Cox TW, Souche JL, Grapel H, Wijst FJV (1983) The control of *Alternaria, Sclerotinia* and *Botrytis* on oilseed rape with spray treatments of a flowable formulation of iprodione. In: Proceedings of. 6th international rapeseed conference, Paris, pp 928–933

Crisan A, Pall I (1986) Effect of infection by *Alternaria alternata* on cauliflower seed germination and effectiveness of some fungicides. Contributii Botanice Gradina Botanica Universitatea Babes Bolyai din Cluj-Napoca, pp 213–218

Daebeler F, Amelung D, Seidel D (1981) The most important fungal diseases of rape and possibilities to reduce them. Nachrichtenblatt fur den Pflanzenschutz in der DDR 35:249–251

Daebeler F, Seidel D, Makowski N (1987) Phytosanitory aspects of designing crop rotations with rape. Nachrichtenblatt fur den Pflanzenschutz in der DDR 41:30–32

Danielsson J, Reva O, Meijer J (2006) Protection of oilseed rape (*Brassica napus*) toward fungal pathogens by strains of plant-associated *Bacillus amyloliquefaciens*. Microb Ecol 54:134–140

Davies JML, Ann DM, Dobson SC, Jones OW, Gladders P, Melville SC, Mc Pherson GM, Popham MD, Price RJ, Wafford JD (1986) Fungicide spray timing on Brussels sprouts. Asp Appl Biol 12:21–28

Dhingra A, Portis AR Jr, Daniell H (2004) Enhanced translation of a chloroplast-expressed RbcS gene restores small subunit levels and photosynthesis in nuclear RbcS antisense plants. Proc Natl Acad Sci U S A 101:6315–6320

Dixon GR (1981) Pathogens of crucifer crops. In: Vegetable crop diseases. AVI Publications Company Inc., Westport, pp 119–123

Domsch KH (1957) The blackening of rape and cabbage siliquae. Z Pflanzenkr Pflanzen schutz 64:65–79

Eastburn D (1989) Disease management of cabbage and broccoli- an IPM approach. Trans Ill State Hortic Soc 123:32–35

Ellis MB (1968a) Alternaria brassicae. In: Commonwealth Mycological Institute. Descriptions of pathogenic fungi and bacteria, vol 162. CMI, Kew/Surrey

Ellis MB (1968b) Alternaria brassicicola. In: Commonwealth Mycological Institute. Descriptions of Pathogenic Fungi and Bacteria, vol 163. CMI, Kew/Surrey

Evans EJ, Gladders P (1981) Diseases of winter oilseed rape and their control, east and south-east England, 1977–81. In: Proceedings of the 1981 British Crop Protection conference, Brighton, vol 1, pp 505–512

Evans EJ, Davies JM, Gladders P, Hardwick NV, Hawkins JH, Jones DR, Simkin MB (1983) The occurrence and control of diseases of winter oilseed rape in England. In: Proceedings of the 6th international rapeseed congress, Paris, pp 1032–1037

Evans EJ, Gladders P, Davies JML, Ellerton DR, Hardwick NV, Hawkins JH, Jones DR, Simkin M (1984) Current status of diseases and disease control of winter oilseed rape in England. Asp Appl Biol 6:323–334

Evans EJ, Bowerman P, Giltrap NJ (1988) A comparison of fungicide spray programmes on four cultivars of winter oilseed rape. Tests Agrochem Cultiv 9:32–33

Gaur RB, Ahmed SR (1980) Efficacy of different fungicides for the control of Alternaria blight and white rust of mustard. Indian J Mycol Plant Pathol 10:48

Geeson JD, Browne KM (1979) Effect of post harvest fungicide drenches on stored winter white cabbage. Plant Pathol 28:161–168

Gladders P, Rhodes AD (1985) Evalution of chlorothalonil and iprodione for the control of leaf diseases of turnip. Tests Agrochem Cultiv 6, Supplement, Ann Appl Biol 106:72–73

Godika S, Pathak AK, Jain JP (2001) Integrated management of *Alternaria* blight *(Alternaria brassicae)* and white rust *(Albugo candida)* diseases of Indian mustard *(Brassica juncea)*. Indian J Agric Sci 71:733–735

Guleria S, Kumar A (2010) Antifungal activity of *Agave americana* leaf extract against *Alternaria brassicae*, causal agent of Alternaria blight of Indian mustard (*Brassica juncea*). Arch Phytopathol Plant Protect 42:370–375

Guleria S, Kumar A, Tiku AK (2010) Toxicity of *Solanum xanthocarpum* fruit extract against *Alternaria brassicae*, causal agent of Alternaria blight of Indian mustard (*Brassica juncea*). Arch Phytopathol Plant Protect 43:283–289

Gupta RC, Saxena A (1984) Efficacy of some fungicides on incidence of seed-borne fungi of *Eruca sativa* grown in Tarai region of Nainital. Madras Agric J 71:512–515

Gupta JJ, Sharma BS, Dalela GG (1977) Control of white rust and Alternaria leaf spot of mustard. Indian J Mycol Plant Pathol 7:163–164

Gupta BD, Chatterjee BN, Chowdhury S, Ghosh RK, Baral K (1985) Fungicidal control of Alternaria blight of mustard. J Oilseeds Res 2:148–151

Gupta SK, Gupta PP, Yadav TP, Kaushik CD (1990) Metabolic changes in mustard due to Alternaria leaf blight. Indian Phytopathol 43:64–69

Gupta K, Saharan GS, Singh D (2002) Effective selection criteria to assess disease tolerance in Indian mustard (*Brassica juncea* L. Czern. & Coss.). J Mycol Plant Pathol 32:72–80

He Y, Palevitz BA, Wetzstein HY (1996) Pollen germination, tube growth and morphology, and microtubule organization after exposure to benomyl. Physiol Plant 96:152–157

Ho WC, Su HJ, Li JW, Ko WH (2006) Effect of extracts of Chinese medicinal herbs on spore germination of *Alternaria brassicicola*, and nature of an inhibitory substance from gallnuts of Chinese sumac (*Rhus chinensis*). Can J Plant Pathol 28:519–525

Ho WC, Wu TY, Su HJ, Ko WH (2007) Effect of Oriental medicinal plant extracts on spore germination of *Alternaria brassicicola* and nature of inhibitory substances from speed weed. Plant Dis 91:1621–1624

Holtzhausen MA (1978) Seed-borne fungal pathogens and diseases of Japanese radish and their control in South Africa. Phytophylactica 10:107–113

Howlidar MAR, Meah MB, Jalaluddin M, Rahman MA (1985) Effect of fungicides in reducing intensity of Alternaria blight of mustard. Bangladesh J Agric 10:41–46

Huang R, Levy Y (1995) Characterization of iprodione-resistant isolates of *Alternaria brassicicola*. Phytopathology 79:828–833

Humpherson-Jones FM, Maude RB (1982) Studies on the epidemiology of *Alternaria brassicicola* in *Brassica oleracea* seed production crops. Ann Appl Biol 100:61–71

Humpherson-Jones FM, Maude RB, Kennedy SC (1980) Gray leaf spot of Brassicas. In: 30th annual report for 1979, National Vegetable Research Station, Wellesbourne, Warwick, p 65

Humpherson-Jones FM, Maude RB, Ainsworth LF (1981) *Alternaria* disease of *Brassica* seed crops. In: 31st annual report for 1980, National Vegetable Research Station, Wellesbourne, Warwick, pp 69–70

Humpherson-Jones FM, Ainsworh LF, Maude RB, Bambridge JM, Spencer A (1983) Seed studies. In: 33rd annual report for 1982, National Vegetable Research Station, Wellesbourne, Warwick, pp 64–65

Humpherson-Jones FM, Ainsworth LF, Bambridge JM, Spencer A, Gott KA, Maude RB, Thomas TH (1984) Seed studies. In: 34th annual report for 1983, National Vegetable Research Station, Wellesbourne, Warwick, pp 71–72

Hussaini SH, Singh H (1989) Efficacy of different fungicides against Alternaria blight of radish seed crop. Plant Dis Res 4:105–107

Iacomi-Vasilescu B, Avenot H, Bataille-Simoneau N, Laurent E, Guenard M, Simoneau P (2004) *In vitro* fungicide sensitivity of *Alternaria* species pathogenic to crucifers and identification of *Alternaria brassicicola* field isolates highly resistant to both dicarboximides and phenylpyrroles. Crop Prot 23:418–488

Jain A, Dhingra HR, Mehta N, Sangwan MS (2000) Effect of fungicides on *in vitro* pollen germination and tube growth in *Brassica juncea*. Indian Phytopathol 53:329–330

Jayant M, Sinha SK (1981) Control of leaf spot disease of brown sarson caused by *Alternaria brassicae* and *Alternaria brassicicola* by the antibiotic substance produced by a strain of *Streptomyces hygroscopicus*. In: 3rd international symposium of plant pathology, New Delhi, p 21 (Abstr.)

Jennrich H (1985) Ronilan-smoke, a new vinclozolin formulation for control of Botrytis cinerea and other diseases in greenhouses and storage rooms. Med fac Landbouww Rijksuniv Gent 50:1227–1233

Kamble A, Bhargava S (2007) b-Aminobutyric acid-induced resistance in *Brassica juncea* against the necrotrophic pathogen *Alternaria brassicae*. J Phytopathol 155:152–158

Kanrar S, Venkateswari JC, Kirti PB, Chopra VL (2002) Transgenic expression of hevein, the rubber tree lectin, in Indian mustard confers protection against *Alternaria brassicae*. Plant Sci 162:441–448

Kanwar ZS, Khanna PK (1979) Mustard seed mycoflora in Central India-their effect on the health of seed, seedling and pod, and their control. Int Pest Control 21:83, 86–87, 92

Kaspers H, Siebert R (1989) Folicur (tebuconazole) for the control of rape diseases. Pflanzen schutz- N achrichten Bayer 42:121–137

Kaushik CD, Kaushik JC, Saharan GS (1983) Field evaluation of fungicides for the control of Alternaria leaf blight of *Brassica juncea*. Indian J Mycol Plant Pathol 13:262–264

Kaushik CD, Saharan GS, Kaushik JC (1984) Magnitude of losses in yield and management of Alternaria blight in rapeseed-mustard. Indian Phytopathol 37:398. (Abstr.)

Kear RW, Williams DJ, Stevens CC (1977) The effect of iprodione on the fungal deterioration of stored white cabbage. In: Proceedings of the British Crop Protection conference, pests and diseases, Brighton, pp 189–195

Keyworth WG (1969) Plant pathology. In: Report of National Vegetable Research Station, vol 20, Warwick, pp 102–113

Khan MR, Khan MM (2010) Effect of intermittent exposure of SO_2 on the leaf blight caused by *Alternaria brassicicola* on Indian mustard. Agric Ecosyst Environ 139:728–735

Khan RU, Kolte SJ (2002) Some seedling diseases of rapeseed-mustard and their control. Indian Phytopathol 55:102–103

Khan MM, Khan RU, Mohiddin FA (2007) Studies on the cost-effective management of Alternaria blight of rapeseed-mustard (*Brassica* spp.). Phytopathol Mediterr 46:201–206

Kharbanda P, Tewari JP (1996) Integrated management of canola diseases using cultural methods. Can J Plant Pathol 18:168–175

Khurana AK, Mehta N, Sangwan MS (2005) Variability in the sensitivity of *Alternaria brassicae* isolates to plant extracts. J Mycol Plant Pathol 35:76–77

Kolte SJ (1985) Diseases of annual edible oilseed crops, vol 2. CRC Press Inc, Boca Raton, p 135

Kolte SJ (2005) Tackling fungal diseases of oilseed Brassicas in India. Brassica 7:07–13

Kolte SJ, Tewari AN (1978) Efficacy of certain chemicals for the control of Alternaria blight of yellow sarson. Indian Phytopathol 31:81–84

Kolte SJ, Sharma KD, Awasthi RP (1981) Third international symposium on plant pathology, New Delhi, 14–18 December, pp 70–71 (Abstr.)

Krishnia SK, Saharan GS, Singh D (2000) Genetics of Alternaria blight resistance in inter and intraspecific crosses of *Brassica juncea* and *B. carinata*. Ann Biol 16:212–216

Kudela V, Novak O, Skorpik M (1978) Effect of chemical control on seed yield of cauliflower. Ochr Rostl 14:275–284

Kumar B, Kolte SJ (2001) Progression of Alternaria blight of mustard in relation to components of resistance. Indian Phytopathol 54:329–331

Kumar B, Kolte SJ (2006) Development of Alternaria blight in genotypes of Indian mustard (*Brassica juncea* (L.) Czern & Cross) under field. Indian Phytopathol 39:314–317

Kumar N, Kumar A (2006) Effect of cultural practices on *Alternaria* blight in *Brassica juncea* and *B. napus*. Indian J Agric Sci 76:389–390

Kumar K, Singh DP (1986) Control of *Alternaria brassicae* infection in mustard and rapeseed. Pesticides 20:22–23

Kupryanova MVK (1957) Measures for the control of cabbage diseases. Plant Prot Moscow 5:28–29

Lapis DB, Ricaforte ET (1974) Alternaria leaf spot of crucifers in the Philippines. Philipp Agric 58:121–137

Latif AM, Abu Koser SM, Khan AI, Habibur M, Rahman M, Anwar HM (2006) Efficacy of some plant extracts in controlling seed borne fungi infection of mustard. Bangladesh J Microbiol 23:168–170

Lou J, Fu L, Peng Y, Zhou L (2013) Metabolites from *Alternaria* fungi and their bioactivities. Molecules 18:5891–5935

Madej T (1986) Incidence of diseases affecting cruciferous plants in Poland. Wissenschaftliche Zeitschrift der Wilhelm-Pieck-Universitat Rostock, Natur wissenschaftliche Reihe 35:50–51

Mahapatra S, Das S (2013) Evaluation of fungicides and botanicals against Alternaria leaf blight of mustard. Indian J Plant Prot 41:61–65

Marmath KK, Giri P, Sharma S, Taj G, Kumar A (2013) *In-silico* interaction studies of *Alternaria brassicae* toxin destruxin B and potential partners of MAPK4 cascade. Int J Agric Environ Biotechnol 6:203–210

Marshall J, Harris RI (1984) Broad spectrum disease control in oilseed rape with prochloraz. In: Proceedings of the British crop protection conference, pest and diseases, Brighton, pp 729–734

Maude RB (1976) Dark leaf spot of Brassicas. In: 26th annual report for 1975, National Vegetable Research Station, Wellesbourne, Warwick, p 98

Maude RB (1978) Vegetable seed treatments. In: Jeffs K (ed) Seed treatments, CIPAC monograph-2. Collaborative International Pesticides Analytical Council, Harpenden, pp 91–101

Maude RB (1983) Seed treatment control of *Phoma* and *Alternaria* infections of *Brassica* seed. In: Proceedings of 10th international congress of plant protection, vol 3, p 1202

Maude RB (1986) Seed treatment control of Phoma and Alternaria infections of *Brassica* seed. British Crop Protection Council monograph no. 33 symposium on healthy planting material, 1986, pp 209–210

Maude RB, Dudley CL (1972) Dark leaf spot (*Alternaria brassicicola*) of cabbage. In; 22nd annual report for 1971, National Vegetable Research Station, Wellesbourne, Warwick, p 76

Maude RB, Humpherson-Jones FM (1980a) Studies on the seed-borne phase of dark leaf spot (*Alternaria brassicicola*) and grey leaf spot (*Alternaria brassicae*) of Brassicas. Ann Appl Biol 95:311–319

Maude RB, Humpherson-Jones FM (1980b) The effect of iprodione on the seed- borne phase of *Alternaria brassicicola*. Ann Appl Biol 95:321–327

Maude RB, Humpherson-Jones FM (1984) Importance and control of seed- borne diseases of oilseed rape. Asp Appl Biol 6:335–341

Maude RB, Shuring CG (1968) Tests of the thiram soak treatment for the control of other seed-borne fungal diseases. In: Proceedings of 18st annual report for 1967, National Vegetable Research Station, Wellesbourne, Warwick, pp 80–81

Maude RB, Suett DL (1986) Application of pesticide to brassica seeds using a film coating technique. In: Proceedings of the Brighton crop protection conference – pests and diseases, vol 1, pp 237–242. BCPC, Farnham

Maude RB, Vizor AS, Shuring CG (1969) The control of fungal seed-borne diseases by means of a thiram seed soak. Ann Appl Biol 64:245–257

Maude RB, Kyle AM, Moule CG, Dudley CL (1972) Seed dressing with systemic fungicides for the control of seed-borne fungal pathogens. In: 22nd annual report for 1971, National Vegetable Research Station, Wellesbourne, Warwick, p 75

Maude RB, Humpherson-Jones FM, Shuring CG (1984) Treatments to control *Phoma* and *Alternaria* infections of *Brassica* seeds. Plant Pathol 33:525–535

Meah MB, Howlider MAR, Alam MK (1988) Effect of fungicide spray at different time and frequencies on Alternaria blight of mustard. Thailand J Agric Sci 21:101–107

Mee E (1991) Evaluation of *Alternaria alternata* as abiological agent against *Alternaria brassicae*. Tests Agrochem Cultiv 12:142–143

Meena PD, Sharma P (2012) Antifungal activity of plant extracts against *Alternaria brassicae* causing blight of *Brassica* spp. Ann Plant Prot Sci 20:256–257

Meena PD, Meena RL, Chattopadhyay C, Kumar A (2004) Identification of critical stage for disease development and biocontrol of Alternaria blight of Indian mustard (*Brassica juncea*). J Phytopathol 152:204–209

Meena PD, Meena RL, Chattopadhyay C (2008) Eco-friendly options for management of Alternaria blight in Indian mustard *(Brassica juncea)*. Indian Phytopathol 62:65–69

Meena PD, Chattopadhyay C, Kumar A, Awasthi RP, Singh R, Kaur S, Thomas L, Goyal P, Chand R (2011) Comparative study on the effect of chemicals on Alternaria blight in Indian mustard–a multi-location study in India. J Environ Biol 32:375–379

Meena PD, Kumar VR, Chattopadhyay C (2012) Epidemiology and biocontrol of Alternaria blight in oilseed *Brassica*. LAP Lambert Academic Publishing GmbH & Co. KG, Saarbrücken, Germany 141p

Meena PD, Gour RB, Gupta JC, Singh HK, Awasthi RP, Netam RS, Godika S, Sandhu PS, Prasad R, Rathi AS, Rai D, Thomas L, Patel GA, Chattopadhyay C (2013) Non-chemical agents provide tenable, eco-friendly alternatives for the management of the major diseases devastating Indian mustard (*Brassica juncea*) in India. Crop Prot 53:169–174

Mehta N (2014) Epidemiology and forecasting for the management of rapeseed-mustard diseases. J Mycol Plant Pathol 44:131–147

Minami H, Kim JS, Ikezawa N, Takemura T, Katayama T, Kumagai H, Sato F (2008) Microbial production of plant benzylisoquinoline alka-loids. Proc Natl Acad Sci U S A 105:7393–7398

Mishra A, Pandey D, Goel A, Kumar A (2010) Molecular cloning and *in silico* analysis of functional homologues of hypersensitive response gene(s) induced during pathogenesis of *Alternaria* blight in two genotypes of *Brassica*. J Proteomics Bioinformatics 3:244–248

Mondal SN, Rashid MA, Monowar SM, Abdullah AM (1989) Efficacy of fungicides in controlling Alternaria blight of radish seed crop. Thailand J Agric Sci 22:191–196

Mondal KK, Bhattacharya RC, Koundal KR, Chatterjee SC (2007a) Transgenic Indian mustard (*Brassica junicea*) expressing tomato glucanase leads to arrested growth of *Alternaria brassicae*. Plant Cell Rep 26:247–252

Mondal L, Bhunia CK, Mukhopadhyay A, Biswas A (2007b) Bioefficacy of some new fungicides in management of Alternaria blight disease of rapeseed-mustard. Pestology 31:33–41

Morner J (1980) Co-ordinated oilseed cropping-problems and possibilities. Vaxtkyddsrap-porter 12:21–29

Mridha AU, Safa N (1985) Prevalence and control of seed-borne *Alternaria brassicae* in mustard. Bangladesh J Plant Pathol 1:9–12

Mukerji KG, Upadhyay RK, Saharan GS, Sokhi SS, Kangura RK (1999) Diseases of rapeseed mustard and their integrated management. In: Rajeev K, Upadhyay KG, Mukerji RL, Rajak RL (eds) IPM system in agriculture. Aditya Books, Noida, pp 91–135

Mukherjee I, Gopal M, Chatterjee SC (2003) Persistence and effectiveness of iprodione against *Alternaria blight* in mustard. Bull Environ Contam Toxicol 70:586–591

Muto M, Mulabagal V, Huang HC, Takahashi H, Tsay HS, Huang JW (2006) Toxicity of black nightshade (*Solanum nigrum*) extracts on *Alternaria brassicicola*, causal agent of black leaf spot of Chinese cabbage (*Brassica pekinensis*). J Phytopathol 154:45–50

Nehrash AK (1961) The antimicrobial properties of cultivated radish. Report I. The anti-microbial activity of extracts and essential oil from cultivated and wild radish. J Microbiol Kiev 23:32–37

Nielsen O (1936) Hot water treatment of cabbage seed. Tidsskr Planteavl 41:450–458

Noon RA, Northwood PJ, Brown MC, Montury A, Charlet C (1988) Flutriafol-based formulations for control of oilseed rape diseases. In: Proceedings of the British Crop Protection conference, vol 3, pests and diseases, Brigton, pp 947–952

Ogilvy SE (1984) Disease control in oilseed rape, with particular reference to *Alternaria brassicae*. Crop Prot s1:210–215

Oh IS, Ae RP, Min SB, Sun JK, Young SK, Ji EL, Na Young KS, Lee HC, Ohkmae KP (2005) Secretome analysis reveals an *Arabidopsis* lipase involved in defense against *Alternaria brassicicola*. Plant Cell 17:2832–2847

Pace MA, Campbell R (1974) The effect of saprophytes on infection of leaves of *Brassica* spp. by *Alternaria brassicicola*. Trans Br Mycol Soc 63:193–196

Pathak AK, Godika S (2010) Effect of organic fertilizers, biofertilizers, antagonists and nutritional supplements on yield and disease incidence in Indian mustard in arid soil. Indian J Agric Sci 80:652–654

Patni CS, Kolte SJ (2006) Effect of some botanicals in management of Alternaria blight of rapeseed-mustard. Annu Plant Prot Sci 14:151–156

Patni CS, Kolte SJ, Awasthi RP (2005) Efficacy of botanicals against Alternaria blight (*Alternaria brassicae*) of mustard. Indian Phytopathol 58:426–430

Pedras MS, Chumala PB, Jin W, Islam MS, Hauck DW (2009) The phytopathogenic fungus *Alternaria brassicicola*: phytotoxin production and phytoalexin elicitation. Phytochemistry 70:394–402

Peruch LAM, Michereff SJ (2007) Saprophytic survival of *Alternaria brassicicola* and management of broccoli leaf debris. Ciência Rural Santa Maria 37:13–18

Prasada R, Khandelwal GL, Jain JP (1970) Morphology, physiology and control of *Alternaria brassicae* on taramira. Indian Phytopathol 23:105–110

Rai B, Singh DB (1980) Antagonistic activity of some leaf surface microfungi against *Alternaria brassicae* and *Drechslera graminea*. Trans Br Mycol Soc 75:363–369

Randhawa HS, Aulakh KS (1982) Comparative evaluation of performance of sixteen fungitoxicants as seed dressers for the control of seed-borne fungi of raya (*Brassica juncea* L. Coss.). J Res Punjab Agric Univ Ludhian 19:343–350

Randhawa HS, Aulakh KS (1984) Efficacy of hot water treatment to control seed-borne fungi of Raya (*Brassica juncea* Coss.). Indian J Plant Pathol 2:73–76

Reshu, Khan MM (2012) Role of different microbial-origin bioactive antifungal compounds against *Alternaria* spp. causing leaf blight of mustard. Plant Pathol J 11:1–9

Saharan GS (1984) A review of research on rapeseed-mustard pathology in India. Paper presented in the Annual Rabi Oilseed Workshop held at Jaipur

Saharan GS (1991) Assessment of losses, epidemiology and management of black spot disease of rapeseed-mustard. In: Proceedings of GCIRC 8th international rapeseed congress, Saskatoon, 9–11 July, p 84 (Abstr.); vol 2, pp 465–470

Saharan GS (1992) Disease resistance. In: Labana KS, Banga SS, Banga SK (eds) Breeding oilseed brassicas. Narosa Publishing House, New Delhi, pp 181–200

Saharan GS, Chand JN (1988) Chapter 3: Diseases of rapeseed and mustard. In: Diseases of oilseed crops (In Hindi). Directorate of Publication, Haryana Agric. Univ. Press, Hisar, pp 84–91

Saharan GS, Kadian AK (1983a) Analysis of components of horizontal resistance in rapeseed and mustard cultivars against *Alternaria brassicae*. Indian Phytopathol 36:503–507

Saharan GS, Kadian AK (1983b) Physiologic specialization in *Alternaria brassicae*. Cruciferae Newsl 8:32–33

Saharan GS, Krishnia SK (2001) Multiple disease resistance in rapeseed and mustard. In: Nagarajan S, Singh DP (eds) Role of resistance in intensive agriculture. Kalyani Publication, New Delhi, pp 98–108

Saharan GS, Mehta N (2002) Fungal diseases of rapeseed-mustard. In: Gupta VK, Paul YS (eds) Diseases of field crops. Indus Publishing Company, New Delhi, pp 193–228

Saharan GS, Mehta N, Sangwan MS (2003) Nature and mechanism of disease resistance to Alternaria blight in rapeseed-mustard system. Annu Rev Plant Pathol 2:85–125

Sandhu KS, Singh H, Kumar R (1985) Effect of different nitrogen levels and dates of planting on Alternaria blight and downy mildew diseases of radish seed crop. J Res Punjab Agric Univ Ludhiana 22:285–290

Sasode RS, Sweta P, Gupta A, Pandya RK, Yadav A (2012) *In vitro* study of some plant extracts against *Alternaria brassicae* and *Alternaria Brassicicola*. J Phytol 4:44–46

Schimmer FC (1953) *Alternaria brassicicola* on summer cauliflower seed. Plant Pathol 2:16–17

Sellam A, Dongo A, Guillemette T, Hudhomme P, Simoneau P (2007) Transcriptional responses to exposure to the brassicaceous defence metabolites camalexin and allyl-isothiocyanate in the necrotrophic fungus *Alternaria brassicicola*. Mol Plant Pathol 8:195–208

Sharma SK, Gupta JS (1978a) Biological control of leaf blight disease of brown sarson caused by *Alternaria brassicae* and *A. brassicicola*. Indian Phytopathol 31:448–449

Sharma SK, Gupta JS (1978b) Effect of brown sarson leaf leachates on the germination of the conidia of *Alternaria brassicae* and *Alternaria brassicicola*. Proc Indian Natl Sci Acad B 44:57–58

Sharma SK, Gupta JS (1979) Role of surface microorganisms of brown sarson in relation to *Alternaria brassicae* and *Alternaria brassicicola*. Agra Univ J Res Sci 28:109–111

Sharma SK, Gupta JS (1980) *Streptomyces rochei* in relation to *Alternaria brassicae* and *A. brassicicola* on the surface of brown sarson. J Indian Bot Soc 59:161–163

Sharma SR, Kolte SJ (1994) Influence of nutritional factors on phytotoxic effects of *Alternaria brassicae*. Indian Phytopathol 47:186–187

Sharma N, Sharma S (2008) Control of foliar diseases of mustard by *Bacillus* from reclaimed soil. Microbiol Res 163:408–413

Sharma R, Sinha SK (1989) A pigmented, xylose-utilizing strain of *Streptomyces bobili*. Curr Sci 58:1405–1406

Sharma AK, Gupta JS, Maheshwari RK (1984) The relationship of *Streptomyces arabicus* to *Alternaria brassicae* (Berk.) Sacc. and *Alternaria brassicicola* (Schew.) Wiltshire on the leaf surface of yellow sarson and taramira. Geobios 3:83–84

Sharma AK, Gupta JS, Singh SP (1985) Effect of temperature on the antifungal activity of *Streptomyces arabicus* against *Alternaria brassicae* (Berk.) Sacc. and *Alternaria brassicicola* (Schew.) Wiltshire. Geobios 12:168–169

Sharma S, Singh J, Munshi GD, Munshi SK (2010a) Biochemical changes associated with application of biocontrol agents on Indian mustard leaves from plants infected with Alternaria Blight. Arch Phytopathol Plant Protect 43:315–323

Sharma S, Singh J, Munshi GD, Munshi SK (2010b) Effect of biocontrol agents on lipid and protein composion of Indian mustard seed from plants infected with *Alternaria* spp. Arch Phytopathol Plant Protect 43:589–596

Shivpuri A, Siradhana BS, Bansal RK (1988) Management of Alternaria blight of mustard with fungicides. Indian Phytopathol 41:644–646

Singh DB (1986) Evaluation of fungicides for control of *Alternaria brassicae* and *Drechslera graminea*. Iranian J Plant Pathol 22:39–43

Singh RS (1987) Chapter 5: Alternaria leaf spot or black spot of crucifers. In: Diseases of vegetable crops. Oxford and IBH Publ. Co, New Delhi, pp 159–163

Singh A, Bhowmik TP (1985) Persistence and efficacy of some common fungicides against *Alternaria brassicae*, the causal agent of leaf blight of rapeseed and mustard. Indian Phytopathol 38:35–38

Singh DB, Rai B (1982) Effect of leaf extracts of mustard and barley on growth behavior of some phylloplane microfungi. Bull Torrey Bot Club 108:419–421

Singh RB, Singh RN (2005) Fungicide management of foliar disease of mustard in mid-eastern India. Indian Phytopathol 58:51–56

Singh HG, Sobti AK (1980) Effect of fungicides on Alternaria blight of mustard. Indian J Mycol Plant Pathol 10:49

Singh BP, Singh SP, Abu M, Mishra GM (1989) Relative efficacy of different fungicides for the control of leaf spot of cabbage. Veg Sci 16:62–65

Singh S, Godara SL, Gangopadhyay S (2013) Studies on antifungal properties of plant extracts on mustard blight caused by *Alternaria brassicae*. Indian Phytopathol 66:172–176

Sinha PP, Prasad RK (1989) Chemical management of Alternaria blight of cauliflower seed crop. Indian J Mycol Plant Pathol 19:204–205

Stovold GE, Mailer RJ, Francis A (1987) Seed-borne levels, chemical seed treatment and effects on seed quality following a severe outbreak of *Alternaria brassicae* on rapeseed in New South Wales. Plant Prot Quart 2:128–131

Sultana NA, Khan MAH, Islam MN, Kamrun N (2009) Evaluation of appropriate time for the application of rovral against Alternaria blight incidence and yield of mustard. Int J Sustain Agric 1:20–23

Tahvonen R (1982a) Preliminary experiments into the use of *Streptomyces* spp. isolated from peat in the biological control of soil and seed-borne disease in peat culture. J Sci Agric Soc Finl 54:357–369

Tahvonen R (1982b) The suppressiveness of Finnish light coloured sphagnum peat. J Sci Agric Soc Finl 54:345–356

Tahvonen R (1985) Mycostop-a biological formulation for control of fungal diseases. Vaxtskyddsnotiser 49:86–90

Tahvonen RT (1988) Microbial control of plant disease with *Streptomyces* spp. Bull OEPP 18:55–59

Tahvonen R, Avikainen H (1987) The biological control of seed-borne *Alternaria brassicicola* of cruciferous plants with a powdery preparation of *Streptomyces* sp. J Agric Sc Finl Helsinki 59:199–208

Tapio E, Pohto-Lahdenpera A (1989) Interaction between antagonists and pathogenic fungi. Vaxtskyddsnotiser 53:12–18

Tasca G, Trandaf F (1984) Efficacy of pre-harvest treatment with fungicides in reducing cabbage decay during storage. Lucrari stiintifice Institutulde cercetari si Proiectari Pentru Valorificarea si Industrializarea Legumelor si Fructelor 15:89–93

Tewari JP, Skoropad WP (1979) The effect of polyoxin B and polyoxin D on *Alternaria brassicae* and the black spot of rapeseed. Can J Plant Sci 59:1–6

Thind TS, Jhooty JS (1988) Differential behaviour of fungicides against Alternaria blight of some cruciferous hosts. Indian J Mycol Plant Pathol 18:122–127

Thomas P (1984) Alternaria blackspot (grey leaf spot). In: Canola growers manual. Canola Council of Canada, Winnipeg, pp 1056–1058

Thomma BPHJ (2003) Pathogen profile- *Alternaria* spp., from general saprophyte to specific parasite. Mol Plant Pathol 4:225–236

Tripathi NN, Singh H, Kaushik CD, Singh M (1985) Compatibility of fungicides with insecticides for the control of Alternaria leaf-spot and aphid on Brown Sarson (*Brassica campestris*). Pesticides 19:21–23

Tripathi NN, Saharan GS, Kaushik CD, Kaushik JC, Gupta PP (1987) Magnitude of losses in yield and management of Alternaria blight of rapeseed and mustard. Haryana Agric Univ J Res 17:14–18

Tsuneda A (1977) Mycoparasitism of *Alternaria brassicae* by *Nectria inventa*. Ph. D. Thesis, Univ. of Alberta, Edmonton

Tsuneda A, Skoropad WP (1977a) Formation of micro sclerotia and chlamydospores from conidia of *Alternaria brassicae*. Can J Bot 55:1276–1281

Tsuneda A, Skoropad WP (1977b) The *Alternaria brassicae-Nectria inventa* host- parasite interfaces. Can J Bot 55:448–454

Tsuneda A, Skoropad WP (1978a) Behavior of *Alternaria brassicae* and its mycoparasite *Nectria inventa* on intact and exicised leaves of rapeseed. Can J Bot 56:1333–1340

Tsuneda A, Skoropad WP (1978b) Nutrient leakage from dried and rewetted conidia of *Alternaria brassicae* and its effect on the mycoparasite *Nectria inventa*. Can J Bot 56:1341–1345

Tsuneda A, Skoropad WP (1978c) Phylloplane fungal flora of rapeseed. Trans Br Mycol Soc 70:329–334

Tsuneda A, Skoropad WP (1980) Interactions between *Nectria inventa*, a destructive mycoparasite, and fourteen fungi associated with rapeseed. Trans Br Mycol Soc 74:501–507

Tsuneda A, Skoropad WP, Tewari JP (1976) Mode of parasitism of *Alternaria brassicae* by *Nectria inventa*. Phytopathology 66:1056–1064

Umalkar GV, Mukadam DS, Nehemiah KMA (1977) Fungistatic properties of some deproteinized leaf extracts. Sci Cult 43:437–439

Valkonen JPT, Koponen H (1990) The seed-borne fungi of Chinese cabbage (*Brassica pekinensis*), their pathogenicity and control. Plant Pathol 39:510–516

Vannacci G, Harman GE (1987) Biocontrol of seed-borne *Alternaria raphani* and *A. brassicicola*. Can J Microbiol 33:850–856

Verma PR, Saharan GS (1994) Monograph on Alternaria diseases of crucifers, Saskatoon Research Centre, Technical Bulletin 1994–6E. Agriculture and Agri-Food Canada, Saskatoon. 162p

Verma SS, Yajima WR, Rahman MH, Shah S, Liu JJ, Ekramoddoullah AK, Kav NN (2012) A cysteinc-rich antimicrobial peptide from *Pinus monticola* (PmAMP1) confers resistance to multiple fungal pathogens in canola (*Brassica napus*). Plant Mol Biol 79:61–74

Walker JC (1927) Diseases of cabbage and related plants, US Dept Agric Fanners' bulletin, 1439. U.S. Dept. of Agriculture, Washington, DC, 30p

Walker JC (1952) Chapter 6: Diseases of crucifers. In: Diseases of vegetable crops. McGraw Hill Book Co, New York/London, pp 150–152

White JG, Linfield CA, Lahdenpera ML, Uoti J (1990) Mycostop a novel biofungicide based on *Streptomyces griseoviridis*. In: Proceedings of the British Crop Protection conference, pests and diseases, vol 1, Brighton, pp 221–226

Williams RR, Child V, Copaz L, Holgate ME (1987) The mechanism of yield suppression by a triadimefon fungicide programme on the apple cv Cox's Orange. Pippin J Hortic Sci 62:291–294

Wu WS, Lu JH (1984) Seed treatment with antagonists and chemicals to control *Alternaria brassicicola*. Seed Sci Technol 12:851–862

Yadav MS (2003) Efficacy of fungitoxicants in the management of Alternaria blight and white rust of mustard. J Mycol Plant Pathol 33:307–309

Yasmeen, Saxena SK (1990) Effect of fern extracts on growth and germination of fungi. Curr Sci 15:798–799

12 Techniques

12.1 Introduction

Techniques are methods, procedures and tools to be utilized by the Brassicologists to do research on various aspects of crucifer's diseases. These techniques are also helpful to the students, extension workers, farmers and all those who wish to handle *Alternaria*–crucifers system. The techniques given in this chapter are standardized, reproducible and developed by renowned scientists and practical *Brassica* workers.

12.2 Stem Explant Culture Inoculation

The technique consists of cutting 6–15-cm terminal shoots bearing leaves and/or flowers and pods at 3–5.3 growth stage of the plants. The cut end of such shoots is immediately immersed in tap water in Erlenmeyer flasks or test tubes. Care is taken to see that cut ends of the shoots remain submerged in water. The leaves and/or pods are spray inoculated with *A. brassicae* conidial suspension. Inoculated shoots are incubated in a polyethylene moist chamber (90–95 % RH) at 20–30 °C maximum and 6–14 °C minimum temperatures in diffused light for 72 h. Special care is taken to make up the loss of water in flasks or test tubes by slowing down the transpiration through shaded moist conditions. Symptoms of the disease appear within 4–5 days of inoculation. In 15–20 days, the cut ends of test shoots initiate root formation with profuse rooting in 30 days. New shoots may generate from the axillary buds. Such stem explants can be transferred to soil in pots where pod formation with good seed development can be possible. This technique is advantageous over others in its simplicity and obtaining the seeds of test genotypes in the same season for further use (Kolte and Yadav 1990).

12.3 Leaf Disc Inoculation

Test plants are grown under controlled conditions. From the excised healthy leaves of such plants, leaf discs 2.5–3.0-cm diameter are cut and washed thoroughly by rinsing 5–6 times with distilled water in test tube of wider diameter (>3 cm). To ensure elimination of surface contaminants, fresh aliquots of sterilized water are used for each rinse. Surface-sterilized discs are placed on a 4-mm-thick layer of 1 % plain sterilized agar Petri dishes containing 60 ppm benzimidazole. The adaxial surface is placed against the agar, and the abaxial surface is drop inoculated with 0.01–0.02 ml of conidial suspension (Humpherson-Jones and Phelps 1989). Suitable controls with distilled sterilized water drop-inoculated discs are maintained. Excess moisture from the leaf discs is removed with sterilized absorbent paper. The inoculated discs are

G.S. Saharan et al., *Alternaria Diseases of Crucifers: Biology, Ecology and Disease Management*,
DOI 10.1007/978-981-10-0021-8_12

incubated under continuous cool white fluorescent light (5.5 µEM^2 S^{-1}) at 18–24 °C temperature for *A. brassicae* and at 20–30 °C for *A. brassicicola* for 4–5 days. The symptoms in the form of lesions are recorded to test the reaction of the genotypes.

12.4 Detached Leaf and Pod Inoculation

Third or fourth leaves of the test plants (at 3.3–4.2 growth stages) and pods (at 5.3–6.5 growth stages) are detached, an approximately 3 mm^2 of the upper surface is gently scratched with the tip of a Pasteur pipette, and 50-µl drop of spore suspension (5×10^4 spores ml^{-1}) is placed on the scratched area (Bains and Tewari 1987); five leaves/pods from each test genotype are inoculated at four points at random. Suitable checks with sterilized water droplet inoculation are kept. The inoculated leaves/pods are incubated for 72 h on a moist filter paper in Petri dishes at 25–26 °C temperature under continuous cool white fluorescent light (5.5 µEM^2 S^{-1}). The symptoms in the form of lesions develop within 4–5 days after inoculation.

12.5 Detached Leaf Inoculation

In a detached leaf inoculation technique under laboratory study, Bansal et al. (1990) screened *Brassica* species against *A. brassicae* on the basis of the size of lesions produced on detached wounded leaves. The leaves are wounded artificially to eliminate the protective effect of the wax layer in order to examine the reaction of the underlying tissues. Ten plants from each genotype are grown individually in 5×5-cm pots containing soil-free growth medium in the greenhouse at 20/16 °C day/night temperature and 18-h photoperiod achieved with supplementary illumination provided by 400-W high-pressure sodium lamps.

Aqueous spore suspension ($4–5 \times 10^5$ spores ml^{-1}) from 10- to 14-day-old cultures of *A. brassicae* grown on V-8 agar supplemented with 0.4 mg/l^{-1} rose bengal at room temperature used 18 photoperiod was used to inoculate fully expanded detached fourth leaf (ca. 18 days after sowing) placed in a clear plastic container lined with water-soaked paper towels. The leaves are punctured of both sides of the midrib with a number 24-needle inoculator to induce circular wounds of ca. 3-mm diameter and inoculated each wound with 2.5-µl drop of spore suspension with an Eppendorf micropipette. Control leaves are inoculated with sterilized distilled water. The plastic container sealed with clear plastic wrap was incubated at room temperature under continuous light. Four days after inoculation, lesion diameter, including the chlorotic zone, was recorded.

12.6 Greenhouse Method for Testing Resistance

Small differences in infection of *A. brassicae* on *Brassica* species are difficult to detect in the field. Therefore, a more detailed method under greenhouse conditions was developed by Grontoft and O'Connor (1990). Plants of test genotypes are grown in pots. When the first true leaf emerges, the plants are thinned to one healthy plant per pot. From the time of sowing to the time of evaluations of genotypes, the temperature is kept constant at approximately 20 °C. The plants are inoculated at the three-leaf stage using *Alternaria* inoculum grown on PDA for 3 weeks at 18 °C. Inoculation is performed using small agar discs (10 mm) cut from just behind the active mycelial growth front with a cork borer. The discs are placed with the fungus growth sides against the upper side of the second leaf over a 4-mm hole. A second hole of similar size is cut on the other side of the midvein. This hole is used to control if the size of necrotic lesions is in any way influenced by leaf surface area growth. Immediately after inoculation, each plant is covered with plastic bag in order to increase relative humidity and enhance fungal growth. The genotypes are evaluated 7 days later by measuring the diameter of the necrotic lesions on the leaves.

12.7 *Brassica* Germ-Plasm Screening for Resistance through AB Toxin

Sensitivity of test genotypes to the *Alternaria* blight toxin is correlated with the disease reaction in this technique assuming that the level of sensitivity of different *Brassica* species to the toxin is similar to the in vivo susceptibility of these to *A. brassicae* (Bains and Tewari 1987). It can cause symptoms of varying severity, depending upon the host genotype, which range from severe chlorosis and necrosis to almost no visible chlorosis (Tewari and Bains 1988). Test genotypes are grown under controlled conditions, 3 mm^2 area is gently scratched with a Pasteur pipette on third or fourth detached leaf (3.3–4.2 growth stage), and 20 µl of the purified AB toxin from the single-spore culture of *A. brassicae* is applied (Bains and Tewari 1987). The leaves are incubated on a moist filter paper in a Petri dish on laboratory benches under continuous fluorescent light, and the reaction of genotypes is recorded 48 h of incubation.

In vitro screening of protoplasts of different *Brassica* genotypes, or screening of secondary embryoids, originally derived from a single anther embryoid through AB toxin, has been employed since plant protoplasts are known to be more sensitive to host-specific toxins (MacDonald and Ingram 1985; 1986; MacDonald et al. 1986). Secondary embryogenic culture lines are maintained on MS medium (Murashige and Skoog 1962) containing 0.8 % Difco agar and 2 % sucrose but without plant growth regulators. When regenerants are required, secondary embryoids are transferred for 3–7 days to MS medium supplemented with kinetin and then transferred to fresh medium without growth substances for root development. Plantlets are potted in peat balls and placed in high humidity in the greenhouse until they are established. Leaves from such plants are surface sterilized and placed in plastic boxes on moist filter paper. Two drops (10 µl each of suspension of approximately 5×10^5 spores ml^{-1} of *A. brassicicola*) are placed on each detached leaf and incubated in low continuous light (approximately 1.7 m W cm^{-2}) at 25 °C. Secondary embryoids have exhibited a wide range of reactions, including a number of resistant plants (MacDonald and Ingram 1985; 1986; MacDonald et al. 1986).

In vitro pollen selection (microspore culture) in *Brassica* for resistance to toxins from *A. brassicae* and *A. brassicicola* has been applied (Hodgkin 1990; Tewari and Bains 1988). The AB toxin from *A. brassicae* at 12.5 µg/ml completely inhibits the germination of pollens of highly susceptible *B. rapa* var. Yellow Sarson, whereas pollens of highly resistant *C. sativa* germinate even at 100 µg/ml of the toxin (Tewari and Bains 1988). Hodgkin (1990) described in vitro pollen selection in *B. napus* for resistance to phytotoxic compounds from *A. brassicicola*. Pollen samples from *B. napus* cv. 'Arran' and 'Hercules' are incubated for 1 h in a germination medium or in a medium to which 20 mg ml^{-1} of an *A. brassicicola* toxic extract has been added. The pollen samples are then used to pollinate cultivar Primar. A number of plants, obtained from such pollinations, produce pollen with a significantly increased ability to germinate in medium containing 10 mg ml^{-1} of the extract. Evidence that some selection for resistance to the toxic compounds produced by *A. brassicicola* has been obtained (Hodgkin and MacDonald 1986).

12.8 Semi-Selective Medium for Detecting Seed-Borne *A. brassicicola*

A semi-selective medium containing benomyl, sodium propionate, streptomycin sulfate and chloramphenicol in peptone–dextrose agar detected seed-borne *A. brassicicola* (Wu and Lu 1984). This medium is more sensitive than the blotter method to detect *A. brassicicola* in cabbage seeds. Seeds incubated on this semi-selective medium at 22 °C for 7 days produced velvety olivaceous colour colonies of *A. brassicicola*, with abundant sporulation. There is a positive correlation between the detached amount of seed-borne *A. brassicicola* infection on this medium and the percentage of diseased seedlings produced in autoclaved field soil or in peat moss.

12.9 Radish Root Extract Agar for *A. brassicae* Sporulation

Preparation of a radish root extract filtrate by cooking 200 g of grated radish roots in 500 ml distilled water and 20 g in 500 ml agar solution supplemented with 1 % sucrose and 1 % mannitol proved very effective in producing large colonies and profuse sporulation of *A. brassicae* (Thakur and Kolte 1985).

12.10 Inducing Sporulation of *A. brassicae*

Single-conidium stock cultures of *A. brassicae* stored at room temperature on oxoid potato dextrose agar (PDA) slopes under sterile paraffin oil was used to initiate Petri dish cultures of individual isolates for 7 days at 25 °C on PDA; a 2-cm^2 mycelium piece is placed in a sterile 25-ml masticator with 10 ml sterile distilled water, macerated at maximum speed for 1 min. and 0.2-ml portion of the macerate is transferred to 9-cm diameter triple vented polystyrene Petri dishes containing 10 ml of diluted autoclaved V-8 juice. Three replicates of each of eight field isolates are incubated separately under near ultraviolet (UV) radiation (two 40-W Philips 310, 420-nm bulb), white light (two 85-W Sylvania cool white (310, 330, 350, 750 nm) fluorescent lamps) and darkness at 25 °C for 7 days. Two to 3 days after incubation, the mycelium forms a loosely interwoven mat of hyphae throughout the liquid medium, producing a solid gelatinous matrix. As the cultures slowly dehydrate, the pinkish-white mycelia turn dark olivaceous brown, which can be correlated with the production of melanized conidiophores and conidia. After 7 days, the cultures are dehydrated to 90 % of their original mass (Senior et al. 1987).

Conidia are removed by flooding individual plates with 10 ml of a 0.01 % (V/V) sterilized Tween 20 solution while rubbing the mycelial mat with a surface-sterilized glass spreader for 1 min. The loosely interwoven mycelial mat and its contents of the Petri dish are transferred to sterilized 30-ml universal bottles and mixed using a whirlimixer for 3 min., and suspension is filtered through two layers of sterile lens tissue (Whatman no. 105) and washed further with a 5 ml of 0.01 % sterilized Tween 20 solution. The conidial suspension is centrifuged for 10 min at 2750 g (rav, 7.5 cm), supernatant is discarded, and conidia are resuspended in 5 ml of 0.01 % sterilized Tween 20 solution and counted using haemocytometer. The mean of six counts is taken for each replicate. Conidial germination is assessed by preparing spread plates (0.1 ml aliquots of conidial suspension spread over PDA plates) and incubating them for 24 h at 20 °C. For each isolate, 200 conidia are examined, and the percentage germination is calculated (Senior et al. 1987).

12.11 *Brassica* Callus Culture to Induce Sporulation in *Alternaria brassicae*

Twenty-five surface-sterilized *B. juncea* seeds placed into a 250-ml conical flask containing 30 ml of sterilized Murashige–Skoog (MS) medium (Murashige and Skoog 1962) are incubated in the dark for 3 days and in the light (300 lux) for 2–3 days until seedlings reach to the neck of the flask. Each cotyledon is excised into two pieces and transferred to a 100-ml conical flask containing 30 ml modified MS replacing IAA and kinetin with 2 mg/l NAA, 3 mg/l 2,4-D and 0.2 mg/l of 2-iPr (2-isopentenyl-purine riboside). The *A. brassicae* inoculated flasks are incubated at 25 °C (±2) under continuous light (300 lux). Callus cultures are subcultured every 3 weeks. A non-sporulating isolate of *A. brassicae* is grown in culture tubes in radish–mannitol agar (RMA) medium, which is prepared in the same way as PDA medium except that potato and dextrose are replaced by radish and mannitol, respectively. The fungus is subcultured every 15 days. Two-week-old calluses are transferred onto 10 g/l plain agar. Further, modified MS medium is supplemented with 100 mg/l Captafol. Petri dishes containing one callus per dish incubated at 25 °C (±2) under continuous light (300 lux) were used to cut approximately 1-mm diameter. Small mycelial disc from the growing edge of the fungus culture was placed at the top of each callus (Joshi et al. 1988).

12.12 Method of Estimating *Alternaria brassicicola* in Seed

Naturally infected cabbage seeds are incubated for 4 days at 18–21 °C with alternating light and dark periods on filter papers moistened with 2000 ppm aqueous solution of Dow sodium salt of 2,4-dichlorophenoxyacetate (2,4-D). When seeds are placed closer than 6 mm apart, or incubated longer than 4 days, higher pathogen counts occur. Two thousand ppm 2,4-D efficiently inhibits seed germination and does not reduce pathogen count. Incubation temperature above the optimum temperature range of 20–25 °C increases problem with the contaminant fungi. Continuous light or alternate light and dark are preferable to continuous dark. Excessive moisture is detrimental. Pretreating seeds with ethanol (EtOH) or sodium hypochlorite (NaOCl) effectively reduces rapidly growing contaminant fungi but significantly reduces pathogen count (Bassey and Gabrielson 1983a, b). However, other common methods in use for detection of seed-borne inoculum are (a) deep-freezing blotter method, (b) germination test, (c) blotter method and (d) agar plate method (Vannacci 1981).

12.13 Identification of Fungicide Antagonists in Leaf Exudates

Attempts have been made to identify the fungicide antagonists in exudate extracts of leaves of wallflower (*Cheiranthus cheiri*) (Beynon and Brown 1969). In extracts, prepared by immersing wallflower leaves in water, the antagonistic activity of the fungicide ethylene thiuram disulphide (ETD) against the spores of *A. brassicicola* has been shown to be due to the presence of glucose and fructose. These compounds are present in the extracts at concentrations corresponding to 0.015 μg of glucose and 0.002–0.005 μg of fructose for every square centimetre of leaf extracted. Dunn et al. (1969) showed that water-soluble leaf exudates extracted from apple, vine, tomato, potato and wallflower reduced the fungitoxicity of ETD to *A. brassicicola.* The spectrum of the leaf exudate effect is wide, and the magnitude of the effect can vary in each host–parasite–fungicide relationship. It seems that a large proportion of fungicide efficacy may be reduced by antagonistic materials, which either occur on the plant surface or are readily and quickly available from intact plant tissue. It is likely that the active principle affects the fungal spore, either by reducing its permeability to fungicides or by influencing some part of the metabolism involved in spore germination. Stimulation of the spore metabolism could activate systems capable of detoxifying alien chemicals. The major part of the exudate effect observed is due to glucose.

12.14 Ovary and Ovule Culture

Four- to 12-day-old ovaries are excised and sterilized with 0.25 % aqueous $HgCl_2$ for 8–10 min. followed by three washings with sterilized distilled water. The base of the pedicel is cut off, and ovaries are cultured on MS medium (Murashige and Skoog 1962) supplemented with various concentrations and combinations of growth regulators. Three to five ovaries are cultured in each tube containing 15 ml of medium solidified with 0.8 % agar and incubated at 25 °C (±1) in an illuminated room (700 lux). Thirty-five days after culture, the ovaries are cut open to collect hybrid seeds. The F_1 seeds are germinated in Petri dishes, and the seedlings transferred to pots in glass house (Sharma and Singh 1992). For ovule culture, 12–17 days after pollination, the ovaries are surface sterilized and cut open aseptically under dissecting microscope. Ten to 12 excised ovules are cultured in test tubes containing MS and modified PC-L2 medium (Phillips and Collins 1979). Observations on seed formation are recorded 15 days after culture. The seeds are first germinated in Petri dishes and transferred to 8″ diameter pots in the glass house. The F_1 hybrids are evaluated for their reaction to *A. brassicae* by inoculating with spore suspension (2000–2500 spores ml^{-1}) in the glass house (Sharma and Singh 1992).

12.15 Method for Evaluating Partial Resistance to *Alternaria brassicicola*

Three inoculation methods (cotyledon, detached leaf and seedling) were evaluated by Doullah et al. (2006). The detached leaf inoculation test is the most suitable for screening *B. rapa* genotypes because clear symptoms are observed on the leaves in less than 24 h. There is a significant positive correlation between the results from the detached leaf inoculation test and seedling inoculation test, an established method considered to yield reliable results. In addition, it is very easy to screen plants for resistance on a large scale and to maintain standard physical conditions using detached leaves. For successful infection, inoculum concentration should be adjusted to 5×10^4 conidia ml^{-1}, and incubation temperature should be between 20 °C and 25 °C. The third/fourth true leaves from 30-day-old plants are optimal for inoculation.

For this test, two leaves (third/fourth true leaves) are collected from 30-day-old plants having approximately six leaves. A replicate for the detached leaf tests consists of thee randomly selected leaves from a set of third/fourth true leaves collected from several plants that are pooled. *Alternaria brassicicola* inoculum (5×10^4 conidia ml^{-1}) is sprayed on the upper surface of the collected leaves to run-off, and inoculated leaves are placed in a plastic box (length × width × height = 32 × 23 × 5 cm) where strips of moist tissue paper are put on four vertical sides to maintain humidity at 98 %. The plastic box is inoculated at 22 ± 1 °C for 3 days in the dark in an incubator or controlled temperature room (Doullah et al. 2006).

12.16 Tissue Culture

Alternaria brassicae is isolated aseptically from the diseased *Brassica* leaves and multiplied on potato dextrose agar (PDA) for further studies. About 5-mm-diameter small mycelia of *A. brassicae* are transferred to the 250-ml Erlenmeyer flask containing 30 ml of sterilized potato dextrose broth (PDB) and incubated at 22 ± 1 °C for 15 days in dark. Fungal mycelium is separated by filtering through sterilized Whatman no. 1 filter paper, and filtrate further is sterilized by passing through millipore filter of 0.22-μm size; the pH of the filtrate is adjusted to 5.8 before filter sterilization. The sterilized filtrate is added to the autoclaved MS medium containing 1 mg l^{-1} NAA and 1 mg l^{-1} BAP (MSN_1B_1) during cooling and dispensed in sterilized conical flasks. The medium is stored at 25 ± 1 °C in dark and used for inoculation within a week of its preparation.

Seedlings of different species and cultivars are raised aseptically on hormone-free MS medium with 0.5 % sucrose and 0.8 % agar. Hypocotyl segments measuring 5–6 mm are excised from 6- to 7-day-old seedlings and implanted on MSN_1B_1 medium with or without fungal CF. The inoculated flasks are incubated at a temperature 25 ± 1 °C for 1 month under 16-h light (1500 lux) and 8-h dark. After 1 month, per cent callus induction and fresh and dry weights of individual callus are recorded (Kiran et al. 2002).

12.17 Inoculation Methods for Pathogenesis of *Alternaria brassicae*

Five inoculation methods including spraying, infiltration, wounding, spore suspension drop and spore suspension drop along with agarose method were compared. Among the five inoculation methods compared, spore suspension drop along with agarose inoculation method proved most ideal as this fixes the inoculum on the target site. Compared to other four inoculation methods, the highest mean numbers of initial disease lesions in drop plus agarose recorded were 312.2, 484.2, 664.2, 734.2 and 799.2, respectively, at 24, 48, 72, 96 and 120 h after inoculation. Besides easy to handle the inoculated plants, plus agarose method has the advantage of being accurate and precise.

A single conidium of *A. brassicae* culture is derived from a conidia produced on a diseased *B. juncea* leaf and maintained on potato dextrose agar (PDA). For conidiation, fungus is grown on

V-8 agar (10 % V-8 juice, 0.02 % $CaCO_3$ and 2 % agar) at 24 °C for 3–4 days under cool white fluorescent light (2000 lux), followed by further 2-day incubation at 18 °C (Aragaki 1964). A conidial suspension is prepared by scraping mycelia and conidia from plates of actively growing cultures into autoclaved water and filtering through four layers of cheese cloth to remove most of the mycelia. The filtered spore suspension is centrifuged at 2000×g for 5 min. and resuspended in to deionized water. This centrifugation is repeated once more in order to ensure a clear conidial suspension-free of metabolites. After the final wash, supernatant is discarded, conidia resuspended in water containing 0.05 % Tween 20, conidia counted using haemocytometer, and the concentration adjusted to 10^4 spores/ml.

Seeds of brown mustard (*B. juncea*) are sown in plastic inserts (7.5×5 cm; two seeds per insert) consisting a mixture of soil, sand and vermicomposts (2:1:1) in the green house (22/18 °C day/night, 16-h photoperiod). Pathogenicity test is carried out with spore suspension on detached leaves' seedling. Equal amount of inoculums (500 μl) is used with a sterile micropipette, and ten replicate plants are used for each method.

In the spore suspension drop method, inoculum is placed on each detached leaf in the form of a drop, whereas in the spore suspension drop plus agarose method, the inoculum drops are sterilized on a target area to prevent any adverse effect on spores due to high temperature; 5 ml of 0.08 % agarose in a test tube is heated in a microwave oven at high power for 40 s to melt the agarose, and the test tube is then placed in a water bath at about 42 °C in a 250-ml beaker to keep the agarose in the liquid state at this temperature. Each inoculum site on the leaf is covered with agarose, and the agarose drop appears as a clear dome on the plant surface. The agarose solidifies and fixes the inoculum on the target site within few minutes.

In the wounding method, inoculation is performed by gently wounding the detached leaves with a pipette tip. In the spraying method, inoculum is sprayed on detached leaves with the help of an atomizer. In syringe infiltration method, the detached leaf is carefully inverted, exposing the abaxial side. A 1-ml needleless syringe containing a spore suspension is used to pressure infiltrate the leaf intracellular space(s), but the vascular system of the leaf is avoided from injection.

Detached leaves are kept in sealed Petri dishes with 1 % agar and placed in growth chambers at 25 °C with 70 % relative humidity. The Petri plates are observed for *A. brassicae* initial symptoms at intervals of 24 h up until 120 h after inoculation. The numbers of disease lesions are counted on the detached leaves in all the inoculation methods at 24, 48, 72 and 120 h after inoculation (Giri et al. 2013).

12.18 PCR-Based Assay for Detecting *Alternaria brassicae* in Cruciferous Seed

Current detection methods, based on culture, and morphological identification of the fungus are time-consuming, laborious and not always reliable. Therefore, a polymerase chain reaction (PCR)-based assay was developed with *A. brassicae-specific* primers designed on the basis of sequence of two clustered genes potentially involved in pathogenicity by Guillemette et al. (2004). Two sets of primers were selected for conventional and real-time PCR, respectively. In both cases, *A. brassicae* is specifically detected using DNA extracted from seed. The real-time PCR-based method can be automated easily, and preliminary results indicate that it is efficient for quantitative estimation of seed infection.

12.18.1 Preparation of Seed Samples

Seed samples are prepared from contaminated seed of radish and cabbage as described by Iacomi-Vasilescu et al. (2002). For artificial contaminations, seeds are disinfected with 1 % sodium hypochlorite, rinsed with sterilized water, soaked for 1 h in a calibrated spore suspension (10^6 spores/ml) and dried at room temperature on sterilized filter paper. Disinfection and inoculation accuracy is confirmed by placing 20 seeds on the surface of malt agar plates for 1 week at 25 °C. Seed batches showing *A. brassicae* 0 % (disinfected seed) or 100 % *A. brassicae* grown (inoculated seed) are selected and mixed together to prepare seed samples with different levels of contamination (0, 5, 10, 50 and 100 %). Samples are produced by placing 20 seeds (artificially contaminated) or 100 seeds (naturally contaminated) in a sterile tube and covering them with 0.5 and 3 ml, respectively, of liquid culture malt extract–dextrose–peptone (2 % malt extract, 2 % dextrose, 0.1 % peptone) (MDP) culture medium. The tubes are incubated at 25 °C for 48 h with occasional shaking. This incubation step is an enrichment phase that allows an optimal increase of the fungal biomass from seed. The tubes are then vortexed briefly to separate mycelia and conidia from seed. The seeds are discarded, and the fungal structures collected on microcentrifuge filter (pore size 0.45 μm; Ultrafree-Mc Millipore) by centrifugation at 5000 × g for 30 min.

12.18.2 DNA Manipulation

DNA is extracted from fungal mycelia and conidia by scraping the surface of Petri plate cultures (extraction from fungal culture) or collected on microcentrifuge filters (extraction from seed samples). Nucleic acids are isolated according to microwave miniprep procedure described by Goodwin and Lee (1993). Alternatively, DNA is extracted using the NucleoSpin food kit according to the manufacturer's instructions.

12.18.3 PCR-Based Assay

Specific oligonucleotides AB Csens (5′CTGGTGAAAAGGTT GCGATCGT 3′) and AB Crev (5′GTGACTTCATGAAATGACA TTGATG 3′), complementary to 3′ end of the open reading frame (ORF) 2 and 115 sens (5′A ACCCTATAGACC CACGTCGACTA-3′), and 115rev (5′GATGGTACGCAAGGCTTGGT-3′), complementary to a portion of ORFI, are designed for the standard and real-time PCR assays, respectively, using the appropriate software (Primer 3 online and ABI PRISM Primer express). The two ORFs are deduced from the nucleotide sequence of an *A. brassicae* genomic DNA fragment identified after screening a cosmid library with a probe corresponding to a portion of the non-ribosomal peptide synthase (NRPS) gene. In order to test the specificity of the primer pairs against *A. brassicae* isolates, amplifications are performed using DNA extracted from pure fungal cultures from a range of *Alternaria* spp. and other fungi isolated from seed. Then the PCR procedure is applied to detect seed contamination. The universal primer pair ITS1/ITS2 is used as a positive control to assess the quality of the extracted DNA. Conventional PCR is performed using 2 μl of undiluted DNA preparation under the following conditions: Tris HCL, pH 9.0, 20 mM $(HN_4)_2SO_4$, 0.01 % (w/v) Tween 20, 1.5 mM $MgCl_2$, 200 μM each deoxyribonucleotide triphosphate and 1 unit of thermostable DNA polymerase. A thermo jet thermo cycler is used with an initial step of 3 min at 95 °C, followed by 35 cycles of 30s at 95 °C, 50s at 60 °C and 1 min at 72 °C and a final step of 10 min at 72 °C. The amplification results are visualized after electrophoresis of an aliquot (10 μl) of the reaction mixture on a 1.2 % agarose gel. The real-time PCR reactions are carried out in a 25-μl final volume containing 2.5 μl of DNA, 12.5 μl of SYBR Green PCR Master Mix 2X, 300 nM forward and reverse primers and H_2O up to 25 μl. An ABI PRISM 700 sequence Detection System (Applied Biosystems) is used with the following steps: 2 min at 50 °C, 10 min at 95 °C and 40 cycles consisting in one step at 95 °C for 15 s followed by one step at 60 °C for 1 min. Nucleic acids are quantified in unknown samples by direct comparison to a standard. The DNA used as standard is extracted from fungal cultures, with the concentration measured by fluorometric assay before dilution.

12.19 Quantitative Inoculation Method

Alternaria brassicae is derived from a conidium produced on diseased leaf of mustard. Conidiation of *A. brassicae* is induced by growing the fungus on V-8 agar (10 % V-8 juice, 0.02 % $CaCO_3$ and 2 % agar) at 24 °C for 3–4 days under cool white fluorescent light (2000 lux) followed by 2-day incubation at 18 °C (Aragaki 1964). A conidial suspension is prepared by placing three pieces of culture blocks (ca 5×5×3 mm) in 5 ml wheat grain extract in a test tube and by agitating the test tube for 1 min with a vortex mixer. Wheat grain extract is prepared by autoclaving whole wheat grains in distilled water at the ratio of 1:6 by volume. Only the top layer of clear extract is used.

Seedlings of black mustard are established from seeds planted in Supersoil potting mix in rectangular plastic pots (6.5×6×6 cm). All the test plants are fertilized with Osmocote 14-14-14 and used after 1–2 months. The spore concentration in the suspensions is adjusted by approximating the desired number per µl. Five 1-µl drops of the spore suspension are then placed on a glass slide and examined under the microscope. Those drops with exact number of spores are needed individually, transferred with microlitre pipette to leaves or stems about 3–5 cm above the pots and temporarily positioned horizontally during inoculation. The emptied spots on the glass slide are examined again with microscope to assure that all the spores in each drop are transferred. One drop per leaf or stem is used.

Sea plaque agarose, which gels at about 30 °C, is used to stabilize inoculum drops on target areas to prevent injuring to spores due to high temperature. Five ml of 0.8 % agarose in a test tube is heated in a microwave oven at high power for 40 s to melt the agarose. The test tube is then placed in water bath at 32 °C in a 250-ml beaker to keep the agarose in the liquid state at this temperature. Each inoculum site on the leaf or stem is covered with 10 µl agarose. The agarose drop appeared as a clear dome on the plant surface. The agarose solidifies and fixes the inoculum on the target site within 5 min. Inoculated plants are placed in plastic bag moist chamber (50-cm diameter, 90-cm high), each with 100 ml water on the bottom, and kept on the laboratory floor at 24 °C for observation at different time intervals. Three replicate plants are used for each treatment, and experiments are repeated at least once with similar results (Xu and Ko 1998).

12.20 Assessment of Methods of Inoculation for Resistance to *Alternaria*

Out of four methods of inoculation (seed inoculation, cotyledonary leaf inoculation, detached true leaf inoculation and detached green pod inoculation), detached true leaf inoculation is more efficient and reliable for screening germ plasm for resistance to *Alternaria* blight of rapeseed and mustard.

12.20.1 Method of Inoculation

A conidial suspension of *A. brassicae* containing 1.5×10^4 conidia/ml is prepared, and then with the help of Eppendorf micropipette, three separate 10-µl drops, each containing 4–6 conidia, are placed on the upper surface of true leaves at three equidistant places. Similarly, distilled water drops (without conidia) are placed on other cotyledons, which serve as checks. All the treatments are replicated three times for each cultivar with each isolate. Observations on disease severity are recorded on 0–5 rating scale as described at Sect. 12.20.3.

12.20.2 Detached True Leaf Inoculation

True leaves are collected from 30-day-old plants, surface sterilized with sterilized distilled water with the help of sterilized cotton, and leaves inoculated following the method described above. Inoculated leaves are kept in Petri plate moist chamber and incubated at 20±2 °C for 7 days, and leaves scored for disease severity.

12.20.3 Infection Score and Disease Index

A 0–5 rating scale is followed, where 0 = no visible symptom, 1 = 1–10 % leaf or pod area covered with small spot, 2 = 11–25 % leaf or pod area covered with bigger (more than 3 mm) spot, 3 = 26–50 % leaf or pod area covered with bigger (more than 3 mm) spots with initiation of coalescing on leaves and deep lesions on pods, 4 = 51–75 % leaf or pod area covered with bigger commonly coalescing spot and 5 = 76–100 % leaf or pod area covered with very large spots giving blighting appearance (Vishwanath and Kolte 1999).

12.21 Image-Based Disease Identification

Image-based rapeseed–mustard disease expert system – an effective extension tool – has been developed by Kumar et al. (2008). The identification of disease is the difficult task. If the disease is identified timely, the control measures can be applied effectively. Nowadays, the use of digital technology can produce high-quality digital images including photos and clips of healthy and infected plants easily that can play very important role in disease identification. Digital images can be seen and shared easily among the experts. Image can be examined on camera screen literally within the second; they are captured and downloaded to a computer for closer inspection within minutes. A computerized expert tool image-based rapeseed–mustard disease expert system has been developed to help extension personals, researchers and farmers in identification and management of these diseases. The expert system uses a hierarchical classification and a mix of the text description, photographs and artistic pictures. The system involves two main subtasks, namely, diagnosis and management. The system designed and developed using Visual Basic as front end and Microsoft Access 2000 as back end software.

12.22 ELISA Diagnostic Kits

A number of disease detection kits have been developed for use at the site where a disease is suspected. These kits, which in most cases do not require laboratory equipment, are especially useful to growers. Some tests only take 5 min. to perform. The diagnostic kits are based on a method that uses proteins called antibodies to detect disease causing organisms of plants (plant pathogen). The technique used is called ELISA (enzyme-linked immunosorbent assay). This assay is based on the ability of an antibody to recognize, and bind to a specific antigen, a substance associated with a plant pathogen. The process is described in Fig. 12.1. The antibodies used in the diagnostic kits are highly purified proteins produced by injecting a warm-blooded animal (like rabbit) with an antigen associated with one particular plant disease (Fig. 12.1(1)). The animal proteins react to the antigen and produce antibodies (Fig. 12.1(2)). The antibodies produced recognize and react only with proteins associated with the causal agent of plant disease (Reddy 2014).

These antibodies are bound to a plastic plate or similar detector unit in a test kit. The individual running the test prepares the plant sample by grinding it between two pieces of abrasive paper. The ground plant sample is placed in a bottle filled with an extraction liquid. This liquid is then placed in wells in the plastic plate (Fig. 12.1(3)). If the disease-causing organism is present in the sample, the specific antibodies in the plastic plate will bind to pathogen-associated proteins and adhere to the unit (Fig. 12.1(4a)). A second antibody is added that also reacts with the pathogen-associated proteins. This antibody is special because it can react with colour-producing chemicals called reagents (Fig. 12.1(5a)). Colour changes on the unit's surface indicate a positive reaction. If no pathogen-associated proteins are present (Fig. 12.1(4b)), the detector antibodies cannot bind to the colour-carrying reagents and are washed away (Fig. 12.1(5b)).

Presently, a limited number of these onsite disease detection kits are available. Many ELISA

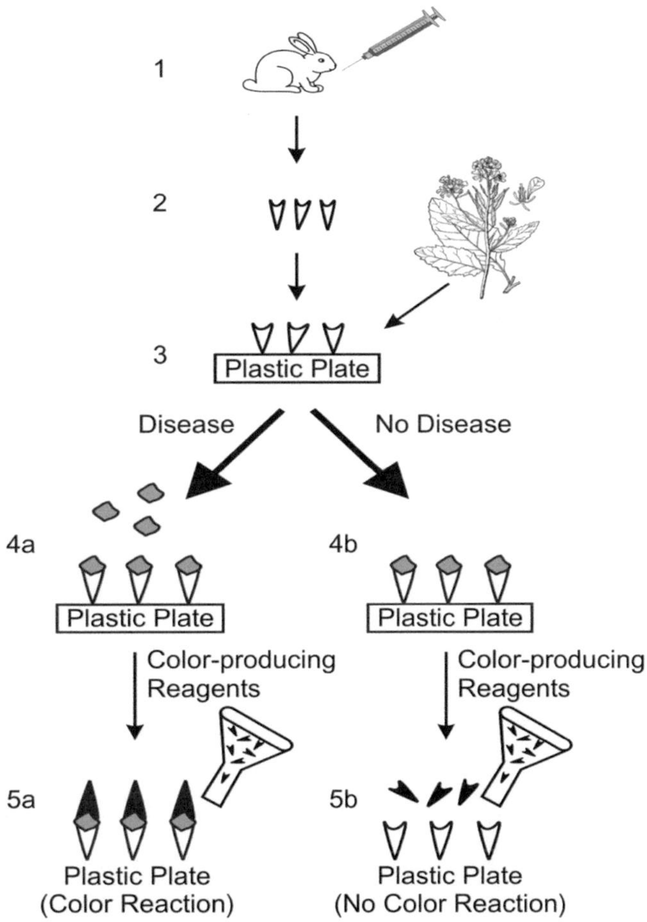

Fig. 12.1 ELISA diagnostic kit (Courtesy, P. Parvatha Reddy 2014)

diagnostic kits are available for use in the laboratory, however, and can test for a wide range of plant pathogens.

12.23 Direct Tissue Blotting

Another diagnostic assay that also uses specific antibodies as a detection tool is known as direct tissue blotting. With this technique, the location of a disease-causing pathogen within the host plant can be determined, allowing earlier detection and a better understanding of how a disease progresses through a plant. Host plant tissue is placed onto a special piece of paper. Antibodies that bind only to the disease-causing pathogen in question are then introduced to this paper. A colour change indicates a positive result and shows the location of the pathogen in the hot tissue (Fig. 12.2) (Reddy 2014).

12.24 Nucleic Acid Probes

Another set of tools that can be used in plant disease diagnostics is nucleic acid probes. These probes are fragments of nucleic acid arranged in a sequence complementary to that of the DNA or RNA of the disease-causing pathogen. Because the sequences complement

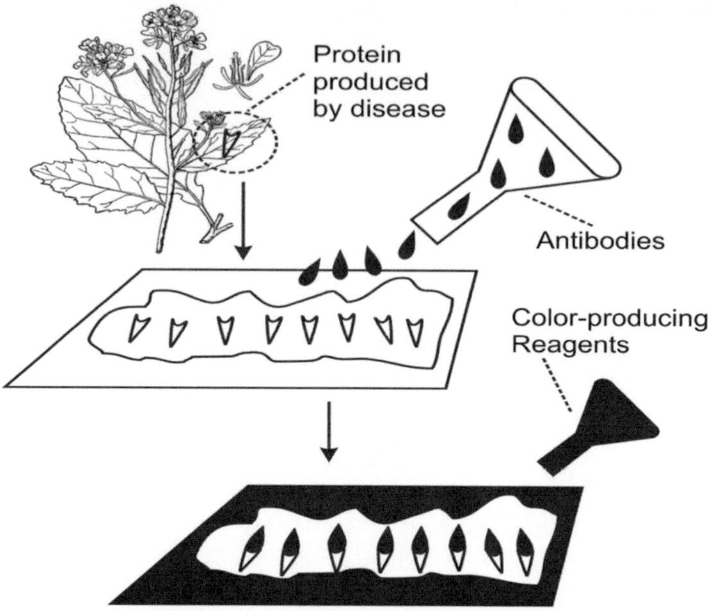

Fig. 12.2 Direct tissue blotting (Courtesy, P. Parvatha Reddy 2014)

each other, the probes can be used to identify specific diseases.

12.24.1 Squash Blot Method

A method that uses nucleic acid probe technology is known as the squash blot method (Fig. 12.3). This technology is similar to direct tissue blotting. Tissue from a plant that is suspected to be diseased is 'squashed' onto a special piece of paper, called a membrane (Fig. 12.3(1)). This membrane is then treated with a probe (Fig. 12.3(2)) that can bind or hybridize with nucleic acid of the plant pathogen suspected to be in the plant tissue. Hybridization or binding will occur when alike sequences are present (Fig. 12.3(3)). After adding several more substances to the membrane, a colour reaction (Fig. 12.3(4)) indicates that the probe and pathogen nucleic acid sequences have hybridized and the disease is present. No colour reaction means the test for the disease is negative (Reddy 2014).

12.24.2 Polymerase Chain Reaction (PCR)

A new technology, PCR, has great potential for raising the sensitivity of various assays that use nucleic acid probes. PCR is used to produce enormous numbers of copies of a specified nucleic acid sequence. This technique can allow the detection of very small amounts of a pathogen in a sample by amplifying the pathogen sequences to a detectable level. PCR is especially useful in plant quarantine point of view owing to its fastness.

The specificity of PCR, be it conventional or real time, depends upon the designing of proper PCR primers that are unique to the target organism. Highly conserved gene resigns are often the target for designing primers. Closely related microbial species often differ in a single (single-nucleotide polymorphism (SNPs)) to few bases in such genes. PCR allows detection of such SNPs (Papp et al. 2003). With the advancements in high-throughput DNA sequencing, more and more genomes of plant pathogens are sequenced,

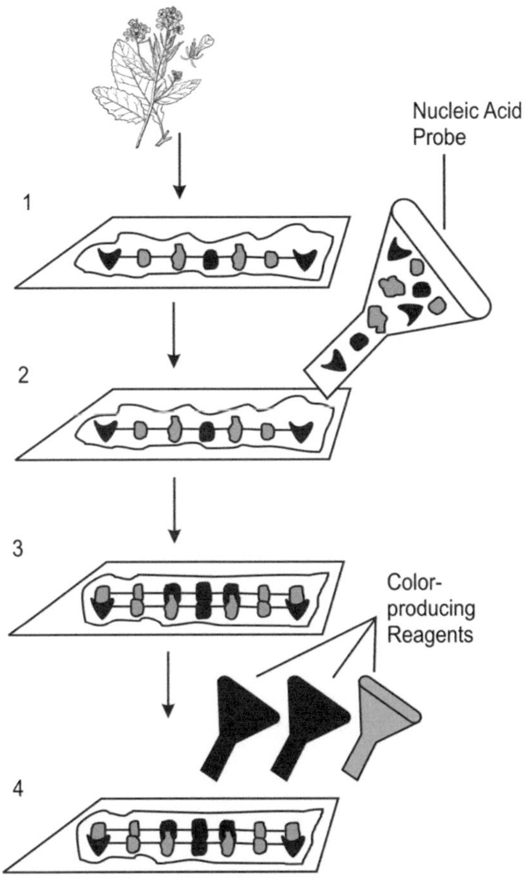

Fig. 12.3 Nucleic acid probe squash blot method (Courtesy, P. Parvatha Reddy 2014)

and nucleotide sequence data will be available increasing the possibility for designing unique primers and probes for specific detection of pathogens.

PCR is highly sensitive technology. However, its sensitivity is greatly affected by the presence of inhibitors, which prevent or reduce amplification. A wide range of inhibitors are reported. Although their mode of action is not clear, these inhibitors are believed to interfere with polymerase activity for amplification of the target DNA. On the other hand, it is worth mentioning that the high sensitivity of PCR also causes one of the limitations of PCR, i.e. detection sensitivity exceeding threshold levels or clinical significance and false-positive results from slight DNA contamination (Yang and Rothman 2004). Hence, stringent conditions are necessary in conducting the assay, and proper negative controls must be included in the test. It is also recommended to have separate dedicated areas for pre- and post-PCR handling.

Due to the advancement of fluorogenic chemistry, a second-generation PCR known as real-time PCR has become an emerging technique for the detection and quantification of microorganisms in the environment. In PCR, the target DNA sequence is amplified over a number of denaturation annealing extension cycles. In a conventional PCR, only the final concentration of the amplifications may be monitored using a DNA-binding fluorescent dye. However, in the quantitative real-time PCR, the concentrations of the amplicons are monitored throughout the amplification cycles using a group of fluorescent reagents. The fluorescence intensity emitted during this process reflects the amplifications' concentration in real time. Undoubtedly, most of the future tests will be quantitative in nature, and the real-time detection system will be a method of choice. The real-time data will serve as useful basis for establishing inoculum threshold levels and detailed analysis of disease epidemics.

12.24.3 DNA Microarray Technology

The DNA microarray technology, originally designed to study gene expression and generate single-nucleotide polymorphism (SNP) profiles, is currently a new and emerging pathogen diagnostic technology which in theory differs a platform for unlimited multiplexing capability. The principle of microarray is the hybridization of fluorescently labeled sequences or targets to their complementary sequences spotted on solid surface, such as glass slides, serving as probes. Tens of thousands of such DNA probes can be spotted in a defined and addressable configuration on the glass slide forming the chip. The unlimited capability for simultaneous detection of pathogen makes microarray to be an approach with a potential capacity of detecting all relevant pathogens of a specific crop. Development of microarray for diagnostic applications is a recent history. In plant pathology, the method was applied for identifying oomycetes, nematode and bacterial and fungal DNA from pure cultures (Fessehaie et al. 2003; Lievens et al. 2003). However, for application in practice, pathogens should be detected from environmental samples (plants, soil, etc.). Recently, the possibilities of parallel detection of pathogens from such environments were shown (Nicolaison et al. 2005). In contrast with studies using pure cultures, microarray-mediated analysis from environmental samples presents several challenges that must be addressed (Whittle et al. 2005).

References

Aragaki M (1964) Relation of radiation and temperature to the sporulation of *Alternaria* tomato and other fungi. Phytopathology 54:565–569

Bains PS, Tewari JP (1987) Purification, chemical characterization and host-specificity of the toxin produced by *Alternaria brassicae*. Physiol Mol Plant Pathol 30:259–271

Bansal VK, Seguin-Swartz G, Rakow GFW, Petrie GA (1990) Reaction of *Brassica* species to infection by *Alternaria brassicae*. Can J Plant Sci 70:1159–1162

Bassey EO, Gabrielson RL (1983a) Factors affecting accuracy of 2, 4-D assays of crucifer seed for *Alternaria brassicicola* and relation of assays to seedling disease potential. Seed Sci Technol 11:411–420

Bassey EO, Gabrielson RL (1983b) The effects of humidity, seed infection level, temperature and nutrient stress on cabbage seedling disease caused by *Alternaria brassicicola*. Seed Sci Technol 11:403–410

Beynon KI, Brown KF (1969) The identification of fungicide antagonists in leaf exudate. Phytopathol Z 64:213–220

Doullah MAU, Meah MB, Okazaki K (2006) Development of an effective screening method for partial resistance to *Alternaria brassicicola* (dark leaf spot) in *Brassica rapa*. Eur J Plant Pathol 116:33–43

Dunn CL, Brown KF, Montagne JTW (1969) Antagonism between fungicides and water soluble exudates from leaves of plants. Phytopathol Z 64:112–118

Fessehaie A, De Boer SH, Levesque CA (2003) An oligonucleotide array for the identification and differentiation of bacterial pathogenic on potato. Phytopathology 93:262–269

Giri P, Taj G, Kumar A (2013) Comparison of artificial inoculation methods for studying pathogenesis of *Alternaria brassicae* (Berk.) Sacc on *Brassica juncea* (L.) Czern. (Indian mustard). Afr J Biotechnol 12:2422–2426

Goodwin DC, Lee SB (1993) Microwave mini-prep of total genomic DNA from fungi, plants, protists and animals for PCR. Biotechniques 15:438–444

Grontoft M, O'Connor D (1990) Greenhouse method for testing of resistance of young *Brassica* plants to *Alternaria brassicae*. Plant Breed 105:160–164

Guillemette T, Iacomi-Vasilescu B, Simoneau P (2004) Conventional and real-time PCR-based assay for detecting pathogenic *Alternaria brassicae* in cruciferous seed. Plant Dis 88:490–496

Hodgkin T (1990) *In vitro* pollen selection in *Brassica napus* L. for resistance to phytotoxic compounds from *Alternaria brassicicola* (Schw.) Wilts. Sex Plant Reprod 3:116–120

Hodgkin T, MacDonald MV (1986) The effect of a phytotoxin from *Alternaria brassicicola* on *Brassica* pollen. New Phytol 104:631–636

Humpherson-Jones FM, Phelps K (1989) Climatic factors influencing spore production in *Alternaria brassicae* and *Alternaria brassicicola*. Ann Appl Biol 114:449–459

Iacomi-Vasilescu B, Blancard D, Guénard M, Molinero-Demilly V, Laurent E, Simoneau P (2002) Development of a PCR-based diagnostic assay for detecting pathogenic *Alternaria* species in cruciferous seeds. Seed Sci Technol 30:87–95

Joshi MK, Singh US, Garg GK (1988) Use of *Brassica* callus culture to induce sporulation in *Alternaria brassicae* (Berk.) Sacc. Plant Cell Tissue Organ Cult 14:59–62

Kiran, Dhingra HR, Mehta N, Sangwan MS (2002) Evaluating relative resistance of *Brassica* to *Alternaria* blight through tissue culture technique. J Mycol Plant Pathol 32:231–233

Kolte SJ, Yadav DS (1990) Stem explants culture and inoculation technique for screening for resistance to white rust and Alternaria blight of rapeseed and mustard. In: First annual report (1989–90), ICAR/IDRC, Directorate of Experimental Station, Pantnagar

Kumar V, Haldar S, Pandey KK, Singh RP, Singh AK, Singh PC (2008) Cultural, morphological, pathogenic and molecular variability amongst tomato isolates of *Alternaria solani* in India. World J Microbiol Biotechnol 24:1003–1009

Lievens B, Brouer M, Vanachter ACRC, Levesque CA, Cammue BPA, Thomma BPHJ (2003) Quantitative assessment of phytopathogenic fungi in various substrates using DNA macroarray. Environ Microbiol 7:1698–1710

MacDonald MV, Ingram DS (1985) *In vitro* selection for resistance to *Alternaria brassicicola* in *Brassica napus* spp. *oleifera* (winter oilseed rape) using partially purified culture filtrates. Cruciferae Newsl 10:97–100

Macdonald MV, Ingram DS (1986) Towards the selection *in vitro* for resistance to *Alternaria brassicicola* (Schw.) Wilts., in *Brassica napus* ssp. *oleifera* (Metzg.) Sinsk., winter Oilseed rape. New Phytol 104:621–629

MacDonald MV, Ingram DS, Hodgkin T (1986) Selection *in vitro* for resistance to *Alternaria brassicicola* in *Brassica napus* spp. *oleifera*, winter oilseed rape. Proceedings of Crucifer Genetic Workshop 3, Crucifer Genetic Co-operative, University of Wisconsin, Madison, Wisconsin, p 66

Murashige T, Skoog F (1962) A revised medium for rapid growth and bioassays with tobacco tissue cultures. Physiol Plant 15:473–497

Nicolaison M, Justesen AF, Thrane V, Skouboe P, Holmstrom K (2005) An oligo-nucleotide microarray for identification and differentiation of trichothecene producing and non-producing *Fusarium* species occurring on cereal grains. J Microbiol Methods 62:57–69

Papp AC, Pinsonneault JK, Cooke G, Sadec W (2003) Single nucleotide polymorphism genotyping using allele-specific PCR and fluorescence melting curves. Biotechniques 34:1068–1072

Phillps GC, Collins GB (1979) *In vitro* tissue culture of selected legumes and plant regeneration from callus cultures of red clover. Crop Sci 19:59–64

Reddy PP (2014) Molecular diagnosis of plant diseases. In: Genetic engineering for crop protection. Studium Press LLC, Houston, pp 209–215

Senior DP, Epton HAS, Trinci APJ (1987) An efficient technique for inducing profuse sporulation of *Alternaria brassicae* in culture. Trans Br Mycol Soc 89:244–246

Sharma TR, Singh BM (1992) Transfer of resistance to *Alternaria brassicae* in *Brassica juncea* through interspecific hybridization among brassicas. J Genet Breed 46(4):373–378

Tewari JP, Bains PS (1988) Mode of action of the host-specific toxins produced by *Alternaria brassicae*. In: Proceedings of 5th international congress of plant pathology, Kyoto, 20–27 August 1988, p 234

Thakur R, Kolte SJ (1985) Radish root extract agar, a suitable medium for the growth and sporulation of *Alternaria brassicae*. Cruciferae Newsl 10:117–118

Vannacci G (1981) Seed born *Alternaria brassicicola*: detection by means of symptoms on seedlings. Acta Horticult 111:123–129

Vishwanath K, Kolte SJ (1999) Methods of inoculation for resistance to Alternaria blight of rapeseed and mustard. J Mycol Plant Pathol 29:96–99

Whittle IHF, Klammer SH, Insam H (2005) Design and application of an oligo-nucleotide microarray for investigation of compost microbial communities. J Microbiol Methods 62:37–56

Wu WS, Lu JH (1984) A semi-selective medium for detecting seed-borne *Alternaria brassicicola*. Plant Prot Bull (Taiwan, RO.C.) 26:67–72

Xu XL, Ko WH (1998) A quantitative confined inoculation method for studies of pathogenicity of fungi on plants. Bot Bull Acad Sin 39:187–190

Yang S, Rothman RE (2004) PCR-based diagnostics for infectious diseases: uses, limitations, and future applications in acute-care settings. Lancet Infect Dis 4:337–348

13 Future Strategies and Priorities for the Management of *Alternaria* Diseases of Crucifers

With the increasing demands of a growing global population, there has been a significant change in cropping patterns of edible and industrial oil-yielding crucifers as well as vegetables. In the absence of host resistance, intensive cultivation has resulted in perpetuation, build-up and dissemination of *Alternaria* species on crucifers in areas where these crops are grown. The main factors responsible for the increase of inoculum of pathogenic *Alternaria* to the present level include the lack of genetic sources of resistance, cultivation of susceptible crops with high plant density, use of irrigation with high rates of nitrogen fertilizers, contiguous areas under monoculture, staggered sowing dates, poor weed management and poor plant protection strategies. The information gathered in this book on *Alternaria* diseases of crucifers indicates that some gaps and bottlenecks still exist in our comprehensive understanding of various dimensions of *Alternaria–host* pathosystem; some of these are highlighted below to be resolved priority wise in areas of crucifer production.

13.1 Disease Epidemiology and Forecasting

Factors governing disease development and consequent progression are not completely understood. Deeper insight with regard to various epidemiological aspects will help to develop strategies to curb the progression of these potential diseases. There is an urgent need to carry out intensive investigation with regard to the environmental parameters responsible for causation and spread of diseases under field conditions. Multilocational trials with staggering dates of planting could be helpful in analysing disease development in relation to environment by computing disease progression at regular intervals. These efforts could be undertaken to study various *disease prediction models* (geophytopathological, bioclimatic, simulation system analysis, etc.) which could play a meaningful role in developing an effective disease control strategy.

13.2 Physiological Specialization

Identification and standardization of host differentials are fundamental prerequisites to the gathering of meaningful information on races/pathotypes. This is a high-priority area of investigation. Designation and nomenclature of pathotypes should be uniform, logical and internationally acceptable. Virulence genes may be identified.

13.3 Genetics of Resistance

Although we have some information on genetic sources of resistance, the information on the nature, mechanism and inheritance of resistance

G.S. Saharan et al., *Alternaria Diseases of Crucifers: Biology, Ecology and Disease Management*,
DOI 10.1007/978-981-10-0021-8_13

are still not satisfactory and/or lacking. The results of this area of investigation are an important prerequisite to incorporate stable and effective sources of resistance in breeding programmes. In diseases caused by *Alternaria* species, where major gene resistance is not known, the potential of minor genes/horizontal resistance, tolerance and slow blighting should be exploited. Feasibility of induced resistance needs attention. Biotechnology or molecular biology and genetic engineering techniques can be used to transfer resistant genes from distantly related sources. Identification and sequencing of resistant genes along with identification of suitable gene combinations may be taken up. Hybrids and GM cultivars with multiple disease resistance is the need of the hour.

13.4 Genetics of Virulence

Characterization of secondary metabolite biosynthetic genes along with signal transduction and their role in pathogenicity and fungal development are important areas of investigation. Disruption of *Aso-1*, a gene required for hyphal fusion (anastomosis) which is important for *Alternaria* spores to become pathogenic, needs attention. Studies on the nature and mechanism of inheritance of virulence genes in different species of *Alternaria* are required. Determination of virulence spectrum of pathotypes will reveal use of number of resistance genes in breeding of disease-resistant brassicas.

13.5 Exploitation of Morphological, Structural and Biochemical Basis of Resistance

Exploitation of morphological, structural and biochemical basis of resistance needs in-depth study in crucifer since a lot of genetic variability is available in related and distantly related crucifers all over the world.

13.6 Comparative Studies on All Aspects of Host–Parasite Interaction

Comparative studies on all aspects of host–parasite interaction in crucifers with all the four species of *Alternaria* may reveal some fruitful results for planning comprehensive management strategies of these pathogens.

13.7 Phytotoxins

Relatively few genes and/or gene products have been identified that contribute to or are required for pathogenicity. More research is required to further characterize secondary metabolites biosynthetic genes and their role in pathogenicity and fungal development along with fungal signal transduction mechanisms.

13.8 Genome Sequencing

Alternaria species genome sequencing will allow the identification of putative secreted pathogen effectors' proteins in silico that can be used in a variety of functional analyses.

13.9 Disease Control Strategy

Some useful information is available on chemical control of various *Alternaria* diseases; however, much more is needed in order to develop a viable disease control strategy under field conditions. Efforts should also continue to search for low-cost effective chemicals, which can provide economically significant disease control. Active ingredient of plant extracts should be identified.

Control of *Alternaria* diseases in rapeseed–mustard by microbial antagonists should also be explored. Seed treatment with *Trichoderma* spp., *Gliocladium* spp., *Penicillium* spp. and Mycostop-like preparations of *Streptomyces* spp. should be further investigated to determine their level of control and economic feasibility under field conditions. Similarly, control by foliar applications of *Streptomyces rochei*, *S. hygroscopicus*, *S. arabicus* and *Nectria inventa* should also be evaluated under field conditions.

13.10 Integrated Management

No single methods or approach can be effective and economical in dealing with any biological systems. Therefore, all means of plant disease control including chemical, biological, cultural, host resistance and biotechnology tools for early detection of disease should be integrated to manage the disease at early stage of plant growth. Considering the role of insects like aphids (virus transmission) and flea beetles (transmission of *A. brassicicola* to cabbage), insecticides should be one of the components in integrated disease management package and practices. Disease forecasting system should be strengthened using modern information technology available at village level to warn farmers in advance for timely action of proper control strategy.

Index

G.S. Saharan et al., *Alternaria Diseases of Crucifers: Biology, Ecology and Disease Management*, DOI 10.1007/978-981-10-0021-8

MIX
Papier aus verantwortungsvollen Quellen
Paper from responsible sources
FSC® C105338
FSC
www.fsc.org

If you have any concerns about our products,
you can contact us on
ProductSafety@springernature.com

In case Publisher is established outside the EU,
the EU authorized representative is:
Springer Nature Customer Service Center GmbH
Europaplatz 3, 69115 Heidelberg, Germany

Printed by Libri Plureos GmbH
in Hamburg, Germany